Virtual Screening
in
Drug Discovery

Virtual Screening
in
Drug Discovery

edited by

Juan Alvarez and Brian Shoichet

CRC Press
Taylor & Francis Group
Boca Raton London New York

CRC Press is an imprint of the
Taylor & Francis Group, an **informa** business
A TAYLOR & FRANCIS BOOK

CRC Press
Taylor & Francis Group
6000 Broken Sound Parkway NW, Suite 300
Boca Raton, FL 33487-2742

First issued in paperback 2019

ISBN-13: 978-0-8247-5479-2 (hbk)
ISBN-13: 978-0-367-39318-2 (pbk)
Library of Congress Card Number 2004059300

Library of Congress Cataloging-in-Publication Data

Virtual screening in drug discovery / edited by Juan Alvarez, Brian Shoichet.
    p.   cm.
Includes bibliographical references and index.
ISBN 0-8247-5479-4 (alk. paper)
    1. Drug development--Computer simulation. 2. Drugs--Structure-activity relationships--Computer simulation. 3. Ligand binding (Biochemistry)--Computer simulation. 4. Pharmaceutical chemistry--Computer simulation.
    [DNLM: 1. Drug Evaluation, Preclinical--methods. 2. Computational Biology--methods. 3. Molecular Biology--methods. 4. Pharmaceutical Preparations--analysis. QV 771 V819 2005] I. Alvarez, Juan, 1965- II. Shoichet, Brian, 1963- III. Title.
RM301.25.V576 2005
615′.19—dc22   2004059300

**Visit the Taylor & Francis Web site at
http://www.taylorandfrancis.com**

**and the CRC Press Web site at
http://www.crcpress.com**

Preface

Identification of a good lead is a critical first step in drug discovery. The qualities of the lead set the stage for subsequent efforts to improve therapeutic efficacy through potency against its target, selectivity against related targets, adequate pharmacokinetics, and minimizing toxicity and side effects. Correspondingly, there has been considerable effort to improve technologies in lead discovery.

Over the last 15 years, the largest source of novel leads has been high throughput screening (HTS) of compound libraries. These libraries typically consist of hundreds of thousands of compounds, and combinatorial chemistry has increased the number of compounds available for screening well into the millions. Recent advances in genomics have identified thousands of gene products as potential targets for therapeutic intervention. This embarrassment of candidates, hits, and targets has led investigators to question traditional medicinal chemistry routes to discovery, as well as structure-based or rational design. Why bother with these information-hungry methods when one can simply make and test all the likely candidates against all the likely targets?

This view, perhaps always something of a strawman, has recently been reconsidered as the rate of actual drug introduction has fallen in the last decade. Few, perhaps no, drugs have been introduced that directly come out of HTS, and the pharmaceutical industry has moved away from large combinatorial chemistry libraries as sources of initial hits (the technique is still widely used for lead elaboration). This has prompted a reconsideration of more information-rich methods of lead discovery, including structure-based design and virtual screening (VS).

VS uses computational tools to identify biologically active molecules against specific targets. Two primary approaches are used — methods that look for similarity to known ligands and docking methods that require the use of the three-dimensional (3D) structure of the target. Both sorts of calculations are typically followed by experimental testing. Because far fewer compounds are tested, virtual screens must hold out the promise of higher hit rates, or more druglike or leadlike molecules, than HTS. The increased robustness of computational algorithms and scoring functions, the availability of affordable computational power, and the potential for timely structural determination of target molecules, have provided new opportunities for VS and made it more practical.

Thus, whereas VS remains less widely used than HTS for lead discovery, the popularity and feasibility of VS have risen significantly in the last 5 years. There is no question that HTS will eventually be replaced by VS as the primary source of lead molecules. The speed at which a virtual screen can be completed makes it attractive for projects with few leads or for which new molecules are sought or at least as a first attempt to identify suitable leads. The technique is relatively inexpensive because it circumvents the need for robotics, reagent acquisition or production, and compound storage facilities. Indeed, there is no requirement for the compounds being virtually screened to be a part of an in-house compound repository; screening of external compound collections and virtual libraries is increasingly popular. This latter point in itself means that the number of chemical entities capable of being virtually screened can currently exceed even large, which in most cases actually also means chemically redundant, corporate libraries, by an order of magnitude, a number that will undeniably increase. It is these greater opportunities that will offset some of the algorithmic limitations of VS and reliance on HTS serendipity. Finally, information-based methods, such as VS, rely on the existence of knowledge, which inevitably grows over time. The linking of an increased number of gene sequences to known protein structures; the greater number of proteins for which a structure has been solved; the identification of a wide variety of ligands, natural and synthetic, for a wealth of targets, will result in fewer truly novel targets for which there are no known ligands and thus the opportunity for a knowledge-based approach as a starting point. HTS, or some other type of random benchtop screening, will be reserved for only those targets for which there is little existing knowledge or for which VS yielded no useful starting points.

Why then, is not everyone using VS? There are four important barriers to entry:

1. The algorithms remain inaccurate.
2. The programs require expert decisions to be effective.
3. The compound databases that are being screened require considerable curation.
4. The techniques require a great deal of information about a target.

For those that lack a 3D structure, which is often the case, docking will not be possible. Similarly, when little is known about ligands, small-molecule-based methods will not be appropriate. In these cases, HTS will certainly be the best choice for novel lead discovery.

In this book, we consider where VS is with regard to the first two of these barriers — the algorithms and how to actually use them. Part I offers perspectives on both ligand-based and docking-based virtual screens. The authors of these chapters frame many of the challenges currently facing the field. Part II considers the choice of compounds that are best suited as drug leads. Part III discusses ligand-based approaches, including descriptor-based similarity, traditional pharmacophore searching, and similarity based on 3D-pharmacophore fingerprints. The final two parts are devoted to molecular docking. Part IV outlines some important and practical considerations relating to the energetics of protein–ligand binding and target-site topography, whereas specific docking algorithms and strategies are discussed in Part V.

Notwithstanding this list of subjects, the book is by no means comprehensive in its coverage of even the popular programs being used in the field, though many of the strategies outlined will transcend the specifics of a given method. Nor does it purport to offer single best ways to use the programs: The practical recipes that are offered for using the programs are what active practitioners have found useful in their own day-to-day work. We hope that these snapshots will nevertheless be useful to both skilled investigators and those new to the field. Part of our motivation for drawing this collection together was that we have often felt the need for such a resource for ourselves and for our research groups.

Juan C. Alvarez, Ph.D.
Transform Pharmaceuticals
Lexington, Massachusetts

Brian Shoichet, Ph.D.
University of California
San Francisco, California

The Editors

Juan C. Alvarez, Ph.D.

Juan Alvarez received a B.Sc. in chemistry from Massachusetts Institute of Technology (MIT) in 1986. In 1992, he received his Ph.D. in pharmaceutical chemistry from the University of California — San Francisco (UCSF) for research focused on the oxidative mechanisms carried out by hemoproteins as well as the structure-based discovery and design of inhibitors of the HIV-1 protease in the lab of Dr. Paul Ortiz de Montellano. Alvarez joined the Small Molecule Drug Discovery Department at the Genetics Institute in Cambridge, Massachusetts in 1992 and later established its computational chemistry group. After the acquisition of Genetics Institute by American Home Products in 1996 and its subsequent integration into Wyeth Research, he became the Associate Director of Computational Chemistry. In addition to supporting therapeutic projects, his group's research focused around molecular docking, scoring functions, automated structure-based lead optimization, pharmacophore elucidation, and high throughput screening data analysis. In 2004, Alvarez joined Transform Pharmaceuticals in Lexington, Massachusetts, where he is currently Senior Director of Informatics.

Brian Shoichet, Ph.D.

Brian Shoichet received a B.Sc. in chemistry and a B.Sc. in history in 1985 from MIT. He received his Ph.D. for work with Tack Kuntz on molecular docking in 1991 from UCSF. Shoichet's postdoctoral research was largely experimental, focusing on protein structure and stability with Brian Matthews at the Institute of Molecular Biology in Eugene, Oregon, as a Damon Runyon Fellow. Shoichet joined the faculty at Northwestern University as an Assistant Professor in 1996 and became an Associate Professor in 2002. Brian Shoichet is an Associate Professor in the Department of Pharmaceutical Chemistry at UCSF. His group uses a mixture of computational and experimental techniques to investigate enzyme structure, function, stability and inhibition, and the links among them. The research is supported by the National Institute of General Medical Sciences, including GM59957 and GM71630.

List of Contributors

Edmond J. Abrahamian
Tripos, Inc.
St. Louis, Missouri

Juan C. Alvarez
Transform Pharmaceuticals
Lexington, Massachusetts

Konstantin V. Balakin
Chemical Diversity Labs, Inc.
San Diego, California

Hugues-Olivier Bertrand
Accelrys SARL
Paris, France

Cristian Bologa
Office of Biocomputing
University of New Mexico School of
  Medicine
Albuquerque, New Mexico

Hans Briem
Schering AG
CDCC/Computational Chemistry
Berlin, Germany

Amedeo Caflisch
Department of Biochemistry
University of Zurich
Zurich, Switzerland

Marco Cecchini
Department of Biochemistry
University of Zurich
Zurich, Switzerland

Paul S. Charifson
Vertex Pharmaceuticals, Inc.
Cambridge, Massachusetts

Robert D. Clark
Tripos, Inc.
St. Louis, Missouri

Jason C. Cole
Cambridge Crystallographic Data
  Centre
Cambridge, England

Michael Dooley
Accelrys, Inc.
San Diego, California

Peter C. Fox
Tripos, Inc.
St. Louis, Missouri

Richard A. Friesner
Department of Chemistry
Columbia University
New York, New York

Marcus Gastreich
BioSolveIT GmbH
St. Augustin, Germany

Anne Goupil-Lamy
Accelrys SARL
Paris, France

Osman F. Güner
Accelrys Inc.
San Diego, California

Thomas A. Halgren
Schrödinger, LLC
New York, New York

Remy Hoffman
Accelrys SARL
Paris, France

Danzhi Huang
Department of Biochemistry
University of Zurich
Zurich, Switzerland

Diane Joseph-McCarthy
Wyeth Research
Cambridge, Massachusetts

Gerhard Klebe
Institute for Pharmaceutical Chemistry
Philipps-University Marburg
Marburg, Germany

Peter Kolb
Department of Biochemistry
University of Zurich
Zurich, Switzerland

Irwin D. Kuntz
Department of Pharmaceutical
  Chemistry
University of California
San Francisco, California

Christian Lemmen
BioSolveIT GmbH
St. Augustin, Germany

Marguerita S.L. Lim-Wilby
Accelrys, Inc.
San Diego, California

Teresa A. Lyons
Accelrys SARL
Paris, France

Eric S. Manas
Department of Chemical and Screening
  Sciences
Wyeth Research
Collegeville, Pennsylvania

Iain J. McFadyen
Wyeth Research
Cambridge, Massachusetts

Susan L. McGovern
Department of Molecular
  Pharmacology and Biological
  Chemistry
Northwestern University
Chicago, Illinois

Demetri T. Moustakas
UCSF/UCB Joint Graduate Group in
  Bioengineering
University of California
San Francisco, California

Robert B. Murphy
Schrödinger, LLC
New York, New York

J. Willem M. Nissink
Cambridge Crystallographic Data
  Centre
Cambridge, England

Marius Olah
Office of Biocomputing
University of New Mexico School of
  Medicine
Albuquerque, New Mexico

Tudor I. Oprea
Office of Biocomputing
University of New Mexico School of
  Medicine
Albuquerque, New Mexico

Sunil Patel
Accelrys SARL
Cambridge, England

Scott C.H. Pegg
Department of Biopharmaceutical
  Sciences
University of California
San Francisco, California

Emanuele Perola
Vertex Pharmaceuticals, Inc.
Cambridge, Massachusetts

Matthias Rarey
Center for Bioinformatics
University of Hamburg
Hamburg, Germany

Thomas S. Rush III
Department of Chemical and Screening
  Sciences
Wyeth Research
Cambridge, Massachusetts

Nikolay P. Savchuk
Chemical Diversity Labs, Inc.
San Diego, California

Christoph Schneider
GmbH
Munich, Germany

Kim A. Sharp
The Johnson Research Foundation
Department of Biochemistry and
  Biophysics
University of Pennsylvania
Philadelphia, Pennsylvania

Gregory J. Tawa
Department of Chemical and Screening
  Sciences
Wyeth Research
Monmouth Junction, New Jersey

Robin Taylor
Cambridge Crystallographic Data
  Centre
Cambridge, England

John H. van Drie
Vertex Pharmaceuticals, Inc.
Cambridge, Massachusetts

Gary Walker
Wyeth Research
Cambridge, Massachusetts

W. Patrick Walters
Vertex Pharmaceuticals, Inc.
Cambridge, Massachusetts

Jinming Zou
Wyeth Research
Cambridge, Massachusetts

Contents

Chapter 3 An Analysis of Critical Factors Affecting Docking and Scoring
Emanuele Perola, W. Patrick Walters, and Paul S. Charifson

PART II: Compound and Hit Suitability for Virtual Screening 87

Chapter 4   Compound Selection for Virtual Screening

Tudor I. Oprea, Cristian Bologa, and Marius Olah

Chapter 5   Experimental Identification of Promiscuous, Aggregate-Forming Screening Hits

Susan L. McGovern

Chapter 8   Using Pharmacophore Multiplet Fingerprints for Virtual High
Throughput Screening
Robert D. Clark, Peter C. Fox, and Edmond J. Abrahamian

PART IV: *Important Considerations Impacting Molecular Docking* 227

Chapter 9   Potential Functions for Virtual Screening and Ligand Binding
Calculations: Some Theoretical Considerations
Kim A. Sharp

Chapter 10 Solvation-Based Scoring for High Throughput Docking

Thomas S. Rush III, Eric S. Manas, Gregory J. Tawa, and Juan C. Alvarez

Chapter 11 Classification of Ligand–Receptor Complexes Based on Receptor Binding Site Characteristics

Marguerita S.L. Lim-Wilby, Teresa A. Lyons, Michael Dooley, Anne Goupil-Lamy, Sunil Patel, Christoph Schneider, Remy Hoffmann, Hugues-Olivier Bertrand, and Osman F. Güner

Chapter 13 Pharmacophore-Based Molecular Docking: A Practical Guide

Diane Joseph-McCarthy, Iain J. McFadyen, Jinming Zou, Gary Walker, and Juan C. Alvarez

Chapter 14 Fragment-Based High Throughput Docking

Peter Kolb, Marco Cecchini, Danzhi Huang, and Amedeo Caflisch

Chapter 15 Protein–Ligand Docking and Virtual Screening with GOLD
Jason C. Cole, J. Willem M. Nissink, and Robin Taylor

Chapter 16 A Brief History of Glide: A New Paradigm for Docking and Scoring
in Virtual Screening

Thomas A. Halgren, Robert B. Murphy, and Richard A. Friesner

Part I

Perspectives on Virtual Screening

1 Virtual Screening: Scope and Limitations

Gerhard Klebe

1.1 INTRODUCTION

In a retrospective view, the last 30 years of drug research have witnessed a series of fundamental developments stimulated by the introduction of new methodologies [1–4]. Frequently enough, these new approaches or concepts have been announced as "a change in paradigm." Undoubtedly, drug discovery has moved toward more rational concepts based on our increasing understanding of the fundamental principles of protein–ligand interactions [5,6]. Nowadays, serendipity, as a key element in drug research [7], plays a less important role, even though recent examples such as the discovery of sildenafil (Viagra®) might suggest the opposite [8].

Common to all these "revolutionary changes in paradigm" is the increased knowledge about the spatial structure of the molecules involved and the improved expertise to manipulate and modify their properties in a well-planned fashion at the molecular level. However, in contrast to the expectations that these developments will provide a dramatic increase of new chemical entities reaching the market, a constant decrease of the number of launched products is observed [9]. Accordingly, the statement "a kingdom for an innovative and also tractable lead" is still essential for drug companies because novel leads remain the most important prerequisite for success and survival on the market.

In this context, the question must be allowed whether virtual screening (VS) [10–15] for new leads by computational means is just one of these trendy new technologies coming up and disappearing again or whether this approach will be established as a major tool of future lead finding. Certainly, experimentalists and synthetic medicinal chemists regard this method with some doubts. In contrast to high throughput screening (HTS), which provides the medicinal chemist with hits in terms of real chemical compounds that actually bind to the target of interest, VS simply suggests computer hits. Possibly these are even not yet synthesized compounds and they are selected on the basis of a sophisticated docking algorithm using an approximate energy scoring function to mutually rank them as putative hits [16].

Both approaches, HTS and VS, are conceptionally rather controversial [17]. Hits discovered by HTS, even though they represent real molecules, do not contribute to our understanding of why and how they act upon the target. Any increase in knowledge is only obtained once structural biology or molecular modeling come into play and detect structural similarities or possible binding modes among the discovered hits. Often enough the hits from HTS are quite diverse in chemical

3

structure, thus preventing any simple intuitive comparison. In contrast, VS requires a key element knowledge about the criteria responsible for binding to a particular target. Accordingly, inevitable prerequisite for its application is the availability of the three-dimensional (3D) structure of the target either determined by crystallography or nuclear magnetic resonance (NMR) or constructed on the basis of a homologous protein. Provided the nature of protein–ligand interactions is sufficiently well understood, the computer can then help to select or even better generate candidate molecules considering the restrictions and requirements imposed by the binding pocket. Furthermore, solvation/desolvation characteristics can be taken into account along with conditions for druglikeness. Even the principles of combinatorial chemistry or concepts for parallel synthesis can be employed in the context of VS by starting with well-selected molecular skeletons reoccuringly found to bind to particular target families [9].

Clearly, the methods have not yet matured enough to answer the question about the relevance of VS for the entire lead finding process. However, it is cheap and fast and can be applied at an early stage of the drug development process. Furthermore, recent efforts in structural genomics are expected to gradually remove the bottleneck of nontimely available structures of the target proteins.

The following contribution will briefly outline our views on the present scope of VS based on some of our own research results. Subsequently, we will discuss by means of some instructive examples from recent research where, in our opinion, major limitations are still given that impede a broad and routine application of VS in lead discovery, at least in the hands of nonexperts.

1.2 STRATEGIES TO VIRTUAL SCREENING

As mentioned, VS is still a developing field and by no means yet a mature and well-established tool for routine work. Accordingly, the number of concepts and strategies proposed is nearly as large as the number of laboratories involved in the development of VS techniques. However, taking only the truly predictive examples into account, which actually suggested new hits from scratch and were subsequently confirmed by experiment, yet a rather small number of success stories have been communicated [18–25]. This selection excludes theoretical studies that calibrate their VS success rates in terms of enrichment factors that discriminate known binders from randomly selected so-called nonbinders (their presumed nonbinding has never been tested). In our own research, we have applied VS to five different protein targets and succeeded in all cases to retrieve micro- to nanomolar hits in a truly predictive manner. In all studies, we tested experimentally only a small set of computer hits retrieved from a large data sample. Out of these small sets, convincingly high hit rates (usually 60 to 100%) could be achieved, though it must be considered that we are dealing with rather small numbers and statistically significant rates are difficult to define.

What are the essential steps in our VS concept and where, in our opinion, are the advantages found? First, we depart from the 3D structure of the target protein. In four of our case studies, appropriate crystal structures had been available. Two of these examples, carbonic anhydrase II [26,27] and thermolysin [28] represent rather rigid enzymes, thus any complications imposed by induced-fit adaptations of the protein are hardly to be expected (only histidine-64 or His64 (carbonic anhydrase) and asparagnine-112 or Asn112 (thermolysin) show conformational changes upon ligand binding). In case of the transfer ribonucleic acid (tRNA) modifying enzyme, tRNA guanine transglycosylase (TGT), some important adaptations of the binding site have been detected through multiple crystal structure analyses of different substrate and inhibitor complexes [29,30]. Once these adaptations are known and related to functional properties of the enzyme, they can be exploited for the design of new inhibitors. As a fourth example we selected aldose reductase (AR), an enzyme operating on a broad range of structurally diverse substrates. The enzyme achieves this promiscuity via pronounced induced-fit adaptations of its binding site. It occurs with various active site conformations in equilibrium. Upon ligand binding, the conformational equilibrium is shifted toward the binding-competent conformations with open pocket and the relevant conformational state becomes predominant in complex formation [31]. Such a manifold system with several alternative binding-site conformers requires new concepts in VS. We followed two strategies, which both suggested new micromolar inhibitors. In a first attempt, we concentrated on one preferred conformer and performed VS assuming that the binding pocket of AR adopts only this conformation [32]. In a second study, we followed a ligand super-positioning method [33]. From three different crystal structures we extracted ligands in their bound conformations, all accommodating the ligands with distinct binding-site conformations. After merging these ligands into one "supermolecule reference," mutual alignments with candidate molecules from the search database were performed. This strategy also retrieved several structurally new AR inhibitors of micromolar potency. As a further example, we constructed a model of the Neurokinin 1 (NK1) receptor as a prominent representative of the class of G protein-coupled receptors (GPCRs) [34]. For the modeling of the receptor we applied a new method recently developed by us [35]. In this approach, knowledge-based contact potentials are used in a combinatorial fashion to suggest optimal side-chain orientations of the active-site residues. Also in case of this homology-built protein receptor, our strategy for the retrieval of antagonists by VS was successful and proposed a compound of submicromolar affinity.

1.3 DEVELOPMENT OF A RELIABLE PHARMACOPHORE HYPOTHESIS

In all of the mentioned examples, we followed a strategy using a series of consecutive hierarchical filters. As an initial data sample, we considered about 800,000 compounds available in public domain and accessible through commercial product suppliers. Most crucial for VS is an initial analysis of the binding pocket of the target protein (Figure 1.1). As a first step, all accessible structural information is evaluated using the tools for molecular comparison available in Relibase [36,37]. Once the conformational properties of the target protein are analyzed and a detailed picture about the possible binding modes of known ligands along with their solvation patterns in the binding site is collected, a protein-based pharmacophore is derived. This step follows the idea that the protein binding-site environment imposes the requirements to be met by a putative ligand. To discover the most favorable areas likely to accommodate a particular type of ligand functional group, we map the binding pocket in terms of hot spots of binding (Figure 1.1). For this analysis, we have applied three different methods: the force-field-based approach GRID, originally developed by Goodford [38], and the two knowledge-based approaches Super-Star [39,40] and DrugScore [41]. The latter ones are both based on crystal data as input. SuperStar calculates preferred interaction sites by mapping contact distributions in terms of scatter plots or propensity contour maps onto active site-exposed residues. These maps are obtained as composite crystal-field environments extracted from the crystal packing of small organic molecules or taken from protein–ligand complexes in the Protein Data Bank (PDB). DrugScore uses statistically derived pair potentials using the so-called "inverse Boltzmann principle" [42]. In a large sample of protein–ligand complexes, the protein-to-ligand interface is analyzed in terms of distance-dependent occurrence frequencies by which a particular ligand atom type is found in contact with a protein atom type. Whether for GRID, SuperStar, or DrugScore, a regularly spaced grid is embedded into the binding site. Subsequently, for different ligand atom types interaction energies or contact preferences are calculated by systematically placing these probe atoms at the various grid intersections. As functional form to describe these interactions, the force-field potential of GRID, the propensity distribution of SuperStar or the knowledge-based contact preferences of DrugScore are applied. To allow for an intuitive graphical interpretation of the hot spots elucidated by these programs, the obtained grid values are contoured according to a predefined level above the detected global minimum.

In case of the flexible protein AR, we followed an alternative concept to consider simultaneously multiple binding modes of putative ligands. In this approach, we took advantage of the successful superimpositioning of ligand molecules in terms of similarities of their spatial physicochemical properties. Different ligand molecules, representing alternative binding modes with the protein, are extracted in their crystallographically determined protein-bound conformation and merged into one supermolecule. This unified reference is then used to define a ligand-based pharmacophore by evaluating the distribution of physicochemical properties of the superimposed ligands in space.

1.4 SERIES OF CONSECUTIVE HIERARCHICAL FILTERS TO MATCH WITH THE PHARMACOPHORE HYPOTHESIS

In a subsequent step, the information of the protein- or ligand-based pharmacophore is examined and used to define a list of essential functional groups to be present in putative candidate ligands. Usually this comprises a certain number of donor- and acceptor-functional groups and a set of hydrophobic groups. These criteria are defined in a search query for Tripos's search engine UNITY® [43], along with limits considering the molecular weight and the number of rotatable bonds (Figure 1.1). In addition, at this filter level the requirements imposed by Lipinski's rule-of-five are applied to reduce the initial sample set of candidate molecules [44].

FIGURE 1.1 VS starts with a detailed analysis of a protein binding site (a). A hot spot analysis is performed using different probe atoms to map the physicochemical properties of the binding-site exposed residues (b). The hot spots indicated by the various probes are translated into a protein-based pharmacophore (c) that serves as input for database searches using UNITY (d). Through a hierarchical filtering, a large initial data sample is reduced to a set of several thousand candidates that are finally docked into the binding site (e) and scored in terms of the expected binding affinity. The most prospective hits are either purchased or synthesized (f) and the binding properties are determined in a binding assay (g). Finally, crystal structures of the best hits together with the protein are determined (h).

In the next step, information from the protein- or ligand-based pharmacophore is considered in terms of an approximate spatial arrangement of the detected hot spots. These are translated into a 3D pharmacophore hypothesis. Such a hypothesis is then used in a UNITY query. At a later stage, this search query can be reformulated in a more stringent manner by adding excluded volume areas in UNITY. These restraints are placed in space to approximately describe the shape of the binding pocket. In the case of our searches for thermolysin inhibitors, we also consulted FeatureTrees [45] at this search level. At all filter steps the search queries can be defined in a more stringent or relaxed fashion. The actual settings are adjusted in a way that a dataset of reasonable size for the following step is retrieved. Furthermore, if known ligands have already been described for the protein under consideration, we examined how these known binders performed at the various filter levels.

In the following filter step, further reduction of the remaining data sample is attempted. If a significant number of diverse ligands have already been characterized, it is advisable to use this dataset as a spatial reference to score the similarity of candidate molecules with known binders. Methods computing molecular superpositionings and spatial similarities of ligands can be applied. We have used either FlexS™ [46] or SEAL in combination with MIMUMBA [47,48] at this level [27,32].

Also here, approximate information about the protein can be included by considering the binding-site environment in terms of repulsive Gaussian functions.

1.5 DOCKING, SCORING, AND VISUAL INSPECTION

As a final step, we employ a docking algorithm to retrieve the most prospective hits out of the remaining data sample (Figure 1.1). We applied the programs FlexX™ [49], GOLD (Genetic Optimization for Ligand Docking) [50], DOCK [51,52] or AutoDock [53,54]. AutoDock provides the opportunity to replace the original Lennard-Jones and Coulomb potential maps by other property maps, such as those derived from DrugScore [55]. In our recent VS study on the NK1 receptor, we performed docking with the FlexX-Pharm™ program [56] that allowed us to consider pharmacophore features during docking. The scoring of the produced docking solutions is essential. In our studies, DrugScore was applied as a routine tool, but as an extension we have also used the models obtained from CoMFA (comparative molecular field analysis) [57] and CoMSIA (comparative molecular similarity index analysis) [58] or the tailored scoring model produced by AFMoC (adaptation of fields for molecular comparison) [16] to rank and score the different docking solutions. Nevertheless, the best scoring function cannot replace the experience of a modeler. Accordingly, the finally obtained solutions should always be subjected to visual inspection. For this analysis, preclustering of the retrieved hits is supportive [32]. In a graphical analysis, the actually generated conformations of the remaining best-scored hits are analyzed simultaneously considering the mutual complementarities of protein and ligand surfaces. In particular, the analysis studied whether any voids in the protein–ligand interface remained for the well-scored docking solutions (Figure 1.1).

In our opinion, this final visual inspection is one of the most crucial steps. It is also used to select those hits that are subsequently purchased and subjected to the binding assay (Figure 1.1). Such a visual inspection can only be applied to a data sample of perhaps 300 to 500 compounds. Accordingly, a reliable filtering has to be performed at the previous steps. The advantage of our approach is that we can keep control over the retrieval rates and selection steps at any level. We refrain from high throughput docking (HTD), mainly because we believe that our presently available docking tools are not precise enough to produce sufficiently accurate binding geometries and that the established scoring and ranking schemes are not reliable and discriminative enough to operate as sole criterion for the final selection of the most prospective hits from a list of several 10,000 solutions. To circumvent this bottleneck, postfilters have been described in literature to analyze the results of HTD runs [59,60]. Clearly, it is a matter of opinion whether postfiltering or forward filtering in a stepwise fashion using consecutive hierarchical filters should be preferred. We prefer the latter strategy, which allows for better control during the search and is computationally less expensive. However, it requires a higher amount of human intervention along the various steps. We believe this latter aspect is essential, at least as long as VS is still a matter of methodological developments.

For our case studies, we tried to resolve a crystal structure of the corresponding target protein together with the obtained hits from VS (Figure 1.1). Successful structure determinations helped us to detect either agreement or deviation between predicted and observed binding modes [27,29]. They also indicated unexpected changes in the binding modes resulting from induced-fit adaptations of the protein or occurrence of water interstitial molecules in the protein–ligand interface. Usually, hits from VS (similarly to HTS) correspond to micromolar binders. Frequently, we were faced with problems of limited solubility. Accordingly, we could not obtain sufficiently concentrated solutions for these low-affinity ligands to succeed in co-crystallization or soaking with the protein. As a result, low population at the binding site, significant disorder, or distribution over several binding modes are detected that hamper a conclusive interpretation of the electron density. Nevertheless, such control studies, performed subsequently to VS, are extremely valuable for analyzing the performance of the presently available VS tools and help to further improve their reliability.

1.6 VIRTUAL SCREENING: MATURED AS A ROUTINE TOOL?

Considering these successful examples, can we assume that VS has matured to be a routine tool providing us in the future with new leads in drug discovery? In our opinion, we are still apart from such an optimistic situation. Presently, we still have a too superficial understanding of essential contributions resulting from induced-fit adaptations of both ligand and protein. Furthermore, the role of water in ligand binding is hardly understood. Water is not only involved in solvation/desolvation of the reaction components, but it is also a frequent partner in generating appropriate binding modes. A statistical analysis of the protein–ligand complexes stored in Relibase reveals that in 2/3 of the complexes a water molecule is involved as binding partner in ligand recognition, mediating an interaction between ligand and protein. Upon binding, ligand and protein functional groups can change their protonation states. In general, these effects are determined by modulations of the local electrostatic properties of the formed complexes. This can easily transform a donor functional group into an acceptor group or vice versa. In docking, such transitions correspond to a dramatic change. Accordingly, we have to maintain control over such effects in VS, provided we are aware of their occurrence and possess the concepts of how to handle them in the screening process. Protein plasticity is a further complication that hampers straightforward application of computer methods to the lead finding process. Likely, most of the uncomplexed proteins show equilibrium of several conformational states involving different orientations of residues occasionally buried or exposed to the binding site. However, experience increasingly shows that usually only those induced-fit adaptations are observed upon ligand binding that correspond to the inherent conformational mobility a protein experiences to achieve its functional properties. Accordingly, the comprehensive study of the functional properties of a protein might be the best but also most elaborate approach to tackle the difficult issue of predicting protein plasticity and induced-fit adaptations upon ligand binding.

By use of some instructive examples from our own recent research, we want to illustrate the importance of the above-described effects.

1.7 PEPTIDE BOND FLIP AND INTERSTITIAL WATER MOLECULES

The enzyme TGT catalyzes the base-exchange reaction of guanine by several modified bases such as $preQ_1$ or $preQ_0$ (Figure 1.2, Figure 1.4) leading to anticodon modification of certain tRNAs [61]. Interestingly enough, the TGT enzymes of species originating from the three kingdoms of life differ in their substrate selectivity. In the prokaryotic enzyme, a peptide-bond flip, triggered by pH conditions and induced by ligand binding, modulates the recognition properties of the substrate-binding site, switching between donor and acceptor functionality [61]. The first protein–ligand complex with the prokaryotic TGT from *Zymomonas mobilis* could be solved together with $preQ_1$ (Figure 1.2). Recently, the structure of the ternary complex of TGT together with a stem-100p RNA substrate and 9-deazaguanine has been determined [62]. In this complex, the RNA is tethered to TGT through the side chain of aspartic acid-280 (Asp280), proving this residue as intermediate nucleophile. Cocrystallized deazaguanine occupies the substrate-base recognition pocket of the enzyme. In the crystal, the exchange reaction to $preQ_1$ can be followed unraveling details of the single-step-base exchange reaction. Important for our drug design considerations is that the observed binding geometries of $preQ_1$ in the binary and ternary complex are identical, suggesting the relevance of our conclusions based on the geometry of binary complexes of TGT with inhibitors. In the TGT-$preQ_1$ complex, the exocyclic methylene amino group of the base forms a hydrogen bond (H-bond) to the binding-site exposed backbone carbonyl group of leucine-231 (Leu231) [63]. Accordingly, in our initial inhibitor design program, we focused on this H-bond as an assumed essential prerequisite to anchor potent inhibitors in the binding pocket of TGT. Surprisingly, in subsequent studies, a new class of inhibitors with a common pyridazinedione skeleton could be discovered [29] that lacks the potential to form a direct H-bond with the backbone carbonyl of Leu231. Subsequent crystal structure analysis revealed the above-mentioned peptide bond flip resulting in the exposure of the NH donor functionality of the peptide bond toward the binding pocket (Figure 1.2). Furthermore, an interstitial water molecule is accommodated, which mediates an H-bond between the inhibitor and the protein. Both observations, the bond flip and the incorporated water molecule would have been hardly predictable, at least at the stage of knowledge in those days.

In the mean time, more structural information has been collected [61] and the detected versatile adaptations of the protein involving also an adjacent glutamate residue as general acid/base to stabilize the peptide switch could be related to its functional role. As mentioned, the enzyme catalyzes the base exchange of guanine by a modified base. Accordingly, also guanine must be recognized as substrate. Compared to $preQ_1$, this base lacks the basic side chain and an acceptor functionality is exposed toward the protein. Taking the binding modes of the pyridazinediones into consideration, the nitrogen in position 7 of guanine forms an H-bond to

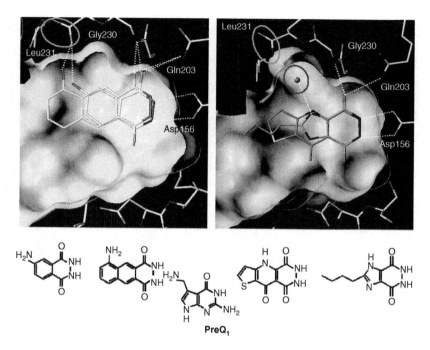

FIGURE 1.2 In the binding pocket of TGT, a peptide bond flip at Leu231 (white circle) is observed upon ligand binding. Either H-bond acceptor (left) or donor properties are exposed toward the binding pocket. In addition, both inhibitors, shown on the right, interact via an interstitial water molecule (black circle) with the exposed NH group. In the center the formula of the natural substrate preQ₁ is shown. It binds to the enzyme conformer that exposes the Leu231CO toward the binding site.

Leu231NH mediated via the interstitial water molecule. Thus, the occasionally detected binding mode of the pyridazinediones most likely resembles a binding-site conformer of prokaryotic TGT suitable to recognize guanine as substrate. This would correspond to one of its intrinsic functional properties. In the recently described ternary TGT-RNA-deazaguanine complex [62] the peptide flip is not reported. However, in contrast to guanine, deazaguanine exposes a donor functionality toward the Leu231 peptide bond, thus supposedly reinforcing the same orientation of the peptide bond as in the complex with bound preQ₁. With respect to drug design, this conformational multiplicity of the peptide bond can be exploited. First, in our hot spot analysis, we considered both binding-site conformers as alternatives suggesting a more sophisticated pharmacophore [29]. Due to its bifunctionality, the interstitial water molecule can operate as either a donor or an acceptor of an H-bond at this site. This versatility has been taken into account in our design attempts. Accordingly, we suggested and synthesized two different molecular scaffolds: both bind with comparable binding affinity, but they expose either a donor or an acceptor functionality toward the interstitial water molecule (Figure 1.3) [30].

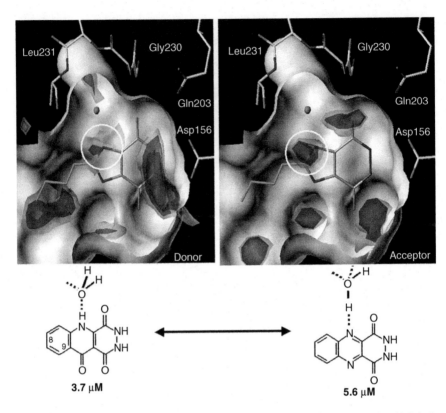

FIGURE 1.3 Using an H-bond donor (left) or an H-bond acceptor as a probe (right), hot spots are indicated in similar areas of the binding pocket (white circle). For orientation, the crystallographically determined binding mode of an imidazopyridazinedione is shown. The equally potent inhibitors of different scaffold expose either a donor or acceptor functionality into this area. Their binding is mediated via the interstitial water molecule to the protein.

1.8 CHANGES OF PROTONATION STATES INDUCED UPON LIGAND BINDING

An additional complex of TGT with another bound substrate points to a further problem that needs to be discussed in VS concepts: the protonation state of functional groups at the binding site. The above-described peptide switch is further stabilized by an adjacent glutamate present in the local environment [61]. It obviously operates as general acid/base. In the Leu231NH-exposed conformer showing the interstitial water molecule as neighboring binding partner, the terminal acid group of glutamic acid-235 (Glu235) forms an H-bond to the Leu231 backbone carbonyl, strongly suggesting a protonated state for this terminal acid group. Prokaryotic TGT also recognizes preQ$_0$, the usual substrate of archaebacterial TGTs. In the crystal structure of *Zymomonas mobilis*, TGT with this substrate the nitrogen of the exocyclic cyano group occupies the position of the interstitial water molecule in related complexes (Figure 1.4). This nitrogen forms an H-bond to the exposed Leu231NH functionality.

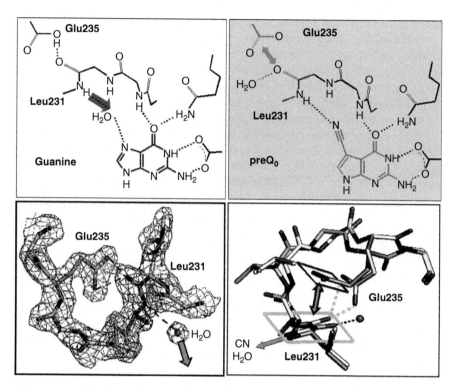

FIGURE 1.4 Guanine (top left) and preQ$_0$ (top right) are both recognized as substrates at the binding site of TGT (upper row, schematic binding mode). In case of guanine, the binding is mediated via an interstitial water molecule, whereas the cyano group of preQ$_0$ binds directly to Leu231NH (lower row, guanine case, grey frame, preQ$_0$ case, black frame). In the guanine complex, the peptide bond is stabilized by the neighboring Glu235COOH through a direct H-bond to Leu231CO. In this case the binding partner for Leu231NH is a water molecule (lower left). Once the cyano group is bound, the Glu235COO is deprotonated and instead of forming a H-bond, the terminal carboxylate group establishes a φ stacking with the neighboring peptide bond. Furthermore an additional water molecule is observed binding to Leu231CO.

Thus, the protein seems to adopt a conformation similar to that required to bind guanine as substrate. However, the remaining part of the structure indicates a more complex situation. Surprisingly, the adjacent acid group of Glu235 is not involved in a short H-bond to Leu231CO, but the carboxylate group stacks with its hydrophobic π-face on top of the neighboring peptide bond. This type of hydrophobic interaction appears to be a compromise to accommodate the obviously deprotonated terminal Glu235COO group next to the flipped peptide bond. In addition, a further water molecule is now incorporated stabilizing the Leu231CO in the direction opposite to the TGT binding site. Obviously, the replacement of the water molecule as binding partner of Leu231NH by the highly polarizing cyano group has drastic effects on the local dielectric conditions in the binding pocket. In consequence, the terminal Glu235 carboxylate does not pick up a proton to directly stabilize the flipped peptide bond via a short H-bond but remains deprotonated. It achieves short-range local interactions through hydrophobic π/π stacking. This observation is highly surprising and would never be predictable by any state-of-the-art computer method. Nevertheless, such functional adaptations of proteins do occur upon ligand binding. In the long term, successful VS must be capable to handle also such effects of locally induced dielectric adaptations to avoid unsatisfactory mispredictions.

Another surprising dependence of the observed binding mode could be discovered in the case of a serine protease inhibitor (for chemical formula see Figure 1.6) originally developed by Zeneca as potent factor Xa ligand [64]. Depending on the applied pH and crystallization conditions, this molecule binds differently to the binding site of trypsin, with virtually reverse orientation. In the cubic crystal form, grown at pH7, the inhibitor binds with its pyridinyl portion into the specificity pocket. Outside the S1 pocket, the remaining part of the inhibitor wraps around the side chain of glutamic acid-192 (Glu192) so that its chloronaphthyl group reaches toward the S1 site. Contrary to this situation, in the orthorhombic crystal form obtained at pH8 the ligand orients its hydrophobic chloronaphthyl moiety deeply into the S1 pocket with its C-Cl bond pointing toward the aromatic portion of tyrosine-228 (Tyr228). The chlorine atom occupies the position of water molecule W416 usually found in inhibitor complexes of trypsin. A new (or shifted) water position appears in the neighborhood on the external surface of the specificity pocket. Concomitant to this displacement is a reorientation of the side chain of serine-190 (Ser190) and the adjacent peptide bond serine-217–glycine-219 (Ser217-Gly219). An almost 90° bend at the sulfonamide group causes the remaining piperidinyl/pyridinyl portion of the inhibitor to accommodate the S2/S3 pocket located above the aromatic moiety of tryptophan-215 (Trp215). The described example demonstrates the dependence of binding modes on local dielectric conditions, likely paralleled by changes in the protonation state. The observed, supposedly isoenergetic binding orientations adopted by a single ligand might appear as a major disadvantage to drug discovery. On the other hand, however, these multiple binding modes can also serve as a guideline to map favorable alternative sites in a binding pocket.

Another remarkable case of change in protonation has been discovered by isothermal titration calorimetry [65]. In a congeneric series of serine protease inhibitors, we detected an unexpected pick up or release of protons upon ligand binding. The 2-carboxy derivative of *N*-(2-naphthylsulfonyl)-3-amidino-phenylalanine and napsagatran (Figure 1.5) acquire a proton during binding, whereas the piperazine derivative releases a proton. The closely related 4-carboxy derivative remains deprotonated at the acid group during binding. The same holds for CRC220: also here the central aspartate remains deprotonated (Figure 1.5). The apparent pK_a shifts of the carboxylate groups involved, all of which show similar values in water, can be explained by examining the corresponding ligand binding modes. In case of CRC220, the carboxylate group orients away from the binding pocket. Accordingly, its carboxylate group remains partially solvated and the local dielectric conditions resemble those in the bulk water phase. In the 2-carboxy derivative and napsagatran, the acid group becomes buried upon binding and points toward the catalytic serine and histidine. This local change has a strong impact on the pK_a value of this acidic group. Consequently, a proton is picked up and H-bonding occurs to this functional

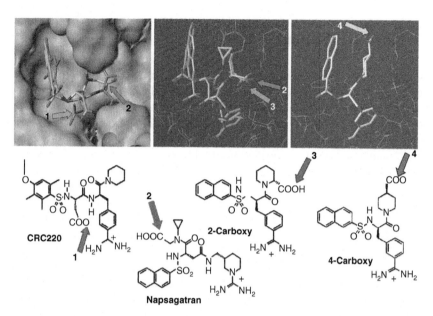

FIGURE 1.5 The carboxylate groups of CRC220 (light grey arrow 1) and napsagatran (dark gray arrow 2) adopt different binding modes at the binding site of thrombin. The 2-carboxy derivative of N-(2-naphthyl-sulfonyl)-3-amidino-phenylalanine (grey arrow 3) places its carboxylate group similarly to napsagatran. The 4-carboxy derivative orients its acidic group in a stacking fashion next to the catalytic histidine (light grey arrow 4). Isothermal titration calorimetry reveals that napsagatran and the 2-carboxy derivative pick up a proton upon binding, whereas CRC220 and the 4-carboxy derivative remain in the protonation state also observed in aqueous solution.

group. The proton can either stay at this acidic group or move on to the adjacent His residue. In the 4-carboxy derivative the acid functional group packs with parallel orientation against the hydrophobic face of the catalytic histidine. The induced pK_a shift, caused by this environment, is not sufficient to result in a net protonation of this acid group or the neighboring histidine residue. The terminal amino group of the piperazine derivative binds into a similar area. The local hydrophobic protein environment reinforces a pK_a shift for this basic nitrogen toward smaller values. Accordingly, the release of a proton is observed. Similar observations could be collected for the binding of carbonic acids or hydantoins as inhibitor of AR [66]. Here, most likely, a tyrosine residue present in the active site changes its protonation state upon inhibitor binding. Interestingly, this change seems to depend furthermore on the oxidation state of the bound cofactor. In the oxidized state, the NADPH/NADP+ places an additional positive charge into the active site. This has direct consequences for the protonation state of the residues in the direct neighborhood and as well as for protonation changes upon ligand binding. Again, such effects are of utmost importance for rational drug design approaches, as changes in protonation state require different concepts in handling donor or acceptor properties, as for example in a docking program.

1.9 WATER, THE NASTY, FREQUENTLY IGNORED BINDING FACTOR

The importance of water molecules as additional partners in protein–ligand inter-actions has already been addressed in case of TGT. The interstitial water molecule mediates an important contact between both interaction partners [29]. Proper treatment of water molecules in docking is a challenging problem [37]. Neglecting tightly bound water molecules in the binding site can result in a high desolvation penalty and unfavorable contributions to binding affinity. Taking such water molecules into account is possible in most docking programs. This is usually performed by considering them as integral parts of the binding pocket. However, such doing requires reliable criteria whether to classify a water molecule as tightly or loosely bound. Furthermore, also a so-called loosely bound water molecule can mediate important favorable interactions between ligand and protein, as seen for example in human immunodeficiency virus (HIV) protease. Besides strict conservation of water molecules at predefined sites or full replacement upon binding, the displacement to a shifted position also has to be considered. These different options have to be regarded as alternatives when setting up an adequate docking protocol. As long as the role of water in ligand binding is so poorly understood [67], the most promising approach to appropriately incorporating water molecules is the sound analysis of solvation patterns as observed in crystal structures of the target protein with a series of structurally diverse ligands. To support such analyses, Relibase has been equipped with appropriate tools to superimpose and characterize binding pockets in terms of their solvation patterns [36,37]. It detects all water molecules being replaced upon ligand binding, or being buried or locked in the binding pocket. Relibase provides a quantitative measure for the positional replacement of reoccurringly observed water mole-cules, either with respect to the structure of the uncomplexed protein or among a series of ligand-bound complexes. A statistical analysis considering several thousand water molecules reveals that about 50% of the water molecules observed in the uncomplexed situation are found at roughly the same position in the complex [68]. About 11% of the waters present in the ligand-free situation are shifted by more than 1.2 Å upon ligand binding. Focusing on water molecules that are hardly accessible, the conservation probability increases to nearly 70%. Thus, treating highly buried water molecules observed in the ligand-free situation as entrenched parts of the protein during docking appears as a reasonable concept for docking protocols including water molecules. If the structure of the uncom-plexed protein has not been characterized, the analysis of reoccurringly detected water molecules in a series of protein–ligand complexes can be used as a rea-sonable indicator to decide on the preplacement of water molecules in a docking experiment. However, in such an analysis, the diversity of the considered ligand series is of significant importance and will have pronounced influence on the discovered correlations.

1.10 PROTEIN PLASTICITY OR "HOW TO HIT A MOBILE TARGET?"

A further limiting aspect in VS is the intrinsic plasticity of proteins. This property is exhibited to a varying extent depending on the function of a protein. Some proteins have to operate under conditions that require one particular preorganized binding-site geometry to achieve optimal substrate recognition or turnover. The serine proteases such as trypsin and factor Xa apparently resemble such situations. For other proteins, such as the above-described AR, binding-site plasticity is an intrinsic property, required to adapt its recognition features to a broad range of substrates. Quite similar arguments hold for the large family of protein kinases [69]. In a comprehensive study to investigate selectivity and specificity determining features in protein–ligand interactions, we have undertaken the reconstruction of the binding pocket of human factor Xa in the structurally related trypsin by site-directed mutagenesis [70–72]. Three sequential regions — the 99-, the 175-, and the 190-loops — were chosen to represent the major structural differences between the ligand binding sites of the two enzymes. Both rat and bovine trypsin have been selected as a starting point. The different chimeric enzymes were characterized with respect to different inhibitors by enzyme kinetics and crystal structure analysis. In particular, the introduction of the 175-loop results in a surprising reorientation and partial unwinding of a short intermediate helix [70]. This reorientation is accompanied by an isomerization of a cysteine–cysteine (Cys-Cys) disulfide bond and the burial or exposure of the critical phenylalanine-174 (Phe174) side chain toward the rim of the binding pocket. In the presence of some inhibitors, a major rearrangement of this helical section yields a geometry virtually identical to that found in factor Xa showing Phe174 in an exposed conformation. In the case of binding of benzamidine, an inhibitor that leaves the aromatic box next to Phe174 unoccupied, a geometry is produced that either strongly resembles factor Xa or shows complete burial of the phenylalanine side chain (Figure 1.6).

The observed structural rearrangements of the intermediate helix, artificially introduced into trypsin by our mutational studies, has been implicated in the functional cofactor dependency of many trypsin-like serine proteases [70]. The blood coagulation factor VIIa, for example, responsible for the initiation of the extrinsic coagulation cascade, exhibits low reactivity toward its substrates as long as its cofactor, the tissue factor, is not bound. A structural comparison of factor VIIa in the presence or absence of tissue factor reveals that the crucial intermediate helix is disordered in the uncomplexed enzyme. Upon complexation of tissue factor, the helix becomes ordered and the active site is conformationally altered to achieve a geometry fully competent with enzyme catalysis [73,74].

Obviously, plasticity of the intermediate helix triggers an important functional role of some members of the trypsin-like serine proteinase family. In factor Xa and trypsin, this functional trigger is not required and accordingly their binding pockets are virtually rigid. Upon mutating this crucial portion, we apparently elicited plasticity next to the trypsin binding pocket. This behavior is well established by other members of the trypsin family. They require such plasticity to properly evolve their functional role. As for the TGT example, detailed knowledge of the functional

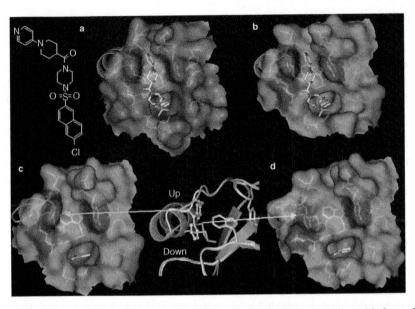

FIGURE 1.6 The binding of a potent factor Xa inhibitor is shown together with factor Xa (a) and a trypsin mutant (b). In the trypsin mutant, the binding site of factor Xa was reconstructed and a similar binding mode is observed, although the inhibitor binds with substantially reduced binding affinity. The mutant could also be crystallized with benzamidine, an inhibitor that leaves the aromatic binding pocket next to Phe174 unoccupied. In one crystal form (d) the phenylalanine is buried in the protein (down conformer) whereas in the other polymorph (c) this residue is exposed (up) to produce a geometry strongly reminiscent of factor Xa. The switch from the down to the up conformation involves partial unwinding of a helical portion and requires an energy contribution to activate the described conformational adaptation. This energy contribution is reflected in the reduced binding affinity of the potent factor Xa inhibitor toward the mutated trypsin, even though a similar binding geometry is produced.

properties of members of a protein family can help to elucidate putative induced-fit adaptations of active site residues to be expected upon ligand binding.

At present, protein plasticity provides significant limitations to VS. First of all, the correct binding site geometry to be considered in a VS run is difficult to define, in particular if the formation of one binding-competent conformer is induced by the shape and physicochemical properties of a bound ligand. Furthermore, estimation of binding affinity is complicated because the conformational adaptations of the protein involve additional energy contributions. In our comparative study of kinetic and structural data recorded for a series of structurally diverse serine protease inhibitors, we could assign individual contributions to the conformational adaptation, in particular with respect to the rigid reference proteins trypsin and factor Xa [75]. However, besides the inherent problem of how to consider the energy contribution of the conformational adaptation in a scoring function, we also observed that to varying extent local interaction geometries are formed between individual portions of the ligands and the structurally adapting binding-site residues. Exactly these ligand induced adaptations give rise to distinct selectivity profiles of structurally diverse inhibitors. With respect to the scoring of solutions obtained in a VS run, we would have to predict rather reliably the adopted binding modes, including protein adaptations. At present, this is beyond the scope of any state-of-the-art simulation technique.

1.11 CONCLUSIONS AND OUTLOOK

VS is a complementary but conceptionally controversial alternative to experimental high throughput screening to discover novel lead structures. Recent examples have demonstrated that VS is able to suggest meaningful and innovative leads for further lead optimization. Nevertheless, VS is still a developing field and by no means an already mature and well-established technology. This arises from two aspects, first, methodological developments are still required to improve and speed up existing tools for database retrieval and ligand docking. Second and perhaps more serious is the fact that our current knowledge and understanding of protein–ligand recognition is not yet profound enough to allow for a fully rational structure-based approach. Any approach of this kind, such as VS, starts with structural knowledge about the target under consideration. With the help of computer simulations we try to depart from the known structural grounds to estimate and predict new properties, such as the binding of novel ligands to our protein target. Thus, we use the computer to push the scope of our understanding further ahead. However, in doing so, we have to critically assess when we are leaving solid grounds for our predictions and when it would be advisable to perform a new and more conclusive experiment. The computer always provides an answer to a simulated problem, but this answer will depend on the data used to calibrate the applied method. These data also decide whether a method can only interpolate or actually allows to extrapolate into terra incognita. We also have to keep in mind, and this explains to some extent the good performance of many scoring functions or quantitative structure–activity relationship (QSAR) methods, that we often correlate on the basis of some indirect and possibly even irrelevant parameters, simply because for the given data sample, they correlate with those properties that are the physically relevant ones. And, because we usually consider only relative differences among a series of compounds, many of the insufficiently understood and thus disregarded effects simply cancel out in this relative comparison. Any further improvements of computer methods require to a large extent the support of well-planned experiments that provide us with a more comprehensive insight into the essential foundations of protein–ligand recognition.

ACKNOWLEDGMENTS

The author is grateful to many present and former members of his research group for their joint efforts in developing various methodological aspects of VS. The computational and experimental work on five different VS targets has been performed by Ruth Brenk (University of California — San Francisco), Sven Grüneberg (Aventis, Frankfurt, Germany), Katrin Silber (University of Marburg, Germany),

Oliver Krämer (Boehringer Ingelheim, Vienna, Austria), and Andreas Evers (Aventis, Frankfurt, Germany). Christoph Sotriffer (University of Marburg, Germany) applied improved docking and MD simulations to better understand the dynamics and binding-competent conformers of proteins. Holger Gohlke (University of Frankfurt, Germany) developed DrugScore, AFMoC, and the hot spot analysis on the basis of knowledge-based potentials. Judith Günther (Schering, Berlin, Germany) and Andreas Bergner (CCDC, Cambridge, England) equipped Relibase with many tools important to embark into a VS project. Many experimental insights into the thermodynamics and plasticity of proteins were collected by Milton Stubbs (University of Halle, Germany), Sabine Reyda (University of Mainz, Germany), Daniel Rauh (University of Halle, Germany), and Frank Dullweber (Altana, Konstanz, Germany).

REFERENCES

[1]   J. Drews, *In Quest of Tomorrow's Medicines,* New York: Springer-Verlag, 1998.
[2]   J. Drews, Drug discovery: a historical perspective, *Science* 287:1960–1964, 2000.
[3]   J. Drews, Drug discovery today — and tomorrow, *Drug Discov. Today* 5:2–4, 2000.
[4]   J. Drews, Strategic trends in the drug industry, *Drug Discov. Today* 8:411–420, 2003.
[5]   H.J. Böhm and G. Klebe, What can we learn from molecular recognition in protein–ligand complexes for the design of new drugs? *Angew. Chem. Int. Ed. Engl.* 35:2588–2614, 1996.
[6]   H. Gohlke and G. Klebe, Approaches to the description and prediction of the binding affinity of small-molecule ligands to macromolecular receptors, *Angew. Chem. Int. Ed. Engl.* 41:2644–2676, 2002.
[7]   G. de Stevens, Serendipity and structured research in drug discovery, *Fortschr. Arzneimittelforsch.* 30:189–203, (1986).
[8]   N.K. Terrett, S.A. Bell, D. Brown, and P. Ellis, Sildenafil (Viagra), a potent and selective inhibitor of Type 5cGMP phosphodiesterase with utility for the treatment of male erectile dysfunction, *Bioorg. Med. Chem. Lett.* 6:1819–1824, 1996.
[9]   G. Mueller, Medicinal chemistry of target family-directed masterkeys, *Drug Discov. Today* 8:681–691, 2003.
[10] W.P. Walters, M.T. Stahl, and M.A. Murcko, Virtual screening — an overview, *Drug Discov. Today* 3:160–178, 1998.
[11] B. Waszkowycz, Structure-based approaches to drug design and virtual screening, *Curr. Opin. Drug Discov. Devel.* 5:407–413, 2002.
[12] A. Good, Structure-based virtual screening protocols, *Curr. Opin. Drug Discov. Devel.* 4:301–307, 2001.
[13] J. Bajorath, Virtual screening in drug discovery: methods, expectations and reality, *Curr. Drug Discov.* 24–28, 2002.
[14] B.K. Shoichet, S.L. McGovern, B. Wei, and J.J. Irwin, Lead discovery using molecular docking, *Curr. Opin. Chem. Biol.* 6:439–446, 2002.
[15] P.D. Lyne, Structure-based virtual screening: an overview, *Drug Discov. Today* 7:1047–1055, 2002.
[16] H. Gohlke and G. Klebe, DrugScore meets CoMFA: adaption of fields for molecular comparison (AFMoC) or how to tailor knowledge-based pair-potentials to a particular protein, *J. Med. Chem.* 45:4153–4170, 2002.
[17] G. Klebe, Virtual screening: an alternative or complement to high throughput screening? Boston: Kluwer Academic Publishers, 2000, 1997.

[18]   M. Schapira, B. Raaka, S. Das, L. Fan, M. Totrov, and Z. Zhou et al., Discovery of diverse thyroid hormone receptor antagonists by high-throughput docking, *Proc. Natl. Acad. Sci. USA* 100:7354–7359, 2003.

[19]   A.V. Filikov, V. Mohan, T.A. Vickers, R.H. Griffey, P.D. Cook, and R.A. Abagyan et al., Identification of ligands for RNA targets via structure-based virtual screening: HIV-1 TAR, *J. Computer-Aided Mol. Des.* 14:593–610, 2000.

[20]   T.N. Doman, S.L. McGovern, B.J. Witherbee, T.P. Kasten, R. Kurumbail, W.C. Stallings et al., Molecular docking and high-throughput screening for novel inhibitors of protein tyrosine phosphatase-1B, *J. Med. Chem.* 45:2213–2221, 2002.

[21]   G. Rastelli, A.M. Ferrari, L. Costantino, and M.C. Gamberini, Discovery of new inhibitors of aldose reductase from molecular docking and database screening, *Bioorg. Med. Chem.* 10:1437–1450, 2002.

[22]   K.E. Lind, Z. Du, K. Fujinaga, B.M. Peterlin, and T.L. James, Structure-based computational database screening, *in vitro* assay, and NMR assessment of compounds that target TAR RNA, *Chem. & Biol.* 9:185–193, 2002.

[23]   E. Vangrevelinghe, K. Zimmermann, J. Schoepfer, R. Portmann, D. Fabbro, and P. Furet, Discovery of a potent and selective protein kinase CK2 inhibitor by high-throughput docking, *J. Med. Chem.* 46:2656–2662, 2003.

[24]   S. Flohr, M. Kurz, E. Kostenis, A. Brkovich, A. Fournier, and T. Klabunde, Identification of nonpeptidic urotensin II receptor antagonists by virtual screening based on a pharmacophore model derived from structure-activity relationships and nuclear magnetic resonance studies on urotensin II, *J. Med. Chem.* 45:1799–1805, 2002.

[25]   A. Aronov, N. Munagala, I. Kuntz, and C. Wang, Virtual screening of combinatorial libraries across a gene family: in search of inhibitors of Giardia lamblia guanine phosphoribosyltransferase, *Antimicrob. Agents Chemother.* 45:2571–2576, 2001.

[26]   S. Grüneberg, B. Wendt, and G. Klebe, Sub-nanomolar inhibitors from computer screening: a model study using human carbonic anhydrase II, *Angew. Chem. Int. Ed.* 40:389–393, 2001.

[27]   S. Grüneberg, M.T. Stubbs, and G. Klebe, Successful virtual screening for novel inhibitors of human carbonic anhydrase: strategy and experimental confirmation, *J. Med. Chem.* 45:3588–3602, 2002.

[28]   K. Silber, Ligand- und Rezeptor-basierte Suchstrategien zum Auffinden neuer Protein-inhibitoren, Institut für Pharmazeutische Chemie, Philipps-University Marburg, PhD thesis, 2002.

[29]   R. Brenk, L. Naerum, U. Gradler, H.D. Gerber, G.A. Garcia, and K. Reuter et al., Virtual screening for submicromolar leads of tRNA-guanine transglycosylase based on a new unexpected binding mode detected by crystal structure analysis, *J. Med. Chem.* 46:1133–1143, 2003.

[30]   R. Brenk, H.D. Gerber, J. Kittendorf, G.A. Garcia, K. Reuter, and G. Klebe, From hit to lead: *de novo* design based on virtual screening hits of inhibitors of tRNA-guanine transglycosylase, a putative target of shigellosis therapy, *Helv. Chim. Acta* 86:1435–1452, 2003.

[31]   C. Sotriffer, O. Krämer, and G. Klebe, Probing flexibility and "induced-fit" phenomena in aldose reductase by comparative crystal structure analysis and molecular dynamics simulations, *Proteins* 56:52–66, 2004.

[32]   O. Krämer, I. Hazemann, A.D. Podjarny, and G. Klebe, Virtual screening for inhibitors of human aldose reductase, *Proteins* 55:814–823, 2004.

[33]   G. Klebe, O. Krämer, and C. Sotriffer, Strategies for the design of inhibitors of aldose reductase, an enzyme showing pronounced induced-fit adaptations, *Cell Mol. Life Sci.* 61:783–793, 2004.

[34] A. Evers and G. Klebe, Ligand-supported homology modelling of g-protein coupled receptor sites: models sufficient for successful virtual screening, *Angewandte Chemie*, 43:248–251, 2004.

[35] A. Evers, H. Gohlke, and G. Klebe, Ligand-supported homology modelling of protein binding sites using knowledge-based potentials, *J. Mol. Biol.* 334:327–345, 2003.

[36] M. Hendlich, A. Bergner, J. Günther, and G. Klebe, Relibase: design and development of a database for comprehensive analysis of protein–ligand interactions, *J. Mol. Biol.* 326:607–620, 2003.

[37] J. Günther, A. Bergner, M. Hendlich, and G. Klebe, Utilizing structural knowledge in drug design strategies: applications using Relibase, *J. Mol. Biol.* 326:621–636, 2003.

[38] P.J. Goodford, A computational procedure for determining energetically favorable binding sites on biologically important macromolecules, *J. Am. Chem. Soc.* 28:849–857, 1985.

[39] M.L. Verdonk, J.C. Cole, and R. Taylor, SuperStar: A knowledge-based approach for identifying interaction sites in proteins, *J. Mol. Biol.* 289:1093–1108, 1999.

[40] M.L. Verdonk, J.C. Cole, P. Watson, V. Gillet, and P. Willett, SuperStar: improved knowledge-based interaction fields for protein binding sites, *J. Mol. Biol.* 307:841–859, 2001.

[41] H. Gohlke, M. Hendlich, and G. Klebe, Predicting binding modes, binding affinities and "hot spots" for protein–ligand complexes using a knowledge-based scoring function, *Persp. Drug Discov. Des.* 20:115–144, 2000.

[42] H. Gohlke, M. Hendlich, and G. Klebe, Knowledge-based scoring function to predict protein–ligand interactions, *J. Mol. Biol.* 295:337–356, 2000.

[43] UNITY: Unity Chemical Information Software (ed 4.1.1.), Tripos Inc., St. Louis, MO.

[44] C.A. Lipinski, F. Lombardo, B.W. Dominy, and P.J. Feeney, Experimental and computational approaches to estimate solubility and permeability in drug discovery and development settings, *Adv. Drug Delivery Rev.* 23:3–25, 1997.

[45] M. Rarey and J.S. Dixon, Feature trees: a new molecular similarity measure based on tree matching, *J. Computer-Aided Mol. Des.* 12:471–490, 1998.

[46] C. Lemmen, T. Lengauer, and G. Klebe, FlexS: a method for fast flexible ligand superposition, *J. Med. Chem.* 41:4502–4520, 1998.

[47] G. Klebe, T. Mietzner, and F. Weber, Different approaches toward an automatic structural alignment of drug molecules: applications to sterol mimics, thrombin and thermolysin inhibitors, *J. Computer-Aided Mol. Des.* 8:751–778, 1994.

[48] G. Klebe, T. Mietzner, and F. Weber, Methodological developments and strategies for a fast flexible superposition of drug-size molecules, *J. Computer-Aided Mol. Des.* 13:35–49, 1999.

[49] M. Rarey, B. Kramer, T. Lengauer, and G. Klebe, A fast flexible docking method using an incremental construction algorithm, *J. Mol. Biol.* 261:470–489, 1996.

[50] G. Jones, P. Willett, R.C. Glen, A.R. Leach, and R. Taylor, Development and validation of a genetic algorithm for flexible docking, *J. Mol. Biol.* 267:727–748, 1997.

[51] T.J.A. Ewing and I.D. Kuntz, Critical evaluation of search algorithms for automated molecular docking and database screening, *J. Comp. Comput. Chem.* 9:1175–1189, 1997.

[52] S. Makino and I.D. Kuntz, Automated flexible ligand docking method and its application for database search, *J. Comput. Chem.* 18:1812–1825, 1997.

[53] D.S. Goodsell and A.J. Olson, Automated docking of substrates to proteins by simulated annealing, *Proteins* 8:195–202, 1990.

[54] F. Osterberg, G.M. Morris, M.F. Sanner, A.J. Olson, and D.S. Goodsell, Automated docking to multiple target structures: incorporation of protein mobility and structural water heterogeneity in AutoDock, *Proteins: Structure, Function, and Genetics* 46:34–40, 2002.

[55] C.A. Sotriffer, H. Gohlke, and G. Klebe, Docking into knowledge-based potential fields: a comparative evaluation of DrugScore, *J. Med. Chem.* 45(10):1967–1970, 2002.

[56] S.A. Hindle, M. Rarey, C. Buning, and T. Lengauer, Flexible docking under pharmacophore type constraints, *J. Computer-Aided Mol. Des.* 16(2):129–149, 2002.

[57] R.D. Cramer, III, D.E. Patterson, and J.D. Bunce, Comparative molecular field analysis (CoMFA). 1. Effect of shape on binding of steroids to carrier proteins, *J. Am. Chem. Soc.* 110:5959–5967, 1988.

[58] G. Klebe, U. Abraham, and T. Mietzner, Molecular similarity indices in a comparative analysis (CoMSIA) of drug molecules to correlate and predict their biological activity, *J. Med. Chem.* 37:4130–4146, 1994.

[59] M. Stahl and H.J. Boehm, Development of filter functions for protein–ligand docking, *J. Mol. Graphics Modelling* 16:121–132, 1998.

[60] H. Jansen, Docklt and Magnet: fast docking with magnetized selection and scoring, (2001). http:www.daylight.com/meetings/mug01/Jansen.

[61] R. Brenk, M.T. Stubbs, A. Heine, K. Reuter, and G. Klebe, Flexible adaptations in the structure of the tRNA modifying enzyme tRNA-guanine transglycosylase and its implications for substrate selectivity, reaction mechanism and structure-based drug design, *Chembiochem*, 4(10):1066–1077, 2003.

[62] W. Xie, X. Lin, and R.H. Huang, Chemical trapping and crystal structure of a catalytic tRNA guanine transglycosylase covalent intermediate, *Nature Struct. Biol.* 10:781–788, 2003.

[63] C. Romier, K. Reuter, D. Suck, and R. Ficner, Crystal structure of tRNA-guanine transglycosylase: RNA modification by base exchange, *Embo. J.* 15:2850–2857, 1996.

[64] M.T. Stubbs, S. Reyda, F. Dullweber, M. Möller, G. Klebe, and K. Stabe et al., pH-dependent binding modes observed in trypsin crystals — lessons for structure-based drug design, *ChemBioChem* 2002:246–249, 2002.

[65] F. Dullweber, M.T. Stubbs, D. Musil, J. Sturzebecher, and G. Klebe, Factorising ligand affinity: a combined thermodynamic and crystallographic study of trypsin and thrombin inhibition, *J. Mol. Biol.* 313:593–614, 2001.

[66] O. Krämer, Rationales Wirkstoff-design am Beispiel der Aldose-Reduktase, Philipps-University Marburg, PhD thesis, 2003.

[67] J.E. Ladbury, Just add water! The effect of water on the specificity of protein–ligand binding sites and its potential application to drug design, *Chem. & Biol.* 3:973–980, 1996.

[68] J. Günther, Entwicklung einer Datenbank und Wissensbasierter Vorhersagemethoden zur Untersuchung von Wassermolekülen in Proteinstrukturen sowie ihrer Rolle in der Protein–Ligand-Bindung, Philipps-University Marburg, PhD thesis, 2003.

[69] M. Huse and J. Kuriyan, The conformational plasticity of protein kinases, *Cell* 109:275–282, 2002.

[70] S. Reyda, C. Sohn, G. Klebe, K. Rall, D. Ullmann, and H.-D. Jakubke et al., Reconstructing the binding site of factor Xa in trypsin reveals ligand-induced structural plasticity, *J. Mol. Biol.* 325:963–977, 2003.

[71] D. Rauh, S. Reyda, G. Klebe, and M.T. Stubbs, Trypsin mutants for structure-based drug design: expression, refolding and crystallisation, *Chem. & Biol.* 383:1309–1314, 2002.

[72] D. Rauh, G. Klebe, J. Stürzebecher, and M.T. Stubbs, ZZ made EZ. Influence of inhibitor configuration on enzyme selectivity, *J. Mol. Biol.* 2003:761–770, 2003.

[73] D.W. Banner, A. D'Arcy, C. Chene, F.K. Winkler, A. Guha, and W.H. Konigsberg, The crystal structure of the complex of blood coagulation factor VIIa with soluble tissue factor, *Nature* 380:41–46, 1996.

[74] A.C. Pike, A.M. Brzozowski, S.M. Roberts, O.H. Olsen, and E. Persson, Structure of human factor VIIa and its implications for the triggering of blood coagulation, *Proc. Natl. Acad. Sci. USA* 96:8925–8930, 1999.

[75] D. Rauh, G. Klebe, and M.T. Stubbs, Understanding protein–ligand interactions: the price ligand binding pays for, *J. Mol. Biol.* 835:1325–1341, 2004.

2 Addressing the Virtual Screening Challenge: The Flex* Approach

Marcus Gastreich, Christian Lemmen,
Hans Briem, and Matthias Rarey

2.1 INTRODUCTION

Due to the increase in target protein structures available as well as the increased quality of homology-based protein models, molecular docking and structure-based virtual screening (VS) are two key technologies in current drug discovery (see [1–4] for recent reviews). Their clear advantage is the direct elucidation of binding properties without any bias of the chemical structure of the small molecules. Solving the docking problem, namely the prediction of the geometry and binding affinity of a protein–ligand complex given a three dimensional (3D) structure of the protein and the ligand molecule, is at the heart of structure-based design. Several variants of the base problem exist that often allow, sometimes request, differentiated method and tool development [4,5]. Some targets show significant protein flexibility in the active site. Frequently, models have a high degree of uncertainty concerning particular atom positions. Water molecules and metal ions might play an important role. When it comes to the docking of multiple molecules, the efficient handling of combinatorial libraries as well as the incorporation of prior knowledge such as enforcing a certain interaction pattern gain importance.

Unfortunately, a protein structure is not always accessible. For some targets and target families like G protein-coupled receptors (GPCRs), virtually no progress has been made to unravel sizable numbers of protein structures for practical docking applications. In such cases, the ligands known to bind to the target of interest are the most valuable source of information for a VS project. Then molecular similarity-based approaches become the method of choice for mining through compound databases for other potentially superior drug candidates.

Small molecule alignment is among the predominantly used methods for the detailed comparison of molecules considering their 3D-conformation and flexibility [6,7]. Descriptor-based methods usually focus in some way on the topology of the small molecule and thereby gain in speed but only allow for a much cruder similarity assessment.

However, VS is more than mere molecular docking or alignment calculations. Input data preparation for ligand molecules must be automated because hundreds of thousands of compounds are typically processed [3]. The overall workflow of the screening process must be supported by an information technology (IT) infrastructure: databases of input molecules must be accessible and the utilization of parallel compute resources should be implemented in a user-friendly way [8,9]. The output of a VS run — a ranked list of compounds with scores—needs to be analyzed in various ways. A structure–activity relationship (SAR) must be found, important binding modes should be determined, known active compounds may be compared and scoring outliers need to be explained. Because several tens of alternative poses are usually calculated, this task requires a data management concept including tools for VS data analysis.

In this chapter, we will summarize the Flex*-approach to VS. We start by surveying the physicochemical models and basic algorithms of the FlexX docking [10] and FlexS alignment engine [11]. Subsequently, modules that allow for combinatorial library handling, the consideration of pharmacophore constraints, and protein flexibility are explained. In VS, the details on the ligand side usually matter: We will therefore explain how we deal with flexible ring systems and unresolved stereo centers. Finally, we will come to VS workflows, introducing a docking database concept for screening results, the job scheduling system, and the analysis software. As application scenarios, we will describe several virtual screening exercises on the basis of data on cyclin-dependent kinase 2 (CDK2) utilizing various components of the Flex* software.

2.2 ELEMENTARY MODELS FOR DOCKING AND STRUCTURAL ALIGNMENT CALCULATIONS

The two major ingredients of a VS tool are the underlying physicochemical models and its algorithms and data structures. Although frequently stated, these two components can be considered independent from each other only if simple, generic optimization algorithms are used. Flex* and several other VS tools make use of the properties of a model to focus the search for optimal poses in promising regions of the search space. Model and method development are linked, which implies that the models have to follow two goals, on the one hand being close to reality such that useful predictions can be made, on the other hand having a structure that allows for the development of efficient algorithms. Here we summarize the Flex* models [10,12] that were developed having these two goals in mind.

2.2.1 CONFORMATIONS

Good coverage of the ligand's conformational space within a docking calculation is extremely important for success in structure-based VS. It is a well-known fact that small molecule conformations in bound states can differ substantially from the low-energy conformations in solution [13]. Because high-energy conformations are unlikely to occur in protein–ligand complexes, the aim is to represent the space of low-energy conformations. To this end, the Flex* tools, follow the MIMUMBA approach [14]. For a set of about 1000 molecular fragments with a central rotatable bond, a statistical analysis of crystallographic data [15,16] augmented by Protein Data Bank (PDB) data [17] was employed to calculate the distribution of torsional angles for each individual fragment. For the molecule of interest, the distributions for the central bond of all matching fragments from the MIMUMBA database are assigned. In case of multiple matches, the distributions are merged into a single one. From these, a set of frequently occurring torsional angles is derived applying a grid following the idea of an inverse Boltzmann law (see for example [18]). Figure 2.1 displays the number of principally accessible conformations if all conformers were enumerated and the root-mean-square deviation (rmsd) of the conformation closest to the crystal structure in relation to the number of rotatable bonds. The figure illustrates that low rmsd values can be achieved with the MIMUMBA approach, however only with a sizeable number of conformations. Efficient data structures and search procedures are needed for handling and navigating in this vast search space.

2.2.2 MOLECULAR INTERACTIONS

In principle, modeling the relative orientation and conformation of a ligand and applying a scoring function is sufficient for the development of a simple VS tool. It turns out however, that a good geometric model for the representation of molecular interactions helps to limit the search space to low-energy regions. Basically every successful structure-based VS software available today relies on a model for molecular interactions and bases its search engine on this model.

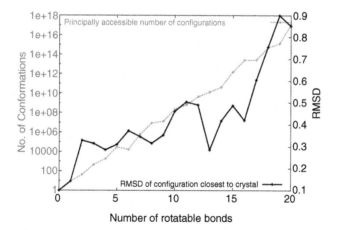

FIGURE 2.1 The diagram shows the principally accessible number of conformations (logarithmic y-axis, dashed line, axis labeling on the left-hand side) and the rmsd value of the conformation closest to crystal structure (solid line, linear y-axis, labeling on the right-hand side) in relation to the number of rotatable bonds. Data calculated from the FlexX-200 dataset using the MIMUMBA conformation model.

The Flex* tools use spherical interaction surfaces, a model which is closely related to the one developed for the *de novo* design software LUDI [19,20]. For an interacting group, an interaction type, a center, and a spherical surface patch (called the "interaction surface") around the center is defined. An interaction can be formed if another interacting group exhibiting a compatible type is placed such that its interaction center A lies on the surface of interaction B and vice versa (see Figure 2.2A).

This model allows the representation of interaction distance and directionality at both interacting groups. It is sufficiently generic to represent all kinds of interactions such as hydrogen bonds, salt bridges, metal coordination, π-π-interactions, as

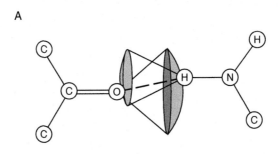

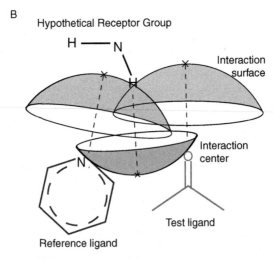

FIGURE 2.2 FlexX interaction surface matching between a carbonyl group and a primary amine function. The interaction centers lie mutually on the partner's surface. B) A FlexS interaction situation: The surface of the pyridine-N acceptor surface of the reference ligand overlaps with the acceptor surface of a carbonyl group. Because both are acceptor groups, they are considered compliant; a hypothetical receptor amine-H may come to lie on either interaction surface, its respective surface matching both N (pyridine) and O (carbonyl).

well as interactions generally referred to as hydrophobic interactions. For the search algorithms described later, the interaction surfaces are approximated by interaction points on the interaction surfaces.

Comparing FlexX (docking) with FlexS (alignment) with respect to (w.r.t.) their interaction models, it is obvious that the interaction partner on the protein side is missing. It was found insufficient to match either the ligand interaction centers or some specific site point along the main axis of the interaction. Rather, the compatibility of the interacting groups and some overlap of the respective interaction surfaces was found appropriate for the modeling of matching interactions in an alignment scenario (see Figure 2.2b for an illustration).

Even though the generic interaction model is directly implemented in the Flex* tools, the full parameterization is well separated. All interaction types, interacting groups, and interaction geometries are defined separate from the code in specific data files. This enables individual adjustments. Interaction geometries can be added (for example, C–H groups as H-bond donors) and can be made more specific (for example a special interaction geometry for a furan-O H-bond acceptor).

2.2.3 SCORING

The molecular docking tool FlexX supports several scoring functions. The original FlexX scoring function (termed "FlexX score") [10] is a variant of the empirical function Score developed by Böhm [21]. The FlexX score contains a hydrogen bonding term, a penalty for protein–ligand overlap, a pairwise hydrophobic potential (instead of a grid-based one), and additional terms for specific hydrophobic contacts. As alternatives, FlexX comes with PLP (Piecewise Linear Potential) scoring [22] and all components necessary to parameterize ChemScore [23]. Also, a generic interface to PMF-scoring functions [24,25] is available.

The scoring function of FlexS on the other hand has two parts. First, the abovementioned matching of interactions is assessed similarly to the Böhm scoring [21], including additive contributions for every match and penalties for geometric deviations. Second, the overlap of similar physicochemical properties of the molecules is considered. Multiple sets of Gaussian (field) functions represent the typically considered properties such as the electron density and charge distribution, hydrophobicity, and the hydrogen-bonding potential. The counterpart to a protein–ligand clash test in docking is the constraint of some minimum overlap requirement of the van der Waals (vdW) surfaces of the molecules to enforce proper overlays. As for the molecular interaction model, scoring terms and functions are parameterized outside the Flex* code.

For both tools, individual scoring terms can easily be combined and weighted relative to each other. Furthermore, for each scoring term, one may distinguish the usage during the docking procedure from the final ranking of compounds. The ScreenScore function [26] is a good example demonstrating the advantage of such an approach.

2.3 BASIC ALGORITHMIC CONCEPTS

Structure-based VS tools can roughly be divided into three classes based on the way they deal with ligand flexibility [27]. The *enumerators* calculate a set of low-energy conformations in advance and dock or align them rigidly. The *randomized searchers* use generic, randomized search engines such as genetic algorithms or simulated annealing to optimize the placement by score. Finally, the *constructors* try to build the conformation of molecules during the actual calculation. Examples for docking tools belonging to all three classes can be found in this book (enumerators: FRED [28] and FRISCO [29]; randomized searchers: GOLD [30,31], GASP™ (Genetic Algorithm Similarity Program) [30], and AutoDock [32,33]; constructors: DOCK 4.0 [34], FlexX [10], and FlexS[11]).

Flex* is based on an incremental construction concept. First, the ligand is decomposed, anchor pieces are placed independently from the rest of the ligand molecule, and finally the molecules are completed piece by piece simultaneously considering the conformational degrees of freedom. The advantage of this technique is that the conformational search can be restricted to conformations which fit into the active site (respectively, fitting into some reference shape) while still being deterministic and in this way preserving an advantage in computing time over randomized approaches. The disadvantage is that the whole solution can be reasonably scored only at the end of this optimization procedure. Algorithmically, the two most challenging steps are the placement of molecular fragments and the efficient implementation of the incremental construction procedure.

2.3.1 PLACING MOLECULAR FRAGMENTS

For fragment placement, Flex* uses an algorithmic technique called pose clustering [10,35]. The algorithm consists of three phases. In the first phase, a hash table of interaction triangles within the protein active site (representing the reference compound) is created. As mentioned earlier, interaction surfaces are approximated by discrete interaction points. All pairs of interaction points are stored in a hash table that can be addressed by interaction types and the distance between the points. This procedure is independent from the ligand and has to be done only once in a VS run. In the second phase, for each triplet of interaction centers (interaction points in FlexS) of the fragment to be placed, corresponding triplets of interaction points in the hash table are searched. This is done by a routine that uses the hash table to quickly create these corresponding triangles on the fly. From every matching triangle, a transformation of the fragment can be computed by superimposing the fragment triangle onto the triangle of interaction points. In the third phase, all transformations are subjected to clustering to come up with a final list of unique fragment placements. More details on all algorithmic phases can be found in [10] and [36].

Flex* uses all kinds of interactions—from salt bridges and hydrogen bonds to hydrophobic interactions—for this fragment process [37]. Due to an efficient implementation, several thousand interaction points can be processed this way, creating a few hundred energetically favorable fragment placements within a few seconds.

FlexS utilizes the same algorithmic machinery of matching triangles of interaction points. However, oftentimes no sufficient number of interaction partners can be found on either molecule to find decent overlays. In such cases, FlexS uses a field-based alignment method for the placement of fragments, focusing only on the overlap of the Gaussians that represent the fields. This optimization-based method called RigFit [38] operates partially in Fourier space, which facilitates an especially efficient translational optimization.

2.3.2 THE INCREMENTAL CONSTRUCTION PHASE

Based on a set of placements for a molecule fragment, a pose for the whole molecule can be created incrementally [12,36]. The ligand fragment is extended by an adjacent piece of the molecule. Correspondingly, the set of poses extends. For every pose, a new set of poses is calculated containing the extended ligand, one entry for each possible conformation of the newly added moiety. The pose for the old part of the ligand is taken from the predecessor in the calculation. For all new poses, an overlap test and filtering is performed, matching interactions between the protein and the new fragment are searched for, the overall orientation can be optimized, and finally scored.

Without any further limitations, this process would end up with a number of poses which is exponential in the number of fragments added (see Figure 2.1). To keep the incremental construction within certain computing time and space limits, a selection of good scoring poses as well as a clustering of poses is performed. For an efficient implementation, it is extremely important to use hierarchical data structures avoiding recalculations for overlap tests and scoring for the already placed part of the ligand in every extension step. Using these data structures enables handling of about 800 partial placements. In every iteration this number is inflated up to 10,000 partial solutions and shrunk down again to about 800 partial solutions. This is performed in less than 2 seconds computing time per incremental construction step. The larger this set can be, the smaller the chance to get trapped in a local (score) minimum during the whole optimization process. User-definable parameters allow for adjusting the optimum balance between speed and accuracy of the computation.

2.3.3 BRINGING IT ALL TOGETHER: BASE SELECTION—BASE PLACEMENT—INCREMENTAL CONSTRUCTION – POSTOPTIMIZATION

Fragment placement and the construction phase are the two major components for an incremental construction docking or alignment algorithm. There is one question left: how to decide where to start the calculation, or, how to choose a base fragment. A good base fragment should not be too flexible, yet bind specifically with respect to the protein–ligand complex. The second criterion, however, is difficult to find out in advance. FlexX and FlexS circumvent the problem by selecting up to four base fragments in different portions of the ligand molecule [39]. In this way, we can be reasonably sure that if the ligand binds, we will have a base fragment interacting with the protein. The fragment placement routine described above is then applied independently for each base fragment. In the incremental construction phase, the situation is more difficult. First, scores for partial ligand molecules need to be comparable even if differently sized and decorated parts of the ligand are in competition. This is performed by certain normalization techniques in combination with a score estimate for the rest of the ligand (see [39] for details). Second, at least a small fraction of solutions from each anchor fragment is forced to survive the construction process. This is accomplished by a two phase selection process, where 50% of the poses are picked individually for each base fragment and 50% are picked from the entire set.

At the end, it is sometimes desirable to leave the discrete conformational search space and energetically relax the placement in a postoptimization step. Flex* facilitates this final numerical optimization for any of the implemented scoring functions. Both methods also allow for defining base fragments by substructure matching, which facilitates a semimanual base selection in cases of a recurring molecular entity but with unknown atom numbering.

2.4 ADVANCED CONCEPTS FOR VS TOOLS

2.4.1 COMBINATORIAL LIBRARIES

Combinatorial libraries contain an internal structure in the molecule set that can be exploited for speeding up the VS procedure. Due to the common core and the fact that every R-group instance occurs multiple times in the enumerated library, a significant speedup can be achieved. To deal with combinatorial libraries, the Flex*C module was developed [40].

For large libraries, Flex*C can evaluate core–R-group instance pairs. These fragments of combinatorial library molecules are placed and scored. Low scoring R-groups can then be removed in order to shrink the size of the library.

For medium-sized libraries, Flex*C uses a combinatorial extension of the incremental construction algorithm. Here, the library molecules are constructed one by one re-using the placement information of the common parts between two molecules in this sequential process. The advantage of this technique is that still every molecule of the library is fully enumerated inside the active site (respectively into the reference shape) such that dependencies between the placement of R-groups can be considered. Depending on the size of the library, speedup factors of up to 50 can be achieved in principle without much deteriorated quality of the VS result.

Here the implementation rationales of the Flex*-software pay off in that these enhancements could be added with moderate extra effort to FlexS, too: With the latest version of the code, there is now also a module (FlexS^C) available for the efficient handling of combinatorial libraries in the small molecule alignment area.

2.4.2 DOCKING UNDER PHARMACOPHORE CONSTRAINTS

A frequently occurring scenario in structure-based VS is an *a priori* definition of
protein atoms that should be directly involved in binding a ligand. For example, an
important metal ion in the active site should be coordinated, a specific hydrogen
bond network should be established, or a certain hydrophobic pocket should be
filled. In principle, this simplifies the docking problem because only a constrained
set of ligand orientations has to be considered. To fully exploit this simplification,
the constraints have to be applied as early as possible such that poses that cannot
fulfill them are ruled out during and not only after the construction procedure. For
constructor-type docking algorithms such as FlexX, the problem occurs that during
the calculation, placement information is available only for a certain part of the
ligand. For each such partial ligand placement, a decision has to be made whether
the constraints can be potentially fulfilled or not. The extension module, FlexX-
Pharm [41] distinguishes three levels of tests with increasing complexity to address
this question. The answer of each test is either "no" or "could be." The *logical* test
ascertains that for each constraint to be fulfilled, a corresponding moiety in the ligand
is available. The *static distance* test checks whether the distances between involved
locations in the active site are compatible with a precomputed distance matrix
containing upper and lower bounds for all ligand interatom distances. The *directed
tweak* test determines whether conformations exist that are able to fulfill all constraints
simultaneously. Due to the narrowing of the search space, a significant speedup, and
chemically even more important, a more intense search within the relevant parts of
the search space can be achieved [42]. An application example of the usage of FlexX-
Pharm will be given below.

2.4.3 PROTEIN FLEXIBILITY

Dealing with conformational flexibility of the target proteins is one of the
most challenging problems in molecular docking. Besides the drastic
increase in the number of degrees of freedom, the complexity of protein
movements and the estimation of free energy differences between protein–ligand complexes with different protein conformations are the reasons
to cause this extra level of complexity. The FlexX extension module FlexE™
considers protein flexibility using an ensemble approach [43]. As input, FlexE
uses a set of alternative structures or configurations for the protein of interest.
For example, this set can be derived from crystal structures of the protein
cocrystallized with different ligands, from an analysis of nuclear magnetic
resonance (NMR)-chemical shift data or molecular dynamic calculations. In
a first step, FlexE determines conserved and moving parts of the protein. Also
it constructs a compatibility graph for all elements of the ensemble (such as
different side-chain conformers). During the docking calculation, for every
pose of a partially placed ligand, FlexE constructs the best fitting protein
structure by combining conformations from different compatible members of
the ensemble. A quite efficient implementation allows this to be taken into

account each time the scoring function is evaluated. This way, only valid protein–ligand complexes are generated. Example applications of FlexE can be found in [43,44].

2.4.4 MULTIMOLECULE ALIGNMENT

A drawback of alignment methods is that they rely extremely on the presence of a suitable reference molecule that has to be more or less a negative imprint of the active site it binds to. However, often different ligands populate different subpockets in the binding site and so, only a set of molecules forms this negative imprint. To take this observation into account, FlexS allows for loading multiple fields describing multiple molecules in addition to the single distinct reference molecule. This has proven to boost the performance in VS studies beyond the capabilities of using only a single reference molecule [45].

2.4.5 DETAILS TO BE TAKEN INTO ACCOUNT

Basic physics and chemistry tell us there are a variety of other factors exhibiting influence on bonding situations than the ones previously assessed in Section 2.2. Here, we would like to discuss a few models relevant in this context and explain to what extent FlexX and FlexS are able to deal with them.

2.4.5.1 Covalent Binders

A prediction of whether and how a ligand binds covalently to a given target protein is out of reach for tools based on fast-to-compute empirical scoring functions. Fortunately, covalent binding is fairly rare in drug design studies. For the few exceptions, the functional groups forming the covalent bond are usually known in both ligand and protein. In this scenario, FlexX offers a manual covalent docking mode. The user specifies the covalent bond, and the base placement algorithm is replaced by an algorithm that forms this bond considering all possible conformations of the base fragment (including rotations around the covalent bond itself). In this way, FlexX can be used to rank order a series of compounds, all of which exhibit the desired covalent bond.

2.4.5.2 Flexible Ring Systems

Flexible ring systems occur frequently in organic compounds and therefore in VS libraries. Although small cycles of up to eight atoms can be reasonably described by a small set of discrete conformations, this is not the case for macrocycles. These are kept rigid in their input configuration. For small rings, Flex* use either Confort™ [46,47] or the ring conformer generator within the 3D structure generator software CORINA [48,49] to enumerate the respective conformations on the fly. The resulting ring-conformers are taken into account as additional degrees of freedom during the placement procedure.

2.4.5.3 Water Molecules and Ions

Water molecules as well as ions located in the interface between a protein and a ligand can have a significant influence not only on the binding affinity, but also on the geometry of the protein–ligand complex [50]. In FlexX calculations, water molecules and ions can manually be placed within the active site of the protein. If this approach is too inflexible—considering the likelihood of presence and exact location of a water molecule to change from ligand to ligand—there is an alternative approach called the particle concept [51]. Here, preferred locations for water molecules in the active site are computed automatically in a preparation step. Depending on the ligand and the actual pose, water molecules may or may not be kept at these positions. The particle concept can also be used for ions or other objects that can be approximated by a spherical interaction model.

2.4.5.4 Stereoisomerism

Another problem that occurs frequently in VS runs is the potentially undetermined stereo chemistry of a ligand. Sometimes stereo centers are left unspecified in a library with good reason (e.g., if racemates are available for testing only). Representing different stereoisomers as different molecules in the database is a quite time-consuming solution for this problem. Flex* allows for considering R/S stereo-chemistry centers at carbon atoms inversion at nonplanar nitrogens and E/Z isomerism at double bonds as additional degrees of freedom during the incremental placement. In this way, the best fitting stereoisomer can be selected by the docking algorithm in the same way as the best fitting conformer.

2.5 SUPPORTING THE WORKFLOW—FROM AN ALGORITHMIC ENGINE TO A VS MACHINE

2.5.1 Scripting the Screening Process

A basis for automated workflows is a powerful scripting language. All Flex* tools come with a simple and originally proprietary, yet versatile batch scripting facility that incorporates loops and branching statements as well as variable assignment and Boolean comparison. Data stream redirection mechanisms allow for flexible I/O (input/output) handling. More and more, Python [52] establishes itself as a leading scripting language. Python is an object-oriented, full featured, open-source programming language that is free to the public. A rapidly increasing number of extensions, called modules (e.g., PyMOL, OElib, Qt, Relibase, etc.), allows for rapid prototyping by easily "plugging together" novel applications. C source code can be wrapped in a Python layer. This way it is possible to turn also native C applications into Python modules. We did this with the Flex* suite of programs thus facilitating the full functionality of this software by a simple `import pyflexx` statement in a Python script. Evidently, this opens an extra degree of flexibility and power for scripting purposes. Scripting entire workflows from assembling the input data over a cascade of VS applications to the statistical analysis and subsequent visualization become an easy task by importing the required statistical and graphical modules (e.g., the gnuplot Python module [53]).

2.5.2 PARALLEL VS

The second main ingredient for setting up an efficient computing workflow is to seamlessly distribute the compute task across several machines in a balanced way—parallelizing. The Flex* suite accomplishes this task by means of the parallel virtual machine (PVM) approach. PVM is "a software package that permits a heterogeneous collection of UNIX® and/or Windows® computers hooked together by a network to be used as a single large parallel computer" [54]. PVM has been ported to various platforms and, as Python is, PVM is open source and therefore freely available. In Flex* scripts, splitting up a compute task for distributed processing including collecting and merging the outputs is automatically taken care of by the extension module PVM. This comprises cross-platform usage; the user can utilize different architectures within the same batch job.

2.5.3 A DATABASE OF VS RESULTS

Despite all comfort when dealing with parallelized scripts, one usually deals with flat file in- and output. For a detailed analysis of docking results (including pharmacophore verification and rescoring), storing computed results in databases is certainly more appropriate. Subsequent queries to the database facilitate all sorts of analyses of the results. Filtering, as well as the assembly of complex statistical data derived from the solution space become straightforward tasks. Examples of such queries for determining a receptor-based pharmacophore have been described previously [8]. By interfacing FlexX to a Python database module and setting up an appropriate database-application in Oracle® 9.2, uploading docking data becomes a routine task. In addition to getting rid of a lot of tedious flat file handling, there is another major advantage for the user: Apart from structured storage of the results, one can organize configuration details and even executables together with the calculated results in so-called VS projects—thus achieving maximum reproducibility of the experiments.

2.5.4 MINING VS RESULTS

The final ingredient for setting up a workflow consists of making it transparent to the user and allowing for easy interaction with the respective computing facilities. In a cooperation project [8], we recently combined the Python-based database interface with an intuitive graphical user interface (GUI) called Seeker. Seeker fulfills three main functions:

1. Setting up experiments—the data-tree type fashion input related to protein, ligand, and configuration-related information can be collected and arranged to stay on top of the flood of data.

2. Result analysis—postprocessing of the outcomes of jobs is feasible through an integrated spreadsheet-like area within Seeker. Here, the user may not only analyze FlexX related data, but for example, insert his own scoring or molecule property data, or define filters and new scoring schemes based on employing user-definable functions, for example a linear combinations of a variety of scoring terms, etc.

3. Molecule visualization—because nothing replaces visual inspection, Seeker has been interfaced with FlexV, the 3D visualization facility developed around the Flex* suite of tools.

2.6 A PRACTICAL EXAMPLE—VS ON A CYCLIN-DEPENDENT KINASE TARGET

2.6.1 An Initial Test—Reproducing and Cross-Docking Crystal Structures

The ability of the docking tool to predict the correct binding mode of an active compound reasonably well can be considered a prerequisite for a successful structure-based VS. In high throughput docking, there is an additional challenge to find a good placement on rank 1 of the list of solutions, as in such experiments one typically has to rely only on the best score of each docked ligand. It has been demonstrated previously [55] that FlexX can reasonably reproduce the correct binding mode (rmsd ≤ 2.5 Å) on top rank with a probability of about 50% on a sizeable benchmark set. In that study, Kramer et al. utilized a diverse set of 200 protein–ligand complexes from the PDB [56] and redocked each ligand into its native protein (i.e., the one in which it was crystallized). In VS though, one is faced with the problem of finding correct placements in binding pockets alien to the respective ligand. Therefore, we were interested not only in redocking but rather in cross-docking (i.e., also docking into nonnative crystal structures of the given target protein). As a test case, we selected 16 PDB complexes of CDK2.[2] CDK2 belongs to the Ser/Thr protein kinase superfamily. A remarkable degree of protein flexibility around the adenosine triphosphate (ATP)-binding pocket, which is the binding site for all currently known inhibitors, has been observed. As such, it is particularly challenging to find good docking solutions in a cross-docking experiment.

Each ligand was docked into each of the 16 protein structures with FlexX, using default parameters and the ScreenScore scoring function [57]. Table 2.1 summarizes the rmsds of the top ranking docking solutions. The left-hand side of the table shows the cross-docking matrix and the right-hand side shows the results of various analyses of the matrix.

The following conclusions can be drawn from this experiment: Redocking led to rmsds ≤ 2.5 Å in 12 of 16 cases (75%). If one considers "any" of the protein structures (Best rmsd All column), there is at least one good docking solution for every ligand. Even if we omit the redocking results, i.e., the diagonal of the docking matrix (Best rmsd No Redock column)—there is a good solution in at least one

TABLE 2.1

rmsd Values of CDK2 Ligands (rows) Docking into Different CDK2 Binding Pockets (columns). Only Rank 1 Solutions Considered.

	1aq1	1b38	1di8	1dm2	1e1v	1e1x	1fin1	1fin2	1fvt	1fvv1	1fvv2	1g5s	1gih	1gii	1gij	1jvp	Redocking	Best RMSD All	Best rmsd No Redock	MRGD by Score All	MRGD by Score No Redock
1aq1	2.27	15.87	11.48	20.93	14.71	11.79	3.17	2.47	9.13	14.79	10.92	6.12	10.84	10.10	14.27	21.57	2.27	2.27	2.47	2.27	6.12
1b38	5.28	1.80	9.19	3.00	2.20	4.10	8.12	3.90	5.20	3.87	2.05	2.32	3.98	4.43	9.86	1.97	1.80	1.80	1.97	3.00	3.00
1di8	7.55	7.60	1.28	8.45	7.15	7.25	7.30	7.92	4.48	8.34	4.48	7.46	8.02	7.41	7.44	8.24	1.28	1.28	4.48	7.25	7.25
1dm2	6.01	4.92	1.37	1.31	8.03	5.96	1.64	1.96	1.41	1.39	1.46	1.27	1.65	6.82	5.36	1.69	1.31	1.27	1.27	1.46	1.46
1e1v	3.72	2.12	3.62	6.73	3.27	1.44	4.57	6.32	1.45	2.51	3.57	2.13	2.65	2.52	2.38	2.22	3.27	1.44	1.44	1.44	1.44
1e1x	2.16	6.13	5.76	5.70	1.77	2.17	6.99	6.10	6.36	6.56	5.49	4.21	5.51	5.90	5.24	6.10	2.17	1.77	1.77	1.77	1.77
1fin1	5.29	2.35	9.78	2.96	2.93	4.31	7.88	3.16	2.32	2.87	2.37	2.94	4.14	5.33	11.35	2.52	7.88	2.32	2.32	2.35	2.35
1fin2	6.46	3.05	9.77	3.30	4.42	2.26	8.32	3.43	2.47	3.00	8.82	5.30	4.78	4.99	9.57	10.44	3.43	2.26	2.26	3.30	3.30
1fvt	2.73	2.72	2.08	1.05	3.62	2.26	2.08	1.79	1.66	1.47	2.23	2.50	1.74	2.56	2.61	4.33	1.66	1.05	1.05	2.23	2.23
1fvv1	5.10	11.05	2.55	3.88	6.53	7.68	3.46	3.11	3.48	1.30	1.15	4.45	2.75	4.31	4.45	3.95	1.30	1.15	1.15	1.15	1.15
1fvv2	4.40	11.29	3.07	3.01	8.14	4.15	3.00	3.13	3.03	2.54	2.15	3.98	2.44	3.71	2.51	2.47	2.15	2.15	2.44	2.15	2.44
1g5s	8.81	2.76	2.35	2.16	8.06	7.93	1.87	1.91	3.94	2.14	2.19	2.34	7.45	7.42	2.32	2.25	2.34	1.87	1.87	2.47	2.47
1gih	3.25	1.55	1.11	1.15	6.52	3.26	2.24	7.79	1.04	1.32	1.37	1.66	1.11	1.06	1.08	1.75	1.11	1.04	1.04	1.37	1.37
1gii	6.01	1.47	1.12	6.15	4.77	2.99	2.03	1.88	1.08	1.50	1.39	1.16	1.26	1.00	1.04	1.04	1.00	1.00	1.04	1.39	1.39
1gij	7.24	7.30	2.26	5.54	7.06	6.99	3.52	1.92	2.03	9.47	1.83	2.15	6.34	8.21	2.53	2.11	2.53	1.83	1.83	1.83	1.83
1jvp	1.47	1.38	1.07	7.30	5.46	7.01	1.76	1.44	1.07	1.45	1.37	1.17	1.56	1.18	1.06	0.74	0.74	0.74	1.06	0.74	1.07

of the alien protein structures, with only one exception (the ligand from 1di8). By merging the docking solutions of a ligand with all of the protein structures and considering the rmsd of the best-scoring solution (MRGD by Score All column*)*, a good pose was found in 13 out of 16 cases (81.3%). Interestingly, the picture does not change much when we omit the redocking results (MRGD by Score No Redock column). At this point, it is interesting to note that only 3 out of 16 ligands (1aq1, 1fvv2, 1jvp) achieve their lowest docking score in their native protein structure.

In summary, these results suggest that for binding mode prediction of an active compound it is beneficial to dock into as many available structures of the target as possible, while considering the poses with the best overall score. For the following, it is important to note that the discussion is limited to top ranking placements only. For these ($16 \times 16 = 256$), we subtracted for each ligand the lowest (i.e., best) score from the list of all scores the ligand delivered with the 16 target structures. So, the resulting scores are relative scores, and the combination of a ligand with the preferred target structure delivers a zero relative score. All resulting 256 (rank 1) placements were subsequently binned according to these relative scores. As Table 2.2 indicates, there is a clear-cut relationship between the relative docking score and the quality of the docked placement (i.e., the rmsd): by far the highest percentage of reasonable poses are found among only the lowest scoring (i.e., zero relative score) of the placements and a steady decline can be seen for the lower scoring placements.

2.6.2 VS FOR CDK2 INHIBITORS

After the validation phase of cross- and redocking, we have applied the Flex* tools for screening 72 manually selected tight binders[3] against the Bionet screening compound collection [58]. This vendor library is freely available and contains approximately 34,000 compounds that are advertised as being apt for screening purposes.

TABLE 2.2
Summary of Top Ranking Solutions by Score Difference w.r.t the Lowest Overall Score Value of the Respective Ligand

Score Difference to Lowest Scoring Instance	Number of Instances Overall	Number of Instances with rmsd ≤2.5 Å	Percentage of Instances with rmsd ≤2.5 Å
0 (best merged solution)	16	13	81.3
> 0; ≤5	56	26	46.4
> 5; ≤10	87	35	40.2
> 10	97	24	24.7

As before, FlexX version 1.13 was employed in combination with the Screen-Score parameters as published in [26].[4] All screening was carried out on the Bio-SolveIT Linux compute cluster with 11 nodes (each equipped with two Athlon™ AMD 1.6 GHz processors and 2 GB RAM). As target, structure 1di8 from the PDB was chosen.

2.6.2.1 Results

As shown in Figure 2.3, reasonably good enrichments can be achieved using the traditional VS by docking approach on this data set. In the following, this experiment will be referred to as Experiment 1.

An extension of the first experiment comprises the incorporation of *a priori* knowledge about the target. Here, we make use of "receptor-based" pharmacophore information which has been extracted previously [8]. A combination of one "essential" and two "optional" interactions constitutes the pharmacophore: We force H-bond interactions between the ligand and the leucine-83 (Leu83) peptidic NH-group (acting as an H-donor) and required at least one more interaction with either the peptide carbonyl functions of Leu83 or glutamic acid-81 (Glu81). Figure 2.3 shows the improved enrichment with pharmacophore constrained docking.

However, note that approximately 11% of the known active compounds did not meet the pharmacophore and were thus rejected as solutions. It must therefore be stressed that pharmacophore-constrained docking always confines the generality of unconstrained docking. Sometimes it is advisable to weaken a pharmacophore or to employ fewer restrictions. Yet, the restrictions may be an advantage in the search

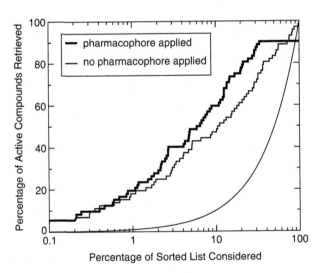

FIGURE 2.3 Enrichment plots (abscissa logarithmically) for 72 active compounds and the 34,000 member Bionet library of screening compounds against the CDK2 target 1di8. Lower line: Result without application of receptor-based pharmacophore information (Experiment 1). The second line represents results with FlexX-Pharm in combination with a derived pharmacophore. The lowest line represents the expectation for a random selection.

for *novel* docking solutions as can be seen from Experiment 2: this quite similar experiment was conducted in collaboration with F. Hoffmann-La Roche AG using the same pharmacophore for screening 773 inhibitors merged into approximately 465,000 in-house compounds of Roche [59].

Analysis of the results revealed that FlexX-Pharm generated novel solutions with the correct binding mode at lower (i.e., "better") scores than FlexX did. This is possible only because FlexX-Pharm checks constraints during the docking procedure and allows for better sampling of the relevant solution space. Note that postdocking filters do not have this advantage. Similar to the case above, FlexX-Pharm enrichments improved significantly (approximately 48 to 60% of the active compounds found among the top 10% of the screened library) compared to the unconstrained computation. Also, because FlexX-Pharm checks for compliance with the pharmacophore constraints and rejects inappropriate placements on the fly, results are obtained faster. Table 2.3 shows this in an overview of the computing times for the experiments.

Another important observation during Experiment 2 was that FlexX exhibits a slight bias toward overestimation of the binding affinity of larger compounds. FlexX-Pharm does not show this effect, which can be advantageous, when screening compound libraries with a large spread of compound sizes (see [59] for details).

2.6.3 ALIGNMENT-BASED SCREENING FOR CDK2 INHIBITORS

Sometimes, one lacks a target protein structure. In these cases, scoring based on similarity or alignment methods can be an alternative approach to docking.

To demonstrate the power of this approach, we have performed an alignment experiment (Experiment 3) as follows: From the set of 72 tight binders (see Table 2.3), we picked the 10 most active compounds for which activity information was available. These 10 were taken as reference compounds (with the 3D coordinates generated by CORINA version 3.0 [48]). Then, the remaining tight binders plus all members of the Bionet set were aligned to each of the ten reference compounds. As usual, a rank-ordering of the compounds comes from sorting by score. Figure 2.4

TABLE 2.3

Computing times for Experiment 1 to Experiment 3. 1: FlexX/-Pharm docking 72 known active compounds vs. the 34,000 member Bionet compound library (22 CPU); 2: screening of 773 known active compounds vs. the 465,000 member subset of the Roche compound library (50 CPU); and 3: FlexS alignment of the ten most active compounds out of the 72 vs. the Bionet library. In parentheses, we have annotated the approximate average computing times per compound on a single CPU.

	Computing Times		
	Experiment 1	Experiment 2	Experiment 3
FlexX/S	10 hrs (23 sec)	4.0 days (37 sec)	2 days 6 hrs (17 sec)
FlexX-Pharm	4.5 hrs (10.3 sec)	2.4 days (22 sec)	—

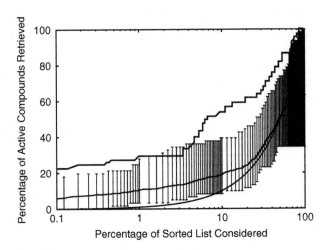

FIGURE 2.4 Enrichment plots after FlexS screening. Fine center line: The setup consisted of superimposing active compounds and the Bionet database (34,000 screening compounds) onto 10 known inhibitors with lowest IC50 (inhibitory concentration 50%) values and subsequent averaging over enrichments; the error bars indicate maximum and minimum values within the 10 experiments. Upper bold line: Enrichment curve derived from the merged list where only the best scores were taken into account. Lower bold line: indicates the expectation for random selection of actives.

shows the average over the 10 individual enrichment curves plus the upper and lower envelope curves as error bars.

Within 2 days and 6 hours on 20 nodes, the approximately 340,000 alignments (10 times the number of compounds in the Bionet library) completed, leading to an average of approximately 17 seconds per alignment (see Table 2.3).

There is significant enrichment in the low percentile of the library, despite the absence of explicit target information. At around 1%, an enrichment factor of 10 may be obtained. At 2%, one gains a factor of around 12 over random.

Taking the result of the cross-docking experiment (see Section 2.6.1) into account, we also applied a merging of the 10 lists into a single one by annotating only the minimum (i.e., best) score for a ligand aligned to all ten reference compounds. Except for few points, the resulting curve is superior to all of the individual enrichments. Note that the enrichments achieved by alignment-based screening without any structural information about the target are in the same ballpark as the docking results and hence constitute an interesting alternative approach.

ACKNOWLEDGMENTS

We express our gratitude to Andreas Steffen at Schering AG and University of Marburg for collecting the 72 active compounds from the literature and to Markus Lilienthal at BioSolveIT for setting up the FlexS screening runs. We extend our thanks to Sally A. Hindle and Holger Claussen for fruitful discussions on FlexX-Pharm and FlexE. Finally, we would like to thank Martin Stahl at Hoffmann-La Roche AG for allowing us to report on his data.

NOTES

1. ScreenScore is a weighted combination of scoring terms from the FlexX-Score and PLP. The weights have been optimized for VS purposes [22].

2. For the PDB entries 1fin and 1fvv, there are two proteins in the unit cell which we emloyed as independent pockets, denoted as 1fin1, 1fin2, 1fvv1, and 1fvv2, respectively.

3. For approximately half of the compounds, there are IC50 values available.

4. Please note that there is a typo in the original publication: Instead of the scaling factor of 1.667 we applied the correct value of 1.766 for weighting charged hydrogen bonds.

REFERENCES

[1] M. Stahl and T. Schulz-Gasch, Practical database screening with docking tools, In H. Waldmann and M. Koppitz, Eds., *Small Molecule-Protein Interaction*, New York: Springer Verlag, pp. 127–150, 2003.

[2] N. Brooijmans and I.D. Kuntz, Molecular recognition and docking algorithms, *Ann. Rev. Biophys. Biomol. Struc.* 32:335–73, 2003.

[3] P.D. Lyne, Structure-based virtual screening: an overview, *Drug Discov. Today* 7(20):1047–1055, 2002.

[4] I. Muegge and M. Rarey, Small molecule docking and scoring. In K.B. Lipkowitz and D.B. Boyd, Eds., *Rev. Comp. Chem.*, pp. 1–60. Wiley-VCH, 2001.

[5] R. Najmanovich, J. Kuttner, V. Sobolev, and M. Edelman, Side-chain flexibility in proteins upon ligand binding. *Proteins* 39:261–268, 2000.

[6] M.G. Bures, Recent techniques and applications in pharmacophore mapping. In P.S. Charifson, Ed., *Practical Application of Computer-Aided Drug Design*, New York: Marcel Dekker, pp. 39–72, 1997.

[7] C. Lemmen and T. Lengauer, Computational methods for the structural alignment of molecules, *J. Computer-Aided Mol. Des.* 14(3):215–232, 2000.

[8] H. Claussen, M. Gastreich, V. Apelt, J. Greene, S.A. Hindle, and C. Lemmen, The FlexX database docking environment — rational extraction of receptor-based pharmacophores, *Curr. Drug Disc. Tech.* 1(1):49–60, 2004.

[9] B. Waszkowycz, T.D.J. Perkins, R.A. Sykes, and J. Li, Large-scale virtual screening for discovering leads in the postgenomic area, *IBM Systems Journal* 40:360–376, 2001.

[10] M. Rarey, B. Kramer, T. Lengauer, and G. Klebe, A fast flexible docking method using an incremental construction algorithm, *J. Mol. Biol.* 261(3):470–489, 1996.

[11] C. Lemmen, T. Lengauer, and G. Klebe, FlexS: a method for fast flexible ligand superposition, *J. Med. Chem.* 41:4502–4520, 1998.

[12] M. Rarey, S. Wefing, and T. Lengauer, Placement of medium-sized molecular fragments into active sites of proteins, *J. Computer-Aided Mol. Des.* 10:41–54, 1996.

[13] M. Vieth, J.D. Hirst, and C.L. Brooks, III, Do active site conformations of small ligands correspond to low free-energy solution structures? *J. Computer-Aided Mol. Des.* 12:563–572, 1998.

[14] G. Klebe, Toward a more efficient handling of conformational flexibility in computer-assisted modelling of drug molecules, *Persp. Drug Disc. Des.* 3:85–105, 1995.

[15] F.H. Allen, S. Bellard, M.D. Brice, B.A. Cartwright, A. Doubleday, H. Higgs, T. Hummelink-Peters, O. Kennard, W.D.S. Motherwell, J.R. Rodgers, and D.G. Watson, The Cambridge Crystallographic Data Centre: computer-based search, retrieval, analysis and display of information, *Acta. Cryst.* B35:2331–2339, 1979.

[16]   F.H. Allen, The Cambridge structural database: a quarter of a million crystal structures and rising, *Acta. Cryst.* B58(3):380–388, 2002.

[17]   F.C. Bernstein, T.F. Koetzle, G.J.B. Williams, E.F. Meyer, Jr., M.D. Brice, J.R. Rodgers, O. Kennard, T. Shimanouchi, and M. Tasumi, The protein data bank: a computer based archival file for macromolecular structures, *J. Mol. Biol.* 112:535–542, 1977.

[18]   M.J. Sippl, Calculation of conformational ensembles from potentials of mean force — an approach to the knowledge based prediction of local structures in globular proteins, *J. Mol. Biol.* 213: 859–883, 1990.

[19]   H.-J. Böhm, The computer program LUDI: a new method for the *de novo* design of enzyme inhibitors, *J. Computer-Aided Mol. Des.* 6:61–78, 1992.

[20]   G. Klebe, The use of composite crystal-field environments in molecular recognition and the *de novo* design of protein ligands, *J. Mol. Biol.* 237:221–235, 1994.

[21]   H.-J. Böhm, The development of a simple empirical scoring function to estimate the binding constant for a protein–ligand complex of known three-dimensional structure, *J. Computer-Aided Mol. Des.* 8:243–256, 1994.

[22]   D.K. Gehlhaar, G.M. Verkhivker, P.A. Rejto, C.J. Sherman, D.B. Fogel, L.J. Fogel, and S.T. Freer, Molecular recognition of the inhibitor AG-1343 by HIV-1 protease: conformationally flexible docking by evolutionary programming, *Chem. & Biol.* 2:317–324, 1995.

[23]   M.D. Eldridge, C.W. Murray, T.R. Auton, G.V. Paolini, and R.P. Mee, Empirical scoring functions: I. The development of a fast empirical scoring function to estimate the binding affinity of ligands in receptor complexes, *J. Computer-Aided Mol. Des.* 11:425–445, 1997.

[24]   I. Muegge and Y.C. Martin, A general and fast scoring function for protein–ligand interactions: a simplified potential approach, *J. Med. Chem.* 42(5):791–804, 1999.

[25]   H. Gohlke, M. Hendlich, and G. Klebe, Knowledge-based scoring function to predict protein–ligand interactions, *J. Mol. Biol.* 295:337–356, 2000.

[26]   M. Stahl and M. Rarey, Detailed analysis of scoring functions for virtual screening, *J. Med. Chem.* 44:1035–1042, 2001.

[27]   M. Rarey, Protein–ligand docking in drug design. In T. Lengauer, Ed., *Bioinformatics —From Genomes to Drugs*, Vol. I, Heidelberg: Wiley-VCH, pp. 315–360, 2001.

[28]   OpenEye Scientific Software, http://www.eyesopen.com/products/applications/fred.html.

[29]   S. Putta, C. Lemmen, P. Beroza, and J. Greene, A novel shape-feature based approach to virtual library screening, *J. Chem. Inf. Comput. Sci.* 42(5):1230–1240, 2002.

[30]   G. Jones, P. Willett, and R.C. Glen, Molecular recognition of receptor sites using a genetic algorithm with a description of desolvation, *J. Mol. Biol.* 245:43–53, 1995.

[31]   G. Jones, P. Willett, R.C. Glen, A.R. Leach, and R. Taylor, Further development of a genetic algorithm for ligand docking and its applications to screening combinatorial libraries. In A.L. Parrill and M.R. Reddy, Eds., Rational Drug Design: Novel Methodology and Practical Applications, No. 719 in ACS Symposium Series, Washington, D.C.: American Chemical Society, pp. 271–291, 1999.

[32]   D.S. Goodsell and A.J. Olson, Automated docking of substrates to proteins by simulated annealing, *Proteins* 8:195–202, 1990.

[33]   D.S. Goodsell, G.M. Morris, and A.J. Olson, Automated docking of flexible ligands: applications of AutoDock, *J. Mol. Recog.* 9:1–5, 1996.

[34]   T.J. Ewing, S. Makino, A. G. Skillman, and I. D. Kuntz, DOCK 4.0: search strategies for automated molecular docking of flexible molecule databases, *J. Computer-Aided Mol. Des.* 15:411–428, 2001.

[35] S. Linnainmaa, D. Harwood, and L.S. Davis, Pose determination of a three-dimensional object using triangle pairs, *IEEE Trans. Pattern Anal. Mach. Intell.* 10(5):634–646, 1988.

[36] C. Lemmen and T. Lengauer, Time-efficient flexible superposition of medium-sized molecules and molecular fragments, *J. Computer-Aided Mol. Des.* 11:357–368, 1997.

[37] M. Rarey, B. Kramer, and T. Lengauer, Docking of hydrophobic ligands with interaction-based matching algorithms, *Bioinformatics* 15:243–250, 1999.

[38] C. Lemmen, C. Hiller, and T. Lengauer, RIGFIT: a new approach to superimpose ligand molecules, *J. Computer-Aided Mol. Des.* 12:491–502, 1998.

[39] M. Rarey, B. Kramer, and T. Lengauer, Multiple automatic base selection: protein–ligand docking based on incremental construction without manual intervention, *J. Computer-Aided Mol. Des.* 11:369–384, 1997.

[40] M. Rarey and T. Lengauer, A recursive algorithm for efficient combinatorial library docking, *Persp. Drug Disc. Des.* 20:63–81, 2000.

[41] S.A. Hindle, M. Rarey, C. Buning, and T, Lengauer. Flexible docking under pharmacophore constraints, *J. Computer-Aided Mol. Des.* 16(2):129–149, 2002.

[42] S.A. Hindle, M. Stahl, and M. Rarey, Flexible docking under pharmacophore type constraints: application to virtual screening. In 14th European Symposium on Quantitative Structure-Activity Relationships, Bournemouth, UK, 2002.

[43] H. Claussen, C. Buning, M. Rarey, and T. Lengauer, FlexE: efficient molecular docking into flexible protein structures, *J. Mol. Biol.* 308:377–395, 2001.

[44] H. Claussen, C. Buning, M. Rarey, and T. Lengauer, Molecular docking into the flexible active site of aldose reductase using FlexE. In H.-D. Höltje and W. Sippl, Eds., Rational Approaches to Drug Design: Proceedings of 13th European Symposium on Quantitative Structure-Activity Relationships, Barcelona: Prous Science, pp. 324–333, 2001.

[45] C. Lemmen, M. Zimmermann, and T. Lengauer, Multiple molecular superpositioning as an effective tool for virtual database screening, *Persp. Drug Disc. Des.* 20:43–62, 2000.

[46] R.S. Pearlman and R. Balducci, Confort: a novel algorithm for conformational analysis, National Meeting of the American Chemical Society, New Orleans, 1998.

[47] R.S. Pearlman, Rapid generation of high quality approximate 3D molecular structures, *Chem. Des. Auto. News* 2(1):5–6, 1987.

[48] J. Gasteiger, C. Rudolph, and J. Sadowski, Automatic generation of 3D-atomic coordinates for organic molecules, *Tetrahedron*, 3:537–547, 1990.

[49] J. Sadowski, J. Gasteiger, and G. Klebe, Comparison of automatic three-dimensional model builders using 639 x-ray structures, *J. Chem. Inf. Comput. Sci.* 34:1000–1008, 1994.

[50] A.Wlodawer, Rational drug design: the proteinase inhibitors, *Pharmacotherapy* 14(6):9S–0S, 1994.

[51] M. Rarey, B. Kramer, and T. Lengauer, The particle concept: placing discrete water molecules during protein–ligand docking predictions, *Proteins* 34(1):17–8, 1999.

[52] http://www.python.org.

[53] Gnuplot.py Home Page: http://gnuplot-py.sourceforge.net.

[54] http://www.csm.ornl.gov/pvm/pvm_home.html.

[55] B. Kramer, M. Rarey, and T. Lengauer, Evaluation of the FlexX incremental construction algorithm for protein–ligand docking, *Proteins* 37:228–241, 1999.

[56] H.M. Berman, J. Westbrook, Z. Feng, G. Gilliland, T.N. Bhat, H. Weissig, I.N. Shindyalov, and P.E. Bourne, The Protein Data Bank, *Nucleic Acid Res.,* 28:235–242, 2000.

[57] M. Stahl and M. Rarey, Detailed analysis of scoring functions for virtual screening, *J. Med. Chem.* 44:1035–1042, 2001.

[58] Bionet Screening Compounds Database, Key Organics Limited UK: http://www. keorganics.ltd.uk/screenin.htm.

[59] S.A. Hindle, M. Stahl, and M. Rarey, Flexible docking under pharmacophore type constraints: application to virtual screening. In 14th European Symposium on Quantitative Structure-Activity Relationships, Bournemouth: UK, 2002.

3 An Analysis of Critical Factors Affecting Docking and Scoring

Emanuele Perola, W. Patrick Walters, and Paul S. Charifson

3.1 INTRODUCTION

Many emerging technologies often overpromise and underdeliver in the early days of their evolution. Recent examples include combinatorial chemistry, ultra-high throughput screening, and even the promise of genomics to impact drug discovery. The field of docking and virtual screening (VS) for drug discovery has been no different in this regard, but has attained a level of maturity and utility based upon incremental improvements in methods and in our understanding of how and when to best apply these methods.

The ability to correctly place ligands of potential interest in protein binding sites has improved significantly in recent years. Some of the successes [1,2] and reasons for failure [3–5] have been summarized in recent publications and there are some good general reviews of currently used docking programs [6–8]. Docking and VS methods, in general, have evolved over the past 20 years to their current state. There are certainly many issues surrounding docking and ranking of docked structures that can still be improved; some of these issues will be discussed in this chapter. In the past few years, the topics of database preparation [9] as well as how to best analyze/filter docking output [10] has also received some attention.

All docking programs require a mathematical function that evaluates whether a given orientation/conformation (pose) is the most favorable relative to all other orientation/conformation combinations sampled. When applied to screening, the process also requires a comparison of the best pose (or top few poses) of a given ligand with those of the other ligands such that a final ranking or ordering can be obtained. For the purposes of this chapter, we will refer to the function used in evaluating the poses of any ligand as a docking function and any function used in either refinement/reranking of docked ligand poses or for comparing different ligands as a scoring function. There is no requirement that both the docking function and scoring function are the same, although this has most often been the case in the early years. In practice, it has been shown that using different scoring functions from those used in the initial docking often provides better results in the screening process [11–14]. It must also be emphasized, however, that the docking function must evaluate a large number of solutions (numbers of poses ranging from 10^4 to 10^5 are typical). Even if one saves a small set of top ranking poses and rescores them with a different function, an initial bias will have been introduced by the docking function. It is often quite challenging for a given search method combined with a given docking function to ensure that the best scoring pose can reproduce crystallographically observed binding modes. This is further complicated by the observation that the ability of the docking function to rank poses is often clouded by the ability of the search method to overcome barriers and obtain adequate sampling [15]. The possibility exists that different types of search algorithms will work best with specific types of docking functions, although an exhaustive study of this kind has yet to be published.

Several studies have shown that rescoring docked poses with a secondary function can improve the selection of poses to be used at the rank ordering stage in database screening [4,9,10]. It has been suggested that a reasonable compromise between speed and accuracy might include using a simplified function (e.g., an empirical or knowledge-based function) as a docking function to save a set of viable poses and then use a more rigorous (e.g., energy-based) function for the final pose selection/ranking of ligands [4,9]. Further, it has been observed that certain types of scoring functions tend to give better predictions in certain types of binding sites [10,12,13], or that combining the best results from two or more scoring functions can lead to a considerable reduction in false positives (consensus scoring) [14]. Two further issues that are often poorly addressed (or not addressed at all) are those of protein flexibility and solvation/desolvation. Such issues can have an effect on the number of false positives and negatives emanating from database screening. Various groups have attempted to describe one or both of these phenomena in docking [16–23], but none have led to a general solution. Fortunately, many biological systems of interest do not exhibit large conformational changes upon binding different ligands, thus allowing for current methods to answer the fundamental question, "Does this ligand have a chance of fitting into this binding site?" Another often overlooked issue in this area is that the desired balance between polar and nonpolar properties in druglike molecules is not necessarily well described by current docking and scoring functions largely because training sets used to derive such functions have not been rigorously chosen with druglikeness as a key parameter.

Database composition is also an issue that has received increased attention over the past few years. Of course, this affects not only docking and VS, but screening in general. Aside from simply removing undesirables from screening databases or culling databases toward druglikeness, little attention has been applied toward database organization. It turns out that this can be an important factor, especially when considered in the context of docking, because even simple procedural changes can have fairly dramatic effects on the outcome. Such examples include comparing compounds of similar formal charge from docking databases [24] when solvation is not considered and docking molecules by families to increase the diversity of hits [25].

The remainder of this chapter will address some of the issues mentioned above while also describing some recent docking studies performed in our group [75]. In the first part of the chapter, we attempt to answer the question, "How does one generate an appropriate and relevant training set for the evaluation of docking and scoring tools used in drug discovery?" We then address two general issues surrounding docking: "Are certain docking programs better than others at predicting the binding modes of druglike molecules to pharmaceutically relevant protein targets?" and "How is the docking accuracy affected by the nature of the binding site?" The third section of this chapter then answers the question, "How good are the existing scoring functions at estimating the affinity of druglike molecules for their targets?" In this section, we also describe an interesting finding concerning subset selection with regard to the scoring functions evaluated in this study. In the fourth and final section of this chapter, we attempt to answer the question, "What is the best combination of docking and scoring methods for docking-based VS?"

3.2 THE IMPORTANCE OF TEST SET SELECTION

A variety of reports have been published on the evaluation of docking programs and on the development and validation of scoring functions [17,26–40]. In each case, a set of protein–ligand complexes of known three-dimensional (3D) structure was selected and used as a test set or training set. In earlier scoring function work, the selection of the complexes was primarily based on the simultaneous availability of x-ray structure and dissociation constant, with diversity and pharmaceutical relevance of the protein targets as secondary criteria [26,33]. As more structural information became available, the test sets were expanded and more restrictive criteria were applied in regards to the quality of the structures included, mostly assessed on the basis of crystallographic resolution [27,28,36]. A more diverse selection was possible for test sets devoted to the evaluation of docking programs, not limited by the availability of binding data [17,34]. In recent scoring function work, test sets have been assembled as a combination of previously reported sets with few additions of new complexes [32,39,40].

The need for the generation of larger test sets, selected with consistent criteria and highly refined, has been highlighted by two recent publications [41,42]. The first describes the generation of a database of 195 complexes with known binding affinity, mostly based on previously published test sets. Criteria for the selection of new complexes are not discussed, although emphasis is given to the characterization and classification of the systems included. The list of complexes is not provided in the paper. The second report describes a database of 305 complexes (the "Astex test set"), which combines the GOLD (Genetic Optimization for Ligand Docking) validation set [17], the ChemScore test set [43] and 123 new complexes selected on the basis of pharmaceutical relevance and structural diversity of the proteins involved. A series of filters are applied to define a higher quality subset, pruned of structures with symmetry related protein units involved in ligand binding, severe clashes between ligand and protein atoms or potential errors. No binding affinities are included.

A careful analysis of the reported test sets shows that, although satisfactory diversity and pharmaceutical relevance of the protein structure has now been achieved, the same cannot be said with respect to the ligands. Structural classes that are of less relevance to drug discovery programs (peptides, sugars, nucleotides) are still overrepresented, with a high degree of redundancy, and the molecular weight (MW) of the ligands generally ranges from 100 to 1000, far beyond the range of interest for a drug discovery program. In general, none of the reported test sets contains more than 20% of truly druglike ligands. Because such test sets are used in the evaluation/calibration of tools for drug design, it is important that the complexes be representative of what is relevant to the process. If the ultimate objective is to predict binding of druglike molecules to pharmaceutically relevant proteins, complexes between such partners should clearly be emphasized. Following this premise, we generated a new test set of complexes of known binding affinity, geared toward druglike ligands and suitable for a variety of tasks: evaluation of docking programs and existing scoring functions, development and calibration of new scoring functions, and analysis of various aspects of protein–ligand binding.

3.2.1 COMPLEX SELECTION

One hundred complexes were selected from the Protein Data Bank and 50 were selected from the Vertex corporate structure collection according to the following criteria:

- General:
  - Ki or Kd available
  - Noncovalent binding between ligand and protein
  - Crystallographic resolution < 3.0 Å
- Ligands:
  - MW between 200 and 600
  - 1 to 12 rotatable bonds
  - Druglike or leadlike
  - Structurally diverse
- Proteins:
  - Multiple classes
  - Diverse within classes
  - Relevant to drug discovery
- Criteria for exclusion (examples in parenthesis):
  - Sugar-containing ligands (4hmg)
  - Steroidal ligands (1a27)
  - Macrocyclic ligands (1mmq)
  - Ligand binding mediated by a complex network of water molecules (1jqe)
  - Uncertain protonation state in the binding region (1k4g)
  - Severe clashes between protein and ligand atoms (1dth)
  - Crystallographically related protein units involved in ligand binding (1bm7)
  - Heme-containing active sites (1phg)
  - Poorly refined water molecules involved in ligand binding (1c4y)
  - Unconventional amino acid residues in the binding site (1hlf)
  - Ligands containing atoms other than C, N, O, S, F, Cl, Br, H (1tha)

Each ligand was included only once, thus avoiding common redundancies like methotrexate bound to different versions of dihydrofolate reductase or the same ligand bound to two closely related proteins. The purpose was to avoid repetitions of almost identical sets of interactions, thus maximizing the diversity of the interactions represented in the test set. These criteria reflect our intention to include the maximum amount of structural information on systems that are of high interest in a structure-based drug design context, and exclude those that are only rarely considered. This explains the exclusion of sugars, steroids, and macrocycles, all present in approved drugs.

3.2.2 Composition of the Test Set

The PDB codes of the 100 complexes selected from the Protein Data Bank are reported in Table 3.1, along with protein names, sources, crystallographic resolutions, and dissociation constants, expressed as pKi ($-\mathrm{Log}_{10}(\mathrm{Ki})$). The test set includes 63 different proteins from a variety of classes, including proteases, kinases, nuclear receptors, phosphatases, oxidoreductases, isomerases, and lyases. Kinases (43 complexes) and proteases (42 complexes) are the most widely represented. There are 24 metalloprotein complexes, all of them with a zinc ion in the active site. Several examples of approved drugs in complex with their targets are also included (e.g., Agenerase®/human immunodeficiency virus (HIV) protease, Aricept®/acetylcholinesterase, lisinopril/ACE).

3.2.3 Complex Preparation

Each of the PDB files of the 150 complexes was processed according to the following protocol: the ligand was extracted, bond orders and correct protonation state were assigned upon visual inspection and the structure was saved to a standard deviation (SD) file. If a cofactor was present, the same procedure was applied and a separate SD file was generated. After removal of the ligand, a clean protein file was generated by removing subunits not involved in ligand binding and far from the active site, solvent, counter ions, and other small molecules located away from the binding site. Metal ions and tightly bound water molecules in the ligand binding site were preserved, and the protein structure was saved to a PDB file.

Hydrogen atoms were then added to the protein, and the structures of protein, ligand, and cofactor were combined in a single MacroModel® file. The active site was visually inspected and the appropriate corrections were made for tautomeric states of histidine residues, orientations of hydroxyl groups, and protonation states of basic and acidic residues. The hydrogen atoms were minimized for 1000 steps with MacroModel in OPLS-AA (Optimized Potentials for Liquid Simulations — All Atoms) force field, with all nonhydrogen atoms constrained to their original positions. Protein (with cofactor if present) and ligand with optimized hydrogen positions were finally saved to separate files.

TABLE 3.1
Composition of the PDF Portion of the Test Set and binding Data Expressed as pK

PDB Code	Resolution	Protein	Organism	pKi
13gs	1.90	glutathione *S*-transferase	human	4.62
1a42	2.25	carbonic anhydrase II	human	9.89
1a4k	2.40	antibody Fab	mouse	8.00
1a8t	2.55	metallo beta-lactamase	*Bacteroides fragilis*	5.80
1afq	1.80	gamma-chymotrypsin	bovine	6.21
1aoe	1.60	dihydrofolate reductase	*Candida albicans*	9.66
1atl	1.80	atrolysin C	*Crotalus atrox*	6.28
1azm	2.00	carbonic anhydrase I	human	6.14
1bnw	2.25	carbonic anhydrase II	human	9.08
1bqo	2.30	stromelysin I	human	7.74
1br6	2.30	ricin	*Ricinus communis*	3.22
1cet	2.05	lactate dehydrogenase	*Plasmodium falciparum*	2.89
1cim	2.10	carbonic anhydrase II	human	9.55
1d3p	2.10	thrombin	human	5.11
1d4p	2.07	thrombin	human	6.30
1d6v	2.00	oxy-Cope catalytic antibody germline precursor	human/mouse hybrid	6.17
1dib	2.70	methylenetetrahydrofolate dehydrogenase	human	7.74
1dlr	2.30	dihydrofolate reductase	human	9.18
1efy	2.20	poly(ADP-ribose) polymerase	chicken	8.22
1ela	1.80	elastase	pig pancreas	6.35
1etr	2.20	thrombin	bovine	7.41
1ett	2.50	thrombin	bovine	6.19
1eve	2.50	acetylcholinesterase	*Torpedo californica*	8.48
1exa	1.59	retinoic acid receptor gamma-2	human	6.30
1ezq	2.20	coagulation factor Xa	human	9.05
1f0r	2.10	coagulation factor Xa	human	7.66
1f0t	1.80	trypsin	bovine	6.00
1f4e	1.90	thymidylate synthase	*Escherichia coli*	2.96
1f4f	2.00	thymidylate synthase	*Escherichia coli*	4.62
1f4g	1.75	thymidylate synthase	*Escherichia coli*	6.48
1fcx	1.47	retinoic acid receptor gamma-1	human	7.12
1fcz	1.38	retinoic acid receptor gamma-1	human	9.22
1fjs	1.92	coagulation factor Xa	human	9.70
1fkg	2.00	FKBP-12	human	8.00
1fm6	2.10	PPAR-gamma	human	7.33
1fm9	2.10	PPAR-gamma	human	8.82
1frb	1.70	FR-1 aldo-keto reductase	mouse	7.77
1g4o	1.96	carbonic anhydrase II	human	8.68
1gwx	2.50	PPAR-gamma	human	7.30

-- *continued*

TABLE 3.1 (continued)
Composition of the PDF Portion of the Test Set and binding Data Expressed as pK

PDB Code	Resolution	Protein	Organism	pKi
1h1p	2.10	cyclin-dependent kinase II	human	4.92
1h1s	2.00	cyclin-dependent kinase II	human	8.22
1h9u	2.70	retinoid X receptor beta	human	8.52
1hdq	2.30	carboxypeptidase A	bovine pancreas	5.82
1hfc	1.56	fibroblast collagenase	human	8.15
1hpv	1.90	HIV-1 protease	HIV-1	9.22
1htf	2.20	HIV-1 protease	HIV-1	8.09
1i7z	2.30	antibody Gnc92H2	human	6.40
1i8z	1.93	carbonic anhydrase II	human	9.82
1if7	1.98	carbonic anhydrase II	human	10.52
1iy7	2.00	carboxypeptidase A	bovine	6.19
1jsv	1.96	cyclin-dependent kinase II	human	5.70
1k1j	2.20	trypsin	bovine	7.68
1k22	1.93	thrombin	human	8.40
1k7e	2.30	tryptophan synthase	Salmonella typhimurium	2.92
1k7f	1.90	tryptophan synthase	Salmonella typhimurium	3.32
1kv1	2.50	p38 map kinase	human	5.94
1kv2	2.80	p38 map kinase	human	10.00
1l2s	1.94	beta-lactamase	Escherichia coli	4.59
1l8g	2.50	protein tryrosine phosphatase 1B	human	6.22
1lqd	2.70	coagulation factor xa	human	8.05
1m48	1.95	interleukin-2	human	5.09
1mmb	2.10	MMP-8	human	9.22
1mnc	2.10	neutrophil collagenase	human	9.00
1mq5	2.10	coagulation factor Xa	human	9.00
1mq6	2.10	coagulation factor Xa	human	11.15
1nhu	2.00	HCV RNA polymerase	Hepatitis C virus	5.66
1nhv	2.90	HCV RNA polymerase	Hepatitis C virus	5.66
1o86	2.00	angiotensin converting enzyme	human	9.57
1ohr	2.10	HIV-1 protease	HIV-1	8.70
1ppc	1.80	trypsin	bovine	6.16
1pph	1.90	trypsin	bovine	6.22
1qbu	1.80	HIV-1 protease	HIV-1 virus	10.24
1qhi	1.90	thymidine kinase	Herpes simplex virus	7.30
1ql9	2.30	trypsin	rat	5.36
1qpe	2.00	lymphocyte-specific kinase	human	8.40

-- continued

TABLE 3.1 (continued)
Composition of the PDF Portion of the Test Set and binding Data Expressed as pK

PDB Code	Resolution	Protein	Organism	pKi
1r09	2.90	human rhinovirus 14	human rhinovirus 14	4.90
1syn	2.00	thymidylate synthase	*Escherichia coli*	9.05
1thl	1.70	thermolysin	*Bacillus thermoproteoliticus*	6.42
1uvs	2.80	thrombin	human	5.40
1uvt	2.50	thrombin	bovine	7.64
1ydr	2.20	c-AMP dependent kinase	bovine	5.52
1yds	2.20	c-AMP dependent kinase	bovine	5.92
1ydt	2.30	c-AMP dependent kinase	bovine	7.32
2cgr	2.20	immunoglobulin-2 beta Fab fragment	mouse	7.27
2csn	2.50	casein kinase-1	*Schizosaccharomyces pombe*	4.41
2pcp	2.20	antibody Fab	mouse	8.70
2qwi	2.00	influenza A neuraminidase	influenza A virus	8.40
3cpa	2.00	carboxypeptidase A	bovine	4.00
3erk	2.10	erk2 kinase	rat	5.12
3ert	1.90	estrogen receptor alpha	human	9.60
3std	1.65	scytalone dehydratase	*Magnaporthe grisea*	11.11
3tmn	1.70	thermolysin	*Bacillus thermoproteoliticus*	5.90
4dfr	1.70	dihydrofolate reductase	*Escherichia coli*	8.62
4std	2.15	scytalone dehydratase	*Magnaporthe grisea*	10.33
5std	1.95	scytalone dehydratase	*Magnaporthe grisea*	10.49
5tln	2.30	thermolysin	*Bacillus thermoproteoliticus*	6.37
7dfr	2.50	dihydrofolate reductase	*Escherichia coli*	4.96
7est	1.80	elastase	pig pancreas	7.60
830c	1.60	collagenase-3	human	9.28
966c	1.90	fibroblast collagenase I	human	7.64

3.3 EVALUATION OF HIGHLY REGARDED DOCKING PROGRAMS

The objective of a docking program is to predict the experimental binding mode of a ligand to a given receptor. To achieve this goal, the program performs an extensive search of the possible conformations, orientations, and positions of the ligand in the putative binding site of the receptor and selects the combination (defined as the ligand *pose*) that minimizes a given function (the docking function). The effectiveness of a docking program is determined by the efficiency of the sampling method and by the accuracy with which the generated poses are ranked. A large number of docking programs have been developed in the last 20 years based on a variety of search algorithms [7]. Analysis of the recent literature seems to indicate that DOCK [29], FlexX [44], and GOLD [17] are the most widely used docking programs [2,45–50]. Direct and indirect comparisons have shown that GOLD consistently outperforms the other two in terms of average docking accuracy on a variety of systems [17,29,34,38,46]. Among recently developed programs, ICM™ (Internal Coordinate Mechanics) [51] and Glide™ (Grid-Based Ligand Docking with Energetics) [52] have been reported to achieve a high degree of accuracy [51,53,54] and they have performed well on internal validation sets at Vertex. Here we report the comparative evaluation of GOLD, ICM, and Glide on a set of pharmaceutically relevant protein–ligand complexes. The objectives of this study were to assess:

- The ability of the three programs to reproduce the experimental binding modes of druglike molecules
- The impact of energy minimization of multiple docking poses and subsequent reranking of the minimized poses on docking accuracy
- The impact of the nature of the binding site on the performance of each tool

The test set of complexes described above was used in the evaluation. Each ligand was docked back into the corresponding binding site, and the accuracy of each prediction was assessed on the basis of the root-mean-square deviation (rmsd) between the coordinates of the heavy atoms of the ligand in the top docking pose and those in the crystal structure. The following paragraphs describe the search algorithm and scoring methods used in the three programs. For each program, details of the calculations performed in this study are provided.

3.3.1 ICM (MoLSoft LLC)

The ICM program is based on a stochastic algorithm that relies on global optimization of the entire flexible ligand in the receptor field (flexible ligand/grid receptor approach) [51]. Global optimization is performed in the binding site such that both the intramolecular ligand energy and the ligand-receptor interaction energy are optimized. The program combines large-scale random moves of several types with gradient local minimization and a history mechanism that both expels from the unwanted minima and promotes the discovery of new minima. The random moves include pseudo-Brownian moves, optimally biased moves of groups of torsions, and single torsion changes. The energy calculations are based on the ECEPP/3 (Empirical Conformational Energy Program for Peptides, version 3) force field [55], with MMFF (Merck Molecular Force Field) partial charges. Five potential maps — electrostatic, hydrogen bond (H-bond), hydrophobic, van der Waals (vdW) attractive, and vdW repulsive — are calculated for the receptor. The location of the receptor binding pocket can be specified by the user or selected by the cavity detection module implemented in the program.

In the present work, the binding pocket of the receptor was defined using the crystallographic coordinates of the ligand as a reference. For each complex, the ligand input structure was generated with CORINA (COoRdINAtes) [56] (Molecular Networks GmbH), and the protein structure, prepared as described in the previous section, was used as a receptor input structure. The Monte Carlo (MC) docking runs were performed using a MC thoroughness setting of 3, which controls the length of the run, and the top 20 poses were generated. Subsequent energy minimization of the ICM-generated poses was performed with MacroModel (v. 8.1) using both MMFF [57–59] and OPLS-AA [60] force fields, with flexible ligand and rigid receptor. Conjugate gradient minimization was performed for 1000 steps. The strain energy of the minimized ligand poses was calculated with a two-step procedure: restrained minimization of the ligand geometry (half-width of flat bottom restraint = 0.5 Å, force constant = 500 kcal/mol/Å) to convergence (0.01 kJ/Å/mol) followed by removal of the constraints and full minimization until convergence (0.01 kJ/Å/mol) into the closest local minimum [14]. The refined poses were reranked based on the calculated interaction energy (vdW and electrostatic) minus the strain energy of the ligand conformation.

3.3.2 GLIDE (SCHRÖDINGER, LLC)

The Glide algorithm [54] approximates a systematic search of positions, orientations, and conformations of the ligand in the receptor-binding site using a series of hierarchical filters that allow for respectable computational speed. The shape and properties of the receptor are represented on a grid by several different sets of fields that provide progressively more accurate scoring of the ligand pose. The fields are computed prior to docking. The binding site is defined by a rectangular box confining the translations of the mass center of the ligand. A set of initial ligand conformations is generated through exhaustive search of the torsional minima, and the conformers are clustered in a combinatorial fashion. Each cluster, characterized by a common conformation of the core and an exhaustive set of side-chain conformations, is docked as a single object in the first stage. The search begins with a rough positioning and scoring phase that significantly narrows the search space and reduces the number of poses to be further considered to a few hundred. In the following stage, the selected poses are minimized on precomputed OPLS-AA vdW and electrostatic grids for the receptor. In the final stage, the 5 to 10 lowest-energy poses obtained in this fashion are subjected to a MC procedure in which nearby torsional minima are examined, and the orientation of peripheral groups of the ligand is refined. The minimized poses are then rescored using the GlideScore function, which is a more sophisticated version of ChemScore [28] with force field-based components and additional terms accounting for solvation and repulsive interactions. The choice of the best pose is made using a model energy score (EModel) that combines the energy grid score, GlideScore, and the internal strain of the ligand.

In the present work, the binding region was defined by a 12 Å × 12 Å × 12 Å box centered on the mass center of the crystallographic ligand to confine the mass center of the docked ligand. Protein and ligand input structures were prepared as described in Section 3.3.1. No scaling factors were applied to the vdW radii. Default settings were used for all the remaining parameters. The top 20 docking poses were energy minimized with MacroModel using both the OPLS-AA and MMFF force fields and reranked as described in Section 3.3.1.

3.3.3 GOLD (CAMBRIDGE CRYSTALLOGRAPHIC DATA CENTRE)

The GOLD program uses a genetic algorithm (GA) to explore the full range of ligand conformational flexibility and the rotational flexibility of selected receptor hydrogens [17,38]. The mechanism for ligand placement is based on fitting points. The program adds fitting points to hydrogen-bonding groups on protein and ligand and maps acceptor points in the ligand on donor points in the protein and vice versa. Additionally, GOLD generates hydrophobic fitting points in the protein cavity onto which ligand CH groups are mapped. The GA optimizes flexible ligand dihedrals, ligand ring geometries, dihedrals of protein OH and NH3+ groups and the mappings of the fitting points. The docking poses are ranked based on a molecular-mechanics-like scoring function, which includes a H-bond term, a 4 to 8 intermolecular vdW term and a 6 to 12 intramolecular vdW term for the internal energy of the ligand.

In the present work, the binding site was defined as a spherical region of 10 Å radius centered on the mass center of the crystallographic ligand. Protein and ligand input structures were prepared as described above. Default GA settings number 4 [38] were used for all calculations, with the exception that 20 GA runs were performed instead of 10. The top 20 docking poses were energy minimized with MacroModel in both OPLS-AA and MMFF force fields and reranked as described above.

3.3.4   Results and Discussion

The results of this study clearly identified Glide as the most accurate of the three docking programs examined, with 61% of the top ranking poses within 2.0 Å of the corresponding crystal structure. Both GOLD and ICM also performed reasonably well, with 48% and 45% of top ranking poses meeting the same criterion. The percentages of top-ranked solutions within a defined rmsd from the experimentally determined structure are reported in Table 3.2.

Analysis of the top 20 solutions produced by each program shows that GOLD was almost as effective as Glide in sampling the correct pose and placing it in the top 20. When the top 20 docking poses were compared to the crystal structure, the percentages of best poses (lowest rmsd) within 2.0 Å from the experimental structure were 79% for Glide and 77% for GOLD (see Table 3.3). Based on this observation, the GOLD algorithm appears to be equally efficient in terms of sampling, but the Glide docking function seems much more accurate than the GOLD fitness function in ranking the sampled poses. The ICM algorithm appears to perform less well than the other two in terms of sampling, although the ICM docking function is better at ranking poses than the GOLD function, but not as accurate as the Glide function.

It is important to point out that, in terms of thoroughness of sampling, the default settings for Glide are clearly defined by Schrödinger, Inc. and extensively validated on many test systems. For GOLD there are four different sets of GA parameters defined as default and corresponding to different degrees of thoroughness and CPU

TABLE 3.2
Percentage of Top-Ranked Docking Poses within a
Defined rmsd from the Corresponding Crystal Structure

RMS	ICM	Glide	GOLD
< 0.5	7	10	3
< 1.0	27	37	22
< 1.5	39	52	37
< 2.0	45	61	48
< 3.0	55	74	62

The rmsd is calculated on the coordinates of the heavy atoms of the ligand.

TABLE 3.3
Top 20 Docking Solutions vs. X-Ray Structures:
Percentage of Closest Pose within a Defined rmsd
from the Corresponding Crystal Structure

RMS	ICM	Glide	GOLD
< 0.5	9	14	8
< 1.0	32	49	39
< 1.5	51	69	65
< 2.0	67	79	77
< 3.0	75	88	86

consumption. ICM allows the user to specify the degree of thoroughness as well, however literature and documentation do not provide a strong indication on what settings to use on a routine basis. To perform the study in an objective manner, we used settings that correspond to similar computing times. The docking studies described in this work averaged 1 to 3 min/molecule depending on processor speed on Linux® operating system (from 900 Mhz Intel® Pentium® III to 2.4 Ghz Intel Pentium IV) for all three programs. Both Glide and ICM require a precalculated set of grid potentials, with average computing times of 30 to 60 minutes per protein for Glide and 5 to 10 minutes for ICM.

The effect of energy minimization of the top 20 poses and reranking of the minimized poses was investigated using two different force fields. Comparison of the percentages of top-ranked structures within 2.0 Å of the crystal structure before and after energy minimization, reported in Table 3.4, shows that minimization and reranking did not affect the accuracy of the Glide poses when the OPLS-AA force field was employed; there was a slight decrease in performance relative to the unminimized poses when MMFF was used. Minimization and reranking marginally improved the accuracy of the ICM poses (from 45 to 49% with either force field); however, there was a more significant improvement on the GOLD poses, especially when OPLS-AA was used (from 48 to 62%). The performance of GOLD equaled

TABLE 3.4
Percentage of Top-Ranking Docked Poses within 2.0 Å
from the Experimental Structure before and after
Minimization and Reranking of the Top 20 Poses

Method	ICM	Glide	GOLD
None	45	61	48
MMFF	49	57	59
OPLS	49	60	62

that of Glide when this additional procedure was applied. The effect of minimization is consistent with the features of the three docking programs examined. In Glide, minimization on an OPLS-AA potential energy grid is already performed in the final stages of docking. It is therefore not surprising that additional refinement with an all-atom minimization using the same force field does not result in an increase of the docking accuracy.

Energy-minimization is also performed as part of the ICM search protocol, but with a different force field. The slight improvement observed upon minimization with either MMFF or OPLS-AA may suggest that either these two force fields provide a more accurate description of the protein–ligand interactions than the ECEPP/3 force field implemented in ICM or simply that the minimization performed by ICM is not as thorough and the docking poses require further refinement. The significant improvement of the GOLD poses after minimization is consistent with the fact that there is no energy minimization involved at the docking stage in this program. Severe clashes between protein and ligand atoms are not uncommon in GOLD-generated poses, partly because of the softness of the repulsive term implemented in the fitness function, and further refinement in a more rigorous fashion appears to be highly beneficial in this respect. In terms of force fields, the performances of MMFF and OPLS-AA were similar, with OPLS-AA achieving a slightly better accuracy in two out of three cases, and equal accuracy in the third (see Table 3.4). This observation further validates the choice of OPLS-AA as the force field used by Glide, which is also the best performing docking program in this study. It is important to mention that the average computing time for the energy minimization step ranges from 30 to 60 sec/pose with the settings used in this work. The cost/benefit ratio should therefore be carefully evaluated when this extra step is considered.

To assess where the difference between Glide and the other programs lies and on what kinds of systems each program performs best, the dependence of the docking accuracy on specific structural descriptors was analyzed. The complexes were classified in a binary or ternary fashion with respect to three structural features — flexibility of the ligand, predominant nature of the interactions between ligand and receptor, and degree of solvent exposure of the binding pocket. Statistical analysis of the docking accuracies was performed with regards to such features.

In terms of flexibility, it is well known that the accuracy of any docking program decreases with the number of rotatable bonds of the ligand. The size of the conformational space to be sampled increases exponentially with ligand flexibility, and the thoroughness of the sampling has to be partially sacrificed to keep the computing time within reasonable limits. Different algorithms use different solutions to circumvent the problem and maximize the efficiency of the conformational sampling. In this study, the test systems were divided in two groups — 87 complexes of ligands with 1 to 6 rotatable bonds and 63 complexes of ligands with 7 to 12 rotatable bonds. The results, summarized in Table 3.5, show that the loss of accuracy going from less flexible to more flexible ligands is relatively small for Glide (from 67 to 52% of correct solutions) and much more dramatic for GOLD and ICM, with the latter losing more than half of its predictive power. This indicates that the multistage systematic algorithm implemented in Glide results in a more extensive coverage of conformational space than both the GA and the stochastic search implemented in GOLD and ICM. This partially explains the relative performances observed on the complete test set

In terms of interactions, H-bonds and hydrophobic interactions are considered the main contributors to protein–ligand binding in the vast majority of complexes. To divide the complexes in our test set between H-bond-driven and hydrophobic-driven, the number of H-bonds between protein and ligand in each complex was determined, and the ratio between number of H-bonds and number of heavy atoms in the ligand was used to define the dominant contributor to binding for each complex. Complexes with a ratio of 0.15 or higher were classified as H-bond-driven, and complexes with a ratio of 0.10 or lower were classified as hydrophobic-driven, with the remaining complexes in the intermediate category. Ligand–metal interactions were counted as H-bonds for their similar nature. The results, reported in Table 3.6, show that all programs perform best on complexes in which there is a relatively even balance between H-bonding and hydrophobic interactions. Interestingly, for both ICM and GOLD, the docking accuracy decreases dramatically when binding is mainly driven by hydrophobic interactions, whereas Glide, which appears to be somewhat less sensitive to the nature of binding, performs better on hydrophobic-driven complexes than on H-bond-driven complexes. The preference of GOLD for complexes rich in H-bonds has been pointed out previously [17], and it can be ascribed to the nature of the algorithm, in which the mapping of H-bond fitting points plays a major role. In the case of ICM, this tendency has not been reported; one possible explanation is that in a MC search, mostly characterized by low energy moves, the presence of a set of H-bonds may lock part of the molecule into its correct orientation during the search, thus allowing for a more efficient sampling of the rest of the molecule. For Glide, the difference in performance is less significant, and this consistency across active sites with various degrees of hydrophobicity/hydrophilicity is another reason for its better performance on the complete test set.

When interactions with metals were specifically considered, no difference in performance was observed among the three programs. On the 24 metal-containing complexes, Glide selected a solution within 2.0 Å of the experimental structure nine times; ICM and GOLD succeeded eight times in the same subset. The success rate of the three programs on such systems was significantly poorer if compared to the overall performance, which points to the necessity of further progress in this area, especially considering the continued interest in zinc metalloproteins as drug discovery targets.

TABLE 3.5
Percentage of Top-Ranking Docking Poses within 2.0 Å from the Experimental Structure on Complexes with Low and High Flexibility of the Ligands

	1 to 6 rotors (87 complexes)	7 to 12 rotors (63 complexes)
ICM	57	27
Glide	67	52
GOLD	57	35

TABLE 3.6
Percentage of Top-Ranking Docking Poses within 2.0 Å from the Experimental Structure on Three Groups of Complexes with Different Degrees of H-bonding

	Low (56 complexes)	Medium (40 complexes)	High (54 complexes)
ICM	34	52	50
Glide	62	70	52
GOLD	34	62	52

Low: HB/HA $\leq$ 0.10, medium: 0.10 < HB/HA < 0.15, high: HB/HA $\geq$ 0.15, where HB = number of H-bonds between ligand and protein and HA = number of heavy atoms in the ligand.

The third aspect analyzed in this context is the impact of the degree of solvent exposure of the binding pocket on the docking accuracy achieved with different search algorithms. It is generally the case that buried binding sites restrict the number of orientations, positions, and conformations accessible to putative binders, but at the same time they require a finer sampling to achieve the proper set of interactions without clashes. On the other hand, solvent-exposed sites require more extensive sampling to cover all the accessible poses, but at the same time they are more tolerant with respect to the combination of pose descriptors required to achieve the proper set of interactions. In this study, the binding sites of the test complexes were divided into three groups — buried, solvent-exposed, and intermediate — and the docking results were dissected accordingly. To assign the complexes to each group, the solvent-accessible surface area of the crystallographic ligand was calculated in the presence and in the absence of the bound protein partner, and the fraction of buried ligand was determined for each complex. If the fraction was 0.75 or lower, the binding site was classified as solvent-exposed. If the fraction was 0.90 or higher, the binding site was classified as buried. Values in between correspond to binding sites with average solvent exposure. The analysis of the performances attained by the three programs on each class, summarized in Table 3.7, shows that all of them achieve the highest degree of accuracy on complexes with buried binding pockets

TABLE 3.7
Percentage of Top-Ranking Docking Poses within 2.0 Å
from the Experimental Structure on Three Groups of
Complexes with Different Degrees of Solvent Exposure
in the Binding Site

	Exposed (41 complexes)	Intermediate (65 complexes)	Buried (44 complexes)
ICM	27	42	66
Glide	46	60	75
GOLD	34	46	64

Exposed: BF ≤ 0.75, intermediate: 0.75 < BF < 0.90, buried: BF ≥ 0.90, where BF is the fraction of the ligand surface area that is buried in the complex.

and consistently lose accuracy with an increase in solvent exposure. Once again, Glide appears to be relatively less sensitive to the features of the binding pockets, while ICM shows the largest decay in performance going from buried to solvent-exposed pockets. These results indicate that all three search algorithms can explore an enclosed binding site much more efficiently than a relatively open one and also points to the obvious observation that, in a more sterically constrained site, the best pose for a given ligand is more unequivocally defined by the shape of the site. As a consequence, the likelihood of generating multiple poses with similar score is much lower and the selection of the best pose is more straightforward. Therefore, it is safe to say that, when docking compounds in a buried binding pocket, an efficient sampling process may be more important than an accurate scoring/ranking method, and in a solvent-exposed pocket, both aspects become equally important.

In addition to the general trends observed, the results of this study highlight some limitations and shortcomings that are common to all docking programs examined. In 12 cases, none of the top 20 poses generated by each of the 3 programs was within 2.0 Å of the experimental structure. Most of these common failures can be ascribed to a combination of structural features that make it especially challenging for any docking program to identify the right solution. Four of the problematic complexes (1cet, 1k1j, 1nhu, 1nhv) are characterized by a dominance of hydrophobic interactions in solvent-exposed sites. In such cases, the shape of the pocket does not help to restrict the number of possible binding orientations, and the lack of a set of specific anchoring points for the ligand makes the selection of the best pose challenging. Moreover, all four ligands are relatively flexible (8 to 9 rotatable bonds), which adds to the sampling problem. Three complexes (1qbu and two HIV-1 protease complexes from the Vertex collection) present highly flexible ligands in almost completely buried binding sites. In these cases, the tightness of the binding pockets and the specific conformational requirements for the ligands to achieve the correct pose call for a thorough sampling process, which is hard to attain within the boundaries of a limited computing time. Another aspect that is sometimes problematic in docking is the presence of charged functionalities in the ligand, because the desolvation energy required for such groups to become available for interaction with the protein is overlooked by most docking functions. In two of the failed complexes (1cet and 1i8z), there is a basic amine in the ligand that does not interact with any protein residue when the crystal complex is analyzed; docking functions tend to favor poses in which such groups form H-bonds or salt bridges. Docking accuracy can also be impaired by the occurrence of unconventional interactions, not properly parameterized in the fitness functions of the programs employed. Two examples are H-bonds between hydrogens of electron-poor aromatic rings and protein acceptors, as observed in one of the Vertex complexes, and H-bonds between the imino form of anilino nitrogens and protein donors, as observed in 1jsv. Both complexes were not reproduced by any of the docking programs. Finally, there are cases in which the interactions between ligand and protein in the experimental pose are tighter than average and predominantly hydrophobic. Imperfect refinement of the crystal structure or the presence of legitimate short-range interactions can introduce apparent clashes that are not compensated by other obvious interactions in the crystallographic pose. When such poses are evaluated in the context of docking, they receive unfavorable scores because they are not properly treated by any known docking functions. This provides a partial explanation for the remaining failures (3ert and one of the kinase complexes from the Vertex collection).

3.3.5 CONCLUSIONS

A comparative evaluation of three cutting-edge docking programs was performed in this study, using a carefully constructed set of pharmaceutically relevant complexes as the test set. The overall accuracy and the dependence of the predictive power of each program on the structural features of the systems studied were assessed, and the impact of force-field-based refinement of the docking poses was investigated. The results of this work provide important guidelines on the use of these tools in drug discovery programs.

The Glide program is shown to have the highest degree of accuracy on a wide and diverse set of systems, which makes it the tool of choice in most cases. Energy minimization of multiple poses is a highly beneficial postprocessing step when docking is performed with GOLD, but the improvement on ICM poses is marginal. Minimization has no impact on the accuracy of the Glide-generated poses, and the combination GOLD docking/OPLS-AA minimization appears to be as reliable a predictor as Glide. The Glide program is more tolerant than both ICM and GOLD of the increase of ligand flexibility, which seems to point to a more effective conformational sampling method. Analogously, Glide appears to be less sensitive to variations in the polarity of the binding pocket, with a slight preference for complexes with prevalent hydrophobic character, but a solid performance across the board. ICM and GOLD can be considered as reliable as Glide when operating on highly polar binding sites, where binding is strongly driven by H-bonding. Comparatively, the ability of these same two programs to predict complexes where binding is driven by hydrophobic interactions is relatively poor. All three programs perform best on buried binding pockets, with a gradual decrease in performance at the increase of solvent exposure. In general, some systems remain a challenge for docking at the current stage, which suggests that there is still a margin for improvement on the existing methods. In particular, the inclusion of properly weighted solvation terms and a more effective representation of metal-mediated interactions in the fitness functions appear to be highly desirable.

3.4 SCORING FUNCTION ANALYSIS

In the previous section of this chapter, we focused on the ability of various docking programs to identify the correctly docked pose for a ligand. In this section, we will focus on another equally important component of docking-based VS, the ability of a scoring function to predict binding affinity. When performing a docking study, one typically docks a set of molecules, optionally rescores, sorts the docked molecules by score, and visually inspects the top scoring molecules. As such, the ability of the scoring function to correctly identify the most active molecules is critical for success. To achieve this goal, a scoring function must be able to reliably rank order collections of molecules in VSs, each performed on a single target. Although scoring functions calibrated on the targets of interest would probably be more effective, the amount of data available at the onset of a drug discovery program is usually insufficient for the task. This reason and the convenience of having tools of general applicability inspired the development of scoring functions designed to predict binding on different systems. Although not all the functions used here have been developed to quantitatively predict binding affinity, all of them have been parameterized on a variety of protein–ligand complexes. Their ability to predict the relative binding affinity on a diverse set of complexes is therefore a reasonable metric for evaluation.

The scoring functions used here provide a sampling of many of the types of functions used in VS studies. We have included empirical functions (PLP [61], ChemScore [8,36,62], GlideScore [54]), molecular-mechanics force fields (MMFF [57–59], OPLS-AA [60]), and knowledge-based potentials (PMF [35] and PMF612 [63]). In the past, we have found that the intermolecular component of the molecular-mechanics interaction energy is a useful metric for ranking molecules in a VS study [14]. As such, we have chosen to evaluate both the total intermolecular energy (vdW plus electrostatics) and the vdW interaction energy produced by the two molecular-mechanics functions. A complete description of the functions used is beyond the scope of this chapter. Here we provide a brief summary of each function. Those interested in additional details should consult the original references or the recent review by Stahl and Bohm [64].

3.4.1 Piecewise Linear Potential (PLP)

The PLP function was developed as a docking function by Gelehaar and coworkers. PLP is a simple function that contains a H-bonding and a lipophilic term. Both terms are flat-well linear potentials that define a repulsive region and ideal region.

3.4.2 CHEMSCORE

The ChemScore function was developed in 1997 by Murray and coworkers as a method of predicting binding affinity. This method is widely used and has been integrated into a number of commercially available software packages. ChemScore consists of a linear combination of four terms: lipophilic, H-bonding, metal binding, and an entropic penalty based on the number of frozen rotatable bonds.

3.4.3 GLIDESCORE

The GlideScore function was developed by the group at Schrödinger, Inc. as a component of the docking program Glide. The function was subsequently optimized to enrich hit rates in VS studies. GlideScore is an extensively modified version of the ChemScore function discussed above. In addition to refitting the ChemScore coefficients, the authors also added terms that differentiate charged and neutral H-bonds. Further terms were also added to account for intermolecular Coulomb and vdW interactions and desolvation.

3.4.4 POTENTIAL OF MEAN FORCE (PMF)

The PMF scoring function was developed by Muegge and Martin as a method of predicting binding affinity. This function is based on a statistical analysis of the contacts found in approximately 800 protein–ligand complexes. The PMF function takes the form

$$E_{ij}(r) = -kT\ln[f_j(r)p_{ij}(r)/p_{ij}]$$

where
k is the Boltzmann constant
T is the absolute temperature
$f_j(r)$ is a factor which corrects for the volume of the ligand
$p_{ij}(r)$ is the number density of pairs of type ij which occur at radius r
p_{ij} is the number density of pairs of type ij where no interaction between i and j occurs

The result of this analysis was a set of pair potentials that describe the interactions between specific protein atom types and ligand types. PMF scores are calculated by looking up the appropriate potentials based on atom type and distance and summing over all protein and ligand atom types.

3.4.5 PMF612

PMF612 is a modified form of the PMF scoring function described above. One drawback to the original PMF function is the presence of irregularities in the pair potentials. These irregularities arise due to coarseness of sampling and the fact that certain pairs of protein and ligand atom types are not commonly found in publicly available crystal structures. The intent of this function was to develop a smoothed potential that would remove many of the irregularities that occur in the PMF pair potential. To generate this function, the authors optimized the width and well depth of a series of vdW 6 to 12 potentials to best fit the potentials used by the PMF function.

3.4.6 MMFF

MMFF is a molecular-mechanics force field developed by Halgren at Merck. MMFF is parameterized for biopolymers and small molecules with a particular emphasis on molecules of pharmaceutical interest. Earlier work published by Holloway and coworkers pointed out the utility of MMFF in predicting the binding affinity of a series of HIV protease inhibitors [65].

3.4.7 OPLS-AA

OPLS-AA is a molecular-mechanics force field developed by Jorgensen. This force field was specifically developed for condensed phase simulations. A great deal of recent work has gone into parameterizing OPLS-AA for molecules of pharmaceutical interest. Recently, components of the OPLS-AA force field have been used to predict binding affinities in linear response calculations [66].

The ChemScore, PLP, PMF, and PMF612 implementations were internally developed and accurately reproduce the results from the original papers. The version of GlideScore used was that implemented in Glide, version 2.5. The MMFF and OPLS-AA energies were obtained using MacroModel, version 8.1. Not all of the 150 complexes described in this chapter were amenable to scoring by each of the functions. Some of the functions (PLP, GlideScore, PMF) were not appropriately parameterized for cofactors. Other functions (PLP, PMF) did not have metal binding terms. To achieve consistent results, we carried out the scoring study using a subset of 111 complexes for which all of the scoring functions had appropriate parameters. The complexes were prepared as described earlier. Each crystallographic ligand pose was then scored in the corresponding protein active site using each of the 7 functions described above. Because we were scoring crystal structures, no scoring function minimization was performed with any of the empirical- or knowledge-based potentials. The energies produced by the molecular-mechanics functions were optimized using 100 steps of conjugate gradient minimization.

3.4.8 CORRELATIONS BETWEEN SCORING FUNCTIONS

Our previous work has shown that combinations of scoring functions can provide an effective means of ranking compounds in a VS study [14]. This method, commonly referred to as consensus scoring, is now widely used in VS studies. One of the main reasons this method is effective is that different scoring functions tend to produce different sets of false positives. Thus compounds that are ranked highly by multiple functions typically have a higher probability of being true actives. One factor not investigated in the previous study was the correlation between scoring functions.

TABLE 3.8
Correlation Coefficients (R²) for the 9 Scoring Functions

	CHEM SCORE	GLIDE SCORE	PLP	PMF	PMF612	MMFF_ VDW	MMFF_ TOT	OPLSAA _VDW	OPLSAA _TOT
CHEM SCORE	1.00	**0.74**	0.52	0.26	0.48	0.28	0.28	0.55	0.51
GLIDE SCORE	**0.74**	1.00	0.52	0.18	0.51	0.40	0.25	0.55	0.40
PLP	0.52	0.52	1.00	0.13	**0.77**	0.54	0.45	**0.71**	0.62
PMF	0.26	0.18	0.13	1.00	0.24	0.03	0.13	0.06	0.15
PMF_ 612	0.48	0.51	**0.77**	0.24	1.00	0.55	0.64	0.69	**0.76**
MMFF_ VDW	0.28	0.40	0.54	0.03	0.55	1.00	0.42	**0.73**	0.42
MMFF_ TOT	0.28	0.25	0.45	0.13	0.64	0.42	1.00	0.39	**0.82**
OPLSAA _VDW	0.55	0.55	**0.71**	0.06	0.69	**0.73**	0.39	1.00	0.68
OPLSAA _TOT	0.51	0.40	0.62	0.15	**0.76**	0.42	**0.82**	0.68	1.00

Table 3.8 shows the correlation (R²) between nine scoring functions for our subset of 111 complexes. Pairs of functions with a correlation greater than 0.7 are shown in bold. It is interesting to note that the scores produced by a number of the functions used in this study are highly correlated. In many cases, these correlations are to be expected. For example, the high correlation between the nonbonded (OPLSAA_VDW vs. MMFF_VDW 0.73) and total (OPLSAA_TOT vs. MMFF_TOT 0.68) molecular-mechanics energies can be attributed to fundamental similarities in the methods and terms used. The correlation between ChemScore and GlideScore can again be attributed to the similarity of the terms used and the fact that ChemScore was the starting point for the GlideScore function. More surprising is the high correlation between the scores produced by PLP and PMF612 and the poor correlation between PMF612 and the original PMF function. The PMF scores produced by PMF612 also correlate well with energies from the molecular-mechanics functions. It appears that by smoothing statistical potentials used by PMF, we have arrived at a function that has characteristics of both a force field and an empirical scoring function.

3.4.9 Scoring Function Accuracy

The plots in Figure 3.1 show the correlation between the scores calculated using each of the scoring functions and the negative log of the experimentally determined binding affinity (pKi). It is immediately obvious that all of the functions used here performed poorly on our test set. Even though the majority of the plots show the appropriate trend, each of the correlations is plagued by a large number of outliers. It is interesting to note that a simple docking function such as PLP performs as well

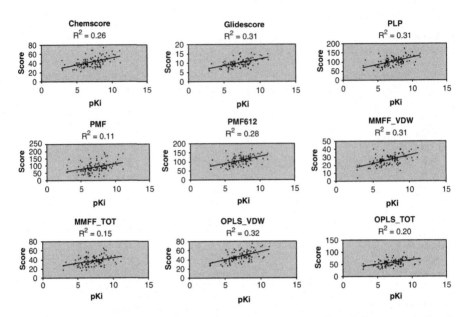

FIGURE 3.1 Correlation between the observed binding affinity (pKi) and the score calculated using 9 scoring functions for a set of 111 diverse protein–ligand complexes. The actual values of the scores are all negative; however, for the purpose of plotting, all the scores have been multiplied by −1.

(or as poorly) as the functions that were carefully tuned to reproduce binding affinities. We were initially surprised that the correlations we observed were significantly lower that those previously published. One would hope that a carefully curated dataset would give results that are at least equivalent to those in the literature. In the original ChemScore paper, the authors obtained an R^2 of 0.71 on their original training set of 82 complexes and an R^2 of 0.63 on a test set of 20 complexes. Similarly, the authors of the PMF function reported an R^2 of 0.61 for a set of 77 complexes. With a few small changes to the composition of the their test set, the authors of PMF were able to increase their R^2 to 0.77.

3.4.10 THE EFFECTS OF TRAINING SET SELECTION

There are number of possible explanations for the poor performance of the scoring functions on our test set. The first has to do with the close link between parameterization and testing of scoring functions. Scoring functions are typically tested on either the same set or a subset of the complexes used for parameterization (the training set). If external test sets are used, these test sets are typically much smaller than the training set and are often composed of structures similar to those in the training set. The set of complexes used here is diverse and contains a number of activity classes that have typically not been used in the parameterization of scoring functions. The set included in the current study also spans a relatively narrow MW range and avoids chance correlations that occur primarily due to molecular size.

We were intrigued by the fact that the performance of a scoring is so highly dependant on the composition of the test set. This phenomenon is not unique to our set of complexes. In the PMF paper, the authors found that R^2 varied between 0.22 and 0.87 depending on the set of complexes that was used. We wondered if it would be possible to select a subset of our complexes that would significantly improve the correlation between score and biological activity.

We chose to treat subset selection as an optimization problem and use a GA [14] to select a subset of complexes for each scoring function that would maximize the correlation between score and biological activity. GAs have been widely used in a variety of computational chemistry applications ranging from docking [17,67] and structure alignment [68] to variable selection for quantitative structure–activity relationships (QSAR) [69]. GAs are probabilistic search techniques based on the principle of evolution and natural selection proposed by Darwin. In a GA, possible solutions are encoded in a chromosome-like data structure. A group of (typically random) chromosomes that make up a population of solutions is allowed to evolve, thereby producing a superior set of solutions. The first step in the GA cycle is the generation of the initial population. Once the initial population is generated, the fitness of each of the chromosomes is evaluated. The most fit members of the population are then chosen to produce the next generation. This is known as the *selection* phase. In the *crossover* phase, two of the selected parents are paired and genetic material is exchanged. To avoid being trapped in local minima, a small fraction of the population undergoes point *mutations*, which effectively increase the gene pool. The new population created by crossover and mutation then completely replaces the current population and the cycle is repeated for a predetermined number of generations.

The use of a GA for subset selection can be best illustrated through the use of a simple example. In this example, we will use the set of 10 complexes shown in Table 3.9. Each complex has a pKi and an associated score. In our GA, each subset of complexes is represented by a bit string of N 1s and 0s, where N is the number of possible complexes. A value of 1 at a position in the bit string indicates that a complex is a member of the subset, a value of 0 indicates that a complex is not. For instance, one potential subset would be represented by 1011011000. This representation indicates that complexes 1, 3, 4, 6, and 7 (13gs, 1aoe, 1atl, 1cet, 1d3p) are members of the subset.

We wish to select a subset of complexes that maximizes two factors: the correlation (R^2) between score and pKi, and the number of complexes used in the correlation. To achieve this goal, we chose a simple objective function for our GA,

$$S = R^2 * N$$

where S is the score, R^2 is the correlation coefficient. Table 3.10 shows the values of S and R^2 for 5 randomly generated subsets of the 10 complexes above. The GA begins by generating a collection (or population) of random subsets and assigning each a score using the method described above. Pairs of high scoring subsets are then randomly selected and combined using a process analogous to biological reproduction known as crossover. In the crossover step, a pair of bit strings is selected

TABLE 3.9
Scores and pKi Values Used to Demonstrate the GA-Based Subset Selection

	Complex	pKi	Score
1	13gs	4.62	30.51
2	1afq	6.21	38.31
3	1aoe	9.66	36.67
4	1atl	6.28	40.42
5	1br6	3.22	37.32
6	1cet	2.89	29.39
7	1d3p	5.11	48.75
8	1d4p	6.30	38.69
9	1dib	7.74	35.74
10	1dlr	9.18	44.90

and each bit string (parent) is divided at an arbitrary position into two segments. The segments from the two parents are then combined to produce a pair of children. For example, the bit string 1111110101 could be divided into two segments — 11111 and 10101. Similarly, the bit string 0001001100 could be divided into 00010 and 01100. Combining the bit strings from the two parents would produce two children **11111**01100 and 00010**10101**. The segments of the two children, which come from the first parent, are shown in bold. The process of randomly selecting high scoring subsets and combining them using crossover is then repeated to produce a new population or generation. The steps above are then repeated for a predetermined number of generations or until the population converges to a single solution.

The correlations between score and biological activity for the optimized subsets of complexes are shown in Figure 3.2 and Table 3.11. It is remarkable that in every case we were able to select a subset of 60 to 80 complexes where the scores are highly correlated with pKi. It is also distressing the extent to which a biased test set can greatly exaggerate the performance of a scoring function. Both the number of complexes selected and the correlations observed here are similar to those observed in the literature. This result underscores the need for extensive external validation of any scoring function.

Because the subsets chosen by the each of the 9 functions were roughly equivalent in size, we wondered whether the different functions were choosing roughly the same subset. One means of addressing this question is to examine the number of subsets in which a particular complex appears. Figure 3.3 shows the frequency of occurrence of complexes in the subsets chosen by the GA. From the graph we can see that 34 complexes were used in all 9 subsets and 24 complexes were used in 8 of 9 subsets. Thus there were 58 complexes that were used in 8 of 9 subsets. This is quite remarkable when we consider that subsets ranged in size from 71 to 88. We can then conclude that 65 to 80% of the complexes selected by the GA were

TABLE 3.10
The Method Used to Assign a Fitness Score to Members of the Population During the GA Optimization

Bit String	Complexes Used	N	R^2	S
1111110101	1,2,3,4,5,6,8,10	9	0.43	3.85
0001001100	4,7,8	4	0.98	3.91
0010111011	3,5,6,7,9,10	7	0.09	0.62
1010101010	1,3,5,7,9	6	0.01	0.05
0101010011	2,4,6,9,10	6	0.73	4.40

The correlation coefficient (R^2) is calculated for only those complexes represented by "1" in the bit string. The R^2 value is then multiplied by the number of complexes used in the correlation (N) to arrive at the score (S).

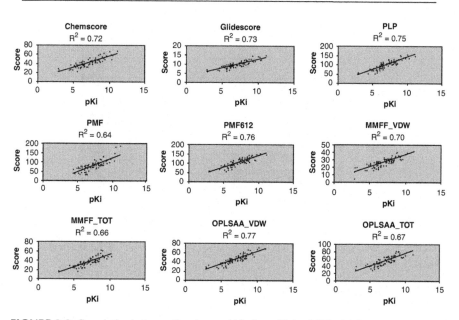

FIGURE 3.2 Correlation between the observed binding affinity (pKi) and the score calculated using nine scoring functions for the complexes selected by the GA for each function.

the same across all 9 scoring functions. This may suggest that there are some complexes that are easy to score and other complexes that are more difficult to score.

We examined simple metrics such as degree of burial and number of H-bonds and were unable to uncover a relationship between the nature of the complex and its frequency of occurrence in the subsets. Table 3.12 lists the frequency of occurrence in subsets as a function of activity class. There is no obvious relationship, with the exception of lyases, the distributions roughly parallel the distributions in the overall dataset.

TABLE 3.11

Correlation between Score and pKi for the Complete Set of Complexes and the Subsets Selected by the GA

	Original R²	N	Subset R²
ChemScore	0.26	79	0.72
GlideScore	0.31	80	0.73
PLP	0.31	79	0.75
PMF	0.11	74	0.64
PMF_612	0.28	81	0.76
MMFF_VDW	0.31	71	0.70
MMFF_TOT	0.15	88	0.66
OPLSAA_VDW	0.32	80	0.77
OPLSAA_TOT	0.19	83	0.67

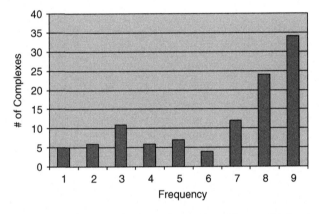

FIGURE 3.3 Frequency of occurrence of complexes selected by the GA for the nine scoring functions used in this study.

3.4.11 CONCLUSIONS

The development of a generally applicable function capable of quantitatively predicting binding affinity from crystallographic complexes has been a goal of computational chemists for more than 20 years. Although a number of published methods purport to be both general and accurate, our results indicate that this may not be the case. We evaluated 9 scoring functions against a set of 111 diverse druglike pharmaceutically relevant complexes and found that, without exception, the observed correlations were poor. Furthermore, the GA-based subset selection demonstrated that, given the appropriate dataset, any of the functions used here can appear to be predictive. We believe that it is important for scoring function developers to follow the examples of practitioners in other model building fields. In fields such as QSAR and machine learning, models are extensively tested using datasets that were not part of the model building process. A careful analysis of complexes that fail this

TABLE 3.12
Frequency of Occurrence of Complexes in the GA-Based Subset Selection
as a Function of Activity Class

	N	8 to 9	5 to 8	0 to 4
serine hydrolase	1	0	1	0
cytokine	1	0	1	0
topoisomerase	8	5	3	0
oxido-reductase	2	1	1	0
rhinovirus coat protein	1	0	1	0
immunoglobuline	1	0	0	1
transferase	46	29	11	6
cis-trans isomerase	1	0	1	0
immunoglobulin	1	1	0	0
serine protease	21	10	6	5
lyase	4	0	0	4
aspartyl protease	9	6	2	1
antibody	3	1	1	1
nuclear receptor	8	4	0	4
hydrolase	4	1	1	2

external validation may lead to additional insights regarding the shortcomings of existing functions.

To predict binding affinity with a higher degree of accuracy, we may have to move beyond simple functions such as those discussed in this section. It is quite likely that factors such as entropy and desolvation, which have been largely ignored, will have to be considered. Fortunately, computers continue to get cheaper and faster. The availability of additional computing is now making it possible for researchers to explore a more rigorous treatment of factors such as protein dynamics and electrostatics [70]. Hopefully, advances in molecular mechanics coupled with the availability of a wider array of experimental data will improve our ability to reliably predict binding affinity.

3.5 EVALUATION OF DOCKING/SCORING COMBINATIONS FOR VIRTUAL SCREENING

VS entails the use of computational methods to rank a collection of compounds with respect to their predicted activity on a given system. The ultimate objective is to select a subset enriched in compounds with the desired activity relative to the entire collection. At the early stage of a drug discovery program, this approach can be used to identify potential lead compounds in corporate collections or in commercial databases. When the structure of a protein target is available, compounds can be evaluated for their ability to bind in the active site using a combination of docking and scoring methods. In the early reports of structure-based VS, the results from docking were directly used to rank different compounds relative to each other, with the docking and scoring function as defined above being the same [24]. As the field matured, computational chemists came to the realization that the function designated to select the best pose for a given compound was not necessarily the most effective at ranking compounds relative to each other. It has therefore become customary to follow up on the initial docking with a postprocessing procedure, with the purpose of ranking compounds in a more reliable manner using a different scoring function [11–13]. To overcome the limitations of individual methods, consensus scoring schemes have been proposed and applied in different occasions [14,71].

The success of a VS is usually reported as *hit rate,* the percentage of the selected compounds that are found to have the desired activity after testing. When the percentage of active compounds in the entire collection is known for a given target, or can be reliably estimated, success can also be described by the enrichment factor, which is the ratio of the percentage of active compounds in the selected subset to the percentage in the entire database.

A variety of docking/scoring combinations have been applied to VS in recent years. In this work, we evaluated the performance of the three docking programs described above in combination with three different scoring functions — Chem-Score, GlideScore, and OPLS-AA interaction energy. ChemScore is the most widely used scoring function for VS: it has been shown to outperform most of the others in comparative studies [14,71]. GlideScore has been specifically designed and refined to maximize enrichment in database screening. It is claimed to be the most effective tool to discriminate between active and inactive compounds on a variety of systems [54]. OPLS-AA is one of the few force fields developed and refined with an eye for both proteins and small organic molecules. The objective of this study was to derive some guidelines on how to select the best protocol for VS on a given target.

Three targets with known high-resolution crystal structure were used in this study — HIV-1 protease, inosine monophosphate dehydrogenase (IMPDH), and p38 mitogen-activated protein (MAP) kinase. Simulated VS was performed on each target using test sets of 10,000 compounds, with N actives selected from Vertex research programs and 10,000 − N decoys selected from commercial databases. The experimental Kis of the active compounds range from low nanomolar to high micromolar, with a few subnanomolar ligands included for p38. The selection of decoys was biased toward druglike molecules using filters for functional groups and cutoffs for MW and number of rotatable bonds. Composition of test sets and cutoffs applied are summarized in Table 3.13. Importantly, active compounds and decoys were selected with a similar distribution of MW to minimize the effects of the notorious tendency of most scoring functions to favor larger molecules.

Each test set was docked into the target crystal structures with the ICM, Glide, and GOLD programs, according to the procedures described in the previous sections. Energy minimization was performed on the docking poses with MacroModel employing the OPLS-AA force field as described in Section 3.3. Both nonminimized and minimized docking poses were rescored with ChemScore, GlideScore, and OPLS-AA interaction energy, the latter corrected by the strain energy of the ligand. Enrichments were finally calculated for all docking/scoring combinations on the top 1%, top 3%, and top 10% of each ranking.

TABLE 3.13

Composition of the Test Sets Used in the Enrichment Studies and Cutoffs Implemented for Rotatable Bonds and Molecular Weight

Target	RB Cutoff	MW cutoff	Number of Actives	Total
HIV-1 protease	12	600	206	10,000
IMPDH	8	500	142	10,000
p38	8	500	247	10,000

3.5.1 RESULTS AND DISCUSSION

The results of this study are reported in Figure 3.4. The performance of each protocol is represented as the percent of the total active compounds identified in the top 1%, top 3%, and top 10% of the corresponding ranking. Whenever the relative performances of different methods are different for different top portions of the ranking examined (e.g., Method A achieves better enrichment than Method B in top 3%, worse enrichment in top 10%), the enrichment in the top 3% will be used as the main indicator of performance in this report. The nature of the three active sites used in this study is significantly different, and the relative performances of the methods examined vary as a function of these differences.

In HIV-1 protease, the binding site is buried and predominantly hydrophobic, with an oblong shape suited for large and flexible ligands. The restrictive size and shape of this binding site make this enzyme a challenging system for docking. Additionally, there is a conserved catalytic water molecule that is an integral part of this active site contributing to the challenging nature of this system because interactions with water are generally not handled accurately by most docking/scoring functions. As a demonstration of this, in the study described in Section 3.3, all three programs performed poorly on the nine HIV-1 protease complexes included in the test set.

In the VS simulation on this system, ChemScore consistently achieved the best enrichment, regardless of the pose generation method (Figure 3.4, first row). The relative insensitivity of this function to repulsive interactions was probably beneficial in a system where a large amount of sampling would be necessary to generate an accurate docking result, and even otherwise correct docking poses may still contain severe clashes between protein and ligand atoms. The performance of ChemScore was largely unaffected by energy minimization of the docking poses, which is consistent with the fact that the attenuation of unfavorable interactions has limited impact on the scores. For this system, the combination ICM/ChemScore achieved the best enrichments, but the performances of Glide/ChemScore and GOLD/ChemScore were comparable.

Energy minimization of the docking poses dramatically improved the enrichments achieved by OPLS-AA and GlideScore on both ICM and GOLD poses, but only OPLS-AA improved on Glide poses and less significantly. Both GlideScore and OPLS-AA performed well on minimized ICM poses, but their performance on

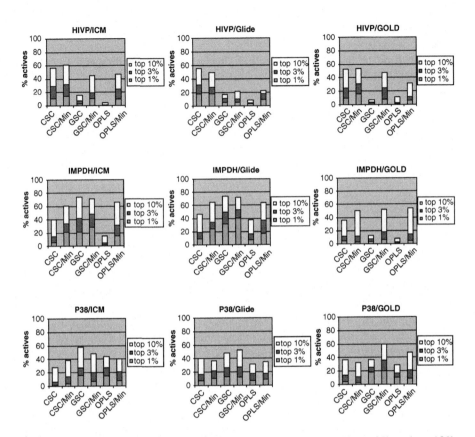

FIGURE 3.4 Percentage of active compounds recovered in the top 1%, top 3%, and top 10% of the rankings with different docking/scoring combinations on the three targets used in this study. CSC: ChemScore on docking poses; GSC: GlideScore on docking poses; OPLS: OPLS-AA force field score on docking poses; CSC/Min: ChemScore on energy-minimized docking poses; GSC/Min: GlideScore on energy-minimized docking poses; OPLS/Min: force field score on energy-minimized docking poses.

both unminimized and minimized Glide poses was surprisingly modest. In terms of pose generation, unminimized poses generated by the three programs achieved similar enrichments, with Glide poses slightly ahead of ICM and GOLD poses. After minimization, the relative performances were reversed with limited benefit for Glide. This last observation is consistent with the results described in Section 3.3.

In IMPDH, the binding site is relatively polar and predominantly solvent-exposed, but it contains a narrow cavity at the bottom that often accommodates hydrophobic moieties of ligands sandwiched between the cofactor and protein residues. The tightness of such cavity requires a fine sampling to correctly place ligands and makes docking challenging. The presence of a cofactor represents an additional challenge, because in many docking functions the interactions involving cofactors are not accurately parameterized. None of the three programs evaluated in this study were particularly effective at reproducing the four IMPDH complexes present in the docking test set. On this system, GlideScore consistently achieved the highest enrichment, regardless of the pose generation method, with the only exception being unminimized GOLD poses, on which ChemScore did better (Figure 3.4, second row). In general, however, all of the scoring functions considered achieved modest enrichments on this particular set of poses. ChemScore consistently outperformed OPLS-AA on unminimized poses. Energy minimization improved the performance of ChemScore on ICM and Glide poses and dramatically improved the performance of OPLS-AA on all sets of poses, raising it to the same level of ChemScore. Minimization dramatically improved the performance of GlideScore on GOLD poses, but only marginally improved the performance of GlideScore on ICM and Glide poses. In terms of pose generation methods, slightly better enrichments were achieved on Glide poses relative to ICM poses. Both docking methods clearly outperformed GOLD, especially in the absence of energy minimization.

The improvement observed in most cases upon minimization highlights the importance of refinement of the docking poses when the active site contains a narrow region important for binding and more extensive sampling may be required. Refinement is more important when the docking protocol does not include a minimization component, as is the case for GOLD. The great benefits of minimization on GOLD-generated poses confirm the observations reported in Section 3.3.3.

The binding site of p38 is relatively buried and mostly hydrophobic and usually accommodates ligands through a combination of shape matching and H-bonds with backbone amide groups. These features and the absence of narrow subpockets make of p38 a good system for docking, as confirmed by the fact that all three docking programs were quite accurate in reproducing the six p38 complexes present in the docking test set.

On this system, once again GlideScore showed the most consistent performance, only matched by OPLS-AA on ICM poses (Figure 3.4, third row). OPLS-AA performed relatively well on all three sets of unminimized poses, with enrichments equivalent to or better than ChemScore. Energy minimization did not significantly affect the enrichments on Glide and ICM poses regardless of the scoring method, although it marginally improved the enrichments achieved with GlideScore and OPLS-AA on GOLD poses. This is consistent with the observation that on this system, the docking poses are generally correct even in the absence of minimization. Interestingly, the best enrichment was achieved in this case by the combination GOLD/GlideScore, although in general the performance of the three pose generation methods can be considered equivalent. To explore all possibilities, the GOLD poses were also ranked on all three systems using the GOLD fitness function. No significant enrichment was achieved in any of the three cases. ICM contains its own empirical function to rescore docked poses in a VS context [72]. The score must be recalculated at the end of the docking stage and it is time-consuming (up to one minute per pose on a Pentium III 1 Ghz). Because the function had not shown satisfactory results when previously tested on internal targets, we decided not to include it in this study.

3.5.2 CONCLUSIONS

The results obtained on the three systems confirm that, as recently stated [12,13,71], a universal docking/scoring combination that outperforms all the others on every system does not exist. Nevertheless this study suggests that some combinations do achieve more consistent performances, and it provides a set of useful guidelines on how to select and use the currently available tools. The three binding pockets used here did not differentiate the performances of the three docking programs as much as the diverse set of complexes used in Section 3.3. Careful analysis of the results indicates that comparable enrichments across different systems were achieved when Glide and ICM poses were rescored with different methods, with the Glide poses performing slightly better on tighter binding sites. Both Glide and ICM clearly outperformed GOLD in the most challenging systems, while GOLD poses achieved comparable enrichments only in the easier binding site of p38. The similar performances of Glide and ICM could be viewed as inconsistent with the results described in Section 3.3, but different aspects come into play when docking programs are engaged in VS. A program must be able to fit real binders to an active site conformation that is not necessarily optimal for them and, at the same time, minimize the occurrence of false positives, i.e., inactive compounds that are docked into the active site and favorably scored. Neither of these issues is present when the crystallographic ligand is docked back into the cognate active site, and the ways each program deals with them contributes to its ability to generate good poses in a docking-based VS exercise.

Analysis of the poses generated by Glide and ICM in different systems shows that, when an ideal fit cannot be achieved, Glide tends to generate more strained ligand conformations to maximize the interactions with the protein, while ICM tends to produce more stable ligand conformations at the expense of less optimal intermolecular contacts. The aggressive approach applied by Glide, which attempts to reconcile induced fit with the rigid receptor approach by softening some of the intramolecular repulsive interactions, increases the probability of finding active compounds, but also the occurrence of false positives. The conservative and more rigorous approach applied in ICM entails a higher risk of missing actives, but also minimizes the incidence of false positives. The interplay of these different tendencies, in conjunction with the intrinsic effectiveness of the search algorithms, the nature of the active site, and the affinity of the actual ligands in the screened collection determine the relative performance of the two programs. The results obtained here indicate that the vigorous approach of Glide can be more successful on tight binding pockets, where the amount of sampling required to achieve the correct set of interactions with the correct conformation of the ligand may be quite large. In more spacious active sites, the conservative approach followed by ICM can be equally or more effective.

Energy minimization of the docked poses in OPLS-AA force field seems to significantly improve the enrichment on systems with tight binding pockets, although it has basically no impact on the enrichments achieved on the more spacious pocket of p38. This suggests that energy minimization should be common practice when docking-based VS is performed on sterically demanding pockets, although in the absence of a clear classification of the binding pocket, it should always be applied. Glide poses seem to generally benefit less from this postprocessing step. This may be ascribed to the fact that the same force field is already implemented in the docking protocol, although in a less rigorous fashion (especially with respect to the intramolecular energy of the ligand).

In terms of scoring, GlideScore appears to be the most general scoring function tested, with the best performance across the board on IMPDH and p38. When the binding site is tight and docking is likely to produce poses with severe intermolecular clashes, a function like ChemScore, more forgiving of repulsive interactions, can be more effective. An alternative approach, suggested by Glide developers, is to reduce the vdW radii of protein and ligand atoms, thus softening the repulsive terms in GlideScore along the lines discussed above.

Overall, exclusive force-field-based scoring, exemplified in this study by the OPLS-AA energy function, appears to be less reliable than empirical scoring when ranking different ligands, although reasonable enrichments can be achieved in some cases, particularly on minimized docking poses and spacious binding pockets. In general, testing different combinations on the target of choice and selecting the best performing one for the real screening is highly desirable whenever sufficient data is available. In the absence of it, the protocol should be selected based on the nature of the active site. This study suggests a choice between Glide and ICM for the docking step, energy minimization in OPLS-AA, and a choice between GlideScore and ChemScore for rescoring, according to the criteria described above.

3.6 OVERALL CONCLUSIONS AND PERSPECTIVES

We have attempted to address some of the fundamental issues, both general and specific, surrounding the use of docking/scoring tools for drug discovery. These types of studies can never address all of the issues and questions as the technology is rapidly changing and access to all of the possible programs is never complete. Nonetheless, we believe that some useful conclusions and insights from this work can be summarized as follows:

- The choice of test set for evaluation of docking/scoring methods should reflect the composition of the ultimate screening sets.
- Careful attention to the details of protein and ligand setup is crucial (e.g., protonation states, bond orders, minimization of hydrogen positions) to enhancing screening hit rates and docking accuracy.
- In general, certain search procedures/scoring schemes perform better in certain types of binding sites due to differences in protocols, search methods, docking functions, and scoring functions.
- ICM, Glide, and GOLD are all reasonable choices for docking/VS in buried binding sites, although Glide appears to perform most consistently with respect to diversity of binding sites, ligand flexibility, and overall sampling. Glide can correctly identify the crystallographic pose approximately 60% of the times within 2.0 Å.

- Saving the top N scoring poses from any docking function followed by energy minimization and reranking can be an effective means to overcome some of the limitations of a given docking function. Because the additional computing cost is no longer prohibitive (e.g., 30 to 60 sec/pose), it may make sense to minimize as a general postdocking step.
- Energy minimization of docked poses can significantly improve the enrichments in systems with sterically demanding binding sites, although Glide poses tend to benefit less than GOLD- or ICM-generated poses.
- GlideScore appears to be an effective scoring function for database screening with consistent performance across several types of binding sites. ChemScore is most useful in sterically demanding sites because it is more forgiving of repulsive interactions.
- Whenever possible, it is advisable to test different docking/scoring combinations on a given system of interest to select the best protocol prior to the full database screen.
- Although some of the functions considered in this study are adequate for selecting the correct pose in docking and in identifying active compounds during database screening, none are acceptable for prediction of binding affinity.
- When attempting to evaluate scoring functions, careful attention must be paid to both the size and composition of the test set as well as the degree of external validation necessary to derive any conclusions about general performance and utility.
- Some complexes are inherently more difficult to score correctly than others, possibly due the fact that a crystal structure is a static entity and in some cases does not adequately describe protein conformational changes and other enthalpic and entropic contributions resulting from the release of water molecules upon ligand binding.

When considering the conclusions drawn above, one is left with the overall sense that some search algorithms and docking functions are adequate. The next large improvements in this field will hopefully be in scoring functions that more accurately describe the physics of binding and allow for not only good discrimination between actives and inactives, but between closely related analogs. In this regard, factors such as protein flexibility and solvation need to be incorporated in a meaningful and efficient manner. Because we recognize some of the current limitations, we have tried to emphasize practices that can sometimes overcome some of these limitations such as saving and reranking poses and using a variety of searching and scoring combinations. To be fair, we should point out that it is common practice to use graphics as part of the final round of selecting compounds in docking-based VS. This can certainly remove some of the shortcomings of various scoring functions (although it tends to introduce other subjective biases). A visualization tool that highlights both favorable as well as unfavorable contributions to binding based upon a variety of scoring functions could be quite useful. Additionally, the judicious use of a variety of postprocessing filters (e.g., Did we maintain a key H-bond?) or the use of constraints during docking are commonly used to enhance the results of docking/VS and should be used whenever possible.

Finally, we would like to comment on how docking methods can best fit into the entire array of screening technologies available in the drug discovery armament. As with many other technologies, docking-based VS has become incorporated over time and currently appears less like a competitor and more like an adjunct. Recent publications have shown that docking-based VS can identify different chemotypes as potential lead classes than those identified via experimental screens on the same target [73,74]. In these reports, the compounds coming out of the virtual screen appeared no less druglike than the hits from the experimental screens. In an environment where diversity of initial lead classes is valued, these approaches can be quite complementary. Another situation in which docking can assist experimental screening is in the analysis of HTS hits. We have found it quite useful to take the results of a HTS screen and perform detailed docking experiments using multiple searching/scoring protocols in an attempt remove false positives identified in the HTS.

The application of docking in current drug discovery has matured over the past 20 years and is generally accepted as a useful tool. We look forward to continued progress in this exciting field.

REFERENCES

[1]   B.K. Shoichet, S.L. McGovern, B. Wei, and J.J. Irwin, Lead discovery using molecular docking, *Curr. Opin. Chem. Biol.* 6:439–446, 2002.

[2]   G. Schneider and H.J. Bohm, Virtual screening and fast automated docking methods, *Drug Discov. Today* 7:64–70, 2002.

[3]   G.M. Verkhivker, D. Bouzida, D.K. Gehlhaar, P.A. Rejto, S.T. Freer, and P.W. Rose, Complexity and simplicity of ligand-macromolecule interactions: the energy landscape perspective, *Curr. Opin. Struct. Biol.* 12:197–203, 2002.

[4]   G.M. Verkhivker, D. Bouzida, D.K. Gehlhaar, P.A. Rejto, S. Arthurs, A.B. Colson, S.T. Freer, V. Larson, B.A. Luty, T. Marrone, and P.W. Rose, Deciphering common failures in molecular docking of ligand–protein complexes, *J. Computer-Aided Mol. Des.* 14:731–751, 2000.

[5]   A.M. Davis and S.J. Teague, Hydrogen bonding, hydrophobic interactions, and failure of the rigid receptor hypothesis, *Angew. Chem. Int. Ed. Engl.* 38:736–749, 1999.

[6]   I. Muegge and M. Rarey, Small molecule docking and scoring, *Reviews in Computational Chemistry*, Vol. 17, New York: Wiley-VCH, 2001.

[7]   R.D. Taylor, P.J. Jewsbury, and J.W. Essex, A review of protein-small molecule docking methods, *J. Computer-Aided Mol. Des.* 16:151–166, 2002.

[8]   I. Halperin, B. Ma, H. Wolfson, and R. Nussinov, Principles of docking: an overview of search algorithms and a guide to scoring functions, *Proteins* 47:409–443, 2002.

[9]   P.S. Charifson and W.P. Walters, Filtering databases and chemical libraries, *J. Computer-Aided Mol. Des.* 16:311–323, 2002.

[10]   M. Stahl and H.J. Bohm, Development of filter functions for protein–ligand docking, *J. Mol. Graphics Modelling* 16:121–132, 1998.

[11]   C. Perez and A.R. Ortiz, Evaluation of docking functions for protein–ligand docking, *J. Med. Chem.* 44:3768–3785, 2001.

[12] C. Bissantz, G. Folkers, and D. Rognan, Protein-based virtual screening of chemical databases, 1. Evaluation of different docking/scoring combinations, *J. Med. Chem.* 43:4759–4767, 2000.

[13] M. Stahl and M. Rarey, Detailed analysis of scoring functions for virtual screening, *J. Med. Chem.* 44:1035–1042, 2001.

[14] P.S. Charifson, J.J. Corkery, M.A. Murcko, and W.P. Walters, Consensus scoring: a method for obtaining improved hit rates from docking databases of three-dimensional structures into proteins, *J. Med. Chem.* 42:5100–5109, 1999.

[15] D.J. Diller and C.L. Verlinde, A critical evaluation of several global optimization algorithms for the purpose of molecular docking, *J. Comput. Chem.* 20:1740–1751, 1999.

[16] A.R. Leach, Ligand docking to proteins with discrete side-chain flexibility, *J. Mol. Biol.* 235:345–356, 1994.

[17] G. Jones, P. Willett, R.C. Glen, A.R. Leach, and R. Taylor, Development and validation of a genetic algorithm for flexible docking, *J. Mol. Biol.* 267:727–748, 1997.

[18] H.B. Broughton, A method for including protein flexibility in protein–ligand docking: improving tools for database mining and virtual screening, *J. Mol. Graphics Modelling* 18:247–257, 302–304, 2000.

[19] H. Claussen, C. Buning, M. Rarey, and T. Lengauer, FlexE: efficient molecular docking considering protein structure variations, *J. Mol. Biol.* 308:377–395, 2001.

[20] A.C. Anderson, R.H. O'Neil, T.S. Surti, and R.M. Stroud, Approaches to solving the rigid receptor problem by identifying a minimal set of flexible residues during ligand docking, *Chem. & Biol.* 8:445–457, 2001.

[21] R.D. Taylor, P.J. Jewsbury, and J.W. Essex, FDS: flexible ligand and receptor docking with a continuum solvent model and soft-core energy function, *J. Comput. Chem.* 24:1637–1656, 2003.

[22] B.K. Shoichet, A.R. Leach, and I.D. Kuntz, Ligand solvation in molecular docking, *Proteins* 34:4–16, 1999.

[23] X. Zou, Y. Sun, and I.D. Kuntz, Inclusion of solvation in ligand binding free energy calculations using the generalized-Born model, *J. Am. Chem. Soc.* 121:8033–8043, 1999.

[24] B.K. Shoichet, R.M. Stroud, D.V. Santi, I.D. Kuntz, and K.M. Perry, Structure-based discovery of inhibitors of thymidylate synthase, *Science* 259:1445–1450, 1993.

[25] A.I. Su, D.M. Lorber, G.S. Weston, W.A. Baase, B.W. Matthews, and B.K. Shoichet, Docking molecules by families to increase the diversity of hits in database screens: computational strategy and experimental evaluation, *Proteins* 42:279–293, 2001.

[26] H.J. Bohm, The development of a simple empirical scoring function to estimate the binding constant for a protein–ligand complex of known three-dimensional structure, *J. Computer-Aided Mol. Des.* 8:243–256, 1994.

[27] H.J. Bohm, Prediction of binding constants of protein ligands: a fast method for the prioritization of hits obtained from *de novo* design or 3D database search programs, *J. Computer-Aided Mol. Des.* 12:309–323, 1998.

[28] M.D. Eldridge, C.W. Murray, T.R. Auton, G.V. Paolini, and R.P. Mee, Empirical scoring functions: I. The development of a fast empirical scoring function to estimate the binding affinity of ligands in receptor complexes, *J. Computer-Aided Mol. Des.* 11:425–445, 1997.

[29] T.J. Ewing, S. Makino, A.G. Skillman, and I.D. Kuntz, DOCK 4.0: search strategies for automated molecular docking of flexible molecule databases, *J. Computer-Aided Mol. Des.* 15:411–428, 2001.

[30] M. Feher, E. Deretey, and S. Roy, BHB: a simple knowledge-based scoring function to improve the efficiency of database screening, *J. Chem. Inf. Comput. Sci.* 43:1316–1327, 2003.

[31] H. Gohlke, M. Hendlich, and G. Klebe, Knowledge-based scoring function to predict protein–ligand interactions, *J. Mol. Biol.* 295:337–356, 2000.

[32] A.V. Ishchenko and E.I. Shakhnovich, SMall Molecule Growth 2001 (SMoG2001): an improved knowledge-based scoring function for protein–ligand interactions, *J. Med. Chem.* 45:2770–2780, 2002.

[33] A.N. Jain, Scoring noncovalent protein–ligand interactions: a continuous differentiable function tuned to compute binding affinities, *J. Computer-Aided Mol. Des.* 10:427–440, 1996.

[34] B. Kramer, M. Rarey, and T. Lengauer, Evaluation of the FLEXX incremental construction algorithm for protein–ligand docking, *Proteins* 37:228–241, 1999.

[35] I. Muegge and Y.C. Martin, A general and fast scoring function for protein–ligand interactions: a simplified potential approach, *J. Med. Chem.* 42:791–804, 1999.

[36] C.W. Murray, T.R. Auton, and M.D. Eldridge, Empirical scoring functions: II. The testing of an empirical scoring function for the prediction of ligand-receptor binding affinities and the use of Bayesian regression to improve the quality of the model, *J. Computer-Aided Mol. Des.* 12:503–519, 1998.

[37] Y.P. Pang, E. Perola, K. Xu, and F.G. Prendergast, EUDOC: a computer program for identification of drug interaction sites in macromolecules and drug leads from chemical databases, *J. Comput. Chem.* 22:1750–1771, 2001.

[38] M.L. Verdonk, J.C. Cole, M.J. Hartshorn, C.W. Murray, and R.D. Taylor, Improved protein–ligand docking using GOLD, *Proteins* 52:609–623, 2003.

[39] R. Wang, L. Liu, L. Lai, and Y. Tang, SCORE: a new empirical method for estimating the binding affinity of a protein–ligand complex, *J. Mol. Model.* 4:379–394, 1998.

[40] R. Wang, L. Lai, and S. Wang, Further development and validation of empirical scoring functions for structure-based binding affinity prediction, *J. Computer-Aided Mol. Des.* 16:11–26, 2002.

[41] O. Roche, R. Kiyama, and C.L. Brooks, III, Ligand–protein database: linking protein–ligand complex structures to binding data, *J. Med. Chem.* 44:3592–3598, 2001.

[42] J.W. Nissink, C. Murray, M. Hartshorn, M.L. Verdonk, J.C. Cole, and R. Taylor, A new test set for validating predictions of protein–ligand interaction, *Proteins* 49:457–471, 2002.

[43] C.A. Baxter, C.W. Murray, B. Waszkowycz, J. Li, R.A. Sykes, R.G. Bone, T.D. Perkins, and W. Wylie, New approach to molecular docking and its application to virtual screening of chemical databases, *J. Chem. Inf. Comput. Sci.* 40:254–262, 2000.

[44] M. Rarey, B. Kramer, T. Lengauer, and G. Klebe, A fast flexible docking method using an incremental construction algorithm, *J. Mol. Biol.* 261:470–489, 1996.

[45] R.D. Clark, A. Strizhev, J.M. Leonard, J.F. Blake, and J.B. Matthew, Consensus scoring for ligand/protein interactions, *J. Mol. Graphics Modelling* 20:281–295, 2002.

[46] N. Paul and D. Rognan, ConsDock: a new program for the consensus analysis of protein–ligand interactions, *Proteins* 47:521–533, 2002.

[47] B. Waszkowycz, Structure-based approaches to drug design and virtual screening, *Curr. Opin. Drug Discov. Devel.* 5:407–413, 2002.

[48] S.D. Pickett, B.S. Sherborne, T. Wilkinson, J. Bennett, N. Borkakoti, M. Broadhurst, D. Hurst, I. Kilford, M. McKinnell, and P.S. Jones, Discovery of novel low molecular weight inhibitors of IMPDH via virtual needle screening, *Bioorg. Med. Chem. Lett.* 13:1691–1694, 2003.

[49]   J.L. Jenkins, R.Y. Kao, and R. Shapiro, Virtual screening to enrich hit lists from high-throughput screening: a case study on small-molecule inhibitors of angiogenin, *Proteins* 50:81–93, 2003.

[50]   E. Vangrevelinghe, K. Zimmermann, J. Schoepfer, R. Portmann, D. Fabbro, and P. Furet, Discovery of a potent and selective protein kinase CK2 inhibitor by high-throughput docking, *J. Med. Chem.* 46:2656–2662, 2003.

[51]   M. Totrov and R. Abagyan, Flexible protein–ligand docking by global energy optimization in internal coordinates, *Proteins* Suppl:215–220, 1997.

[52]   Glide, 2.5 ed., Schrödinger, New York, NY.

[53]   R. Abagyan and M. Totrov, High-throughput docking for lead generation, *Curr. Opin. Chem. Biol.* 5:375–382, 2001.

[54]   http://www.schrodinger.com/docs/fd2_2002_2/pdf/fd25_technical_notes.pdf.

[55]   G. Nemethy, K.D. Gibson, K.A. Palmer, C.N. Yoon, G. Paterlini, A. Zagari, S. Rumsey, and H.A. Scheraga, Energy parameters in polypeptides: 10. Improved geometrical parameters and nonbonded interactions for use in the ECEPP/3 algorithm, with application to proline-containing peptides, *J. Phys. Chem.* 96:6472–6484, 1992.

[56]   J. Gasteiger, C. Rudolph, and J. Sadowski, Automatic generation of 3D-atomic coordinates for organic molecules, *Tetrahedron Comp. Method.* 3:537–547, 1990.

[57]   T.A. Halgren, Merck molecular force field: I. Basis, form, scope, parameterization, and performance of MMFF94, *J. Comput. Chem.* 17:490–519, 1996.

[58]   T.A. Halgren, Merck molecular force field: II. MMFF94 van der Waals and electrostatic parameters for intermolecular interactions, *J. Comput. Chem.* 17:520–552, 1996.

[59]   T.A. Halgren, Merck molecular force field: III. Molecular geometries and vibrational frequencies for MMFF94, *J. Comput. Chem.* 17:553–586, 1996.

[60]   W.L. Jorgensen, D. Maxwell, and J. Tirado-Rives, Development and testing of the OPLS all-atom force field on conformational energetics and properties of organic liquids, *J. Am. Chem. Soc.* 118:11225–11236, 1996.

[61]   D.K. Gehlhaar, G.M. Verkhivker, P.A. Rejto, C.J. Sherman, D.B. Fogel, and S.T. Freer, Molecular recognition of the inhibitor AG-1343 by HIV-1 protease — conformationally flexible docking by evolutionary programming, *Chem. & Biol.* 2:317–324, 1995.

[62]   C.A. Baxter, C.W. Murray, D.E. Clark, D.R. Westhead, and M.D. Eldridge, Flexible docking using Tabu search and an empirical estimate of binding affinity, *Proteins* 33:367–382, 1998.

[63]   D.A. Pearlman and W.P. Walters, manuscript under preparation.

[64]   M. Stahl and H.J. Bohm, The use of scoring functions in drug discovery, *Reviews in Computational Chemistry*, Hoboken, NJ: Wiley-VCH, pp. 41–87, 2002.

[65]   M.K. Holloway, J.M. Wai, T.A. Halgren, P.M. Fitzgerald, J.P. Vacca, B.D. Dorsey, R.B. Levin, W.J. Thompson, L.J. Chen, and S.J. deSolms, A priori prediction of activity for HIV-1 protease inhibitors employing energy minimization in the active site, *J. Med. Chem.* 38:305–317, 1995.

[66]   R. Zhou, R.A. Friesner, A. Ghosh, R.C. Rizzo, W.L. Jorgensen, and R.M. Levy, New linear interaction method for binding affinity calculations using a continuum solvent model, *J. Phys. Chem.* B 105:10388–10397, 2001.

[67]   C.M. Oshiro, I.D. Kuntz, and J.S. Dixon, Flexible ligand docking using a genetic algorithm, *J. Computer-Aided Mol. Des.* 9:113–130, 1995.

[68]   G. Jones, P. Willett, and R.C. Glen, A genetic algorithm for flexible molecular overlay and pharmacophore elucidation, *J. Computer-Aided Mol. Des.* 9:532–549, 1995.

[69]   H. Kubinyi, Variable selection in QSAR studies: I. An evolutionary algorithm, *Quant. Struct. Act. Relat.* 13:393–401, 1994.

[70] P.A. Kollman, I. Massova, C. Reyes, B. Kuhn, S. Huo, L. Chong, M. Lee, T. Lee, Y. Duan, W. Wang, O. Donini, P. Cieplak, J. Srinivasan, D.A. Case, and T.E. Cheatham, III, Calculating structures and free energies of complex molecules: combining molecular mechanics and continuum models, *Acc. Chem. Res.* 33:889–897, 2000.

[71] T. Schulz-Gasch and M. Stahl, Binding site characteristics in structure-based virtual screening: evaluation of current docking tools, *J. Mol. Model.* (Online) 9:47–57, 2003.

[72] M. Totrov and R. Abagyan, Derivation of sensitive discrimination potential for virtual ligand screening, *RECOMB '99: Third Annual International Conference on Computational Molecular Biology*, Association for Computing Machinery, New York: Lyon, France, 1999.

[73] A.M. Paiva, D.E. Vanderwall, J.S. Blanchard, J.W. Kozarich, J.M. Williamson, and T.M. Kelly, Inhibitors of dihydrodipicolinate reductase, a key enzyme of the diaminopimelate pathway of *Mycobacterium tuberculosis*, *Biochim. Biophys. Acta.* 1545:67–77, 2001.

[74] T.N. Doman, S.L. McGovern, B.J. Witherbee, T.P. Kasten, R. Kurumbail, W.C. Stallings, D.T. Connolly, and B.K. Shoichet, Molecular docking and high-throughput screening for novel inhibitors of protein tyrosine phosphatase-1B, *J. Med. Chem.* 45:2213–2221, 2002.

[75] E. Perola, W.P. Walters, and P.S. Charifson, A detailed comparison of current docking and scoring methods on systems of pharmaceutical relevance, *Proteins* 56:235–249, 2004.

Part II

Compound and Hit Suitability for Virtual Screening

4 Compound Selection for Virtual Screening

Tudor I. Oprea, Cristian Bologa, and Marius Olah

4.1 THE ROLE OF LEADS IN DRUG DISCOVERY

Preclinical pharmaceutical research has shifted its focus from high-quality candidate drugs to high-quality leads [1] as a result of the increasing pressure to streamline drug discovery. Part of this pressure is artificially created not by the need to discover and launch new drugs, but by the marketing-driven need to develop single-dose, orally available compounds at low dosages (for example, not exceeding 500 mg/day). Safety (such as a good therapeutic window with respect to the toxic dose[1]), the need to minimize the number of side effects and drug–drug interactions, and the pressure to become best in class are some of the additional constraints imposed on preclinical drug discovery. Another significant constraint relates to competitor activities, as reflected in patents and publications, which are often unquantifiable in the early (for example, lead identification) stage because of the secretive nature of the process. This explains the pressure to explore, as early as possible, large areas of chemical space, as quantified by chemical diversity, by means of virtual screening (VS).

VS has emerged as an adaptive response [2] to the massive throughput drug discovery machine (such as high throughput screening, [HTS] and combinatorial chemistry), which has pressured the computational chemistry community to rapidly evaluate billions of virtual (that is, existing *in silico* only) molecules and to assist synthesis prioritization and HTS analyses. In fairness, VS is just one of the cornerstone technologies of the novel drug discovery paradigm, briefly outlined below:

- Large numbers of preplated compounds are tested, typically under single-dose[2]/single-experiment conditions, in what constitutes a HTS experiment. Compounds that show activity in this process are designated as *HTS hits*. The HTS hits are subsequently retested to confirm activity and structure; the reason for structural confirmation relates to the combinatorial or the commercial history of any particular molecule (or both). The usual success rate for single-dose HTS is under 0.1%.
- Less than 1 in 1000 molecules are likely to be active in a primary HTS run, considering a diverse library and a random target. False positives [3] have been shown to interfere with the biological assay, that is, chemically reactive, unstable, or suspected cytotoxic species (see Figure 4.1), or perhaps dyes and fluorescent compounds are often responsible for interfering with the biochemical assays [4], which in turn reduces the success

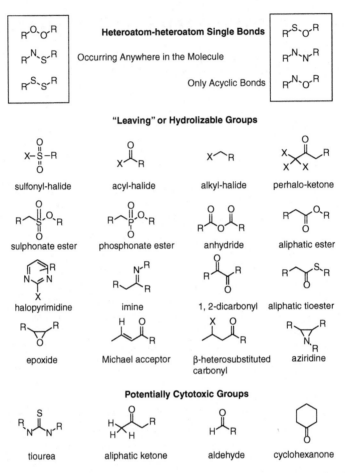

FIGURE 4.1 Examples of chemical substructures that can cause interference with biochemical assays under HTS conditions. (Modified from G.M. Rishton, *Drug Discov. Today* 2:382–384, 1997. With permission.)

rate. *Promiscuous binders,* compounds that sometimes aggregate forming particles with 20 to 400 nm in diameter thus acting as enzyme inhibitors [5], can perhaps be filtered electronically with other frequent hitters [6]. Secondary screening provides the experimental means to weed out HTS hits and provides *HTS actives,* which are compounds with known structure and activity (at this stage, dose-response curves are standard).

- If multiple HTS actives are part of the same chemical family, they become a *lead series* or structures that are amenable for further chemistry optimization [7,8] that may be positioned favorably in terms of intellectual property. Lead series are queried further and derivatized to derive analogs with the appropriate blend of pharmacokinetic and pharmacodynamic (PK/PD) properties.

• After all criteria for lead optimization are met, the molecule becomes a *drug candidate*. Typically, 1 in 10,000 HTS actives reach this stage [9]. Following approval from regulatory authorities, drug candidates enter the development phase and are progressed into clinical trials. Only 1 in 10 candidate drugs [10] is successful in clinical trials. If that happens, and if the marketing situation looks favorable, then the compound is approved and becomes a *launched* (or marketed) *drug*. The progression from HTS active to candidate drug is illustrated in Figure 4.2.

In the current climate of rather low success rates and diminished efforts in examining individual molecules at the early stages, the preclinical discovery paradigm is shifting from optimizing pharmacodynamic properties (e.g., binding potency) first, and worrying about pharmacokinetic properties (e.g., passive intestinal absorption) later, to the simultaneous optimization of PK/PD properties [2], as depicted in Figure 4.3.

By examining Figure 4.2, one notices that (from a numerical standpoint) the critical step is selecting which HTS actives are progressed into lead series. Before this particular step, there is no high-impact decision, because one has only to select which HTS hits are to be retested in secondary screens. At that stage one is constrained only by the number of available slots for retesting, hence small structure–activity relationship (SAR) analyses are conducted to select chemically related molecules and by the number of molecules that are correctly synthesized and recorded in the chemical registry. Thus, the key step occurs after sifting through the HTS hits and confirming the HTS actives according to various company- or user-defined criteria. The definition of a lead, and the leadlike concept, are discussed in the next section, which provides an overview of our current knowledge of what might constitute a high-quality lead in the context of producing orally available

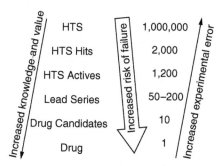

FIGURE 4.2 The compound attrition rate in the current lead and drug discovery paradigm. The numbers estimate of the current success rate as the molecules are progressed from HTS hit to launched drug. As the knowledge and value of the molecule is increasing — often inversely related to experimental error — so does the risk of failure (from a financial standpoint) increase. (Modified from T.I. Oprea, *Chemoinformatics — from Data to Knowledge*. In J. Gasteiger and T. Engels, Eds., Vol. 2, New York: VCH Wiley, pp. 1508–1531, 2003. With permission.)

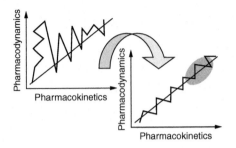

FIGURE 4.3 The progression from HTS hit to candidate drug seen as a number of intermediate steps in preclinical drug discovery. Historically, PD properties were optimized first without concern for PK aspects (left). This often resulted in drawbacks when early compounds were screened *in vivo*, which in turn led to a number of chemical alterations. Thus, properties were optimized sequentially. In the ideal scenario (right), PK properties are optimized simultaneously with PD, because they are considered to be of equal importance. The shaded area in the upper right illustrates the region of (sub)optimality that is likely to be discovered with less effort and time and quite often yields acceptable solutions. (Modified from T.I. Oprea, *Molecules* 7:51–62, 2002. With permission.)

drugs. All these criteria, as well as the simultaneous PK/PD optimization technology can and should be applied not only after physical screening or HTS, but also after VS, because one is confronted with similar issues [2]. One should combine results from both physical screening and VS, as well as other chemoinformatics tools to make the appropriate decisions [12].

4.2 THE LEADLIKE CONCEPT

Chemical aspects of the history of drug discovery history are discussed in the three volume series *Chronicles in Drug Discovery* [13–15], in Sneader's *Drug Prototypes and their Exploitation* [8], and in *Integration of Pharmaceutical Discovery and Development: Case Histories* [16]. These accounts, rich in historical chemical information, do not focus on the choice of lead structures (also called drug prototypes) in drug discovery. This can be explained, in part, by the fact that pharmaceutical industry scientists have not properly documented their reasoning behind certain decisions (and chemical structures) at the time they were taken. In fact, most of the rationale behind drug discovery is done *a posteriori,* making the study of chemical leads one of conjecture and sometimes subjectivity. For example, one of the early accounts of lead discovery highlights chemical leads and their selection as an argument to explain why structured management does not work in the context of preclinical drug discovery [7]. The opinion expressed by DeStevens is, for the most part, related to the role of serendipity in lead discovery, examined by Kubinyi in the context of high throughput technologies [17].

The importance of restricting small molecule synthesis in the property space defined by hydrophobicity (the calculated logarithm of the octanol/water partition coefficient, or CLogP), size (molecular weight, MW), the number of hydrogen bond donors and acceptors (HDO and HAC) was emphasized by Lipinski et al. [18]. The post-HTS analysis of the early (1994) results of HTS and combinatorial chemistry at Pfizer showed that failure to progress to candidate drug stage could be, in part, explained by size (high MW) and hydrophobicity (high CLogP) for the HTS hits. This made their optimization a rather difficult process. Lipinski and coworkers looked at the 90th percentile distribution of MW ($\leq$ 500), LogP ($\leq$ 5), HDO ($\leq$ 5) and HAC ($\leq$ 10), based on 2245 orally available compounds from the World Drug Index (WDI) that had reached phase II clinical trials or higher [18]. Natural products and actively transported molecules were excluded from this analysis. They concluded that if any two of the above conditions are violated, the molecule is unlikely to be an orally active drug. This work became known as the rule-of-five (RO5) and has significantly influenced the current thinking regarding leads, their physicochemical properties, and lead discovery in general. Although there are some notable exceptions in the area of chemical genetics [19,20], such as at Infinity Pharmaceuticals [21] and at the Harvard Institute for Chemistry and Cell Biology [22], 90% of the over 9 million (by 2004) unique chemicals are RO5 compliant, according to the Chem-Navigator Web site [23]. In other words, more than 5 million structures that are now available for purchase observe the property distribution of orally available drugs.

The existence of a druglike space [24,25] was documented based on the ability to discriminate, using chemical fingerprints, between drugs and nondrugs. This working definition considers that a *drug* is any molecule present in a pharmaceutically oriented chemical database — such as Derwent's WDI [26] or MDL's Drug Data Report (MDDR) [27] — regardless of its development phase (preclinical or further). In the same manner, *nondrugs* are defined by their inclusion in a chemical database of buyable chemical reagents, such as the Available Chemicals Directory (ACD) [27]. Although not clear-cut, because there is a small fraction (approximately 1.6%) of drugs included in the ACD dataset [28], this working definition enabled the cheminformatics community to establish the basis for soft filters to further process virtual chemical libraries and even to process commercially available libraries. The drug/nondrug discrimination causes some confusion, as it is often believed that by observing RO5 criteria, chemicals are also druglike. This is not the case, because more compounds in ACD are RO5 compliant, compared to MDDR compounds [28]. In the same report [28], we warned that the reason druglike filters are valid is, in part, due to molecular complexity, because ACD chemicals are on average less complex when compared to MDDR structures: 62.68% of ACD compounds have between 0 and 2 rings and up to 17 rigid (nonflexible) bonds, but 61.22% of MDDR compounds have 3 rings or more and 18 rigid bonds or more. Thus, a compound has a 2:1 probability of being classified as druglike just by having 3 or more rings and 18 or more rigid bonds.

As the aim of VS and combinatorial library design is not to seek drugs, but *leads*, the use of RO5 is not appropriate. Between 1997 and 1999, the vast majority of molecules screened in pharmaceutical discovery were RO5 compliant, regardless of the fact that RO5 guidelines had been derived from analyzing drugs, not leads. Thus, the outcome of many HTS campaigns were only *micromolar hits* [29], which did not prove to be easily amenable to lead optimization, even though they were RO5 compliant, because this optimization would take their property profile outside the RO5 range. Starting from a set of 18 lead–drug pairs, we suggested [29] in 1999 that leadlike libraries should be designed with lower MW (≤ 350) and lower CLogP (≤ 3.0) profiles, as opposed to druglike libraries. In their property profile of 470 lead–drug pairs extracted from Sneader's book [8], Hann and colleagues found [30] that lead's property profiles are left-shifted when compared to the resulting drugs. They concluded that, on the average, leads have lower MW, lower CLogP, fewer aromatic rings, fewer HACs, and lower Andrews binding energy [31], compared to their drug counterparts.

By including lead–drug pairs documented in *Chronicles in Drug Discovery* [13–15] and extending our set from 18 to 96 lead–drug pairs [32], we also found that lead structures exhibit, on the average, lower complexity (lower MW, lower number of rings, RNG, and lower rotatable bonds, RTB), less hydrophobicity (lower CLogP) and are less druglike. In 2001, we revised Lipinski's RO5 in the context of lead discovery and combinatorial library design such that leadlike libraries should not exceed the following property values: 450 for MW, CLogP between −3.5 and +4.5, no more than 4 rings, no more than 10 nonterminal flexible bonds, no more than 5 HDOs, and no more than 8 HACs. However, in 2002, when Proudfoot [33] examined the 25 lead–drug pairs launched in 2000, he found much less significant property differences between leads and drugs. This is, in part, explained by the fact that 4 drugs launched in 2000 were enantiopure drugs of previously marketed racemic mixtures, and another 10 represented extremely minor changes compared to the initial lead.

To further study the differences between leads and drugs, we combined [11] the 3 lead–drug data sets [30,32,33]. We identified 176 unique leads (these are leads only and are not marketed drugs), 186 unique *leadrugs* (the term indicates a dual nature, as leads *and* marketed drugs), and 532 unique drugs that are launched and have not been recorded as leads before. For the sake of clarity, we exclude leadrugs from the distributions depicted in Figure 4.4 and from Table 4.1 and Table 4.2. However, none of three categories is fixed in time. For example, omeprazole and bupivacaine, drugs in two of the analyses [30,32] became leads in Proudfoot's analysis (their enantiopure equivalents are now marketed as drugs) and leadrugs in this analysis [11]. Sometimes drugs are discontinued, for example, Organon's Regonol™ (pyridostigmine) [34], but there are also drugs that are voluntarily withdrawn from the market: Bayer's Baycol® (cerivastatin) [35], Interneuron Pharmaceuticals' Redux™ (dexfenfluramine) [36], and Wyeth's Pondimin™ (fenfluramine) [36].

From the property distribution, it can be observed that leads are smaller than drugs (MW), are likely to be more polar (%PSA, the fraction of polar surface area, is higher in leads) and soluble (LogSw, the logarithm of aqueous solubility), less hydrophobic (left-shifted CLogP and %NPSA, the fraction of nonpolar surface area), and less complex (left-shifted RNG and RTB). The sum of donors and acceptors (SDoAc) does not appear to differ significantly between leads and drugs. These differences are rendered more clearly by comparing the median of the above eight properties (Table 4.1).

For the purpose of property filtering during VS, the differences between drugs and leads should be exploited in terms of compound processing prior to docking or at least in terms of prioritizing postdocking analyses. Combined with the property

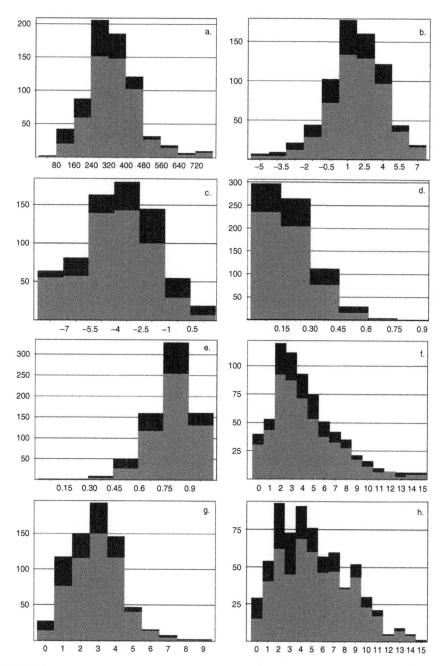

FIGURE 4.4 Property profiling for leads (black) and drugs (grey): (a) MW; (b) CLogP; (c) LogSw; (d) %PSA; (e) %NPSA; (f) SDoAc; (g) RNG; (h) RTB. The vertical axis shows absolute numbers. See text for details.

TABLE 4.1
The Median Properties for Leads and Drugs, as
Depicted in Figure 4.4

Property	Leads	Drugs
MW	295.3	335.7
ClogP	1.79	2.62
LogSw	-2.71	-3.83
%PSA	21.3%	16.6%
%NPSA	78.7%	83.4%
SdoAc	4	4
RNG	2	3
RTB	4	5

distribution analyses [28] from MDDR and the *Physicians' Desk Reference* (PDR), these differences allow us to formulate several property guidelines for leadlikeness:

- MW $\leq$ 460
- $4 \leq$ ClogP ≤ 4.2
- LogSw ≥ -5
- RTB ≤ 10
- RNG ≤ 4
- HDO ≤ 5
- HAC ≤ 9

The minor change in HACs, compared to Lipinski's RO5, is due to a median difference between leads and drugs [11] and does not contradict the "no change" conclusion based on SDoAc (see Figure 4.4 and Table 4.1). Size and hydrophobicity are not necessarily responsible for binding affinity, an observation initially documented by Kuntz et al. [37] and confirmed in our earlier work [1]. Other properties in designing good leads should be observed experimentally. Some of these (this list assumes orally available drugs for long-term dosage) are related to rats or some other mammals:

- Bioavailability (%F) above 30%
- Low clearance (under 30 mL/min*Kg)
- LogD$_{7.4}$ (LogP at pH 7.4) between 0 and 3 (preferably not higher)
- No binding or poor binding (> 100 µM) to drug-metabolizing cytochrome P450 isozymes
- Plasma protein binding below 99.5%

- Lack of acute and chronic toxicity at the expected therapeutic window (e.g., given a 500 mg/day regimen for 7 days)
- No genotoxicity and carcinogenicity (even at doses higher than the therapeutic window)

One word of caution: Unlike the Planck constant, values attributed to leads and used to define and refine the leadlike concept are soft. They are mainly developed with the aim of assisting discovery scientists in filtering through large sets of HTS hits or virtual hits. There are always counterexamples, e.g., tetrahydrofolate (MW = 574.5), which served as lead for methotrexate (MW = 454.4), or tubocurarine (MW = 610.7), the lead for gallamine (MW = 510.8). However, as Rishton points out [4], "most drugs found in the compiled databases were classically discovered and developed using biological assays, selective cytotoxicity assays, and animal models of disease, not using biochemical assays." In other words, these leads were optimized at a time when chemists could modify 1 to 10 molecules, have them screened, and interpret the results before another design/make/test cycle would start. Today, when high throughput experiments substitute sequential (but innovative) thinking as millions of molecules are screened simultaneously, we have to face the fact that the design/make/test cycle occurs only in lead optimization, not in lead identification. This increases the difficulty of the choices we make in the leadlike space, because the critical decision in preclinical discovery remains the choice of the lead compounds.

4.3 COMPOUND PROCESSING PRIOR TO VIRTUAL SCREENING

VS can be a time-consuming process. Assuming a library of 1 million compounds, 100 conformers per molecule, and the ability to dock 100 conformers per central processing unit/second, which is reasonably fast, it still takes two weeks to virtually screen one target. Faster processors, more processors, higher connectivity speed, and more efficient docking algorithms are part of the solution to improve screening efficacy. Our ability to preselect, or to filter, the molecules before the virtual screen is a component of equal importance. One should not needlessly waste resources on molecules that are, ultimately, of no interest to drug discovery.

What follows is a step-by-step procedure on how an existing or acquired collection of compounds should be processed. Its emphasis is on software from Daylight [38], OpenEye [39], and MESA [40], although similar software is available from other vendors, such as Optive Research [41], Accelrys [42], Tripos [43], and Chemical Computing Group [44].

4.3.1 Assembling the Virtual Compound Collection

Large pharmaceutical companies have acquired, in time, databases that contain a significant number of molecules, including marketed drugs and other high-activity compounds. Such compound collections are, by themselves, a valuable resource that is routinely screened against novel targets. It is sometimes argued that these collections reflect the chemistry used to address targets from the past, and that novel targets require novel chemistry, because yesterday's chemistry is by now overpatented. None of these arguments should, however, prevent the inclusion of such molecules into the Virtual Compound Collection (VCC).

By the same token, such databases should be enriched with existing sets of commercially available chemicals, or perhaps with truly virtual molecules (e.g., planned or possible combinatorial chemistry products), thus creating a truly diverse VCC. The largest database of commercially available chemicals is available at the ChemNavigator Web site [23]. As of August 2003, their database contains over 11 million samples, representing over 6 million unique chemicals. This database is available on a subscription basis. If the fees for this database are an issue, one can download databases from the Internet (see Table 4.2). ChemNavigator offers compounds from over 100 companies; therefore, the list in Table 4.2 is far from being exhaustive. The virtual space alternative is best represented by the ChemSpace™ technology: a patented database/software approach from Tripos [43] that routinely explores 10^{12} molecules in virtual space [45]. ChemSpace is available on a collaborative basis.

4.3.2 CLEANING UP THE VCC

This may come as a surprise to the nonspecialist, but there is no perfect chemical database unless it contains simple molecules (e.g., NaCl, H_2O) or a rather small number of molecules. The user needs to spend a significant effort in cleaning up the VCC. Some sites, such as ChemNavigator, provide their own solution to this problem. We prefer FILTER, a program available from OpenEye. Regardless of the method used, the user needs to make some early decisions about his VCC makeup. One obvious suggestion is to remove unwanted chemical structures, such as those depicted in Figure 4.1. Other substructures will be discussed below.

1. Remove garbage from the VCC.
   a. Split covalent salt, remove small fragments (salts), normalize charges — this is clearly an instance where the user is confronted with multiple choices. For example, it is advisable to remove unwanted structures such as those depicted in Figure 4.1.
2. Verify molecular structure integrity.
   a. Get canonical SMILES using Daylight and in-house scripts.
      i. Convert SDF to Daylight's Thor Data Tree (TDT) using mol2smi.
      ii. Extract SMILES and other SDF fields of interest from TDT and place them all in the same line (in-house program) — Daylight's mol2smi may give errors if the cis/trans bonds switch is enabled for ring bonds that are cis/trans. Results should be inspected if this switch was used.
   b. Get SMILES using an alternative approach.
      i. Get SMILES files for each property (e.g., SMILES and molname, SMILES and CLogP, SMILES and LogSw) using

           `alternate_software` (e.g., BABEL, OEChem, MOE, etc.)
for the SMILES conversion.

    ii. Join these fields in one single line per SMILES (Linux command).

    iii. Extract and remove SMILES that give errors at conversion (in-house program, log errors in `VCC_bad`).

3. Obtain canonical SMILES from the two files.

    a. Get canonical SMILES for the set generated using Daylight's `mol2smi` (`cansmi -i` [isomeric SMILES]).

    b. Get canonical SMILES for the set generated using the alternative approach (`cansmi`, log errors in `VCC_bad2`).

4. Compare both sets and analyze the differences — the two sets may not be identical. Error logs should be compared if, and when, the magnitude of the difference warrants such analyses (if under 0.05% and if under time constraints, this issue is best ignored). Compared to other programs, `mol2smi` appears to convert most structures, even some that appear to be problematic.

    a. Remove missing SMILES from Daylight set (Java™ implemented in-house program).

    b. Compare SMILES and remove differences, for analysis (Java) — problematic structures should be inspected individually, if time allows. The number of errors differs significantly among chemical vendors, ranging from under 0.05% to 10% or higher. A totally automated method for error detection and removal of faulty structures should be implemented.

5. Generate unique SMILES — after canonical SMILES are derived, only unique SMILES (not compound IDs or molnames) should be kept. In particular, if the VCC originates from different software vendors, this may remove up to 50% of the initial set. At this step, it is important to list preferred or trusted vendors first. If price is an issue, a script to keep the low-price structures can be used.

6. Process the unique SMILES to explore alternative structures — the user needs to consider various structural modifications [46] of the starting SMILES that are possible in solution or at the ligand–receptor interface:

    a. Tautomerism, which shifts one hydrogen along a path of alternating single/double bonds, mostly involving nitrogen and oxygen (e.g., imidazole).

    b. Protonation states, which should explore different states of the molecule by assigning formal charges to those chemical moieties that are likely to be charged (e.g., phosphate or guanidine) and by assigning charges to some of those moieties that are likely to be charged under different microenvironmental conditions (chargeable moieties such as tetrazole and amine).

    c. Pyramidal ("flappy") nitrogen inversions, which explore noncharged, nonaromatic, pseudochiral nitrogens (three substituents) that can easily interconvert in three dimensions.

TABLE 4.2
Examples of Company Databases Available for Purchase

Company Name	WEB Address	Number of Compounds	Description
4SC	http://www.4sc.de/	3,300,000	virtual library; small molecule drug candidates
ACB Blocks	http://www.acbblocks.com/acb/bblocks.html	90,000	building blocks for combinatorial chemistry
Advanced SynTech	http://www.advsyntech.com/omnicore.htm	170,000	targeted libraries: protease, protein kinase, GPCR, steroid mimetics, antimicrobials
Ambinter	http://ourworld.compuserve.com/homepages/ambinter/Mole.htm	1,700,000	combinatorial chemistry and parallel as building blocks, high throughput screening
Asinex	http://www.asinex.com/prod/index.html	120,000	Platinum collection
		230,000	GOLD collection
		5,009	targeted libraries: GPCR (16 different targets)
		4,307	kinase targeted library (11 targets)
		1,629	ion-channel targeted (4 targets)
		1,200,000	combinatorial constructor
BioFocus	http://www.biofocus.com/pages/drug__discovery.mhtml	100,000	diverse primary screening compounds
Cerep	http://www.cerep.fr/Cerep/Users/pages/ProductsServices/Odyssey.asp	250 chemical families	Odyssey II library: diverse and unique discovery library
Chemical Diversity Labs, Inc.	http://www.chemdiv.com/discovery/downloads/	550,000	leadlike compounds for bioscreening
ChemStar	http://www.chemstar.ru/page4.htm#Bookmark	73,000	high quality organic compounds for screening
COMBI-BLOCKS	http://www.combi-blocks.com/	660	combinatorial building blocks
ComGenex	http://www.comgenex.hu/cgi-bin/inside.php?im=products&l_id=compound	260,000	pharma relevant, discrete structures for multitarget screening purposes
		240	GPCR library
		2,000	cytotoxic discovery library: toxic compounds suitable for anticancer and antiviral discovery research

Company	URL	Number	Description
EMC Microcollections	http://www.microcollections.de/catalogue_compunds.htm	30,000	highly diverse combinatorial compound collections for lead discovery
InterBioScreen	http://www.ibscreen.com/products.shtml	288,000	synthetic compounds
		30,000	natural compounds
Maybridge plc	http://www.maybridge.com/html/m_company.htm	53,000	organic druglike compounds
MicroSource Discovery Systems, Inc.	http://www.msdiscovery.com/download.html	2,000	GenPlus: collection of known bioactive compounds; NatProd: collection of pure natural products
Nanosyn, Inc.	http://www.nanosyn.com/thankyou.shtml	46,715	Pharma library
		18,613	Explore library
Neokimia	http://www.biotechcarecenter.com	35,000	macrocycles library
Pharmacopeia	http://www.pharmacopeia.com/dcs/order_form.html	N/A	target library: GPCR and kinase
Polyphor	http://www.polyphor.com/	15,000	diverse general screening library
Sigma-Aldrich	http://www.sigmaaldrich.com/Area_of_Interest/Chemistry/Drug_Discovery/High_Throughput_Screening/Compound_Libraries.html	130,000	diverse library of small-molecule compounds, including plant extracts and microbial cultures
Specs	http://www.specs.net/	230,000	diverse library
		230,000	focussed library
		10,000	World Diversity Set: preplateled library
		4,000	building blocks
TimTec	http://www.timtec.net/	800	natural products (diverse and unique)
Tripos	http://www.tripos.com/sciTech/researchCollab/chemCompLib/lqCompound/index.html	130,000	compound libraries and building blocks
		80,000	LeadQuest compound libraries

These structural modifications, if not explored, lock the virtual screen to a single state of the parent compound. Such changes, however, are likely to occur because the receptor, the solvent environment, or simple Brownian motion may influence the particular three-dimensional (3D) state that the molecule is sampling. Their combinatorial explosion needs to be, within limits, explored at the SMILES level, before the 3D structure generation step.

4.3.3 FILTER FOR LEADLIKENESS

After cleanup, the VCC is further processed to remove compounds that do not have leadlike properties (see Section 4.2 for details). Compounds that pass this filter (usually between 10 and 50%, depending on the library) are prioritized for VS. It is advisable to cluster the remaining VCC (nonleadlike) set and to include a representative set of these compounds (up to 25%), because they are likely to capture additional chemistry. Because high flexibility will decrease the accuracy of the VS, it is also advisable to eliminate molecules that contain:

- More than 8 connected single bonds not in ring
- More than 6 connected unsubstituted single bonds not in ring
- Macrocycles with more than 22 atoms in a ring
- Macrocycles with more than 14 flexible bonds

Other suggestions of leadlike criteria exclusions are:

- RNG > 4
- RNG > 3 for fused aromatic rings (avoid polyaromatic rings)
- HDO > 4 (HDO ≤ 5 is the RO5 criterion, but 80% of drugs have HDO ≤ 3)
- More than 4 halogens (except fluorine, avoid pesticides)
- More than two CF_3 groups (avoid highly halogenated molecules)

This list is likely to reflect the "cultural" bias that is particular to each company. For example, companies active in contraceptive research, such as Organon and Wyeth, will regard steroids favorably at this stage, whereas other companies may want to actively exclude them from the virtual screen. Similar arguments could be made for the lactam nucleus (penicillins, cephalosporins) and peptides. An additional step may include removal of frequent hitters [6] and promiscuous binders [5], the removal of compounds that have low druglike scores, or that contain chemical fragments believed to be cytotoxic (see Figure 4.1).

The VCC can be seen as the overall collection that can be manipulated one time prior to all virtual screens, assuming that targets are similar and that the drug discovery projects have similar goals (e.g., orally available drugs that should not penetrate the blood-brain barrier). However, the VS set (VSS) may be just a subset of the VCC if the collection needs to be filtered using different criteria for different targets and discovery goals. For instance, targets located in the lung require a different pharmacokinetic profile — aerosols — compared to targets located in the urinary tract (which may require good aqueous solubility at pH 5) or on the skin

(LogP between 5 and 7 is ideal for such topical agents). Such biases should be included at the property filtering stage as much as possible.

Please note that in its current implementation, FILTER rewrites the canonical (Daylight) SMILES. One cannot restore the canonical format post-FILTER by redirecting the output via `cansmi`, because canonicalization using OELib (pre-OEChem product from OpenEye) to generate SMILES can be erroneous for some structures. Instead, the users are advised to restore the original, canonical Daylight SMILES, from the input file (pre-FILTER). Future, OEChem-based, versions of FILTER will not require this fix.

4.3.4 SIMILARITY SEARCH IF KNOWN ACTIVE MOLECULES ARE AVAILABLE

If high-activity molecules are known (from the literature, from patents, from in-house data) it is advisable to perform a similarity search on the entire VCC for similar molecules and to include them in the VSS even though they may have been removed during the other steps. These molecules should serve as positive controls, assuming the similarity principle holds. If the 3D structure of the bioactive conformation is available (e.g., active ligand cocrystallized in the target binding site), a 3D similarity search will perform better than a 2D-based one.

4.3.5 3D STRUCTURE GENERATION

Depending on the VS approach, one or multiple conformers per molecule are generated at this stage. Some software (e.g., FRED [47]) can generate the 3D structures from SMILES. Other docking programs require a separate 3D conversion step [48] using, for example, CONCORD™ [41,43] or CORINA [49]. OpenEye's Omicron [39] has integrated force field and solvation models, allowing the user to explore multiple conformational spaces; other conformational "exploders" are also available: Catalyst® [42], Confort [41,43], and OMEGA [39]. To address the missing or improper chirality information, CONCORD is now coupled with StereoPlex™ [41,43], a software that makes educated guesses about the chiral centers that require 3D explosion. The brute-force approach (exploding all possible chiral isomers) can be handled at the SMILES level, using the `chiralify` routine from the Daylight `contrib` directory [38].

4.4 CONCLUSIONS

The methods described above can be summarized as follows:

1. Assemble the VCC starting from in-house and online databases.
2. Cleanup the VCC by removing garbage, verifying structural integrity, and making sure that only unique structures are screened, with limited exploration of their tautomeric and protonation states.

3. Perform property filtering to remove unwanted structures based on sub-structures, or property profiling, or various scoring schemes. The VCC becomes the VSS at this stage and is target and project dependent.
4. Enrich the VSS by adding molecules similar to known actives, even if they were removed from the VCC in previous steps.
5. Generate the 3D structures in preparation for VS.

Each of the above steps requires active choices from the end user. Following literature trends (e.g., the leadlike concept) is useful as long as the aims of the project are compatible with the "orally available drug, single dose daily" paradigm [50]. In this case, a desirable VSS property profile should observe the following criteria: MW $\leq$ 460, -4 $\leq$ ClogP $\leq$ 4.2, LogSw -5, RTB $\leq$ 10, RNG $\leq$ 4, HDO $\leq$ 5, and HAC $\leq$ 9. Additional criteria may include (predicted) %F above 30; (predicted) low clearance; (predicted) $\log D_{7.4}$ between 0 and 3; poor binding (> 100 μM) to cyto-chrome P450s; (predicted) low or no toxicity, genotoxicity, and carcinogenicity. However, this profile may change when other administration paradigms are considered: higher solubility may be required for intravenous drugs, whereas higher hydro-phobicity is required for topical (skin) administration. Therefore, project chemists and biologists need to be consulted before filtering compounds for VS.

ACKNOWLEDGMENT

These studies were supported in part by New Mexico Tobacco Settlement funds.

NOTES

1.   It's the dose that makes the poison — Theophrastus Bombastus Paracelsus.
2.   Typical concentrations for HTS screening are 5, 10, or 50 μM.

REFERENCES

[1]   T.I. Oprea, Lead structure searching: are we looking for the appropriate properties? *J. Computer-Aided Mol. Des.* 16:325–334, 2002.
[2]   T.I. Oprea, Virtual screening in lead discovery: a viewpoint, *Molecules* 7:51–62, 2002.
[3]   G.M. Rishton, Reactive compounds and *in vitro* false positives in HTS, *Drug Discov. Today* 2:382–384, 1997.
[4]   G.M. Rishton, Nonleadlikeness and leadlikeness in biochemical screening, *Drug Discov. Today* 8:86–96, 2003.
[5]   S.L. McGovern, E. Caselli, N. Grigorieff, and B.K. Shoichet, A common mechanism underlying promiscuous inhibitors from virtual and high-throughput screening, *J. Med. Chem.* 45:1712–1722, 2002.

[6]   O. Roche, P. Schneider, J. Zuegge, W. Guba, M. Kansy, A. Alanine, K. Bleicher, F. Danel, E.M. Gutknecht, M. Rogers-Evans, W. Neidhart, H. Stalder, M. Dillon, E. Sjoegren, N. Fotouhi, P. Gillespie, R. Goodnow, W. Harris, P. Jones, M. Taniguchi, S. Tsujii, W. von Saal, G. Zimmermann, and G. Schneider, Development of a virtual screening method for identification of "Frequent Hitters" in compound libraries, *J. Med. Chem.* 45:137–142, 2002.

[7]   G. DeStevens, Serendipity and structured research in drug discovery, *Prog. Drug Res.* 30:189–203, 1986.

[8]   W. Sneader, *Drug Prototypes and Their Exploitation,* Hoboken, NJ: Wiley, 1996.

[9]   D.B. Boyd, Progress in rational design of therapeutically interesting compounds. In T. Liljefors, F.S. Jorgensen, and P. Krogsgaard-Larsen, Eds., *Rational Molecular Design in Drug Research,* Copenhagen: Munksgaard, pp. 15–29, 1998.

[10]   T. Kennedy, Managing the drug discovery/development interface, *Drug Discov. Today* 2:436–444, 1997.

[11]   T.I. Oprea, Chemoinformatics and the quest for leads in drug discovery. In J. Gasteiger and T. Engel, Eds., *Chemoinformatics — from Data to Knowledge*, Vol. 2, Advanced Topics, New York: VCH-Wiley, pp. 1508–1531, 2003.

[12]   T. Olsson and T.I. Oprea, Cheminformatics: a tool for decision-makers in drug discovery, *Curr. Opin. Drug Discov. Dev.* 4:308–313, 2001.

[13]   J.S. Bindra and D. Lednicer, *Chronicles of Drug Discovery,* Vol. 1, New York: Wiley-Interscience, 1982.

[14]   J.S. Bindra and D. Lednicer, *Chronicles of Drug Discovery,* Vol. 2, New York: Wiley-Interscience, 1983.

[15]   D. Lednicer, *Chronicles of Drug Discovery,* Vol. 3, Washington, DC: ACS Publishers, 1993.

[16]   R.T. Borchardt, R.M. Freidinger, T.K. Sawyer, and P.L. Smith, *Integration of Pharmaceutical Discovery and Development: Case Histories,* New York: Plenum Press, 1998.

[17]   H. Kubinyi, Chance favors the prepared mind — from serendipity to rational drug design, *J. Rec. Signal Transduction Res.* 19:15–39, 1999.

[18]   C.A. Lipinski, F. Lombardo, B.W. Dominy, and P.J. Feeney, Experimental and computational approaches to estimate solubility and permeability in drug discovery and development settings, *Adv. Drug Deliv. Reviews* 23:3–25, 1997.

[19]   H.E. Blackwell, L. Perez, R.A. Stavenger, J.A. Tallarico, E. Cope Eatough, M.A. Foley, and S.L. Schreiber, A one-bead, one-stock solution approach to chemical genetics: part 1, *Chem. & Biol.* 8:1167–1182, 2001.

[20]   P.A. Clemons, A.N. Koehler, B.K. Wagner, T.G. Spriggings, D.R. Spring, R.W. King, S.L. Schreiber, and M.A. Foley, A one-bead, one-stock solution approach to chemical genetics: part 2, *Chem. & Biol.* 8:1183–1195, 2001.

[21]   See the Inifinity Pharmaceuticals, Inc. Web site: http://www.ipi.com.

[22]   See the Harvard Institute for Chemistry and Cell Biology Web site: http://iccb.med. harvard.edu.

[23]   See the ChemNavigator Web site: http://www.chemnavigator.com.

[24]   A. Ajay, W.P. Walters, and M.A. Murcko, Can we learn to distinguish between "drug-like" and "nondrug-like" molecules? *J. Med. Chem.* 41:3314–3324, 1998.

[25]   J. Sadowski and H. Kubinyi, A scoring scheme for discriminating between drugs and nondrugs, *J. Med. Chem.* 41:3325–3329, 1998.

[26]   WDI and ACD are available from Daylight Chemical Information Systems, http://www.daylight.com.

[27]   MDDR and ACD are available from MDL, http://www.mdli.com.

[28]   T.I. Oprea, Property distribution of drug-related chemical databases, *J. Computer-Aided Mol. Des.* 14:251–264, 2000.

[29]   S.J. Teague, A.M. Davis, P.D. Leeson, and T.I. Oprea, The design of leadlike combinatorial libraries, *Angew. Chem. Int. Ed.* 38:3743–3748, 1999; German version: *Angew. Chem.* 111:3962–3967, 1999.

[30]   M.M. Hann, A.R. Leach, and G. Harper, Molecular complexity and its impact on the probability of finding leads for drug discovery, *J. Chem. Inf. Comput. Sci.* 41:856–864, 2001.

[31]   P.R. Andrews, D.J. Craik, and J.L. Martin, Functional group contributions to drug-receptor interactions, *J. Med. Chem.* 27:1648–1657, 1984.

[32]   T.I. Oprea, A.M. Davis, S.J. Teague, and P.D. Leeson, Is there a difference between leads and drugs? A historical perspective, *J. Chem. Inf. Comput. Sci.* 41:1308–1315, 2001

[33]   J.R. Proudfoot, Drugs, leads, and drug-likeness: an analysis of some recently launched drugs, *Bioorg. Med. Chem. Lett.* 12:1647–1650, 2002.

[34]   The list of drugs to be discontinued is continuously updated at the US Food and Drug Administration, Web site: http://www.fda.gov/cder/drug/shortages/#disc.

[35]   Baycol was voluntarily withdrawn by Bayer in August 2001 because of reports of sometimes fatal rhabdomyolysis (a severe muscle adverse reaction). See Web site: http://www.fda.gov/cder/drug/infopage/baycol/default.htm.

[36]   Redux and Pondimin were voluntarily withdrawn in September 1997 because 30% of the patients taking these two drugs had abnormal echocardiograms. See Web site: http://www.fda.gov/cder/news/phen/fenphenpr81597.htm.

[37]   I.D. Kuntz, K. Chen, K.A. Sharp, and P.A. Kollman, The maximal affinity of ligands, *Proc. Natl. Acad. Sci. USA* 96:9997–10002, 1999.

[38]   See the Daylight Chemical Information Systems, Inc. Web site: http://www.daylight.com.

[39]   See the OpenEye Scientific Software Web site: http://www.eyesopen.com.

[40]   See the Mesa Analytics & Computing Web site: http://www.mesaac.com.

[41]   See the Optive Research Web site: http://www.optive.com.

[42]   See the Accelrys Web site: http://www.accelrys.com.

[43]   See the Tripos, Inc. Web site: http://www.tripos.com.

[44]   See the Chemical Computing Group Web site: http://www.chemcomp.com.

[45]   K.M. Andrews and R.D. Cramer, Toward general methods of targeted library design: topomer shape similarity searching with diverse structures as queries, *J. Med. Chem.* 43:1723–1740, 2000.

[46]   P.W. Kenny and J. Sadowski, Structure Modification in Chemical Databases. In T.I. Oprea, Ed., *Cheminformatics in Drug Discovery,* New York: Wiley-VCH, 2004, in press.

[47]   FRED is available from OpenEye Scientific Software, http://www.eyesopen.com.

[48]   J. Sadowski and J. Gasteiger, From atoms and bonds to three-dimensional atomic coordinates: automatic model builders, *Chem. Rev.* 93:2567–2581, 1993.

[49]   CORINA is available from Molecular Networks, http://www.molnet.de.

[50]   T.I. Oprea, Chemical space navigation in lead discovery, *Curr. Opin. Chem. Biol.* 6:384–389, 2002.

5 Experimental Identification of Promiscuous, Aggregate-Forming Screening Hits

Susan L. McGovern

5.1 INTRODUCTION

Many screening methods have been developed to identify novel inhibitors for drug design. Although both virtual and experimental high throughput approaches have successfully discovered new lead compounds [1–8], it has become clear that screening hit lists are plagued by problematic molecules. Often, these compounds display perplexing behaviors, such as a flat structure–activity relationship [2], time-dependent activity, and steep inhibition curves [9]. Because of these undesirable properties, many screening hits end up as developmental dead ends, often after considerable time and resources have been devoted to them.

Well-known mechanisms account for the activity of some false positives from screening. For instance, compounds can be chemically reactive [10]. Other screening hits interfere with colorimetric or fluorimetric assays, thereby creating the appearance of inhibition through experimental artifact [11]. These and related observations [12–14] have spawned a variety of computational and experimental methods [9–11,15] to identify and remove difficult compounds from screening libraries.

Despite these efforts, problematic compounds continue to haunt hit lists, without explanation. To investigate additional mechanisms of nonspecific inhibition, over 100 small molecules, including screening hits, leads, and clinically used drugs, have been studied. At least 50 of these diverse compounds inhibit unrelated model enzymes in a time-dependent manner. These promiscuous compounds are sensitive to enzyme concentration, ionic strength, albumin, and detergent. This collection of properties is not compatible with known mechanisms of enzyme inhibition; to account for these observations, it has been proposed that the nonspecific compounds form aggregates in solution, and the aggregate species is responsible for enzyme inhibition (Figure 5.1) [16]. Consistent with this hypothesis, the promiscuous inhibitors have been observed to form micron or submicron aggregates by light scattering and electron microscopy [16,17]; these particles are absent from solutions of classically behaved inhibitors.

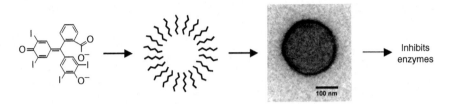

FIGURE 5.1 Model of promiscuous enzyme inhibition. Some compounds form micron or submicron aggregates in solution, and the aggregate species is responsible for enzyme inhibition. (Reproduced from S.L. McGovern et al., *J. Med. Chem.* 45:1712–1722, 2002. With permission.)

It appears that nonspecific inhibitors acting via aggregation may be common among libraries of small molecules [16,17]. Indeed, such compounds may be enriched in high throughput screening hit lists. Unfortunately, their promiscuity makes them undesirable as lead molecules for development. Early identification and abandonment of aggregate-forming compounds would allow more resources to be devoted to specific inhibitors.

Therefore, given a screening hit, how can one determine if it is well-behaved or if it is an aggregate-forming, promiscuous inhibitor? It turns out that the peculiar properties of aggregators can be used to differentiate them from classical inhibitors. This chapter discusses the origin of these properties as described by the aggregate model, recent computational efforts to predict aggregators, and detailed protocols for using the experimental signatures of aggregators to identify these compounds.

5.2 OVERVIEW

To date, over 100 compounds have been explicitly studied for aggregate-based promiscuity. These compounds include screening hits [16,18], leads widely employed as kinase inhibitors [17], and clinically used drugs [19]. At least 53 aggregate-forming, promiscuous compounds (35 hits, 8 leads, and surprisingly, 10 drugs) and 62 non-aggregate-forming, nonpromiscuous compounds (10 hits, 7 leads, and 45 drugs) have been identified.

This growing collection of molecules has allowed the mechanism of aggregate-mediated inhibition to be explored by biophysical and microscopic studies. One possible model is that aggregates sequester substrate from solution, leading to apparent enzyme inhibition. If so, one might expect that the presence of inhibitor alone would decrease the apparent concentration of substrate. However, as judged by spectroscopy experiments, aggregates have no effect on the substrate concentration. Additionally, kinetic assays show that inhibition by aggregates is noncompetitive with respect to substrate. These observations imply that substrate sequestration is not likely responsible for aggregate-mediated inhibition.

Another potential mechanism of nonspecific enzyme inhibition by aggregates is denaturation [20,21]. In kinetic experiments, it has been observed that solvent denaturants do not affect or actually decrease the potency of aggregate-forming compounds [16]. It has also been found that the potency of aggregators does not increase as the thermodynamic stability of the target enzyme decreases [22]. Furthermore, microscopic studies demonstrate that green fluorescent protein retains a fluorescence signal in the presence of aggregates [22]. Together, these results suggest that enzyme denaturation alone does not account for inhibition by aggregate-forming compounds.

Instead, initial results from cosedimentation studies show that aggregates can pull enzyme molecules from solution, consistent with a physical interaction between these species [22]. The nature of this association has been investigated by electron and confocal microscopy, which revealed that enzyme molecules can adsorb onto the surface of aggregates (Figure 5.2), although absorption into the aggregate interior could not be ruled out [22]. Additional experiments demonstrated that cosedimentation, adsorption, and enzyme inhibition were prevented by the addition of the detergent Triton® X-100 [22]. Indeed, addition of 0.01% Triton X-100 was sufficient to reverse, within seconds, enzyme inhibition by aggregate-forming compounds, but not by classically behaved inhibitors [22]. These observations suggest that promiscuous enzyme inhibition by aggregate-forming compounds occurs through the reversible adsorption of enzyme molecules onto the surface of aggregates, although absorption cannot be excluded [22].

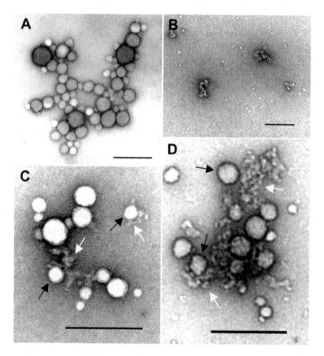

FIGURE 5.2 Images from transmission electron microscopy of aggregate particles and enzyme molecules. A. 100 μM tetraiodophenolphthalein alone. B. 0.1 mg/mL β-galactosidase alone. C. and D. 100 μM tetraiodophenolphthalein with 0.1 mg/mL β-galactosidase. Black arrows mark tetraiodophenolphthalein aggregates; white arrows mark β-galactosidase. Bar = 200 nm. (Reproduced from S.L. McGovern et al., *J. Med. Chem.* 46:4265–4272, 2003. With permission.)

An adsorption mechanism can be reconciled with the peculiar inhibition properties of aggregate-forming compounds. For instance, the model is consistent with the promiscuity of these inhibitors, as a surface interaction might be expected to provide little discrimination on the basis of steric or electrostatic features. The observed time-dependence of inhibition may result from the time required for the interaction of enzyme with aggregates, which is limited by the low, typically nanomolar, concentration of each species. The observation that albumin decreases apparent inhibition by aggregates is consistent with the displacement of enzyme molecules from the aggregate surface by albumin, thereby returning free enzyme to solution. The sensitivity of aggregate-forming inhibitors to enzyme concentration may be analogous to the albumin effect, except that the enzyme molecules are displaced by additional enzyme. Finally, the mechanism is consistent with the ability of detergent to prevent or reverse inhibition by aggregators [21,22]. Low concentrations of Triton X-100 may be sufficient to coat the surface of aggregates to prevent adsorption from occurring; higher concentrations of Triton X-100 are known to dissolve the aggregates themselves and return active enzyme to solution [22].

A reversible adsorption model is consistent with the properties of aggregate-based inhibitors; these properties provide simple experimental methods for distinguishing aggregate-based inhibitors from classical inhibitors.

5.3 CONCEPTUAL ADVANCES

Although the experimental signatures of aggregators are well-known, there is a considerable interest in the development of computational methods for predicting compounds likely to act as aggregate-based inhibitors. Compared to experimental methods, virtual approaches would allow for rapid and inexpensive filtering of large compound collections or hit lists to identify potential aggregate-forming compounds. The most troubling limitation of predictive methods is the possibility of incorrectly labeling a *bona fide* inhibitor as an aggregate-based inhibitor and eliminating it from further consideration for development. As more compounds are explicitly identified as either "aggregator" or "nonaggregator," it is hoped that the molecular properties responsible for aggregation will be more clearly defined and the likelihood of throwing out good compounds will decrease.

One of the first computational algorithms to be developed for predictive purposes is a recursive partitioning model that divides compounds into aggregator or nonaggregator nodes according to properties such as ClogP (the calculated logarithm of the octanol/water partition coefficient), conjugation, and the presence of ionizable groups [19]. The model correctly classified 94% of a test set of 111 molecules studied for aggregation, suggesting that physicochemical characteristics may be useful for predicting aggregation behavior.

Expanding the scope of nonspecificity to include compounds that are promiscuous for any reason, Roche and colleagues trained a neural net to identify frequent hitters [11]. Although this work did not specifically consider aggregators, the 479 frequent hitters in the test set included problematic compounds, such as nonspecific and reactive compounds, and also desirable compounds, such as those containing privileged substructures more likely to recognize receptor binding sites [11]. Despite this variety, the method successfully classified 90% of the test compounds; much future work devoted to the prediction of frequent hitters and aggregators in particular can be anticipated.

Beyond the identification of these compounds, many additional questions regarding aggregators remain outstanding. For instance, what is the effect of aggregate-based inhibitors on membrane bound receptors? Why do these molecules appear in virtual screening hit lists? How does enzyme adsorption lead to inhibition? What is the arrangement of monomers in an aggregate? It is tempting to speculate about possible answers to these questions, but more definitive discussions await future experiments.

5.4 PROTOCOLS

Given a screening hit, how can one determine if it is well-behaved or if it is an aggregate-forming, promiscuous inhibitor? The classification of a small molecule as an aggregate-former is based on the satisfaction of experimental criteria (Table 5.1), many of which involve adaptations of a typical inhibition assay against a single enzyme. The most sensitive of these experiments reflect the peculiar differences between aggregate inhibitors and classically behaved inhibitors. Fortunately, the easiest tests to perform are among the most discriminating. Additionally, these

TABLE 5.1
Experimental Signatures of Promiscuous, Aggregate-Forming Inhibitors

Experimental Property	Well-Behaved Inhibitor	Aggregate-Based Inhibitor
Inhibition of two dissimilar enzymes	Rarely	Yes
Inhibition of three dissimilar enzymes	No	Yes
Decreased inhibition in the presence of 0.01% Triton X-100[a]	No	Yes
Reversal of inhibition upon addition of 0.01% Triton X-100[a]	No	Yes
Decreased inhibition against a 10-fold increase in enzyme concentration[a,b]	No	Yes
Steep dose-response curve[a]	Rarely	Yes
Increased inhibition with incubation[a]	Maybe	Yes
Decreasing inhibition with increasing ionic strength[a]	Maybe	Yes
Decreasing inhibition in the presence of albumin[a]	Maybe	Yes
Formation of particles detectable by light scattering	No	Yes

[a.] In practice, these experiments only need to be performed against one enzyme; repeating with additional enzymes should yield qualitatively similar results.

[b.] Requires that the total inhibitor concentration greatly exceeds total enzyme concentration after the increase in enzyme concentration.

experiments may elicit signs of other pathologies present in the compound, such as chemical reactivity or interference with the assay readout.

Considering experiments most readily adaptable for regular use by academic or industrial screening groups, a rigorous set of criteria for classification of a compound as a promiscuous, aggregate-forming inhibitor includes the following:

- Decreased inhibition in the presence of Triton X-100
- Inhibition of three diverse enzymes
- Decreased inhibition with a 10-fold increase in enzyme concentration
- Increased inhibition with incubation (time-dependent behavior)
- Formation of particles detectable by dynamic light scattering

Such thorough experiments might be reasonable tests for a few (about 10 to 100) compounds of particular interest; if several hundred or thousand compounds are under evaluation, one might consider a reduced set of criteria. The single most useful test is the sensitivity of inhibition to the presence of detergent [21,22]; we have found that to be the easiest method to rapidly differentiate aggregate-forming inhibitors from well-behaved compounds.

Detailed protocols for each of these experiments are described below.

5.4.1 KINETIC ASSAYS

Typically, the most accessible experiments for determining if a compound acts as an aggregate species involve permutations of the original assay used to discover the inhibitor. However, these experiments may also be performed in other assay systems, especially if the original screening assay is particularly laborious, expensive, or difficult to modify. We have found that model systems with colorimetric readouts and commercially available reagents, such as β-lactamase [16,17,19], chymotrypsin [16,17,19], or malate dehydrogenase (MDH) [17,19], are robust systems for the assays described below.

Some general considerations regarding the assay components:

- Buffer — The ionic strength [16] and pH of the buffer can affect the behavior of the aggregate-forming compounds [23]. Buffers such as 50 mM potassium phosphate, pH 7.0 [16] have proven robust in previous work.
- Enzyme concentration — Enzyme concentration can also affect the apparent inhibition by aggregate-forming compounds. Ideally, the final assay concentration of enzyme might be 1 to 10 nM. Low enzyme concentrations allow for greater dynamic range of the assay and increased inhibition by aggregate-forming compounds (see Sensitivity to Enzyme Concentration, below). For the purposes of the experiments described below, avoid assays that include albumin, as this, or any other excess protein, will mask the behavior of aggregate-forming inhibitors.
- Substrate concentration — Ideally, the substrate concentration should at least be greater than the Km, preferably at a concentration that allows for

maximal or near-maximal reaction velocities. This increases the signal and dynamic range of assay, allowing for greater flexibility in later modifications.

- Inhibitor concentration — Potencies of aggregate-forming inhibitors will vary depending on the assay conditions and the particular enzyme, but most compounds show detectable enzyme inhibition at 100 μM or less. Ideally, most of the experiments described below are performed with the inhibitor present at the IC_{50} concentration. The inhibitor is typically diluted from a 10 mM dimethyl sulfoxide (DMSO) stock into aqueous buffer. Previous work suggests that DMSO can affect the behavior of aggregate-forming compounds; to control for possible effects, limiting the final total concentration of DMSO to no more than about 5% has been useful. Because aggregate-forming inhibitors are sensitive to the molar ratio of inhibitor to enzyme [16], the final concentration of inhibitor should be in marked excess of enzyme concentration, roughly 1000 or more individual inhibitor molecules to 1 molecule of enzyme.

A template assay and example assays for initial inhibitor screening against β-lactamase, chymotrypsin, and MDH are given in Protocol 5.1. These are the basic recipes that are modified in the following sections to yield assays that are diagnostic for aggregate-forming, promiscuous inhibitors. Each protocol that follows will be described in a generic template and also in a detailed format for the β-lactamase model system, although the experiments can also be performed with chymotrypsin, MDH, or another robust enzyme system. Repeating these experiments with additional dissimilar enzymes should produce qualitatively similar results.

5.4.1.1 Sensitivity to Detergent

Detergents decrease inhibition by aggregate-forming inhibitors but not by well-behaved inhibitors [19,22]. This observation suggests that the addition of detergents to screening assays is a simple method for experimentally discriminating between aggregate inhibitors and well-behaved inhibitors. Although weak nonionic detergents such as digitonin and saponin [19,21] have been used in previous work in this context, this discussion will be limited to the use of Triton X-100, because it can reverse inhibition by aggregate-forming compounds [22].

When Triton X-100 is added to an inhibition assay, it is simplest to add it to the reaction cuvette before either enzyme or inhibitor (Protocol 5.2). Alternatively, 0.01% (v/v) Triton X-100 can be included in the buffer to decrease the number of components added to the reaction. Indeed, it may be interesting to consider including Triton X-100 in large-scale screening efforts as a prophylactic measure to prevent aggregate-forming inhibitors from showing activity in screening projects. Some caution is warranted, as stocks of Triton X-100 can lose potency over time; ideally, fresh stocks should be made daily. If inconsistent results are obtained, it is worthwhile to repeat the experiment with a fresh stock of Triton X-100.

A few control experiments should be performed at the outset. For instance, does Triton X-100 affect the activity of the target receptor, substrate, or any of the other

PROTOCOL 5.1
Initial Inhibition Screening Assay

Template Assay

1. Add buffer.
2. Add inhibitor or DMSO control.
3. Add enzyme.
4. Mix.
5. Incubate for n minutes.
6. Add substrate.
7. Mix.
8. Monitor reaction progress.

β-lactamase Assay[a] [16,17,19]

1. Add 50 mM potassium phosphate (KPi), pH 7.0.
2. Add inhibitor or DMSO control.
3. Add β-lactamase to 1 nM final concentration from 100 nM stock in 50 mM KPi.
4. Mix.
5. Incubate for 5 minutes.
6. Add nitrocefin to 200 μM final concentration from 20 mM stock in DMSO.
7. Mix.
8. Collect optical density (OD) data at 482 nm for 5 minutes.

MDH Assay[a] [17,19]

1. Add 50 mM KPi, pH 7.0.
2. Add inhibitor or DMSO control.
3. Add MDH to 2 nM final concentration from 200 nM stock in 50 mM KPi.
4. Mix.
5. Incubate for 5 minutes.
6. Add oxaloacetate to 200 μM final concentration from 20 mM stock in 50 mM KPi.
7. Add reduced nicotinamide adenine dinucleotide (NADH) to 200 μM final concentration from 20 mM stock in 50 mM KPi containing 2 mM dithiothreitol (DTT).
8. Mix.
9. Collect OD data at 340 nm for 5 minutes.

Chymotrypsin Assay[a] [16,17,19]

1. Add 50 mM KPi, pH 7.0.
2. Add inhibitor or DMSO control.
3. Add chymotrypsin to 28 nM final concentration from 2.8 μM stock in 50 mM KPi.
4. Mix.
5. Incubate for 5 minutes.
6. Add succinyl-Ala-Ala-Pro-Phe-*p*-nitroanilide to 200 μM final concentration from 20 mM stock in DMSO.
7. Mix.
8. Collect OD data at 410 nm for 5 minutes.

[a.] Performed at room temperature.

PROTOCOL 5.2
Assay for Evaluating Inhibitor Sensitivity to Triton X-100

Template Assay [22]

1. Add buffer.
2. Add Triton X-100 to 0.01% (v/v) final concentration from 1% stock in buffer.
3. Add inhibitor or DMSO control.
4. Add enzyme.
5. Mix.
6. Incubate for n minutes.
7. Add substrate.
8. Mix.
9. Monitor reaction progress.

β-lactamase Assay [22]

1. Add 50 mM KPi, pH 7.0.
2. Add Triton X-100 to 0.01% (v/v) final concentration from 1% stock in 50 mM KPi.
3. Add inhibitor or DMSO control.
4. Add β-lactamase to 1 nM final concentration from 100 nM stock in 50 mM KPi.
5. Mix.
6. Incubate for 5 minutes.
7. Add nitrocefin to 200 μM final concentration from 20 mM stock in DMSO.
8. Mix.
9. Collect OD data at 482 nm for 5 minutes.

---

assay components? Previous work has found that the model enzymes β-lactamase, MDH, and chymotrypsin are unaffected by the presence of 0.01% Triton X-100; this should also be established in other assay systems. If the detergent shows effects independent of the presence of inhibitor, it might be helpful to investigate lower concentrations of Triton X-100 or the use of other detergents, such as saponin or digitonin. Separately, if a well-behaved inhibitor of the target is available, it should be used as a negative control to verify that the detergent will not affect its activity. Alternately, one might consider using a previously identified aggregate-forming inhibitor, such as tetraiodophenolphthalein [16,22], as a positive control to show that the detergent can attenuate the effect of aggregate-forming compounds.

5.4.1.2 Inhibition of Dissimilar Enzymes

Aggregate-forming screening hits have a striking ability to inhibit many diverse enzymes [16,17]. This promiscuity can be detected by testing hits against one or more enzymes that are dissimilar to the screening target.

Many mechanisms are consistent with promiscuous enzyme inhibition, including covalent binding to active site residues and the recognition of privileged substructures [11]. Unlike compounds acting via these mechanisms, inhibition by aggregate-forming compounds is independent of the composition of the enzyme binding site. Therefore, an ideal secondary assay will use an enzyme with a dissimilar structure, activity, and ligand set from the target receptor. For it to be practically applied, this assay should also be straightforward and involve a minimal number of readily available components. β-lactamase, chymotrypsin, and MDH have proven useful as model enzymes for secondary screening; example assays with these enzymes are detailed in Protocol 5.1.

How many secondary assays should be performed to determine that a screening hit is promiscuous? Depending on the degree of certainty desired, one additional assay may be sufficient; testing with two assays in addition to the original screening target will provide greater assurance that any observed promiscuity is not due to enzyme similarity or experimental artifact.

Notably, the same aggregate-forming compound can show 10-fold or greater differences in IC_{50} values against dissimilar enzymes [16]. These differences may be due to differences in assay conditions, enzyme concentrations used in the assay, enzyme diameter, enzyme molecular weight, or other physical properties of the test enzymes. Therefore, potent known aggregate-forming compounds, such as rottlerin [17] or tetraiodophenolphthalein [16], should be used as positive controls to determine that a given assay will detect inhibition by aggregate-forming compounds. Similarly, if they are available, well-behaved inhibitors for the original screening target and each test enzyme should be used to evaluate any cross-recognition of ligands that many occur between systems.

Because there are many mechanisms that can lead to inhibitor promiscuity, demonstrating that a given compound is promiscuous is not sufficient to determine that a compound is an aggregate-former. Other assays described here, such as sensitivity to detergent or enzyme concentration, should be performed in conjunction with promiscuity screening to establish that a compound is an aggregate-former. Depending on the purpose of the secondary screening assays, showing that a particular compound is promiscuous, by any mechanism, may be sufficient to abandon it in favor of more specific screening hits.

5.4.1.3 Sensitivity to Enzyme Concentration

One of the most peculiar characteristics of aggregate-forming inhibitors is that they are exquisitely sensitive to the molar ratio of inhibitor to enzyme [16]. For instance, an increase in the concentration of enzyme from 1 nM to 10 nM can dramatically attenuate inhibition by aggregate-forming compounds present at micromolar concentrations, almost as if the enzyme is titrating out the inhibitor [16]; classically behaved inhibitors do not have this property.

To test the sensitivity of a screening hit to enzyme concentration, perform the original screening assay with a 10-fold increase in the concentration in enzyme (Protocol 5.3). Ideally, this is performed with the inhibitor present at IC_{50} (inhibitory concentration 50%) as determined in the same enzyme system prior to the increase in enzyme concentration. For aggregate-forming inhibitors, the percentage of inhibition should drop from 50% to at least less than 40% against 10-fold more enzyme.

PROTOCOL 5.3
Assay for Evaluating Inhibitor Sensitivity to Enzyme Concentration

Template Assay [16,17]

1. Add buffer.
2. Add inhibitor at IC_{50} concentration[a] or DMSO control.
3. Add enzyme to a 10-fold greater final concentration than in the original assay.
4. Mix.
5. Incubate for n minutes.
6. Add substrate.
7. Mix.
8. Monitor reaction progress.

β-lactamase Assay [16,17]

1. Add 50 mM KPi, pH 7.0.
2. Add inhibitor at IC_{50} concentration[a] or DMSO control.
3. Add β-lactamase to 10 nM final concentration from 1000 nM stock in 50 mM KPi.
4. Mix.
5. Incubate for 5 minutes.
6. Add loracarbef to 100 µM final concentration from 10 mM stock in DMSO.
7. Mix.
8. Collect OD data at 260 nm for 5 minutes.

[a.] As determined in original assay without the increase in enzyme concentration.

Depending on the conditions of the assay, the substrate may be depleted too quickly to allow for accurate measurement of the reaction rate. To improve the dynamic range of the assay, it may be necessary to decrease the concentration of enzyme in the original assay. Alternatively, a poorer substrate for the reaction can be used to decrease the reaction rate. For instance, an ester derivative of cephalothin [16] or loracarbef [19] have been used as substrates for this assay in β-lactamase (Protocol 5.3), instead of the better substrates cephalothin or nitrocefin (Protocol 5.1).

This experiment requires that the concentration of the inhibitor greatly exceeds the concentration of enzyme. Practically, this means that the molar ratio of inhibitor to enzyme should be roughly 1000 to 1, after the 10-fold increase in enzyme concentration. As a negative control experiment, a well-behaved inhibitor should be tested against the increased enzyme concentration to verify that the increase in enzyme does not affect its potency. As with all of the protocols described here, a known aggregate-forming compound, such as tetraiodophenolphthalein [16,22] or rottlerin [17,21], can be used as a positive control to evaluate the assay.

5.4.1.4 Incubation Effect

Apparent inhibition by aggregate-forming compounds increases when they are pre-incubated with the target enzyme for 5 minutes prior to the addition of substrate, compared to the same assay without preincubation [16]. Because other mechanisms can produce time-dependent inhibition, such as covalent binding or a slow on-rate, an incubation effect is not specific for aggregate-formation. However, it is an easy test to perform and, in conjunction with other experiments, the results can be used to differentiate between various modes of inhibition. For instance, covalent inhibitors often have an incubation effect, but are not typically sensitive to increases in enzyme concentration or the presence of detergent.

Many enzyme assays, including those described above, include an incubation period. Therefore, to test for incubation effect, the assay is modified so that there is no incubation period (Protocol 5.4). If the inhibitor shows time-dependent behavior, apparent inhibition should decrease in the absence of an incubation period. Ideally, this assay is performed with the inhibitor at the IC_{50} concentration obtained in the typical assay with incubation. For aggregate-forming inhibitors, the observed inhibition should decrease from 50% to less than 40% without incubation. As a control experiment, classically behaved, nontime-dependent inhibitors of the target should not show an appreciable difference in inhibition with or without an incubation period [16].

5.4.1.5 Other Kinetic Properties

In addition to the properties described above, aggregate-forming promiscuous inhibitors show other peculiar properties in enzyme assays. These include a decreased potency in the presence of albumin [16], decreasing potency with increasing ionic

---

PROTOCOL 5.4
Assay for Incubation Effect

Template Assay [16,17]

1. Add buffer.
2. Add inhibitor at IC_{50} concentration[a] or DMSO control.
3. Add substrate.
4. Mix.
5. Add enzyme.
6. Mix.
7. Monitor reaction progress.

β-lactamase Assay [16,17]

1. Add 50 mM KPi, pH 7.0.
2. Add inhibitor at IC_{50} concentration[a] or DMSO control.
3. Add nitrocefin to 200 μM final concentration from 20 mM stock in DMSO.
4. Mix.
5. Add β-lactamase to 1 nM final concentration from 100 nM stock in 50 mM KPi.
6. Mix.
7. Collect OD data at 482 nm for 5 minutes.

[a.] As determined in original assay with incubation.

strength [16], steep inhibition curves [17], and noncompetitive inhibition [16]. Because these properties are not specific for aggregate-forming inhibitors or are experimentally more laborious than the protocols described above, they are not discussed further.

5.4.2 LIGHT SCATTERING

Light scattering is a rigorous method for determining that a promiscuous compound forms aggregates and may therefore act as an aggregate species. Coupled with the assays described above, light scattering completes the set of experiments necessary to fully evaluate a compound for aggregate-based, promiscuous enzyme inhibition.

5.4.2.1 Overview

Light scattering has become a powerful technique for the analysis of molecular properties. Indeed, a well-designed light scattering experiment can yield a variety of analytical measurements for a broad range of compounds. This discussion will focus on the use of dynamic light scattering (DLS) to address two simple questions: Are aggregates present in an aqueous mixture of a promiscuous compound? If so, how large are the aggregate particles?

In a typical DLS experiment, an aqueous sample containing the test compound is placed in the path of a laser. If particles of a sufficient size, for instance 1 to 1000 nm in diameter, are present in the sample, they will cause the incident laser light to scatter. The scattered photons are recorded by a detector at a fixed angle relative to the sample. The intensity of the scattered light at the detector will undergo microsecond fluctuations because of Brownian motion by the particles in the sample. The time scale of these fluctuations is proportional to the rate of diffusion and therefore to the size of the particles; DLS uses these fluctuations to calculate the hydrodynamic radius of the particles. For more details on the theory and practice of light scattering, see references [24,25].

5.4.2.2 DLS Instruments

Several manufacturers produce DLS instruments suitable for measuring particles in the 1 to 1000 nm diameter size range; a few suppliers are listed in Table 5.2. Basic

TABLE 5.2
DLS Instrument Suppliers

Company	Instrument	Web Site
Brookhaven Instrument Corp.	BI200SM, 90Plus	www.bic.com
Horiba	LA-920, LA-300	www.horiba-particle.com
Particle Sizing Systems	Nicomp 380	www.pssnicomp.com
Precision Detectors, Inc.	PDDLS/Batch	www.precisiondetectors.com
Proterion (Protein Solutions)	DynaPro-MS/X	www.proterion.com
Wyatt Technology Corp.	Dawn EOS, WyattQELS	www.wyatt.com

components of the instrument include a laser source, sample chamber, detector, and an autocorrelator; additional features are often available, such as adaptors for flow-through sample analysis, multiangle detectors, and temperature-controlled sample chambers. The user controls the instrument through an attached personal computer, which typically contains manufacturer-supplied software for data gathering and analysis.

For the purposes described here, the most relevant feature of the instrument is the signal to noise ratio: Is the instrument sensitive enough to differentiate between a solution without particles and a solution with particles? Negative control experiments with solutions known to be particle-free, such as filtered buffer or water, or solutions of compounds known not to form particles, such as 8-anilino-1-naphthalene-sulfonic acid (ANS) [16,26], are useful as baseline measurements for particle-free samples. Positive control experiments should also be performed with solutions known to contain detectable particles, such as latex beads, concentrated bovine serum albumin, or tetraiodophenolphthalein [16]. Ideally, DLS experiments are performed with the inhibitor concentration in the range of the IC_{50}; this often means that particles are present in fairly low concentrations. Does the instrument have adequate sensitivity for low concentrations of particles? This can also be explored by performing control experiments on serial dilutions of solutions known to contain particles and determining the concentration at which a reliable scattering signal disappears for a given compound.

A second issue to consider is the wavelength of the laser used in the DLS instrument. If a compound absorbs light at the wavelength of the laser, particles of that compound will not scatter light from that laser. Many DLS instruments have lasers with wavelengths less than 800 nm to avoid this problem.

A final aspect to consider is the sample volume required for one measurement. An instrument that requires a sample volume of 1 mL or more may be prohibitive if compound quantity is limited; many manufacturers have options for working with sample volumes less than 100 μL. Some instruments can work in flow-through as well as batch mode; this may be an attractive option for some users. Recently, Protein Solutions introduced a plate reader system for high throughput light scattering; this has the added advantage of speed as well as smaller sample volumes.

5.4.2.3 Methods

A basic DLS experiment on a promiscuous compound consists of sample preparation, data collection, and data analysis. Typically, samples are prepared by dilution from a DMSO stock into filtered buffer, often the same buffer used for inhibition assays. Because DMSO can affect the compound distribution, the final concentration of DMSO should be as low as possible, usually less then 5%. Data collection and analysis are performed as described by the instrument manufacturer.

5.4.2.4 Data Analysis

Intensity data from a light scattering experiment are interpreted with an autocorrelation function [24] that is continually updated throughout the course of data collection. The appearance of the autocorrelation function at the end of the run can be a useful tool for initial evaluation of the data.

The presence or absence of particles can be determined by considering the decay of the autocorrelation function. Well-defined decay patterns, seen for the promiscuous compounds rottlerin and K-252c (Figure 5.3A and Figure 5.3B), are consistent with the presence of particles. Poorly defined decay patterns, such as the one seen for suramin (Figure 5.3C), suggest the absence of particles, even in a 400 μM solution of this compound [17,27].

The size of the particles is reflected in the time scale of the decay of the autocorrelation function: smaller particles move more rapidly and yield a shorter decay time; larger particles diffuse more slowly and lead to a longer decay time. For instance, in Figure 5.3A, the autocorrelation function obtained from rottlerin

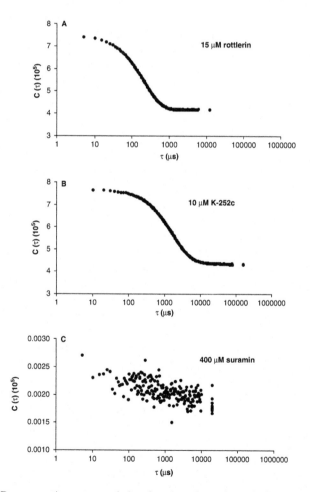

FIGURE 5.3 Representative autocorrelation functions from dynamic light scattering experiments. A. 15 μM rottlerin, B. 10 μM K-252c, C. 400 μM suramin. (Reproduced from S.L. McGovern et al., *J. Med. Chem.* 46:1478–1483, 2003. With permission.)

shows a decay on the 100 μs time scale; in this experiment, this compound has a mean particle diameter of 99 nm [17]. In Figure 5.3B, the signal from K-252c decays on the 1000 μs time scale and has a larger mean particle diameter of 780 nm [17].

Often, it is also helpful to note the average intensity of the scattered light measured at the detector over the time of the experiment. This value is usually reported in counts per second (cps) or kilocounts per second (kcps); it reflects the concentration and size of the scattering species in solution. For particles of a given compound, the scattering intensity increases as the concentration of the particles increases. There is a basal level of scattering from aqueous samples; measured intensities should be compared to those obtained by the testing buffer alone. Because the magnitude of the scattering intensity also depends on sample-independent factors, such as laser power and size of the detector aperture, it is not usually possible to compare scattering intensities from different instruments.

Commercial light scattering devices typically include software for the calculation of particle size. Often, the algorithms make assumptions about the particle shape or the distribution of particle sizes to arrive at a final calculated particle size. It is worthwhile to experiment with standards, such as latex beads or albumin, over various concentration ranges to explore the robustness of the algorithm.

5.4.2.5 Problems and Troubleshooting

Some of the most common problems encountered in a DLS experiment include:

- Dust scatters light independently of the sample particles and can wreak havoc on data analysis. Because dust particles are usually much larger than the test particles, they can be detected by sudden increases or spikes in a plot of scattering intensity over the time of the experiment. Using a filtered buffer may be sufficient to remove most dust most of the time. Other steps include frequent washing and drying of instrument cuvettes or cleaning out the sample chamber as appropriate.
- Air bubbles can also scatter light and interfere with a DLS measurement. Careful pipetting and mixing may be sufficient; otherwise, consider using degassed buffer.
- If the compound concentration is too low, the scattering intensity may not be sufficient to make a reliable measurement, even if particles are present.
- If the compound concentration is too high, especially in the case of aggregate-forming inhibitors, it is possible that aggregates of aggregates will form and the average particle size will be skewed to a higher value than that obtained with a lower concentration of compound.

5.5 CONCLUSIONS

Aggregate-forming, promiscuous compounds are common among screening hits. Because of their nonspecific mode of action, these molecules are poor candidates for lead design efforts. Early experimental identification and abandonment of aggregate-based screening hits allow more resources to be devoted to classically behaved inhibitors.

ACKNOWLEDGMENTS

The author thanks the editors for this opportunity and Beth Beadle, Brian Feng, Pamela Focia, and James Seidler for helpful discussions and critiques of this chapter. This work was supported by T32-GM08061 (Kelly Mayo, PI) and a Northwestern University Presidential Fellowship.

REFERENCES

[1]   J.R. Somoza, A.G. Skillman, Jr., N.R. Munagala, C.M. Oshiro, R.M. Knegtel, S. Mpoke, R.J. Fletterick, I.D. Kuntz, and C.C. Wang, Rational design of novel antimicrobials: blocking purine salvage in a parasitic protozoan, *Biochem.* 37:5344–5348, 1998.

[2]   H.J. Boehm, M. Boehringer, D. Bur, H. Gmuender, W. Huber, W. Klaus, D. Kostrewa, H. Kuehne, T. Luebbers, N. Meunier-Keller, and F. Mueller, Novel inhibitors of DNA gyrase: 3D structure based biased needle screening, hit validation by biophysical methods, and 3D guided optimization: a promising alternative to random screening, *J. Med. Chem.* 43:2664–2674, 2000.

[3]   S. Gruneberg, B. Wendt, and G. Klebe, Subnanomolar inhibitors from computer screening: a model study using human carbonic anhydrase II, *Angew. Chem. Int. Ed.* 40:389–393, 2001.

[4]   R. Brenk, L. Naerum, U. Gradler, H.D. Gerber, G.A. Garcia, K. Reuter, M.T. Stubbs, and G. Klebe, Virtual screening for submicromolar leads of tRNA-guanine transglycosylase based on a new unexpected binding mode detected by crystal structure analysis, *J. Med. Chem.* 46:1133–1143, 2003.

[5]   P.C. Wyss, P. Gerber, P.G. Hartman, C. Hubschwerlen, H. Locher, H.P. Marty, and M. Stahl, Novel dihydrofolate reductase inhibitors: structure-based versus diversity-based library design and high-throughput synthesis and screening, *J. Med. Chem.* 46:2304–2312, 2003.

[6]   R.Y. Kao, J.L. Jenkins, K.A. Olson, M.E. Key, J.W. Fett, and R. Shapiro, A small-molecule inhibitor of the ribonucleolytic activity of human angiogenin that possesses antitumor activity, *Proc. Natl. Acad. Sci. USA* 99:10066–10071, 2002.

[7]   J. Bajorath, Integration of virtual and high-throughput screening, *Nat. Rev. Drug Discov.* 1:882–894, 2002.

[8]   R. Abagyan and M. Totrov, High-throughput docking for lead generation, *Curr. Opin. Chem. Biol.* 5:375–382, 2001.

[9]   W.P. Walters and M. Namchuk, Designing screens: how to make your hits a hit, *Nat. Rev. Drug Discov.* 2:259–266, 2003.

[10]   G.M. Rishton, Reactive compounds and *in vitro* false positives in HTS, *Drug Discov. Today* 2:382–384, 1997.

[11]   O. Roche, P. Schneider, J. Zuegge, W. Guba, M. Kansy, A. Alanine, K. Bleicher, F. Danel, E.M. Gutknecht, M. Rogers-Evans, W. Neidhart, H. Stalder, M. Dillon, E. Sjogren, N. Fotouhi, P. Gillespie, R. Goodnow, W. Harris, P. Jones, M. Taniguchi, S. Tsujii, W. von der Saal, G. Zimmermann, and G. Schneider, Development of a virtual screening method for identification of "frequent hitters" in compound libraries, *J. Med. Chem.* 45:137–142, 2002.

[12]   C.A. Lipinski, F. Lombardo, B.W. Dominy, and P.J. Feeney, Experimental and computational approaches to estimate solubility and permeability in drug discovery and development settings, *Adv. Drug Deliv. Rev.* 23:3–25, 1997.

[13] I. Muegge, S.L. Heald, and D. Brittelli, Simple selection criteria for drug-like chemical matter, *J. Med. Chem.* 44:1841–1846, 2001.

[14] A. Ajay, W.P. Walters, and M.A. Murcko, Can we learn to distinguish between "drug-like" and "nondrug-like" molecules? *J. Med. Chem.* 41:3314–3324, 1998.

[15] W.P. Walters, A. Ajay, and M.A. Murcko, Recognizing molecules with drug-like properties, *Curr. Opin. Chem. Biol.* 3:384–387, 1999.

[16] S.L. McGovern, E. Caselli, N. Grigorieff, and B.K. Shoichet, A common mechanism underlying promiscuous inhibitors from virtual and high-throughput screening, *J. Med. Chem.* 45:1712–1722, 2002.

[17] S.L. McGovern and B.K. Shoichet, Kinase inhibitors: not just for kinases anymore, *J. Med. Chem.* 46:1478–1483, 2003.

[18] S. Soelaiman, B.Q. Wei, P. Bergson, Y.S. Lee, Y. Shen, M. Mrksich, B.K. Shoichet, and W.J. Tang, Structure-based inhibitor discovery against adenylyl cyclase toxins from pathogenic bacteria that cause anthrax and whooping cough, *J. Biol. Chem.* 278:25990–25997, 2003.

[19] J. Seidler, S.L. McGovern, T.N. Doman, and B.K. Shoichet, Identification and prediction of promiscuous aggregating inhibitors among known drugs, *J. Med. Chem.* 46:4477–4486, 2005.

[20] D.W. Miller and K.A. Dill, Ligand binding to proteins: the binding landscape model, *Protein Sci.* 6:2166–2179, 1997.

[21] A.J. Ryan, N.M. Gray, P.N. Lowe, and C. Chung, Effect of detergent on "promiscuous" inhibitors, *J. Med. Chem.* 46:3448–3451, 2003.

[22] S.L. McGovern, B.T. Helfand, B. Feng, and B.K. Shoichet, A specific mechanism for non-specific inhibition, *J. Med Chem.* 4265–4272, 2003.

[23] C. Tanford, *The Hydrophobic Effect: Formation of Micelles and Biological Membranes,* 2nd ed., New York: Wiley, 1980.

[24] N.C. Santos and M.A.R.B. Castanho, Teaching light scattering spectroscopy: the dimension and shape of tobacco mosaic virus, *Biophys. J.* 71:1641–1650, 1996.

[25] P.J. Wyatt, Light scattering and the absolute characterization of macromolecules, *Analytica Chimica Acta* 272:1–40, 1993.

[26] B. Stopa, M. Gorny, L. Konieczny, B. Piekarska, J. Rybarska, M. Skowronek, and I. Roterman, Supramolecular ligands: monomer structure and protein ligation capability, *Biochimie* 80:963–968, 1998.

[27] T. Polenova, T. Iwashita, A.G. Palmer, and A.E. McDermott, Conformation of the trypanocidal pharmaceutical suramin in its free and bound forms: transferred nuclear overhauser studies, *Biochem.* 36:14202–14217, 1997.

Part III

Ligand-Based Virtual Screening Approaches

6 Data Mining Approaches for Enhancement of Knowledge-Based Content of *De Novo* Chemical Libraries

Nikolay P. Savchuk and Konstantin V. Balakin

6.1 INTRODUCTION

In the past decade, the pharmaceutical industry has realized the increasing significance of methods, traditionally referred to as the hit-to-lead optimization phase, in the early stage of the drug discovery process. In particular, knowledge-based approaches emerged and evolved to address a multitude of significant issues such as biological activity profile, metabolism, pharmacokinetics, toxicity, lead- and drug-likeness [1]. In this chapter, we will focus on the development and application of knowledge-based methods related to classification algorithms in virtual screening (VS) programs. Compound classification methods used for correlation of molecular properties with specific activities play a significant role in modern VS strategies. Because the current drug discovery paradigm states that mass random synthesis and screening do not necessarily provide a sufficiently large number of high-quality leads, such computational technologies are of great industrial demand. The most typical application of classification algorithms includes the identification of compounds with desired target-specific activity, which constitutes an essential part of the VS ideology. Statistical methods can be applied to process the results of high throughput screening (HTS) or known literature data and develop predictive models of biological activity. These models can further be used for selection of screening candidates from virtual databases. However, achieving desired specificity and activity alone is not sufficient to produce high-quality clinical candidates. Accordingly, druglikeness, favorable ADME-Tox profile (see Section 6.11), solubility issues, and pharmacokinetic and metabolic characteristics should be taken into consideration as early as possible. This obvious trend seriously influences contemporary VS programs. Usually, it is realized in development and implementation of various special filters in screening tools. Such filters allow the prediction of compound characteristics significant for particular bioscreening purposes.

The aim of our work is to elucidate and comment on some useful approaches to compound classification based on nonlinear mapping techniques. These approaches are integrated on the basis of a corporate chemoinformatics platform, ChemoSoft™ (Chemical Diversity Labs, Inc.). They represent real practical tools for enhancement of informational content of virtual compound selections. All the VS technologies presented here are focused on the small molecular level, as opposed to target structure-based design or docking methodology. The *leitmotif* of our methods is property-based approach realized in the context of classification of quantitative structure–activity relationship (QSAR) modeling. The theoretical discussion will be illustrated by description of practical algorithms in the knowledge-based design of *de novo* chemical libraries.

6.2 CLASSIFICATION QSAR IN VIRTUAL SCREENING

Structural or physicochemical properties of a ligand directly affect its biological activity against a biological target. The quantitative method of assessment of this activity is known as the determination of QSARs, which has been pioneered by Hansch and Leo [2] and further developed in multiple studies. Historically, statistical techniques of regression QSAR were applied to relatively small data sets consisting of several tens to several hundreds of molecules with known activity values. Such activity measurements are usually performed manually in the laboratory and produce relatively accurate experimental results (for example, IC_{50} value or the concentration required for 50% inhibition). Unfortunately, low precision and high error rate HTS data are often incompatible with theoretical assumptions of most regression techniques. In addition, there are a number of practically significant tasks for which the continuous measurement of activity is impossible or problematic, such as the drug-likeness problem, assignment of compounds to a particular therapeutic or target-specific activity group (for example, anticancer, antituberculosis, G protein-coupled receptors [GPCRs], proteases), or studies of structure-metabolism relationships. In all these tasks, classification QSAR approaches represent a viable and sometimes unavoidable alternative to regression techniques. It should also be noted that classification QSAR is ideally suitable for *in silico* HTS of large virtual compound libraries. Because data fitting is not used, the predictive capacity of classification QSAR is not interpolative, but based on generalizations substantiated by the available data. In addition, classification QSAR is less sensitive to measurement errors than regression-based QSAR methodology and has been successfully used in a number of QSAR studies.

Table 6.1 lists some typical examples of classification QSAR modeling studies carried out in the last decade. The tasks range from general problems, such as druglikeness or central nervous system activity of compounds, to specific issues related to assessment of potential target-specific activity, segregation of compounds based on toxicity, metabolic stability, blood–brain barrier (BBB) permeability, etc. The selected concepts and investigations in compound classification and VS have been reviewed in a recent paper [3].

TABLE 6.1
Typical Tasks and Reference Databases Used for Generation of Knowledge-Based Classification Models

Task	Reference Database	Ref.
Discrimination between drugs/nondrugs	38,416 drugs from the World Drug Index and 169,331 nonpharmaceutical compounds from the Available Chemicals Directory	[4]
Discrimination between drugs/nondrugs	Approximately 80,000 drugs from the Comprehensive Medicinal Chemistry database and MACCS-II Drug Data Report and approximately 170,000 nonpharmaceutical compounds from the Available Chemicals Directory	[5]
Discrimination between CNS-active/inactive drugs	Approximately 15,000 CNS-active drugs and approximately 60,000 CNS-inactive drugs; both sets are from the Comprehensive Medicinal Chemistry database and MACCS-II Drug Data Report	[6]
Discrimination between human CYP450 substrates/nonsubstrates	485 compounds described as substrates for human CYP450 isozymes and 523 products of the cytochrome-mediated biotransformations; both sets are from MetaDrug™ database (GeneGo, Inc., www.genego.com)	[7]
Discrimination between genotoxic/nongenotoxic amines	334 aromatic and secondary amine compounds with different levels of genotoxicity; data are from literature	[8]
Discrimination between active cyclic 3'5',-guanosine monophosphate (cGMP) phosphodiesterase V (PDEV) inhibitors and compounds without cGMP PDEV activity	61 cGMP PDEV inhibitors and 1739 compounds inactive with respect to cGMP PDEV inhibition; both sets are from literature	[9]
Discrimination between active ACAT (acyl-CoA:cholesterol O-acyltransferase) inhibitors and compounds without ACAT activity	446 ACAT inhibitors and 1771 compounds without reported ACAT inhibiting activity; both sets are from literature	[9]
Discrimination between compounds active/inactive against estrogen receptors	463 estrogen analogues with different levels of activity; data are from literature	[10]
Discrimination between particular groups of GPCR-specific ligands	3365 GPCR-active agents belonging to 9 different GPCR classes; data are from the Ensemble database of pharmaceutical compounds (Prous Science, 2002; www.prous.com)	[11]
Discrimination between Carbonic anhydrase II inhibitors/noninhibitors	337 compounds with different levels of carbonic anhydrase II inhibiting activity; data are from literature	[12]
Discrimination between antibacterial agents and compounds without antibacterial activity	661 compounds, classified according to whether they had antibacterial activity; data are from literature	[13]

6.3 DATA ANALYSIS AND VISUALIZATION

One of the difficult challenges in data analysis is to be able to represent whatever complexities might be intrinsic to the data in a simple and intuitive form. Traditional methods of data visualization, such as property distribution histograms, are often inadequate to represent the extremely large, high-dimensional data sets common to statistical analyses of large virtual databases. To minimize the complexity and reduce the number of individual plots needed to visualize this sort of data, one must attempt to reduce the dimensionality of the representation. Several different techniques were proposed to achieve dimensionality reduction, while preserving the topology of the original space. That is, points near each other in the high-dimensional space are also near each other in the low-dimensional space.

Self-organizing maps (SOMs) or Kohonen networks [14] belong to a class of neural networks known as competitive learning or self-organizing networks. They were originally developed to model the ability of the brain to store complex information as a reduced set of salient facts without loss of information about their interrelationships. High-dimensional data are mapped onto a two-dimensional rectangular or hexagonal lattice of neurons in such a way as to preserve the topology of the original space. Each object, or molecule, can be represented as a vector, the components of which are variables with a definite meaning (molecular descriptors). The Kohonen neural network automatically adapts itself to the input data in such a way that similar input objects are associated with the topologically close neurons in the neural network. In the Kohonen approach, the neurons learn to determine the location of the neuron in the neural network that is most similar to the input vector. This means that objects on the 2-D map located physically close to each other have similar properties. Kohonen SOMs can be effective for visualizing, comparing, and filtering chemical libraries. In the last decade, Kohonen maps became popular for comparative analysis and visualization of datasets [15]. A study on comparison of benzodiazepine and dopamine datasets was performed with an implementation of a Kohonen network [16]. In another study, a dataset of 31 steroids binding to the corticosteroid binding globulin (CBG) receptor was modeled (17). Kohonen SOMs were used for distinguishing between drugs and nondrugs with a set of descriptors derived from semiempirical molecular orbital calculations [18]. Kohonen map-based algorithm was recently used for clustering and visualization of the National Cancer Instutute's publicly available antitumor drug-screening data [19]. This analysis identified relationships between chemotypes of screened agents and their effect on four major classes of cellular activities — mitosis, nucleic acid synthesis, membrane transport and integrity, and phosphatase- and kinase-mediated cell cycle regulation. In our laboratories, we developed an effective classification scheme for segregation of human cytochrome P450 (CYP450) substrates/nonsubstrates using Kohonen SOMs [7]. The same methodology was applied for discrimination between BBB well- and poorly permeable drugs [20].

Nonlinear Sammon's mapping is an alternative multivariate statistical technique that approximates local geometric relationships on a two- or three-dimensional (2D or 3D) plot [21]. Nonlinear maps were used for visualization of protein sequence relationships in two dimensions [22] and were employed as a means of visualizing and comparing large compound collections, represented by a set of molecular descriptors [23,24].

Both of these mapping techniques are based on unsupervised learning algorithms. However, in most of the reported classification strategies in drug discovery, an alternative supervised learning strategy was used. The choice between the supervised or the unsupervised learning approach depends on the problem and the available data. In both cases, the objects with known answers are needed. In supervised learning, the answers are directly used to influence the learning system; in unsupervised learning, the answers are needed to identify and label the output neurons. Whereas with supervised learning, the system adapts itself to a selected representation of classes, an unsupervised learning method is more flexible due to its many possible outputs. Using the supervised learning, the multivariate objects should be split into three sets — training, control, and test. In unsupervised learning, the control set is not required, because the learning continues until the network stabilization. It should also be emphasized that Kohonen or Sammon map-based classification methods do not depend on the definition of a negative set, and, therefore, the virtual search for compounds belonging to a particular category of interest can be conducted more objectively. This property of unsupervised learning strategies is particularly important in many practical tasks when it is hard to define the negative training set correctly.

In this work, we will describe a practical approach to the virtual selection and *de novo* design of combinatorial libraries with enhanced informational content based on the implementation of the nonlinear mapping techniques. Specifically, three main topics will be discussed — target-specific library design, early prediction of human toxicity, and CYP450-mediated drug metabolism. These tasks represent highly actual problems in modern drug discovery, which attracted serious attention from researchers in the last years. Each of these difficult problems can be considered separately from others, but combined, they provide a good illustration of typical issues that should be taken into account while planning a rational VS strategy.

6.4 PREPARATORY STAGES

Several important preparatory procedures are required prior to initiation of any knowledge-based statistical data mining experiment. Two main steps usually precede the development of a classification model:

1. Collection and processing of the knowledge database aiming at generation of quality training data sets.
2. Calculation of molecular descriptors and selection of an appropriate subset of descriptors for further modeling.

6.5 KNOWLEDGE DATABASE

At the first step, one should collect a comprehensive knowledge database, which represents a set of compounds with defined properties, such as activity against a target of interest, BBB permeability, toxicity, CYP450 substrate properties, etc. Several comprehensive pharmaceutical databases, such as Prous Science's Ensemble® database or CrossFire Beilstein database, as well as proprietary knowledge databases can be used as the source of information about structures and their activities. The key to success of any statistics-oriented predictive modeling is the availability of a large set of quality data used as a training set. Usually, at least several tens of structures belonging to a defined activity category are required for generation of a robust classification model, though this number can vary in a wide interval depending on the task. Sometimes, for an optimal statistical experiment, one needs up to hundreds of thousands of compounds in the training set. Table 6.1 shows the typical sizes of reference knowledge bases recently used to generate various classification models.

Prior to the statistical experiments, the molecular structures should be filtered and normalized to fulfill certain criteria (see Protocol 6.1).

Point 6 of Protocol 6.1 needs some additional explanations. For many practical tasks, it is required that the compounds pass a rejection filter that removes chemically reactive or otherwise not suited compounds. Filtering criteria depend on a particular task. For example, compounds can be filtered based on the presence of chemically reactive or toxicophoric fragments, such as alkylating or acylating groups, strong electrophiles, etc. Care should be taken when using a standard set of reactive fragments for exclusion of hypothetically hazardous compounds. For some specific classes of pharmaceutical chemicals, such as oncolytic or protease-active agents, the chemically reactive "warheads" can be necessary elements of the structure. The rejection filter can also include a number of criteria that ensure the general drug-likeness [25] or leadlikeness [26] of compounds. In addition, it is usually required to filter molecules based on atom type content, because many standard programs for calculation of molecular descriptors correctly work only with C, N, O, H, S, P, F, Cl, Br, and I atom-containing structures.

PROTOCOL 6.1
Preparatory Steps before Statistical Modeling Experiments

1. Remove structures with obvious errors.
2. Remove counterions and solvent molecules to obtain single-compound records.
3. Where possible, neutralize charges at acidic and basic groups by adding or removing protons.
4. Remove duplicates within the reference database.
5. Remove redundant tautomeric compound forms, such as tautomers.
6. Remove compounds that do not pass a rejection filter (specific for each particular task).

6.6 MOLECULAR DESCRIPTORS

At the second step, one should analyze the physicochemical and topological properties of compounds and select a minimal set of key parameters adequately describing the selection. *Physicochemical* properties have long been used to develop structure–activity relationships (SARs). They quantify a large number of molecular features known to determine the pharmacokinetic and pharmacodynamic properties of compounds. Among them are molecular weight (MW), octanol-water partition coefficient (logP), molar refractivity, van der Waals (vdW_vol) volume and surface area, the number of filled orbitals, highest occupied molecular orbital (HOMO) and lowest unoccupied molecular orbit (LUMO) energies, partial atomic charges and electron densities, dipole moment, ionization potential, and many others. *Molecular connectivity* or *topological* indices are numeric values calculated from certain invariants (characteristics) of a molecular graph [27]; these indices encode features such as number of atoms, branching, ring structures, heteroatom content, and bond order. They are attractive for quantifying molecular diversity because they are inexpensive to compute and have been validated through years of use in the field of structure–activity correlation. The available programs calculate indices based on connectivity, shape, subgraph counts, topological equivalence, and electrotopological state. The applicability and use of various topology indices is discussed in several recent papers [28,29].

A thorough overview of different molecular descriptors can be found in *Handbook of Molecular Descriptors,* a fundamental edition presenting a comprehensive collection of molecular descriptors and a detailed review from the origins of this research field up to the present day [30]. Molecular property descriptors have also been extensively reviewed in several excellent reviews [31,32].

Table 6.2 lists some commercially available software packages that can be used to calculate different molecular descriptors.

After available descriptors are calculated, a feature reduction stage should be performed. The high-dimensional data representations that are commonplace in statistical modeling applications pose a number of problems. First, as the number of variables used to describe data increases, the likelihood that some of the variables are correlated dramatically increases. Although certain applications are more sensitive to correlation than others, in general, redundant variables tend to bias the results. Second, the amount of the computational effort needed to perform the analysis increases in proportion to the number of dimensions. The latter fact is particularly significant in VS programs, where the possibility of real-time calculations is a crucial problem.

Therefore, to simplify the analysis and representation of the data, it is often desirable to reduce the dimensionality of the space by eliminating dimensions that add little to the overall picture. Several techniques were developed to perform the dimensionality reduction, such as principal component analysis (PCA) [33,34], multidimensional scaling [35,36], genetic algorithm (GA) [37], and sensitivity analysis [38]. It should be noted, however, that none of the existing methods guarantees to extract the optimal set of important features for the application at hand. Moreover, the underlying molecular features that influence the biological activities of drugs

TABLE 6.2
Commercially Available Programs for Calculation of Molecular Descriptors*

Program	Descriptors	Vendor
Cerius$^{2®}$ Descriptor+	A wide range of molecular descriptors, describing topological, electronic, and structural features	Accelrys, Inc. 9685 Scranton Road San Diego, CA 92121-3752 Phone: 858-799-5000 www.accelrys.com
MOE	A wide range of molecular descriptors including topological indices, structural keys, E-state indices, physical properties, topological polar surface area and others	Chemical Computing Group, Inc. 1010 Sherbrooke Street West, Suite 910 Montreal, Quebec Phone: 514-393-1055 www.chemcomp.com
Dragon	More than 1400 physicochemical, topological, spatial, and other molecular descriptors calculated from 2D structures or 3D structures created by external programs	University of Milano-Bicocca, Department of Environmental Sciences Piazza della Scienza, 1 — 20126 Milano, Italy Phone: +39-02-64482820 http://www.talete.mi.it
Molecular Modeling Pro	Over 100 physicochemical, topological, quantum, and spatial descriptors	Norgwyn Montgomery Software Inc. 216 Lower Valley Rd. North Wales, PA 19454 Phone: 215-378-6274 http://www.norgwyn.com
ChemOffice/C hemSAR	A range of topological, shape, thermodynamic, quantum, and surface property descriptors	CambridgeSoft Corp. 100 CambridgePark Drive, Cambridge, MA 02140 Phone: 800-315-7300 http://www.cambridgesoft.com/

*The shown data reflect the status of a survey completed in mid-2003.

are usually unknown; this fact makes *a priori* feature selection problematic in many cases. It can be concluded that selection of variables for QSAR applications is difficult and, despite many different statistical criteria for the evaluation of the resulting models, a highly subjective and ambiguous procedure [39]. This is why in many practical experiments, particularly those associated with the use of nonlinear learning approaches, computational chemists have to make intuitive decisions upon selecting molecular descriptors. In the modeling studies described here, for reduction of the number of input variables, we used PCA as it is implemented in ChemoSoft software. The principles of PCA have been described many times [33,34] and will not be described here.

After all preparatory procedures are completed, the reference database with selected molecular descriptors is used to develop an *in silico* model with the most appropriate architecture and learning strategy. The following examples represent real computational filtering technologies developed at Chemical Diversity Labs for enhancement of knowledge-based content of exploratory chemical libraries for bio-screening at the stage of combinatorial synthesis planning.

6.7 ENHANCEMENT OF TARGET-SPECIFIC INFORMATIONAL CONTENT OF VIRTUAL COMPOUND SELECTIONS

Choosing structures that are most likely to have a predefined target-specific activity of interest from the vast assortment of structurally dissimilar molecules is a particular challenge in compound selection. This challenge has been tackled with powerful computational methodologies, such as docking available structures into the receptor site and pharmacophore searching for particular geometric relations among elements thought critical for biological activity. The subject is discussed in several recent reviews [40–43]. Both methodologies focus on conformational flexibility of both target and ligand, which is a complex and computationally intense problem. The latest developments in this field pave the way to wide industrial application of these technologies in drug design and discovery, though the limits of computational power and time still restrict the practical library size selected by these methods.

Another popular approach to VS is based on ligand structure and consists of selecting compounds structurally related to hits identified from the initial screening of the existing commercial libraries and active molecules reported in research articles and patents [44–47]. Although broadly used in the development of SAR profiling libraries, these methods usually perform poorly when it comes to the discovery of structurally novel lead chemotypes. In general, the target and ligand structure-based technologies cannot adequately address all real problems of rational drug design, particularly those connected with VS of large compound databases or discovery of novel lead chemotypes.

An alternative design for target-specific libraries is based on statistical data mining methods, which are able to extract information from knowledge databases of active compounds. Here, we describe a practical approach to limiting the size of virtual combinatorial libraries and selection of molecular subsets with enhanced target-specific informational content. In this work, we used two methods of QSAR and data visualization based on Kohonen SOMs and Sammon nonlinear maps.

At the initial stage of this work, we collected a 22,110-compound database of known drugs and compounds entered into preclinical or clinical trials. Each compound in this database is characterized by a defined profile of target-specific activity, focused against 1 of more than 100 different protein targets. The database was filtered based on MW (not more than 800).

Molecular features encoding the relevant physicochemical and topological properties of compounds were calculated from 2D molecular representations and selected by PCA (Table 6.3). These molecular descriptors encode the most significant molecular features, such as molecular size, lipophilicity, H-binding capacity, flexibility,

TABLE 6.3
Descriptors Used for Target-Specific Activity Modeling

Descriptor	Definition
logD	Log of 1-octanol/water partition coefficient at pH 7.4
RotB	Number of rotatable bonds
Hb_acc	Number of H-bond acceptors
Hb_don	Number of H-bond donors
a_aro	Number of aromatic atoms
TPSA	Total polar surface area
K1	Kier first shape index
HI1	First order atomic valence connectivity index
HI2	Second order atomic valence connectivity index

and molecular topology. Taken in combination, they define both pharmacokinetic and pharmacodynamic behavior of compounds and are effective for property-based classification of target-specific groups of active agents. However, it should be noted that for each particular target-specific activity group, another, more optimal set of descriptors can be found, which provides better classification ability.

The procedure in Protocol 6.2 should be carried out for Kohonen network training and map generation.

A Kohonen SOM of 22,110 pharmaceutical leads and drugs generated as a result of the unsupervised learning procedure is depicted in Figure 6.1. It shows that the studied compounds occupy a wide area on the map, which can be characterized as the area of druglikeness. Distribution of various target-specific groups of ligands in the Kohonen map demonstrates that most of these groups have distinct locations in

---

PROTOCOL 6.2
Procedure of Kohonen Map Generation

1. Identify and label molecular descriptors and training set compounds, which will be used for modeling.
2. Set training parameters: number of iterations, starting adjustment radius, decay factor (the default values of these parameters are 2000, 0.01, and 0.001, correspondingly; they can be recommended for most applications).
3. Normalize descriptor values.
4. Set a proper map size to provide the studied molecules with a sufficient distribution space. For most applications, an optimal map contains 10 to 100 objects (molecules) per node.
5. Run the training process. After the training is accomplished, the resulting Kohonen map can be created, the model can be saved and used for further testing and visualizing other objects on the same map.

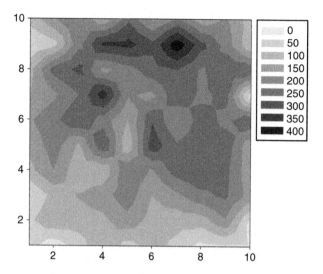

FIGURE 6.1 Property space of 22,110 pharmaceutical leads and drugs visualized using the Kohonen map. The data have been smoothed.

specific regions of the map (Figure 6.2a through Figure 6.2e). A possible explanation of these differences is in that, as a rule, receptors of one type share a structurally conserved ligand-binding site. The structure of this site determines molecular properties that a receptor-selective ligand should possess to properly bind the site. These properties include specific spatial, lipophilic, and H-binding parameters, as well as other features influencing the pharmacodynamic characteristics. Therefore, every group of active ligand molecules can be characterized by a unique combination of physicochemical parameters differentiating it from other target-specific groups of ligands. Another explanation of the observed phenomenon can be related to different pharmacokinetic requirements to drugs acting on different biotargets.

The described algorithm represents an effective procedure for selection of target-biased compound subsets compatible with high throughput *in silico* evaluation of large virtual chemical space. Whenever a large enough set of active ligands is available for a particular receptor, the quantitative discrimination function can be generated allowing selection of a series of compounds to be assayed against the target. Once a Kohonen network is trained and specific sites of location of target-activity groups of interest are identified, the model can be used for testing any available chemical databases with the same calculated descriptors. The Kohonen mapping procedure is computationally inexpensive and permits real-time calculations with moderate hardware requirements. Thus for a training database consisting of 20,000 molecules with 9 descriptors, approximately 15 min are required for a standard personal computer (PC) (Pentium 1.8-GHz processor) on a Windows 2000 platform to train the network. The time increases almost linearly with the size of the database. After the Kohonen network is trained, the 2D map can be created in a short time. For example, for a 1,000,000-compound database with 9 descriptors, less than 10 min are required for the Kohonen map to be generated using the same

(a)

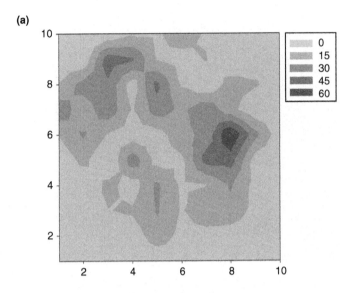

(b)

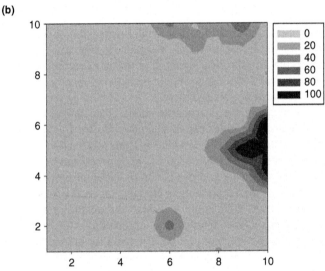

FIGURE 6.2 Distribution of five large target-specific groups of pharmaceutical agents on the Kohonen map: (a) tyrosine kinase inhibitors (1405 compounds); (b) nuclear receptor agonists/antagonists (1021 compounds); (c) GPCR agonists/antagonists (12,512 compounds); (d) potassium channel activators (1060 compounds); (e) calcium channel antagonists (1230 compounds).

computer. It is important to note that focusing on physicochemical rather than structural features makes this approach complementary to any available ligand structure similarity technique.

Our own experience and literature data demonstrate that Kohonen SOMs are efficient clustering, quantization, classification, and visualization tools useful in the

(c)

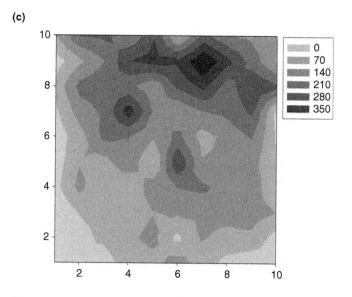

(d)

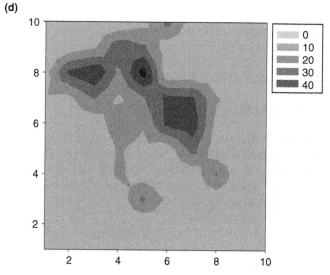

FIGURE 6.2 (continued)

design of chemical libraries. Possible limitations of this approach are related to the fact that the SOM algorithm is designed to preserve the *topology* between the input and grid spaces; in other words, two closely related input objects will be projected on the same or close nodes. At the same time, the SOM algorithm does not preserve *distances*: there is no relation between the distance between two points in the input space and the distance between the corresponding nodes. The latter fact sometimes makes the training procedure unstable when the minor changes in the input parameters lead to serious perturbation in the output picture. As a result, it is often difficult

(e)

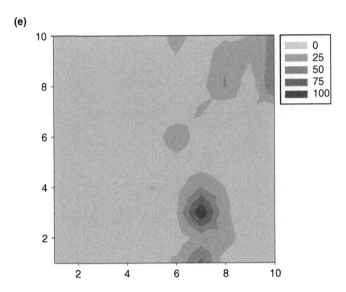

FIGURE 6.2 (continued)

to find the optimal training conditions for better classification. Another potential problem is associated with the quantization of the output space. As a result, the resolution of low-sized maps can be insufficient for effective visualization of differences between the studied compound categories.

6.8 SAMMON MAPPING

Sammon mapping represents a useful alternative or supplement to Kohonen SOM algorithm, because this dimensionality reduction technique often provides much greater detail about the individual compounds and their interrelationships. In contrast to Kohonen SOMs, the Sammon mapping preserves distances between the input and projection spaces, which can facilitate the analysis of relationships between the training parameters and output picture.

To illustrate the application of Sammon mapping to determine the potential target-specific activity profile of compounds, consider the following example. The sites of distribution of tyrosine kinase inhibitors and potassium channel openers (Figure 6.2a and Figure 6.2d, correspondingly) on the Kohonen map are similar, and, therefore, their differentiation using this particular set of molecular descriptors seems to be problematic. To test the classification ability of the alternative nonlinear mapping technique, we processed these two categories of active agents on a 2D Sammon map using the same set of molecular descriptors and Protocol 6.3.

The resulting Sammon map is shown in Figure 6.3. The positioning of the separation line is determined using nonlinear Support Vector Machine algorithm as it implemented in LibSVM-2.4 program [48]. This line provides the largest margin separating the studied classes (margin is defined as a sum of the shortest distances from decision line to the closest points of both classes) and, thus, can serve as an optimal discriminator between the two studied compound categories.

PROTOCOL 6.3
Procedure of Sammon Map Generation

1. Identify and label molecular descriptors and training set compounds, which will be used for modeling.
2. Set training parameters: maximal number of iterations and optimization step (the default values of these parameters which can be recommended for standard applications are 300 and 0.3, correspondingly). Similarity measure based on Euclidean distance is preferred for most of the tasks.
3. Run the training process. After the training is accomplished, the resulting Sammon map is created automatically. The program permits the user to select one best map (the one with minimal stress value) among several maps generated as a result of a specified number of independent trainings.

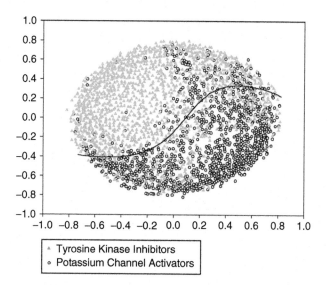

△ Tyrosine Kinase Inhibitors
○ Potassium Channel Activators

FIGURE 6.3 Sammon map of the combined set of tyrosine kinase inhibitors (1405 agents) and potassium channel activators (1060 agents).

As Figure 6.3 and Figure 6.4 show, the studied categories of active agents — tyrosine kinase inhibitors and potassium channel openers — occupy distinctly different regions on the map. It can be concluded, that in this particular case, Sammon mapping algorithm can be used as a complementary to Kohonen map tool, which discriminates the compound categories of interest with greater effectivity. In a real situation of combinatorial synthesis planning, a virtual combinatorial compound database should be processed on the Sammon map simultaneously with the active compounds. The sites of location of these active agents will define the combinatorial subset, which can be recommended for further synthetic development as a target-biased selection.

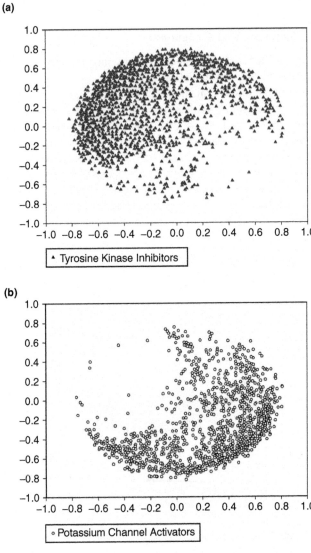

FIGURE 6.4 Areas of distribution of tyrosine kinase inhibitors (a) and potassium channel activators (b) shown separately on the same Sammon map.

Among the other dimensionality reduction techniques that have appeared in statistical literature, Sammon nonlinear mapping is unique for its conceptual simplicity and ability to reproduce the topology and structure of the data space in a faithful and unbiased manner. A major shortcoming of this method is its quadratic dependence on the number of objects scaled, which imposes practical limitations on the size of data sets that can be effectively manipulated. Usually, not more than 10^4 compounds can be processed during one training procedure using a standard PC workstation on the Windows platform. Nevertheless, the method has an undoubted practical value and can be recommended for analysis of medium-sized combinatorial libraries (up to thousands of compounds) aiming at selection of subsets with enhanced knowledge-based informational content.

The mapping techniques presented denote useful approaches to filtering combinatorial libraries for selection of target-specific subsets. They often permit the user to reduce the size of initial chemistry space up to two orders of magnitude and can be recommended as efficient classification and visualization tools for practical combinatorial design. It is important that these property-based methods are complementary to other target and ligand structure-based approaches to VS. Kohonen map-based method is compatible with high throughput VS protocols, and Sammon mapping technique is more relevant to analysis of small-to-medium-sized chemical libraries. The principal limitation of these statistics-oriented techniques is that prior to the experiment, one should have a large enough data set of compounds active against the target of interest.

6.9 MODELING CYP450-MEDIATED DRUG METABOLISM

With the growing use of chemicals as therapeutic agents, food additives, cosmetics, agricultural fertilizers, and pest management agents, people are increasingly exposed to exogenous compounds (xenobiotics). Both the parent agent and the products of its metabolism in the liver and other organs may contribute to the composite toxic effect of an agent on the human organism. The metabolic transformations may profoundly affect the initial bioavailability, the desired activity, the tissue distribution, the toxic action, and the eventual elimination of a compound. The understanding of the possible toxicity and the metabolic fate of xenobiotics in the human body is particularly important in drug discovery, where such early assessment may eliminate the potentially toxic candidates from further development prior to expensive clinical trials [49–52]. With the obvious significance of the early assessment problem, there is a need for *in silico* methodologies to uncover the relations between the structure and metabolism of novel molecules.

We developed a computational algorithm capable of recognizing potential substrates for human CYP450 and evaluating certain metabolism-related effects. A comprehensive set of more than 2200 substrate–product reactions for 38 human cytochromes was assembled from experimental literature [7]. Two overlapping data sets were distinguished within the database, filtered as described in Protocol 6.1, and used in the neural network experiments. The first data set consisted of 485 compounds described as substrates for the human CYP450 enzymes. The second dataset comprised 523 products of the cytochrome-mediated biotransformations for which no data on their further cytochrome-mediated metabolism were found. It was assumed that this dataset models the properties of the nonsubstrates to the whole CYP450 family.

TABLE 6.4
Descriptors Used for Modeling of the CYP450-Mediated Metabolism

Descriptor	Definition
$logD_{7.4}$	Log of 1-octanol/water partition coefficient at pH 7.4
HOMO	Energy of highest occupied molecular orbital
PPSA-1	Partial negative surface area
TPSA	Total polar surface area
MW	Molecular weight
HBA	Number of H-bond acceptors
HBD	Number of H-bond donors

Seven molecular descriptors used for training the model (Table 6.4) were selected from more than 60 initially calculated descriptors using principal component analysis. A 10×10 Kohonen network was generated using Protocol 6.2 and the combined dataset of cytochrome substrates. The cytochrome substrates (Figure 6.5a) are distributed throughout the map as irregularly shaped islands, with a clearly defined trend toward the top of the map. The area occupied by the cytochrome substrates is relatively large, which reflects the broad substrate specificity of the studied set of cytochromes. It can be suggested that the physicochemical properties of a molecule falling into one of the positive regions of the Kohonen map are consistent with the molecule's ability to be a cytochrome substrate. For comparison, we processed the additional data set of 523 products of cytochrome-mediated biotransformations (Figure 6.5b) on the same Kohonen map. These compounds occupy the distinct areas on the map substantially different from the regions of the substrates' localization.

To quantitatively assess the prevalence of CYP450 substrates over the final products in each point of this map, we built a differential map (Figure 6.6), where each node contains a result of division of the percentage of substrates (485 substrates correspond to 100%) on the percentage of products (523 products correspond to 100%) in this node. In the case of randomly distributed compounds, this coefficient approximates to 1 in each node thus indicating equal probability of occurrence of substrates/products everywhere on the map. Using this differential map, we can identify areas, in which the probability of occurrence of CYP450 substrates is up to one order of magnitude higher than in the random case.

Although the general classification power of our model is still moderate, it reasonably discriminates between CYP450 substrates and nonsubstrates when the studied compounds fall into meaningful regions of the map. The enhancement factors for the areas of substrates and nonsubstrates are equal to 7.17 and 4.21, correspondingly. The enhancement factor is a ratio between the fractions of correctly and incorrectly classified compounds within the corresponding areas on the map. Effectively, it shows how many folds the number of CYP450 substrates or nonsubstrates are found in the corresponding areas on the map exceeds the random distribution expectation.

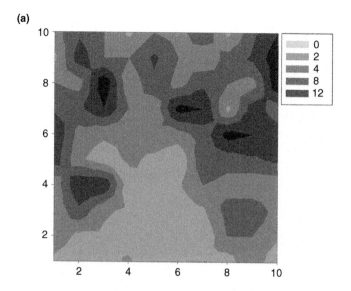

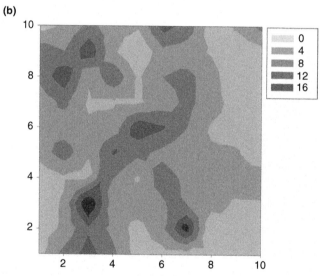

FIGURE 6.5 Kohonen network trained with preselected descriptors for 485 cytochrome substrates (a) and 523 products of metabolic reactions (b). The data have been smoothed.

The described model can be used for *in silico* filtering virtual combinatorial libraries based on compound's ability to be CYP450 substrate, which allows for the selection of compounds with predefined pharmacokinetic behavior. It also allows for the development of an automated computational algorithm for early assessment of possible cytochrome-mediated metabolic transformations that any compound can undergo in the human body.

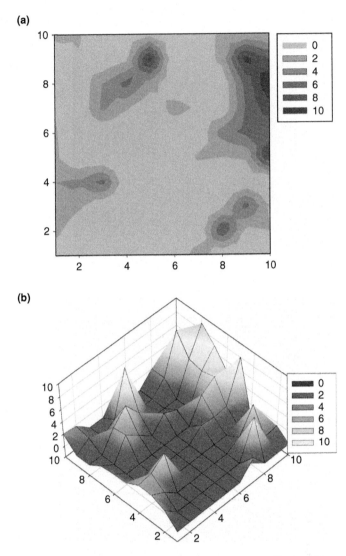

FIGURE 6.6 2D (a) and 3D (b) differential map of preferential distribution of human CYP450 substrates on the Kohonen SOM.

6.10 PREDICTION OF TOXICITY FOR HUMAN FIBROBLASTS

Early assessment of the potential toxicity of organic compounds is critical in today's drug discovery programs. There are numerous limitations, which affect effectiveness of the early toxicity assessment, which can create a significant bottleneck in the drug discovery process. The ability to predict potential toxicity of compounds based on analysis of their physicochemical properties and structural characteristics prior to their synthesis would be economically beneficial when designing new drugs. Although various *in silico* algorithms of toxicity prediction have been reported (reviewed in a series of recent papers [53–56]), in general, the quantitative relationship between the toxicity of structurally diverse compounds and their physicochemical/structural properties has proved to be an elusive goal due, in part, to the complexity of the underlying mechanisms involved.

Here we describe a case study, which demonstrates the ability of neural network classification approach based on Kohonen SOMs to effectively solve the problem. Our goal was to develop a method for automated prediction of human fibroblasts toxicity of compounds based on the extrapolation of knowledge gained from the experimental toxicity data for 10,000 noncongeneric compounds. Successful development of a quantitative model, which would take into consideration relative contributions of various physicochemical features of a molecule into its potential toxicity profile, could permit the estimation of the potential acute toxicity of candidate compounds prior to their synthesis. In addition, such a model would also provide information concerning the modification of structural features necessary to lower the toxicity potential of a drug candidate. Using neural network classification approach, we have developed an effective algorithm for classification of the molecules into toxic and nontoxic applicable in high throughput VS programs.

The analysis used 10,000 compounds, which were representatively selected from 461 combinatorial libraries from the Chemical Diversity collection. Each compound was tested on its cytotoxicity for human fibroblasts. Out of the total 10,000 compounds in the experimental compound set, 2,263 compounds displayed mild to severe cytotoxicity with the cell viability ranging from 0 to 75% at the compounds' concentration of 10 μM. Based on a three-point dose-response curve of the cell viability, the compounds were conditionally divided into 4 classes — Class 1 (highly toxic), Class 2 (moderately toxic), Class 3 (low toxic), and Class 4 (nontoxic) Table 6.5.

For further classification experiments, we used compounds belonging to two principal categories: 868 toxic compounds from the combined set of Class 1 and Class 2 compounds and 4154 nontoxic compounds from Class 4. Among the compounds classified as nontoxic, we selected those that showed no apparent effect on the cell viability (compounds with the cell viability index of 90 to 110%). A total of 5022 compounds from the specified categories were used for Kohonen map generation.

TABLE 6.5
Compounds Toxicity Classification

	Class 1 (High Toxicity)	Class 2 (Moderate Toxicity)	Class 3 (Low Toxicity)	Class 4 (No Toxicity)
Cell viability, %	0–25	26–50	51–75	> 75
LC_{50} (lethal concentration 50%) range, μM	< 3	3–10	> 10	> 50
Number of compounds	414	540	1309	7737

This experiment (Table 6.6) used 10 descriptors that were selected from 60 initially calculated descriptors by principal component analysis. Using Protocol 6.2, we generated a Kohonen map for the entire 5022-compound database. As in the case of CYP450 substrates/nonsubstrates, the studied compound categories, toxic and nontoxic, occupy different sites on the map (data not shown). To quantitatively assess these differences, we generated a differential map (Figure 6.7) similar to that described in the section devoted to classification of CYP450 substrates/nonsubstrates. Each node on this map contains a result of division of the percentage of nontoxic compounds (4154 nontoxic compounds correspond to 100%) on the percentage of toxic compounds (868 toxic compounds correspond to 100%) in this node. The differential map shows the areas where the percentage of nontoxic compounds is 4 to 18 times higher than the percentage of toxic compounds as compared to their random distribution. Using this map, one can select compound subsets in virtual databases for which the probability of occurrence of toxic compounds is significantly reduced as compared to randomly distributed compounds. The described method is computationally inexpensive and permits real-time calculations even for large virtual compound libraries.

6.11   OTHER FILTERS

The data mining approaches based on different classification algorithms can be applied to solving many different tasks, highly actual for contemporary early stage drug discovery. Some of them were mentioned in Table 6.1. Here we will also outline other possibilities related to the use of classification QSAR techniques.

Major reasons preventing many early candidates from reaching market are the inappropriate ADME (absorption, distribution, metabolism, and excretion) properties and drug-induced toxicity. From a commercial perspective, it is desirable thatpoorly behaved compounds are removed early at the discovery phase rather than during the more costly drug development phases [57]. The availability of advanced analytical

TABLE 6.6
Descriptors Used for Toxicity Modeling

Descriptor	Definition
logD	Log of 1-octanol/water partition coefficient at pH 7.4
RotB	Number of rotatable bonds
Hb_acc	Number of H-bond acceptors
Hb_don	Number of H-bond donors
a_aro	Number of aromatic atoms
rgyr	The radius of gyration
ASA_H	Molecular hydrophobic surface area
density	Molecular density
vdw_vol	VdW volume
$logS_W$	Log of water solubility at pH 7.4

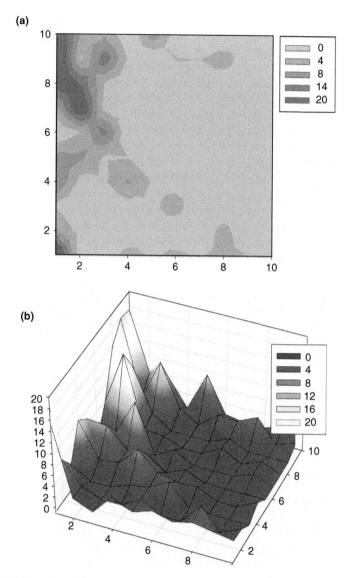

FIGURE 6.7 2D (a) and 3D (b) differential map of preferential distribution of nontoxic compounds on the Kohonen SOM.

and automation technologies, as well as experimental knowledge databases, has dramatically increased our ability to investigate ADME-Tox through increasingly powerful *in silico* methods. Now an understanding of the principles relating chemical structure to ADME-Tox behavior can be based on data from thousands of molecules and applied in VS routines. Use of informatics in ADME-Tox can provide general guidance on properties such as absorption, BBB penetration, plasma clearance, tissue distribution, and metabolic effects. Some advanced approaches in this field are discussed in a series of recent reviews [58–61].

The stability of organic substances under storage and experimental screening conditions is a fundamental issue in early stage drug discovery because it can dramatically impact screening efficiency. The estimation of various factors affecting compound stability has recently attracted serious attention from exploratory chemistry vendors and companies involved in early HTS [62–64]. Recently, we reported a computational classification algorithm, based on published and proprietary data, that is capable of assessing the potential instability of small molecule compounds during their prolonged storage [65]. The algorithm permits high throughput identification of such compounds in large virtual chemical libraries.

The solubility of organic compounds in dimethyl sulfoxide (DMSO) is another important issue in drug discovery, given its widespread use as a storage and screening medium. It is now recognized that DMSO solubility is a problem at least as serious as compound stability in combinatorial libaries, because it may cause artifacts in library screening. On the other hand, preparation of samples in the form of DMSO solutions or thin films, considered as a routine operation, can be complicated in the case of poorly soluble compounds. Chemical Diversity developed an effective *in silico* procedure for estimation of DMSO solubility based on an extensive set of proprietary data (more than 70,000 compounds with experimental data on DMSO solubility) [66,67]. The model represents a robust QSAR algorithm to reveal DMSO-insoluble compounds potentially incompatible with assay protocols prior to screening. Its automated version is now routinely used to filter the large virtual compound collections designed for initial bioscreening programs.

6.12 *DE NOVO* DESIGN OF CHEMICAL LIBRARIES

The described algorithms are effective tools for synthesis planning of *de novo* chemical libraries. Due to a series of specific filters, the properties of a virtual chemical space to be synthesized can be modulated in a wide range of possibilities to optimize them according to the purposes of a particular bioscreening program. Usually, the practical design of target-specific combinatorial libraries also includes elements of other VS approaches, such as selection by structural similarity to known selective ligands (including bioisosteric, topologic, heterocyclic, and substructure similarity), 3D pharmacophore search, flexible docking and so forth. After synthetic feasibility assessment, the combinatorial libraries focused toward particular biotargets are synthesized and used in primary screenings. This general strategy applies for generating the focused libraries toward several protein target classes, such as GPCRs, protein kinases, nuclear receptors, and ion channels. These libraries are available as commercial products at Chemical Diversity Labs, Inc.

All the methods of rational selection of screening candidates mentioned in this chapter belong to a category of product-based approaches. An alternative design principle can be based on reagent-based strategy widely used in the design of combinatorial libraries. Reagent-based approach implies selection of reagent molecules only, which are further used for synthesis of a full combinatorial library of products. Although product-based selection is usually more accurate and provides more information per molecule synthesized [68], reagent-based selection is somewhat faster and the results are easier to implement in the laboratory. The superiority of the product-based design has to be balanced against its increased synthetic cost, which can be unacceptable when the resulting library is intended for repeated screening, using many different primary assays [69]. These considerations suggest that different strategies for manipulation of both product and reagent subsets should be present in the arsenal of a contemporary combinatorial library designer. It can be envisioned that the combination of these strategies will permit the user to dramatically enhance the productivity of the early bioscreening. In our recent work, we discussed a novel knowledge-based method for rational design of target-biased combinatorial libraries [70]. Specifically, we described a multistep procedure of selection of privileged and peripheral structural motifs, which are the key structural elements of target-biased libraries. In the framework of our approach, we demonstrated the practical significance of the target-specific peripheral structural fragments concept and suggested a method for their rational selection based on applying an advanced machine learning algorithm.

6.13 CONCLUSIONS

The knowledge-based strategy to VS seeks to bring together all relevant pieces of information and create a knowledge-oriented process to deploy such information in drug discovery. The computational approaches described in this chapter are based on data collected from literature sources and can be used as inexpensive and efficient filtering tools preceding or limiting the need for more precise and computationally demanding compound selection strategies, such as flexible docking or analysis of 3D conformational space. It can be envisioned that, in a foreseeable future, classification QSAR technologies will have significant applications in various fields of medicinal chemistry. Their increasing popularity can be explained by two main factors: the growing availability of quality experimental data on pharmaceutical agents and their targets and the increasing computational power of today's computers. The further evolution of statistical data mining methods paves the way to fix the development of integrated data mining technology platforms, which will be able to effectively handle the most significant issues related to selection of rational pharmaceutically relevant screening compounds (such as target-specific activity profile, intellectual property [IP] position, favorable ADME-Tox profile, stability and solubility issues, compatibility with assay protocol, and so forth).

Rational VS becomes a multistage process where similarity-based compound clustering techniques, structure-based docking and scoring, and statistical QSAR methods are considered simultaneously in the framework of an integrated platform. The synergy of these powerful technologies will bring great benefit to the industry, with a more efficient production of higher quality clinical candidates.

ACKNOWLEDGMENTS

The authors are grateful to Dr. Yan Ivanenkov and Olesya Tsurgan for invaluable help in preparation of this chapter.

REFERENCES

[1]   V.N. Viswanadhan, C. Balan, C. Hulme, J.C. Cheetham, and Y. Sun, Knowledge-based approaches in the design and selection of compound libraries for drug discovery, *Curr. Opin. Drug Discov. Devel.* 5:400–406, 2002.

[2]   A. Leo, C. Hansch, and C. Church, Comparison of parameters currently used in the study of structure–activity relationships, *J. Med. Chem.* 12:766–771, 1969.

[3]   J. Bajorath, Selected concepts and investigations in compound classification, molecular descriptor analysis, and virtual screening, *J. Chem. Inf. Comput. Sci.* 41:233–245, 2001.

[4]   J. Sadowski and H. Kubinyi, A scoring scheme for discriminating between drugs and nondrugs, *J. Med. Chem.* 41:3325–3329, 1998.

[5]   A. Ajay, W.P. Walters, and M.A. Murcko, Can we learn to distinguish between "drug-like" and "nondrug-like" molecules? *J. Med. Chem.* 41:3314–3324, 1998.

[6]   A. Ajay, G.W. Bemis, and M.A. Murcko, Designing libraries with CNS activity, *J. Med. Chem.* 42:4942–4951, 1999.

[7]   D. Korolev, K.V. Balakin, Y. Nikolsky, E. Kirillov, Y.A. Ivanenkov, N.P. Savchuk, A.A. Ivashchenko, and T. Nikolskaya, Modeling of human cytochrome P450-mediated drug metabolism using unsupervised machine learning approach, *J. Med. Chem.* 46:3631–3643, 2003.

[8]   B.E. Mattioni, G.W. Kauffman, P.C. Jurs, L.L. Custer, S.K. Durham, and G.M. Pearl, Predicting the genotoxicity of secondary and aromatic amines using data subsetting to generate a model ensemble, *J. Chem. Inf. Comput. Sci.* 43:949–963, 2003.

[9]   P. Labute, S. Nilar, and C. Williams, A probabilistic approach to high throughput drug discovery, *Comb. Chem. & High Throughput Scr.* 5:135–145, 2002.

[10]   H. Gao, C. Williams, P. Labute, and J. Bajorath, Binary quantitative structure-activity relationship (QSAR) analysis of estrogen receptor ligands, *J. Chem. Inf. Comput. Sci.* 39:164–168, 1999.

[11]   K.V. Balakin, S.A. Lang, A.V. Skorenko, S.E. Tkachenko, A.A. Ivashchenko, and N.P. Savchuk, Structure-based versus property-based approaches in the design of G-protein-coupled receptor-targeted libraries, *J. Chem. Inf. Comput. Sci.* 43:1553–1562, 2003.

[12]   V.V. Zernov, K.V. Balakin, A.A. Ivashchenko, N.P. Savchuk, and I.V. Pletnev, Drug discovery using support vector machines. The case studies of drug-likeness, agrochemical-likeness and enzyme inhibition predictions, *J. Chem. Inf. Comput. Sci.*, 43:2048–2056, 2003.

[13]   M.T. Cronin, A.O. Aptula, J.C. Dearden, J.C. Duffy, T.I. Netzeva, H. Patel, P.H. Rowe, T.W. Schultz, A.P. Worth, K. Voutzoulidis, and G. Schuurmann, Structure-based classification of antibacterial activity, *J. Chem. Inf. Comput. Sci.* 42:869–878, 2002.

[14]   T. Kohonen, *Self-Organizing Maps,* Heidelberg: Springer-Verlag, 1996.

[15]   S. Anzali, J. Gasteiger, U. Holzgrabe, J. Polanski, J. Sadowski, A. Teckentrup, and M. Wagener, The use of self-organizing neural networks in drug design. In H. Kubinyi, G. Folkers, and Y.C. Martin, Eds., *3D QSAR in Drug Design,* Vol. 2, Dordrecht: Kluwer/ESCOM, pp. 273–299, 1998.

[16]   H. Bauknecht, A. Zell, H. Bayer, P. Levi, M. Wagener, J. Sadowski, and J. Gasteiger, Locating biologically active compounds in medium-sized heterogeneous datasets by topological autocorrelation vectors: dopamine and benzodiazepine agonists, *J. Chem. Inf. Comp. Sci.* 36:1205–1213, 1996.

[17]   S. Anzali, G. Barnickel, M. Krug, J. Sadowski, M. Wagener, J. Gasteiger, and J. Polanski, The comparison of geometric and electronic properties of molecular surfaces by neural networks: application to the analysis of corticosteroid binding globulin activity of steroids, *J. Computer-Aided Mol. Des.* 10:521–534, 1996.

[18] M. Brstle, B. Beck, T. Schindler, W. King, T. Mitchell, and T. Clark, Descriptors, physical properties, and drug-likeness, *J. Med. Chem.* 45:3345–3355, 2002.

[19] A.A. Rabow, R.H. Shoemaker, E.A. Sausville, and D.G. Covell, Mining the National Cancer Institute's tumor-screening database: identification of compounds with similar cellular activities, *J. Med. Chem.* 45:818–840, 2002.

[20] N.P. Savchuk, *In silico* ADME-Tox as part of an optimization strategy, *Curr. Drug Discov.* 4:17–22, 2003.

[21] J.W. Sammon, A non-linear mapping for data structure analysis, *IEEE Trans. Comp.* C-18:401–409, 1969.

[22] D.K. Agrafiotis, A new method for analyzing protein sequence relationships based on Sammon maps, *Protein Sci.* 6:287–293, 1997.

[23] D.K. Agrafiotis and V.S. Lobanov, Nonlinear mapping networks, *J. Chem. Inf. Comput. Sci.* 40:1356–1362, 1997.

[24] D.K. Agrafiotis, J.C. Myslik, and F.R. Salemme, Advances in diversity profiling and combinatorial series design, *Mol. Diversity* 4:1–22, 1999.

[25] D.E. Clark and S.D. Pickett, Computational methods for the prediction of "drug-likeness," *Drug Discov. Today* 5:49–58, 2000.

[26] T.I. Oprea, A.M. Davis, S.J. Teague, and P.D. Leeson, Is there a difference between leads and drugs? A historical perspective, *J. Chem. Inf. Comput. Sci.* 41:1308–1315, 2001.

[27] L.B. Kier and L.H. Hall, *Molecular Connectivity in Structure-Activity Analysis,* New York: Wiley, 1986.

[28] S.C. Basak, A.T. Balaban, G.D. Grunwald, and B.D. Gute, Topological indices: their nature and mutual relatedness, *J. Chem. Inf. Comput. Sci.* 40:891–898, 2000.

[29] D. Bonchev, Overall connectivities/topological complexities: a new powerful tool for QSPR/QSAR, *J. Chem. Inf. Comput. Sci.* 40:934–941, 2000.

[30] R. Todeschini, V. Consonni, R. Mannhold, H. Kubinyi, and H. Timmerman, *Handbook of Molecular Descriptors,* New York: Wiley, 2000.

[31] P.C. Jurs, S.L. Dixon, L.M. Eglof, Representations of molecules. In H. van de Waterbeemd, Ed., *Chemometric methods in molecular design,* Weinheim, Germany: Wiley-VCH, pp. 15–38, 1995.

[32] D.J. Livingstone, The characterization of chemical structures using molecular properties. A survey, *J. Chem. Inf. Comput. Sci.* 40:195–209, 2000.

[33] I.T. Jolliffe, *Principal Component Analysis,* New York: Springer-Verlag, 1986.

[34] W. Cooley and P. Lohnes, *Multivariate Data Analysis*, New York: Wiley, 1971.

[35] W.S. Torgerson, Multi-dimensional scaling: I. Theory and method, *Psychometrika* 17:401–419, 1952.

[36] J.B. Kruskal, Non-metric multi-dimensional scaling: a numerical method, *Psychometrika* 29:115–129, 1964.

[37] D.E. Goldberg, *Genetic Algorithms in Search, Optimization, and Machine Learning,* Boston: Addison Wesley, 1989.

[38] J.P. Bigus, *Data Mining with Neural Networks,* New York: McGraw-Hill, 1996.

[39] H. Kubinyi, Variable selection in QSAR studies. I. An evolutionary algorithm, *Quant. Struct-Act. Relat.* 13:285–294, 1994.

[40] J. Krumrine, F. Raubacher, N. Brooijmans, and I. Kuntz, Principles and methods of docking and ligand design, *Methods Biochem. Anal.* 44:443–476, 2003.

[41] D. Xu, Y. Xu, and E.C. Uberbacher, Computational tools for protein modeling, *Curr. Protein Pept. Sci.* 1:1–21, 2000.

[42] R.D. Taylor, P.J. Jewsbury, and J.W. Essex, A review of protein-small molecule docking methods, *J. Computer-Aided Mol. Des.* 16:151–166, 2002.

[43] P.J. Gane and P.M. Dean, Recent advances in structure-based rational drug design, *Curr. Opin. Drug Discov. Dev.* 10:401–404, 2000.

[44] P. Willett, Chemical similarity searching, *J. Chem. Inf. Comput. Sci.* 38:983–996, 1998.

[45] M.I. Skvortsova, I.I. Baskin, I.V. Stankevich, V.A. Palyulin, and N.S. Zefirov, Molecular similarity. 1. Analytical description of the set of graph similarity measures, *J. Chem. Inf. Comput. Sci.* 38:785–790, 1998.

[46] R.P. Sheridan and S.K. Kearsley, Why do we need so many chemical similarity search methods? *Drug Discov. Today* 7:903–911, 2002.

[47] C. Merlot, D. Domine, C. Cleva, and D.J. Church, Chemical substructures in drug discovery, *Drug Discov. Today* 8:594–602, 2003.

[48] C. Chang and C-J. Lin, LibSVM: a library for support vector machines, 2001. URL: http://www.csie.ntu.edu.tw/~cjlin/libsvm/.

[49] C.M. Masimirembwa, U. Bredberg, and T.B. Andersson, Metabolic stability for drug discovery and development: pharmacokinetic and biochemical challenges, *Clin. Pharmacokinet.* 42:515–528, 2003.

[50] A.M. Palmer, New horizons in drug metabolism, pharmacokinetics and drug discovery, *Drug News Perspect.* 16:57–62, 2003.

[51] J.K. Nicholson, J. Connelly, J.C. Lindon, and E. Holmes, Metabonomics: a platform for studying drug toxicity and gene function, *Nat. Rev. Drug Discov.* 1:153–161, 2002.

[52] J. Langowski and A. Long, Computer systems for the prediction of xenobiotic metabolism, *Adv. Drug Deliv. Rev.* 54:407–415, 2002.

[53] A.G. Wilson, A.C. White, and R.A. Mueller, Role of predictive metabolism and toxicity modeling in drug discovery — a summary of some recent advancements, *Curr. Opin. Drug Discov. Devel.* 6:123–128, 2003.

[54] M.T. Cronin, Computer-aided prediction of drug toxicity and metabolism, *Experientia Supplementa* 93:259–278, 2003.

[55] M.T. Cronin, The current status and future applicability of quantitative structure-activity relationships (QSARs) in predicting toxicity, *Altern. Lab. Anim. Suppl.* 2:81–84, 2002.

[56] J.D. McKinney, A. Richard, C. Waller, M.C. Newman, and F. Gerberick, The practice of structure activity relationships (SAR) in toxicology, *Toxic Sci.* 56:8–17, 2000.

[57] J. Lin, D.C. Sahakian, S.M. de Morais, J.J. Xu, R.J. Polzer, and S.M. Winter, The role of absorption, distribution, metabolism, excretion and toxicity in drug discovery, *Curr. Top. Med. Chem.* 3:1125–1154, 2003.

[58] H. van de Waterbeemd and E. Gifford, ADMET in silico modelling: towards prediction paradise? *Nat. Rev. Drug Discov.* 2:192–204, 2003.

[59] F. Darvas, G. Keseru, A. Papp, G. Dorman, L. Urge, and P. Krajcsi, *In silico* and *ex silico* ADME approaches for drug discovery, *Curr. Top. Med. Chem.* 2:1287–1304, 2002.

[60] S. Ekins, B. Boulanger, P.W. Swaan, and M.A. Hupcey, Towards a new age of virtual ADME/TOX and multidimensional drug discovery, *Mol. Divers.* 5:255–275, 2002.

[61] H. van de Waterbeemd, High-throughput and in silico techniques in drug metabolism and pharmacokinetics, *Curr. Opin. Drug Discov. Devel.* 5:33–43, 2002.

[62] J. Hochlowski, X. Cheng, D. Sauer, and S. Djuric, Studies of the relative stability of TFA adducts vs. non-TFA analogues for combinatorial chemistry library members in DMSO in a repository compound collection, *J. Comb. Chem.* 5:345–350, 2003.

[63] B.A. Kozikowski, T.M. Burt, D.A. Tirey, L.E. Williams, B.R. Kuzmak, D.T. Stanton, K.L. Morand, and S.L. Nelson, The effect of freeze/thaw cycles on the stability of compounds in DMSO, *J. Biomol. Scr.* 8:210–205, 2003.

[64] M. Turmel, R. Spreen, and D. Nie, Preliminary investigation of compound stability under various storage conditions. Presented at the Drug Discovery Technology Congress, Boston, 2002.

[65] N.P. Savchuk, Advanced chemoinformatics tools for determination of optimal storing conditions of drug-like compounds. Presented at the Drug Discovery Technology Congress, Compound Inventory Management Section, Boston, 2003.

[66] K.V. Balakin, DMSO solubility and bioscreening, *Curr. Drug Discov.* 8:27–30, 2003.

[67] K.V. Balakin, Y.A. Ivanenkov, A.V. Skorenko, Y.V. Nikolsky, N.P. Savchuk, and A.A. Ivashchenko, *In silico* estimation of DMSO solubility of organic compounds for bioscreening, *J. Biomol. Scr.,* 9:22–31, 2003.

[68] V.J. Gillet, P. Willett, and J. Bradshaw, The effectiveness of reactant pools for generating structurally diverse combinatorial libraries, *J. Chem. Inf. Comp. Sci.* 37:731–740, 1997.

[69] A.M. Ferguson, D.E. Patterson, C.D. Garr, and T.L. Underiner, Designed chemical libraries for lead discovery, *J. Biomol. Scr.* 1:65–73, 1996.

[70] N.P. Savchuk, S.E. Tkachenko, and K.V. Balakin, Rational design of GPCR-specific combinatorial libraries based on the concept of privileged substructures in T. Oprea, Ed., *Cheminformatics in Drug Discovery*, Weinheim, Germany: Wiley-VCH, pp. 287–313, 2004.

7 Pharmacophore-Based Virtual Screening: A Practical Perspective

John H. van Drie

7.1 INTRODUCTION

"All truths are easy to understand once they are discovered; the point is to discover them."

Galileo

Since its appearance over 15 years ago, the successful applications of pharmacophore-based virtual screening (VS) are appearing in a steady stream. A casual perusal of recent years of *Journal of Medicinal Chemistry* alone indicates that such successes are being reported at the rate of approximately 2 to 3 per year, with publications primarily from the pharmaceutical industry, but increasingly from academic settings.

It is timely to review these successful applications and to distill the lessons learned on how best to employ these methods in drug discovery. This review is intended as a practical guide — a primer on the use of such techniques. In that sense, the aim is more pedagogical than to serve as a comprehensive index into the literature; the aim is to give guidance to both novice and experienced practitioners.

In this spirit of serving these aims, initially some basic terms and concepts will be defined and explained, then the steps of the protocol for pharmacophore-based VS will be detailed, and finally a series of case studies will be considered. Case studies are detailed analyses of published success stories in the literature. The use of case studies is common in the study of law or business, but is less frequently used in a scientific review, where the aim customarily is to cover the literature broadly but shallowly. An in-depth analysis of a select set of publications should better illuminate the key issues surrounding this topic.

7.2 MOTIVATIONS DRIVING THE EARLY EVOLUTION: HISTORICAL DEVELOPMENTS

In the early 1980s, a common complaint about computational approaches to drug design by those outside the field was that these approaches explained after-the-fact why a newly discovered compound had biological activity, but these computational approaches almost never suggested something new. In part, this reflected the origins of computational chemistry in physical chemistry, a discipline that fundamentally tends to be oriented to understanding in greater atomic/molecular detail the discoveries of other branches of chemistry. The leaps of discovery are expected to come from the individual researcher in the wake of this improved understanding.

In response to this complaint, two paths of research emerged, each with the explicit aim to suggest new molecules that may have the desired biological activity. One group of researchers, led by Kuntz and his colleagues at University of California — San Francisco [1] and their work on the DOCK software, took as their starting point the x-ray structure of the drug target, aiming to find novel molecules by identifying those molecules that showed greatest complementarity to the active site of the protein (the subject of many of the chapters of this book).

The other path of research built upon the pioneering work of Marshall [2], taking as their staring point a *pharmacophore:* the spatial orientation of a minimal collection of features on a molecule that tend to confer biological activity. Seminal in this path of research was the collaboration between this author and Martin at Abbott in the development of the ALADDIN software [3], the first of a series of software tools for performing pharmacophore-based VS. Independently and simultaneously, Willett and colleagues at the University of Sheffield in Britain explored this path of research [4], focusing primarily on the algorithmic aspects. Pharmacophore-based approaches are mainly relevant when no x-ray structure of the target is available, as is the case still today for the majority of drug targets.

Either the protein-based approach or the pharmacophore-based approach may be depicted schematically as shown in Figure 7.1. In the case of pharmacophore-based VS, the inputs to the process are a pharmacophore and a database of conformations of molecules. The output is a set of *hits:* a collection of molecules that match the pharmacophore and may possess the desired biological activity. As shown in Figure 7.1, the explicit aim of these methods is to suggest something new. What is new in this case is not the molecules *per se,* it is their association with the specified biological activity. This process has come to be known as *virtual screening;* after the virtual screen is performed, the final step is to submit these hits to biological testing. If any of those hits actually display biological activity, an activity previously unanticipated, that virtual screen may be considered a success. Inevitably, in pharmacophore-based VS, after that activity is found, it looks obvious in retrospect (frequently, the match to the pharmacophore is immediately apparent by visual inspection), but again the motivation is to suggest something new.

Interestingly, the key conceptual breakthrough that converted VS from a research curiosity into one of real utility in the pharmaceutical industry was the realization that the database of conformations did not need to be a database of experimentally determined three-dimensional (3D) structures, but that it sufficed to use computationally determined conformations. With experimental 3D databases, like the Cambridge Crystallographic Database, one is severely limited in the number of molecules and the types of molecules available. With a database of computationally determined conformations — what is now usually meant by the term 3D database — one can access a larger set of molecules, a set of molecules that are more druglike, and, most importantly, a set of molecules which are physically archived and readily accessed

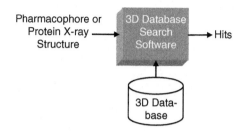

FIGURE 7.1 Overview of the information flow in VS.

for the follow-up biological testing. As most pharmaceutical companies in the late 1980s had compound archives with tens to hundreds of thousands of molecules, it was a straightforward step from this conceptual breakthrough to create 3D databases from those corporate compound collections. Although our initial 3D database at Abbott was based on multiple conformations per compound as determined by molecular mechanics [5], subsequently a rule-based software tool emerged that quickly became the standard for corporate 3D database construction: CONCORD, from Pearlman and his colleagues at the University of Texas [6], a tool for rapidly generating a single high-quality approximation to the low-energy conformation. In 1988, Bures at Abbott was the first to generate a 3D database from a corporate compound collection using CONCORD.

ALADDIN was the prototype of a class of 3D database searching software that emerged in the late 1980s and early 1990s. More on the historical development may be found in a 1995 review [7]. The first commercial 3D database search system developed from the ground up was Catalyst, a product of the BioCAD corporation, a Silicon Valley start-up that this author helped to establish, which existed from 1990 to 1994. That product is still marketed today by Accelrys [8]. In general, pharmacophore-based VS is oriented toward those cases where no x-ray structure of the target is available and where one must infer a pharmacophore indirectly from the structure–activity relationship (SAR). This is a general guideline; there have been exceptions [9].

Independent of the role of computer-aided design tools, one may broadly depict the flow of drug discovery as shown in Figure 7.2. The first step is *target validation*: performing the biological experiments that demonstrate that modulation of protein X in the human body is likely to alter the disease course for disease Y. For example, for the disease hypertension in the early 1970s it was shown that inhibition of angiotensin-converting-enzyme (ACE) would lower blood pressure. The second step is *lead discovery*: discovering a molecule that shows modest biological activity against the target and that shows promise at being amenable to synthetic modification. The third step is the highly iterative process of *lead optimization*: making structural modifications to the lead to improve the biological activity against the target and other important properties. The process is rarely a linear, sequential one, but this depiction in Figure 7.2 illustrates the important steps. VS is generally employed in the lead discovery phase; still in its infancy is its application to lead optimization, where virtual libraries are screened. The term *virtual library* refers to a set of molecules that may be readily synthesized (usually via combinatorial or parallel synthesis).

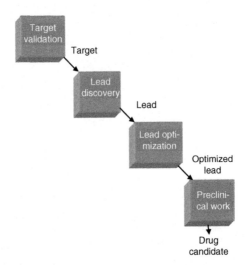

FIGURE 7.2 Highly simplified flow in drug discovery. This process is depicted as linear, though in fact it is highly iterative.

7.3 CONCEPTS AND DEFINITION OF TERMS

It is sometimes daunting for a novice or outsider to read the publications on pharmacophore-based VS, partly because the writer assumes an understanding of some common terminology. To facilitate the future discussion, the definitions will first be clarified.

7.3.1 3D DATABASE

A collection of computationally determined 3D structures of a set of compounds, usually compounds that are readily available for biological testing. There may be multiple conformations per compound (multiconformer database) or there may be only one (single-conformer 3D database). A 3D database may also be constructed from the compounds in a virtual library.

7.3.2 VIRTUAL SCREENING, *IN SILICO* SCREENING, 3D DATABASE SEARCHING

These terms are synonymous, and all refer to the process depicted in Figure 7.1. Pharmacophore-based VS is the computational process of taking a pharmacophore and walking through a 3D database asking of each compound: Can this compound adopt a reasonable conformation consistent with the pharmacophore? Those compounds for which the answer is "Yes" are reported in the output as hits.

7.3.3 HITS, HIT LIST

The output of a virtual screen. The hits are the subset of molecules present in the 3D database for which a conformation matches the pharmacophore. VS is usually used for lead discovery. In this case, the molecules in a hit list are submitted to biological testing; those that show activity may function as the lead for subsequent lead optimization.

7.3.4 PHARMACOPHORE

This may be defined in multiple ways. Practically, one may define it as the spatial arrangement of molecular features essential for biological activity. Mathematically, one may define it as the precise objective description of a class of molecules, a description that correlates with biological activity. Loosely speaking, a pharmacophore represents the essential components of molecular recognition, or it is a "summary of what it takes for a molecule to be active." Note that the first definition explicitly makes reference to 3D properties; this is in contrast to how that term is frequently used by medicinal chemists, who use that term synonymously with *scaffold* (defined below). *Everywhere in this document the term pharmacophore will refer to an entity with 3D properties.* An example of a simple pharmacophore common to many central nervous system (CNS) agents is shown in Figure 7.3: an aromatic ring separated from a basic amine by 5 to 7 Å. Such a simple pharmacophore is highly unselective (it would retrieve about a quarter of the corporate 3D database of the former Upjohn company, a company with a long history in CNS research); nonetheless, using this as a 3D search query against any company's corporate 3D database and sorting by molecular weight (MW) to bring the smallest compounds to the top of the hit list inevitably brings potentially interesting biologically active compounds to one's attention.

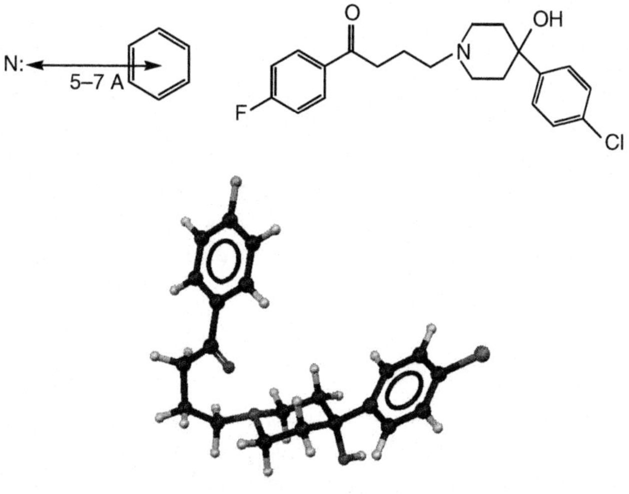

FIGURE 7.3 Fundamental CNS pharmacophore and haloperidol, a D2 antagonist. In a folded conformation this pharmacophore may map to haloperidol in two different ways.

7.3.5 STRUCTURE–ACTIVITY RELATIONSHIP

A SAR is a collection of molecules with associated biological activity against some target. A pharmacophore is usually constructed to be consistent with the SAR and may be thought of as a simple, primitive, intuitively visual model of the SAR.

7.3.6 SEARCH QUERY

In the context of pharmacophore-based VS, this term is synonymous with pharmacophore. The term more generally means the set of criteria one applies to search a database.

7.3.7 FEATURE, MAPPING, MULTIPLE MAPPINGS

A feature is a component of a pharmacophore that specifies a collection of a certain type of atoms. For example, in the simple CNS pharmacophore in Figure 7.3, there are two features:

1. Aromatic ring — In this case, this feature specifies an all-carbon 6-membered aromatic ring; it would not match a pyridine, pyrimidine, etc.
2. Basic amine — In this case, it specifies a nitrogen with a free lone pair; it would not match an aniline nitrogen (a nitrogen attached to an aromatic ring), nor an amide nitrogen.

One wants to be able to specify features with both generality and precision, which necessitates some subtleties in the software used to perform the virtual screen. This subtlety is one of the aspects that differentiates various types of software for pharmacophore-based VS.

As shown in Figure 7.3, note that for a molecule like haloperidol, this simple pharmacophore may match in two ways (i.e., there are two mappings). In a carefully chosen conformation, this pharmacophore can map the aromatic ring feature to the fluorophenyl ring, with the basic amine mapped to the piperazine nitrogen. In that same conformation, the pharmacophore can map to the chlorophenyl ring and the piperazine nitrogen. It is easy to overlook the possibilities for multiple mappings when staring at a graphics screen; therefore, one needs software to systematically evaluate all possible mappings of a molecule to a pharmacophore to ensure that the process is objective (i.e., free of user bias).

7.3.8 DYADS, TRIADS, TETRADS

A dyad is a pharmacophore with only two features. A triad contains only three features. A tetrad contains four. A minimalist pharmacophore is a dyad with one distance constraint, such as the CNS pharmacophore shown in Figure 7.3.

7.3.9 Overlay

In the older literature, it was common to use interactive molecular graphics to superimpose conformations of different molecules in a way that was suggestive of how these molecules might be binding to the receptor in a common way. From such an overlay, people would identify the pharmacophore by inspection of that image. Today, with automated methods for discovering a pharmacophore, the opposite happens: a pharmacophore is used to overlay a set of molecules to suggest a common mode for how they may bind to the receptor. A synonym for overlay is *alignment*.

7.3.10 Scaffold

This term is used in various ways, but here the term will be used to denote the chemical substructure common to a class of molecules. No reference is made to 3D properties. Some examples of well-recognized scaffolds are shown in Figure 7.4: benzodiazepines (ion channel modulators), quinolines and phenyl-oxazolidinones (both antibiotic scaffolds), and the tricyclic scaffold (antidepressants).

7.3.11 Selectivity

The selectivity of a pharmacophore may be defined as the proportion of molecules in the 3D database that are retrieved as hits: it is the ratio of the size of the hit list to the size of the database. This definition depends on the database one chooses. Typical values of selectivity for a database of druglike molecules for good pharmacophores range from 0.1 to 20%; the selectivity may be viewed as a measure of the quality of a pharmacophore.

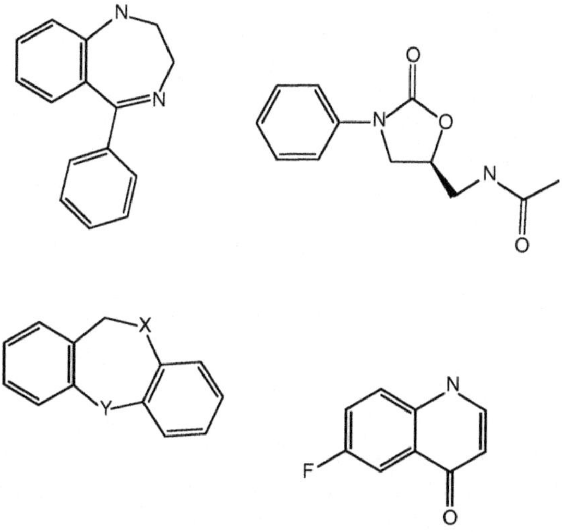

FIGURE 7.4 Common drug scaffolds (clockwise from top left): benzodiazepines (ion channel modulators), phenyl-oxazolidinones and quinolones (both antibiotic scaffolds), the tricyclic scaffold (numerous CNS indications).

7.3.12 SUCCESS RATE

This is the proportion of the hit list that shows interesting biological activity. Obviously this definition depends on the pharmacophore, the database, the threshold for what consititutes interesting, etc. Typical values for a success rate for a good pharmacophore range from 0.5 to 20%. This compares with rates of active hits from random screens typically ranging from 0.1 to 0.5%. Typical values for the threshold of what constitutes interesting are 10 to 100 μM. Typical actives in a hit list are single-digit μM. It is rare to have hits with subμM activity.

7.3.13 ENRICHMENT RATIO

This is the ratio of the VS success rate to the random screening success rate. In other words, if one were to perform biological testing on all the molecules in the database, the enrichment ratio is:

$$\frac{\text{Proportion of actives in hit list}}{\text{Proportion of actives in database}}$$

Basically, the enrichment ratio tells you how much work one saves by performing a virtual screen followed by testing those compounds in the hit list, compared with random screening of the entire collection. The enrichment ratio is another measure of the quality of a pharmacophore (though, of course, such a measure is of limited utility, because one must perform some random screening before one can assess it). Typical values of an enrichment ratio for good pharmacophores range from 3 to 100.

7.3.14 GEOMETRIC OBJECT, GEOMETRIC CONSTRAINT

For the simple pharmacophore shown in Figure 7.3, the geometric constraint is the distance range 5 to 7 Å. The geometric objects are the center of the aromatic ring and the position of the basic nitrogen: these are the points on the conformation from which the distance is measured. Geometric objects are necessary constructions, because it does not make sense physically to always measure things from atomic centers, as in the case of the aromatic ring in Figure 7.3.

Geometric objects are not restricted to points. It frequently makes sense to refer to the orientation of a feature, not merely its position. For example, one might want to elaborate the simple CNS pharmacophore shown in Figure 7.3 to that shown in Figure 7.5. Here, one constructs vectors based on the coordinates of the phenyl ring and on the coordinates of the nitrogen atom and its neighbors to indicate their orientations. One can then impose an additional geometric constraint on these vectors, as indicated, to only match molecules that can adopt a conformation where the orientation of those groups is such that the torsion angle between these two vectors is in the range +33 to +101°. Additional geometric constraints can quickly convert unselective pharmacophores, plagued in many cases by multiple mappings per molecule, into highly selective ones, which almost always map to a molecule in a unique way. Note that, due to the signed torsion angle, the dyad pharmacophore in Figure

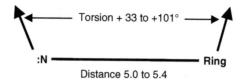

FIGURE 7.5 A more elaborate CNS pharmacophore.

7.5 is enantioselective: it may match a particular chiral molecule, but will not necessarily match that molecule's enantiomer.

7.3.15 PROJECTED POINT, RECEPTOR POINT, DUMMY ATOMS, OUTRIGGERS

The older modeling literature often employs an additional geometric object, where a projected point is envisioned along the vector, the putative site on the receptor where the ligand binds. Figure 7.6 is adapted from reference [10], which shows a projected point along the lone pair vector, 2.8 Å from the atomic center of the basic nitrogen in a CNS pharmacophore. Synonyms for projected point are dummy atoms [11] and outriggers [12]. The observation is that, with modeling based on interactive graphics, these projected points yield improved overlays when the projected points are used to superimpose multiple molecules, rather than superimposing the positions of the nitrogen atoms themselves.

7.3.16 STERIC CONSTRAINT, FORBIDDEN REGION, SHAPE-ENHANCED PHARMACOPHORES

Every medicinal chemist and drug designer is aware of *forbidden regions:* elaborations of a scaffold in a particular direction that inevitably lead to sharp drops in activity. The customary inference is that these represent steric clashes with the putative boundaries of the receptor. Figure 7.7 shows a forbidden region well-known in the design of dopamine D1 agonists [13]; attempts to add substituents to a scaffold in this region tend to lead to inactive molecules.

Forbidden regions can be included in the definition of a pharmacophore. Although this idea dates back to some of the earliest work of Marshall [2], it tends to get neglected, in part because of the limitations in the software used to build 3D

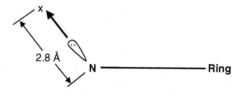

FIGURE 7.6 Projected point along the lone pair of the nitrogen.

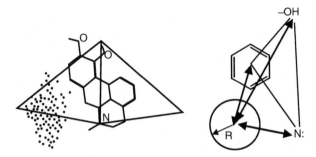

FIGURE 7.7 The presence of a forbidden region in the dopamine SAR (left). Representation of that forbidden region by a sphere radius R centered on a point defined by 3 distances to the three features.

pharmacophores. ALADDIN and similar systems provide the ability to define steric constraints, as a means to define forbidden regions in a rigorous way. Figure 7.7 shows how a steric constraint may be added to a 3-feature D1 pharmacophore: the steric boundary is approximated by a sphere, whose size is specified and where the center is defined by three distance constraints to the geometric objects associated with the three features. Even this prose description sounds complicated.

More recent developments in software methodology [14] allow more facile, easily constructed definitions of steric boundaries, leading to the notion of a *shape-enhanced pharmacophore,* a pharmacophore that additionally contains these steric constraints as inferred from the SAR. Such pharmacophores tend to be exquisitely selective, yielding unusually high success rates and enrichment ratios. An example of such a pharmacophore for agonism to the serotonin 5-HT2a subtype is shown in Figure 7.8.

7.3.17 RIGID MATCH, RIGID SEARCH VS. FLEXIBLE MATCH, FLEXIBLE SEARCHING

The process of determining if a particular conformation of one molecule matches the constraints of the pharmacophore is termed *matching.* If this matching occurs without any modification to the conformation, it is termed a *rigid match.* If adjustments are made to the conformation to test whether the molecule may adopt a conformation that meets the constraints of the pharmacophore, this is termed a *flexible match.* The process of screening a 3D database with a rigid match algorithm is called a *rigid search*; with a flexible match algorithm, it is called a *flexible search.* ALADDIN, MDL's MACCS-3D and Version 1.0 of Catalyst all used a rigid match algorithm, similar to what is described in [15]. Later versions of Catalyst use an unpublished flexible match algorithm, MDL's ISIS (Integrated Scientific Information System) uses a flexible match algorithm [16], Tripos' UNITY uses another flexible match algorithm [17], Chemical Design's Chem-3DBS used an unpublished flexible match algorithm. The tradeoffs between performing a rigid search on a multiconformer database vs. a flexible search on a single conformer database will be discussed later.

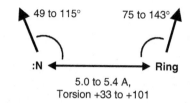

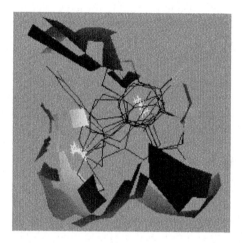

FIGURE 7.8 A shape-enhanced pharmacophore for agonists of the 5-HT2c receptor.

7.3.18 PHARMACOPHORE DISCOVERY

The process of calculating a pharmacophore from a data set is variously termed pharmacophore identification, pharmacophore mapping, or pharmacophore discovery. This process includes both the determination of which features should be used in the pharmacophore, as well as determining the geometric constraints that should be imposed on those features.

7.3.19 2D, 1D

These terms have nonintuitive meanings, in association with the term 3D, which refers to the three dimensions of space. *2D* usually refers to the atoms and bonds of a molecule and how they are connected; this is what a synthesis chemist draws on a piece of paper to represent a structure. This includes *chirality,* that is, the wedges and hashes of a chemical structure drawing. The term *configuration* is a more formal term synonymous with 2D structure. *1D* gets used in a variety of semiserious ways, though here this term will be used to denote calculated attributes (e.g., MW, calculated logP (octanol-water partition coefficient), polar surface area, etc.) and the biological data associated with a compound. Hence, after assaying, a hit list may display 3D, 2D, and 1D information; a SAR may consist of 2D and 1D data.

7.3.20 SIMILARITY, SIMILARITY SEARCHING

It is intuitively clear to a chemist when two molecules are similar (though what one chemist may label similar, another may not). Similarity is something that can be objectively defined, usually based on the 2D attributes of a molecule [18], though in some cases it may be defined based on 3D attributes [19]. It is a powerful concept, and is, in many ways, a complementary one to pharmacophores. *Similarity searching* is the process of VS where one looks simply for molecules in the database that are similar (where similarity is defined as above) to a molecule specified as a query.

7.3.21 CLUSTER, CLUSTERING

A *cluster* is a set of molecules that are all similar to one another. *Clustering* is the process of converting a sequential hit list into clusters. Clustering is an especially useful way to process large hit lists, as one may wish to limit one's attention only to representatives from each cluster.

7.3.22 SMILES, MOL FILES, SD FILES

SMILES is a handy notation for a molecule in a line of text [20]. For example, aspirin may be written in SMILES as OC(=O)c1ccccc1OC(=O)C. Frequently, chemical databases are stored as text files containing thousands of SMILES. An alternative way to store chemical structures in text files is to use the MOL format. In the MOL format, there is one line for every atom in the structure, and one line for every bond [21]. A collection of MOL formatted molecules is called a SD file. Some regard a file of SMILES or a SD file as synonymous with a 3D database, but as will be shown shortly, it is not that simple.

7.4 STEP 1 — CONSTRUCTING 3D DATABASES

A number of discrete issues must be faced, before one can perform a virtual screen against a 3D database as shown in Figure 7.1. Some of these are:

- Where do these lists of molecules come from?
- What types of data-cleaning must one perform on the given structures?
- How does one handle chirality, in particular chiral centers with unspecified chirality?
- How does one generate the 3D information? How does one handle conformational flexibility?

Each of these issues will now be considered in turn.

7.4.1 WHERE DO THESE LISTS OF MOLECULES COME FROM?

The key criterion in choosing molecules for input to 3D database construction is that the compounds should be readily available for testing (no synthesis required, preferably) and, if biologically active, should serve as a suitable lead for lead optimization. In other words, it should be druglike and free of obvious liabilities that cannot readily be fixed, such as polyaromatic systems, which tend to be toxic via deoxyribonucleic acid (DNA) intercalation.

When applied to lead optimization, the lists of molecules are taken from the virtual library. The virtual library is usually constructed based on the needs for the parallel synthesis chemistry and the availability of reagents consistent with that synthetic scheme.

When applied to lead discovery, multiple sources are available for finding lists of thousands of molecules satisfying the above criteria:

- In the pharmaceutical company, the collection of compounds in the company archive. This typically totals 100,000 to 1,000,000 molecules.
- Compound vendors usually make their catalog available as a SD file; researchers frequently compile compound lists from multiple vendors, usually resulting in lists of 1.2 to 2.0 million compounds. Molecular Design Ltd. (MDL) provides their own compilation of vendor lists, the Available Chemicals Directory (ACD), which today exceeds 1,000,000 compounds.
- The National Cancer Institute compiles a list of molecules it has tested for carcinogenicity, molecules that have been submitted from multiple sources [22]. This list today totals roughly 200,000 compounds.
- Other databases of biologically interesting compounds are available, such as the Comprehensive Medicinal Chemistry (CMC) database [23] or the MDL Drug Data Report [24], containing 7,000 and 100,000 compounds, respectively.

When VS first appeared, there was the misguided notion that it sufficed to have databases containing only hundreds of molecules. It was naively assumed that the VS process would find a lead from such a collection. Looking back, it is clear that a prerequisite for success is to have large databases, at least of the order of hundreds of thousands of molecules.

7.4.2 What Types of Data-Cleaning Must One Perform on the Given Structures?

One of the painful, little-discussed steps in 3D database construction is data-cleaning. This catch-all phrase covers a variety of problems:

- Stripping salts — a MOL file may contain multiple molecular entities (e.g., the drug molecule and citrate or a counter-ion, etc.); the entry in the 3D database should contain only the organic component that would give rise to the biological activity.
- Adding hydrogens appropriately, possibly taking into account what the proper protonation state may be at physiological pH. A SMILES usually does not specify hydrogens, and a MOL file may or may not contain them.

- Eliminating compounds that do not belong in the final database, such as compounds with > 20 rotatable bonds, pentapeptides and higher, inorganic complexes, compounds likely to produce false positives in biological assays (e.g., aggregators, molecules with fluorophores, reactive molecules), etc.

Depending on the software used to construct the 3D database, some of these issues may be already handled. Though the details change from version to version, Catalyst's catDB utility automatically handles salt-stripping and hydrogen addition. CONCORD [6] and CORINA [25] both add hydrogens as needed, without reference to pH-dependent protonation states. In addition to CONCORD, Pearlman's suite of software for building 3D databases includes software ProtoPlex for assigning protonation states; however, this is more an issue for protein-structure-based VS, because one can adjust the definition of pharmacophore features to match, for example, carboxyl or carboxylate (as is done in the Catalyst feature definition NEGATIVE_IONIZABLE). REOS [26] provides a systematic way to apply rules for eliminating compounds that must be stripped from the database, based on their potential for reactivity and other measures of nondruglikeness.

In general, it is frequently necessary to build custom scripts to handle many of these data-cleaning tasks. Each database usually presents its own set of idiosyncrasies: the cleaner the input, the less data-cleaning required. The process of data-cleaning is usually iterative, with each cycle cleaning up progressively more problems; it is not infrequent that the final iteration consists of visual inspection and making the final fixes by hand. In some cases, the vendors supply preconstructed 3D databases from those compound databases listed above, which in principle should obviate the painful processes described above.

7.4.3 How Does One Handle Chirality, in Particular Chiral Centers with Unspecified Chirality?

Frequently, the compound databases from which 3D databases are constructed do not specify stereochemistry, either cis/trans double-bond stereochemistry or point-chirality (R/S). In some cases, this accurately represents the compound, which may be a racemate. In many cases, this sloppily misdescribes the actual structure, where the stereochemistry is known, such as representing L-alanine in SMILES as NC(C)C(=O)O. For many years, one of the dirty, little secrets in 3D database searching was that people would frequently ignore this issue and when faced with stereochemical ambiguity would randomly choose one stereoisomer to present to the conformational analysis.

In the development of Catalyst at BioCAD in the early 1990s, we spent a lot of effort ensuring that chirality was handled in a thorough, rigorous, and sensible way. In the wake of that, most tools for constructing 3D databases now handle things comparably. In Catalyst, for each stereocenter or stereobond that is ambiguously defined, both 2D structures are created and each is passed along to conformational analysis. Of course, this means that for a molecule with 8 undefined centers, one will populate the database with 256 2D structures. (There are user-defined thresholds to cap this stereochemical explosion.) CORINA also functions in this way. Both

CORINA and Catalyst have additional rules to prune possibilities at chiral ring systems.

CONCORD does not handle this stereochemical ambiguity in this way and requires preprocessing to handle it appropriately. Originally, users needed to perform that preprocessing themselves with custom scripts; now, that suite of 3D database-building tools from Pearlman's lab has been extended to include a software tool, StereoPlex, to perform this preprocessing.

Chirality is a deep and rich domain of chemistry, and the paragraphs above gloss over a number of subtleties. In Catalyst, we included the capability to recognize point-chirality at centers other than carbon (e.g., sulfoxides); in practice, however, users typically employ the default setting that turns that off. Issues of relative vs. absolute stereochemistry, pseudochirality, centers which defy the Cahn-Ingold-Prelog R/S nomenclature, etc., all complicate matters further.

7.4.4 HOW DOES ONE GENERATE THE 3D INFORMATION? HOW DOES ONE HANDLE CONFORMATIONAL FLEXIBILITY?

The final step in 3D database construction is the conversion of the chirally unambiguous 2D structures into one or more 3D conformations. A variety of conformational analysis tools are available for this task. Described here are ones with which the author has firsthand experience; the list is certainly not exhaustive.

In the early days of pharmacophoric 3D database searching, typically one created a 3D database with a single conformation per compound using CONCORD. Many considered this outrageous, as this ignores the conformational flexibility of the molecule, but the successes reported indicated that this assumption was not altogether unjustified. In the ongoing evolution, two divergent paths emerged: populating databases with multiple conformers per compound [3,7,27] or the use of flexible matching algorithms. Catalyst was the first system to employ both methods simultaneously.

The different options available for constructing 3D databases will now be considered. Timings are not cited, as this experience has stretched over many years, and the timings would not be comparable; with today's fast machines, this is not a crucial issue anymore. The reader is also referred to an older review of this topic [28].

- Molecular-mechanics (e.g., MacroModel) — As mentioned in Section 7.1, the first 3D databases were simply archives of molecular-mechanics runs. However, once it became apparent that one needed hundreds of thousands of compounds in the database, this was deemed impractical. With today's machines, it could be done, for example, using MacroModel's Monte Carlo Multiple Minimum procedure [29] to generate hundreds of minimized, conformationally diverse conformations, but this would be silly. This is overkill and would represent a lack of appreciation for the sources of error that are inherent in the overall process of pharmacophore-based VS.
- CONCORD — This generates a single high-quality approximate conformation. Conformationally flexible portions, such as aliphatic chains, are

created in an extended conformation. Ring conformations are surprisingly good, the results of minimizing a novel strain function [30]. The primary limitation of CONCORD (besides the chirality issues listed in Section 7.4.3) is that it generates only one conformation.

- Catalyst/Fast — This method is unpublished and has undergone constant evolution. It produces multiple conformations per compound. It was originally based on distance-geometry [31], then was extended with exhaustive torsion-driving methods, augmented by a novel algorithm for ensuring conformational diversity [32]. It is significantly faster than the molecular-mechanics-based Catalyst/Best algorithm, which would be overkill for populating a 3D database.

- CORINA — Originally, this generated a single high-quality approximate conformation [25]. It has recently been extended to include some sampling of the ring conformations, though extended acyclic chains are still unsampled. A separate software module, ROTATE, works in conjunction with CORINA, generating conformations of those acyclic sections, based on statistical probabilities of likely torsion angles taken from the Cambridge Crystallographic Database.

- OMEGA® — This generates many conformations, given an input conformation, by sampling torsion angles according to a user-customizable library of angles [33]. It is extremely rapid, but it has no knowledge of energy or statistical likelihood of preferred angles.

- Confort — This generates multiple conformations, according to a highly sophisticated, unpublished algorithm that differentially samples regions of torsion conformational space based on a set of clever rules [34]. It does not have an explicit force field, but nonetheless generates conformers that are energetically sensible. There is a conformational cost for this improved quality: it is significantly slower than OMEGA or Catalyst/Fast.

As a final addendum to this section on treating conformational flexibility in 3D database searching, it should be noted that a mathematical relationship was discovered that allows one to quantify the effects of inadequately treating conformational flexibility [35]. This inequality, which becomes an equality in the limit of perfect treatment of conformational flexibility, allows one to define a quantity δ, which measures the degree to which conformational flexibility is being properly assessed in a 3D database search. For a dyad pharmacophore whose distance constraint is from X to Y Angstroms, |(X,Y)| will be used to denote the number of hits that dyad returns in a database search. With this notation, the inequality may be written as follows:

$$|(a,b)| + |(0,b)| + |(0,b)| - |(0,\infty)|$$

In words, this inequality states that the number of hits returned by the dyad constrained from a to b is less than or equal to the number of hits returned by the similar dyad with distance constraints a to ∞ plus the number of hits returned by the similar dyad with distance constraints 0 to b minus the number of hits returned by the similar dyad with no distance constraints (all dyads use the same two features). One may define the dimensionless quantity δ as a measure of how well this inequality becomes an equality:

$$\delta = \frac{|(a, \infty)| + |(0,b)| - |(0, \infty)| - |(a,b)|}{|(a,b)|}$$

As shown in Figure 7.9, adapted from [35], when δ is plotted for different types of queries, as the number of conformations per compound increases, δ decreases monotonically. Each curve corresponds to a different precision for the pharmacophore: the more precise the pharmacophore (the smaller the difference between a and b), the larger the value of δ. Unpublished observations indicate that δ is also smaller when comparing flexible search algorithms to rigid matching algorithms; δ is smallest when using both multiconformer databases and flexible search algorithms. Thus, the δ inequality allows one to objectively compare different algorithms for conformational analysis and 3D database searching. This is an area that deserves further investigation.

However, the most significant observation resulting from the δ inequality is that there is an interplay between the precision of the pharmacophore and the demands for treating conformational analysis well: more precise pharmacophores demand better treatment; sloppy pharmacophores allow simplistic approaches. Stop reading for a moment and mull this over — this is an observation rich in implication. For example, as most pharmacophores in the literature around 1990 were sloppy, this explains the success of single-conformation 3D database searching in that time frame.

7.5 STEP 2 — PHARMACOPHORE DISCOVERY

The general flow of pharmacophore-based VS depicts two inputs to the process — the 3D database and the pharmacophore. Section 7.4 detailed how one constructs the 3D database; this section will review the processes by which one obtains the pharmacophore.

It is amazing how little attention this issue receives in typical reviews and many publications on pharmacophore-based VS. A casual reader could easily be left with the impression that one simply consults the literature, takes a pharmacophore off the shelf, and uses this to perform the 3D search. This may have been done in the early days, but if there is anything to discover using these literature pharmacophores, it has likely already been discovered. Custom pharmacophores inferred from one's own proprietary data is the path to novel discovery today.

The overall process for pharmacophore discovery is shown in Figure 7.10. A variety of methodologies have been published for performing pharmacophore discovery; these have been recently reviewed [36,37]. In this chapter, some of the crucial issues in the practical application of these methods will be highlighted.

The steps taken to perform pharmacophore discovery are:

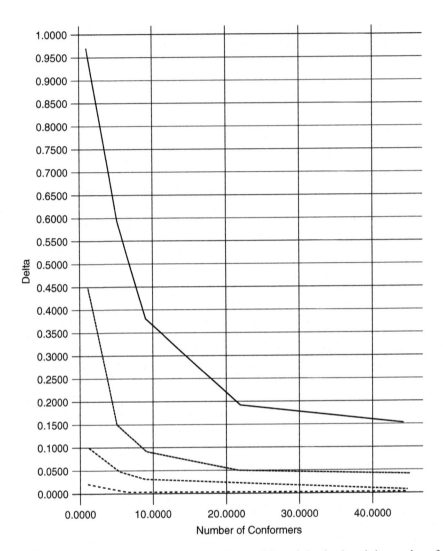

FIGURE 7.9 The relationship between delta, the precision of the dyad, and the number of conformers/molecule in the 3D database. The top curve is for a dyad with precision ± 0.25 Å; the next, ± 0.5 Å, the next ± 1.0 Å, the lowest curve ± 2.0 Å. (Adapted from W.P. Walters et al., American Chemical Society meeting, Spring 1999, unpublished manuscript.)

1. One begins by compiling a dataset, which defines the SAR.
2. Next, one performs exhaustive conformational analysis on all of those molecules.
3. The final computational step is to identify patterns common in the SAR, of which there are usually many, which may be called *candidate pharmacophores*.

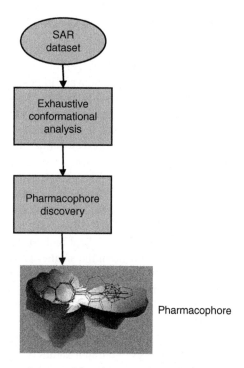

Pharmacophore

FIGURE 7.10 The general protocol for pharmacophore discovery.

4. The ultimate step is usually the user-intensive process of inspecting each of these candidate pharmacophores and sifting through them by a variety of means.

There is a common misperception that this process of pharmacophore discovery should be rapid, of the order of a few seconds. It is not uncommon to witness people taking one low-energy conformation of one molecule, positioning some features on that conformation, measuring distances, throwing in some arbitrary tolerances on those distances, and calling that a pharmacophore. Such a minimalist approach should have accompanying minimalist expectations for the result of its use in a virtual screen. Performing high-quality pharmacophore discovery takes considerable time, but it is time well spent.

7.5.1 How Should the Dataset Be Selected?

There is a misperception that this dataset should be a small, representative subset of the SAR, of the order of 20 compounds. It is impossible to identify a representative subset in an objective way; the insistence on using such a small number of compounds usually stems from limitations imposed by the methodology. It is clear from practical experiences that one should use all the molecules in a SAR to infer a pharmacophore; this may range from a few molecules as in the beginning of a project to thousands of molecules late in the lead-optimization stage.

7.5.2 How Should the Conformational Analysis Be Performed?

Although 3D database construction imposes relatively weak demands on conformational analysis, by contrast pharmacophore discovery imposes high demands. The methods for conformational analysis described in Section 7.4.4 may all be used in pharmacophore discovery. As reviewed in greater detail in [36,37], from practical experience is it apparent that one needs the highest-quality force fields and the highest-quality exhaustive conformational analysis that one can find. The best methods usually are also the most time-consuming, Confort and MacroModel/MCMM/MMFF.

It does not suffice to merely use the low-energy conformations of each molecule. One open issue is how many kcal/mol above the minimum energy must one include in the exhaustive conformational analysis, called τ. This issue has also been discussed in [36]; proposed values for τ range from 3 kcal/mol to 20 kcal/mol. Practical experience suggests the best values for τ lie in the smaller values of that range, and that the upper end is incredible (though this may be a comment more on the weaknesses of the force field used to measure τ). Of course, the optimum value of τ may be problem-dependent; a method for dialing-in the optimum value of τ has been mooted [36]. Overlooked in the utilization of large values of τ is that the conformation that best fits the pharmacophore is therefore often one that has an implausibly high degree of strain energy. Optimum values of τ and the role of strain energy in assessing fits to pharmacophores are issues deserving more study.

7.5.3 How Can One Detect Candidate Pharmacophores?

Consciously or not, virtually all methods for pharmacophore discovery have rediscovered a method for detecting a candidate pharmacophore first proposed by Marshall and coworkers [11]. In effect, their method almost defines what it takes to be a candidate pharmacophore; it will henceforth be denoted the MNMM method.

Their method is indicated schematically in Figure 7.11. After exhaustive conformational analysis on all active molecules, and after selecting three features — F1, F2, and F3 — one measures the distances between all mappings of all features on all conformations of all actives. Each active molecule will span a region of space for the distance between F1 and F2, the abscissa d(F1,F2), and for the distance between F1 and F3, the ordinate d(F1,F3), and the distance between F2 and F3, not shown. One looks for the region of space that is the intersection region among all active molecules; the distance constraints for F1-F2 and F1-F3 can be read off of this intersection region. That is the tightest set of constraints on those features consistent with at least one conformation of each active molecule.

The MNMM method can also be applied to angular constraints. The example in that paper used 30 nM ACE inhibitors, the resulting candidate pharmacophore is shown in Figure 7.12. Exhaustive conformational analysis can vary the distance from the carboxyl to the Zn-binding group from 6 to 10 Å; the optimum distance is that indicated, 8.5 to 8.7 Å, and encompasses at least one energetically reasonable conformation of each ACE inhibitor in their dataset.

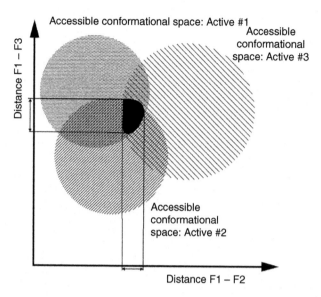

FIGURE 7.11 Those geometries in common to all the active molecules is indicated in black; the distance constraints may be directly read off from that. (Used with permission from Mayer et al., *J. Computer-Aided Mol. Des.* 1:3–16, 1987.)

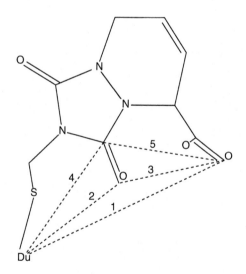

FIGURE 7.12 The ACE pharmacophore of Mayer et al. The optimal values of the distance constraints in order are (all values in Å): 8.5 to 8.7, 4.9 to 5.1, 3.8 to 3.9, 5.2 to 5.4, 4.0 to 4.1. (Used with permission from Mayer et al., *J. Computer-Aided Mol. Des.* 1:3–16, 1987.)

In their ACE example, the MNMM method yields one unique candidate pharmacophore, so they may easily judge that to be the actual pharmacophore. In general, this is not observed: the MNMM method yields multiple candidate pharmacophores consistent with the dataset.

7.5.4 WHAT COMBINATIONS OF FEATURES AND CONSTRAINTS SHOULD BE USED?

With Catalyst's Hypothesis Generation, we were the first to employ a method that was intended as a simple expedient: trying all possibilities from a standard library of features, H-bond-acceptors, -donors, negative-charge, positive-charge, aromatic ring, general hydrophobic. Surprisingly, this approach still remains the state-of-the-art [38].

However, Catalyst's Hypothesis Generation always attempts to find a pharmacophore with the same number of features, four. This approach is suboptimal. The most effective way to handle this is to begin with dyads and to steadily elaborate as needed [36,38]. Many find it astonishing to see that one can sometimes have a high-quality pharmacophore that is only a dyad (one distance constraint, possibly additional angular constraints).

The capability to compute projected points was provided in ALADDIN and Catalyst for composing such pharmacophores in 3D database searching, but practical experience suggests that these tend not to be useful constructs in the context of 3D database searching. It appears that a lever-arm effect tends to amplify noise in the conformational analysis. The use of angular constraints appears to accomplish the same aim, but more robustly. However, this may be an issue deserving more careful analysis.

7.5.5 HOW SHOULD ONE SIFT THROUGH THE CANDIDATE PHARMACOPHORES?

Virtually all datasets are consistent with multiple candidate pharmacophores. Little has been discussed in the literature as to how one sifts through these multiple possibilities systematically.

Nonsystematic approaches predominate. For example, it is sometimes recommended to carefully select a subset of the SAR to teach the computer the salient aspects of the data. Also, sometimes it is recommended to use a reference conformer for one molecule, usually the lowest-energy conformer of one of the more potent molecules. Sometimes it is recommended to visually inspect each candidate pharmacophore and identify the best one.

All of these approaches are prone to the vagaries of subjectivity. Ideally, one would like an objective procedure, which would ensure that different users with the same data and the same methods would arrive at the same conclusion. This author has proposed the *principle of selectivity* as such an objective criterion for ranking candidate pharmacophores [38]. Here, a 3D database search is performed using each candidate pharmacophore, and the proportion of the database that is returned as hits is tabulated, denoted q. The selectivity index is calculated for each pharmacophore, $S = q^N$ where N is the number of molecules in the data set. The candidate pharmacophores are ranked by their values of S; those with the smallest value of S are preferred. The statistical basis for this procedure has been articulated [36,38,39], with a slightly refined formula for S.

7.5.6 ARE STERIC CONSTRAINTS IMPORTANT? HOW SHOULD ONE CONSTRUCT THEM?

Steric constraints added to the pharmacophore are important as they help filter out molecules that have the right spatial arrangements of features but have groups that protrude into the receptor. Marshall's "active-analog approach" [2] specified an approach long ago for calculating such sterically forbidden regions: computing the union volume of the active molecules and the union volume of the inactives and subtracting them. In ALADDIN, capabilities were provided for including such steric constraints into pharmacophores in two different ways. Catalyst and other commercial pharmacophore discovery methods do not allow any automatically generated steric constraints, though one can add them manually in Catalyst. A variation on the Marshall approach was proposed, which solved the weaknesses of the approaches used in ALADDIN, and demonstrated its effectiveness in both pharmacophore discovery and 3D database searching, namely the shrink-wrap methodology [14,39]. After a set of conformers are overlaid, the surface, which is the smallest volume enclosing at least one conformer of each active molecule, is computed. The polygons of this surface are then annotated, depending on whether they provide steric discrimination in the dataset. Figure 7.13 shows a cut-away view of the shrink-wrap surface associated with the gram-negative activity of a set of oxazolidinone antibiotics [40], computed from over 3000 molecules. With that many molecules and high-quality conformational analysis (MacroModel/MCMM/MMFF), this surface may be defined in exquisite detail. This surface was used in lead-optimization mode, in which a 3D database searches against a designed library to discover novel oxazolidinone analogs (proprietary, unpublished results). Figure 7.8 shows a shape-enhanced pharmacophore for the serotonin 5-HT2c receptor, derived from the methods described in this section; unfortunately, this pharmacophore model when used in lead optimization did not lead to the synthesis of any novel analogs, as it suggested variations to the scaffold that were not readily amenable to synthesis.

7.5.7 WHICH COMPUTATIONAL APPROACHES WORK BEST? HOW DO THEY COMPARE?

There are no simple answers to this question. Users are encouraged to perform their own analysis being mindful of the issues outlined in [36,37]. In particular, computational control experiments are vital and enlightening.

An unappreciated aspect of pharmacophore discovery is that different approaches may be more relevant, depending on the goals of the virtual screen. For example, applying the principle of selectivity optimizes for the enrichment ratio. One may wish to optimize for pharmacophores that minimize false positives; to optimize to minimize false negatives, one should skip VS entirely and fall back to random screening.

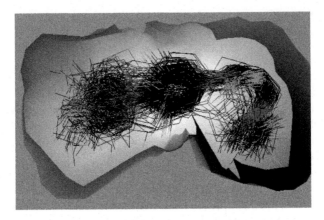

FIGURE 7.13 Cutaway view of the shrink-wrapped pharmacophore of the oxazolidinone antibiotics. The details of the features and distance constraints is still proprietary; what is shown is a cutaway view of the steric constraints associated with gram-negative activity. The oxazolidinone ring is on the right; the fluorophenyl in the middle. Without the cutaway, the surface would envelop the entire molecule. This surface was calculated from over 3000 analogs.

7.6 STEP 3 — WHAT DOES ONE DO WITH THE OUTPUT OF A VIRTUAL SCREEN?

There are three general categories of activities that should be undertaken after performing a virtual screen:

1. Refine the query, after inspecting the hit list combined with additional data, to see if data already exists that tests one's hypothesis.
2. Computational postprocessing of hits.
3. Biological testing.

7.6.1 REFINE QUERY — SEE IF DATA THAT TESTS HYPOTHESIS EXISTS

If one is searching a pharmaceutical company's corporate database, one typically will have available much preexisting biological data on the molecules in the hit list that were not considered in the pharmacophore-discovery process, but which suggest refinements in the query. These refinements may be either practical issues or scientific ones. Typical practical issues are:

- Hits for which no inventory exists may need to be screened out.
- Hits that already show activity against another target may be undesired.
- Too many hits, indicating distance constraints, may need tightening.

Typical scientific issues are:

- A negative charge feature picks up undesired sulfates.
- Hits on aliphatic sulfhydryls may be acceptable, but aromatic sulfhydryls are not.
- The definition of basic amine may not be perfect and some unusual nitrogens that are basic are not recognized as such.

7.6.2 COMPUTATIONAL POSTPROCESSING OF HITS

A number of processes may be followed; these are most frequently used:

- Remove junk — Frequently molecules appear in the hit list that contain reactive functionality, are intrinsically fluorescent, are insoluble, or are promiscuous [41,42]. When these can be identified, they are usually best removed prior to biological screening.
- Sorting — Frequently, the first step after running a 3D database search is sorting by MW, to bring to the top of the list those molecules that most parsimoniously present the pharmacophore. If other data accompanies the hits from the corporate database (e.g., activity against another target), this may also be useful in sorting.
- Clustering — Frequently, when faced with large hit lists (> 1000 compounds), one may wish to cluster the hit list and then work with a subset of the hit list, one that includes representatives of each cluster. One can also remove entire clusters from the hit list, if they represent a scaffold that is undesired (e.g., not novel). At Pharmacia & Upjohn, we used an approach we called *consistent clustering* [43]; in that approach, a hit list is clustered, and clusters which appear to show trends in the SAR within that cluster are given preference for further analysis.
- Molecular modifications — Long ago, a method was reported using custom routines to trim down molecules in a hit list to identify common scaffold among hit lists [44]. Mestres and Veeneman recently reported using custom routines to modify the molecules in a hit list; this is a means to identify *latent hits* — existing molecules that almost have the right structural properties [45].

7.6.3 BIOLOGICAL TESTING

This is the *sine qua non* of VS. One of the most interesting observations about biological testing in the context of VS is that on occasion one finds things that were overlooked or ignored via random screening. One of our early users of Catalyst reported that a hit from a virtual screen had already been identified as active in a random screen, but was sidelined as an artifact based on visual inspection; when presented with how that molecule fit the pharmacophore as part of a virtual screen, it was then followed up as a real hit. In another example during the author's experience at Pharmacia & Upjohn, a random screen against an exciting new target turned up nothing. A virtual screen was then done, but none of the hits from the virtual screen showed activity. We were so confident in the virtual screen that those hits were reassayed from fresh powder. A series of active compounds were then found, the common scaffold of which became the low-single-digit-μM lead for that project. (It was later determined that these compounds had decomposed in the dimethyl sulfoxide, DMSO, solution in the screening plates, which led to their being missed in the original screen.)

7.7 CASE STUDIES

The following case studies have been carefully selected. They are all VS successes and exemplify many best practices; they should help to establish expectations for what one may reasonably expect from pharmacophore-based VS. Each case study distinguishes itself for different reasons, which will be clarified at the end of the discussion of each. The overall lessons to be extracted from these successful case studies will be discussed in the following section.

Although it might serve the pedagogical aim to include among these successful case studies some examples that highlight practices to avoid, that will not be done here. In part, this acknowledges that failures are rarely reported in the literature.

7.7.1 D1 Agonists (Abbott)

The events of this case study occurred in 1987, but they were not described in the literature until 1992 [46]. The biological hypothesis was that a drug that was the agonist for the D1 receptor, selective against D2, would be a useful therapeutic for Parkinson's disease and free of the side effects of unselective dopamine agonists. (At the time, in the era before cloned receptors, it was still relatively new that dopamine receptor subtypes existed at all [47]; now, it is understood that there are five subtypes.)

One molecule was known at the time as a D1-selective agent, SKF-38393, shown in Figure 7.14 next to the endogenous ligand dopamine. A pharmacophore model was constructed, by hand, using a custom-built interactive 3D graphics system, based on the model and data published by Seeman et al. [13]. The actual pharmacophore used to perform the 3D database search was never published, but it resembled that shown in Figure 7.14 with the primary constraint being a distance between the projected point off of the basic amine and that off of the catechol hydroxyl.

ALADDIN was used to perform the 3D database search, software developed by this author, building upon the routines of the Pomona Med Chem software developed by Weininger [48]. In the summer of 1987, the following events happened in close succession:

- The first version of ALADDIN was completed.
- In its first run, it was used by Cathy May to discover a novel lead for the D1 project using a multiconformer 3D database built by Danaher.
- The first draft of the methodology paper [3] was written.
- ALADDIN was described briefly by Martin at the end of her talk at the QSAR Gordon Conference (now called the Computer-Aided Drug Design Gordon Conference).

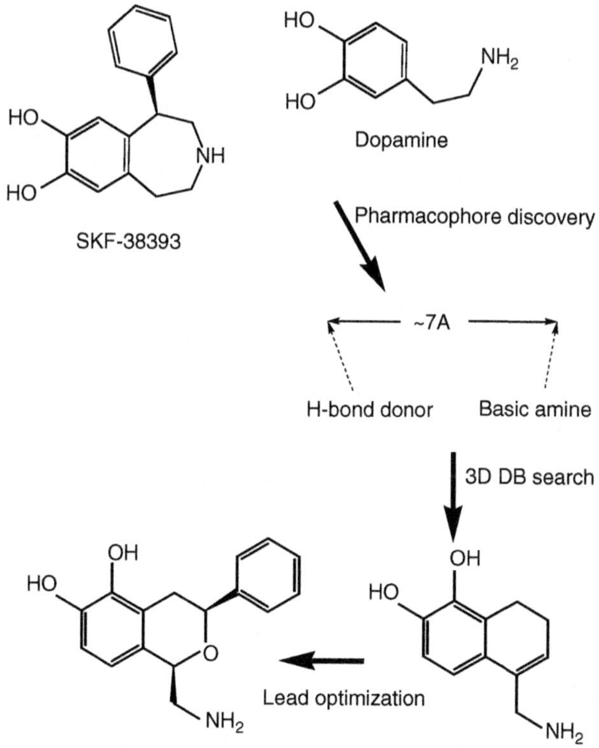

FIGURE 7.14 SKF38393 (a dopamine agonist) and dopamine; a crude D1 pharmacophore; the lead discovered at Abbott; and A-68930, the clinical candidate optimized from that lead.

The 3D database search was run against a multiconformer database of a few thousand compounds that had already tested positive in some assay against a CNS receptor. The multiple conformers were by-products of molecular-mechanics runs in the course of modeling efforts. Compounds that had already been tested for D1 or D2 activity were excluded. Compounds with insufficient inventory for testing were also excluded. Of the 17 hits from the virtual screen, 9 showed activity against the D1 receptor, and 2 of those 9 were selective for D1 with affinities in the low-double-digit nM range. One of those, the aminotetralin shown in Figure 7.14, was quickly adopted as the lead and optimized by the lead chemist, Schoenleber; A-68930 was taken into the clinic in 1989 [49].

One of the reasons details of this story were slow to become public is because of concerns by Abbott's lawyers: because a computer discovered this lead, then *ipso facto* it must be obvious and hence not amenable to patent protection. As VS became commonplace, these concerns subsided.

Note especially the following aspects of this case study:

- The discovery team was in a situation where they were desperate for a new lead, as the one they had been pursuing was not panning out. It is a rare circumstance when a team is in such a situation and has the *élan* to pounce upon a new lead.

- Although the pharmacophore may appear simple and unsophisticated, it distilled the essence of much careful modeling of the D1 receptor, exploiting data on competitors' compounds, data from the academic literature, etc.
- The database was small, but represented a focused database of molecules with known CNS activity. Also, the database consisted of compounds available for testing and included 1D data indicating the amount of inventory available for testing.

7.7.2 FIBRINOGEN ANTAGONISTS (MERCK)

In 1992, a group at Merck reported the discovery of nonpeptide fibrinogen receptor antagonists (GP IIb/IIIa antagonists) [50]. The biological hypothesis was based on the observation that the final step in platelet aggregation is the binding of fibrinogen to GP IIb/IIIa. The key epitope in fibrinogen interacting with that receptor is a Rat Genome Database (RGD) sequence arginine-glycine-aspartic acid (Arg-Gly-Asp). A series of peptidic compounds based on RGD, some cyclic-constrained peptides, had been reported as fibrinogen antagonists [51].

The publication is coy on their lead discovery strategy:

A directed search of the Merck Sample Collection for compounds that possessed amino and carboxylate functionalities that were separated by through-bond distances of 10-20 Å (approximating the distance between the basic guanidine and the acidic carboxylate of the Arg-Gly-Asp) proved the lead compound 1.... [50]

That lead compound is shown in Figure 7.15.

Amazingly, there is no detail in this report, though the remaining lead-optimization was reported in detail. It is conceivable that legal nervousness underlay this as it had earlier at Abbott.

Nonetheless, from these few words some things can be reconstructed, based on what is known publicly about the Merck modeling environment at that time. The 3D database search was likely done using the MDL MACCS-3D rigid search system, likely on a single-conformer database constructed using CONCORD. That database size was likely of the order of 100,000 compounds.

This pharmacophore — a positive charge separated from a negative charge by 10 to 20 Å — is among the crudest pharmacophores ever published. This should not serve as a source of embarrassment; it is a testament to the power of VS that even such a crude dyad pharmacophore may be useful. At roughly that same time, a pharmacophore based on the SAR in [51] could be constructed using the MNMM algorithm; that was a dyad, with a distance constraint of 13 to 14 Å. Later improvements in pharmacophore discovery methodology led to a RGD dyad pharmacophore constrained by 11.9 to 12.2 Å [38].

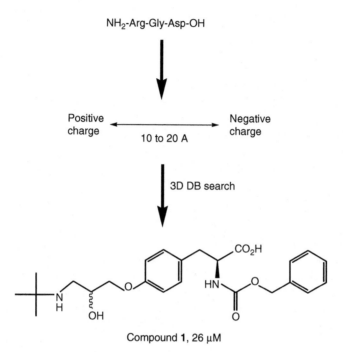

NH₂-Arg-Gly-Asp-OH

Positive charge ←————→ Negative charge
10 to 20 Å

3D DB search

Compound 1, 26 μM

FIGURE 7.15 Merck's fibrinogen antagonist pharmacophore and lead. (From G.D. Hartman et al., *J. Med. Chem.*, 35:4640–4642, 1992. With permission.)

The Merck group did not publish the number of hits that emerged from this virtual screen. Despite the imprecision of the pharmacophore, it is likely that the enrichment ratio was sufficient to make this a useful exercise.

Key observations about this case study:

- While competitors were pursuing a classic peptidomimetic strategy to convert the RGD peptide into a druglike molecule, Merck was able to exploit this information in a different way using VS to discover a lead in their compound collection. As in the Abbott case study, timing is crucial: the discovery team was in the ideal situation to respond to the results from the virtual screen.
- The pharmacophore may seem absurdly broad, but in fact any dyad with the lower bound of the distance constraint greater than 9 Å will tend to be selective. The crudeness of their methodology (single-conformer database, rigid search) actually complemented the imprecision of the pharmacophore.

7.7.3 PROTEIN KINASE C AGONISTS (NCI)

A series of superb VS studies were reported during the 1990s from the laboratory of Milne at the National Cancer Institute (NCI). One of the first of these was by Wang et al., describing the discovery of novel PKC agonists [52]. The biological hypothesis was that modulators of PKC, due to PKC's role in cellular proliferation pathways, would be useful therapeutics in cancer.

Their database totaled 207,000 structures, "open" compounds contained within the NCI's Drug Information System. The 3D search software was Chem-3DBS, then commercially available from Chem-X, with multiconformer database construction occurring via a torsion-driving algorithm, with chirality when ambiguous being arbitrarily assigned, and with compounds with more than 15 rotatable bonds excluded. The pharmacophore is quite precise and is shown in Figure 7.16. The three features were chosen based on known SAR around the known natural-product ligand, phorbol dibutyrate (PDBu). The small-molecule crystal structure of phorbol, a close analog of PDBu, was used to model its structure and to serve as a template for constructing a pharmacophore. Conformational searches in QUANTA® were used, along with other PKC SAR, to assign the three distance ranges in the pharmacophore.

The search yielded 535 hits; of those, 286 were physically available for testing. Their software did not allow them to specify a particular type of hydrophobic group that their modeling suggested was critical; therefore, the hit list was scrutinized visually, and those compounds missing that functionality were discarded. After that visual filtering, they had 125 compounds, which were submitted to biological testing. Eleven were found to be active (> 10% of [³H]-PDBu displaced); hence their success rate was a respectable 9%. They do not present the data needed to calculate an enrichment ratio, but if one assumes a typical random-screening hit rate of 0.2%, this implies a 40-fold enrichment — a testament to the quality of their pharmacophore.

Ki's were measured on the five compounds most potent in the radioligand displacement assay; these ranged from 7.8 to 26.6 μM. The most potent compounds are shown in Figure 7.16. Both of these compounds displayed multiple mappings to the pharmacophore, despite the precision of that pharmacophore. The strain energy of the actives fit to the pharmacophore ranged from < 1 to 9.7 kcal/mol, using QUANTA, version 3.3 with CHARMm® 2.2 force field. They report that the root-mean-square (rms) fit to the pharmacophore does not correlate with biological activity as most would anticipate.

Their conclusion reflects a common sentiment:

> It is our opinion that the main purpose of a 3D-database pharmacophore search is not to find compounds with ultra-potent activity. A realistic goal of a 3D-database pharmacophore search should be the discovery of new leads which can be subsequently used as the basis for further synthetic modifications [52].

There are a number of salient features in this case study:

- Their database was huge and consisted largely of druglike molecules for which inventory was on hand and available for testing.
- Their search methodology included flexible matching, which complemented the comparative precision of their pharmacophore.

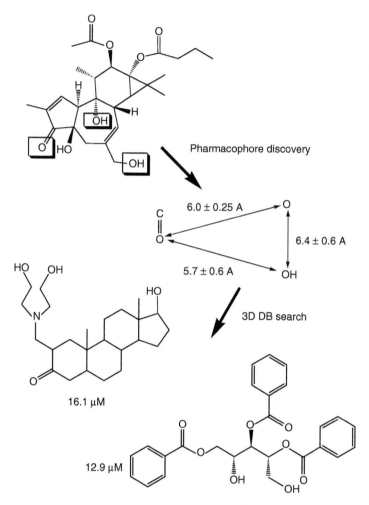

FIGURE 7.16 The protein kinase C pharmacophore of Wang et al. and some active hits. (Used with permission from Wang et al., *J. Mol. Chem.* 37:4479–4489, 1994.)

- The pharmacophore, though based on a small dataset, is precise and effective judging from their success rate.
- As this was an academic setting, one should not expect that the active hits served as leads for a discovery effort; the only detraction from this effort is that these active hits would likely not be compelling for such an application.

7.7.4 HIV INTEGRASE INHIBITORS (NCI)

Another VS success story from Milne's lab at the NCI dealt with HIV-1 integrase inhibitors by Nicklaus et al. [53]. The biological hypothesis for this target as a therapy for acquired immunodeficiency syndrome (AIDS) is that HIV-1 integrase is an essential enzyme for the integration of the viral DNA into the infected cell chromosome.

The 3D database was essentially the same as that described above for the PKC agonist lead discovery. The pharmacophore used in the search, shown in Figure 7.17, was derived from two known inhibitors of the target, CAPE and NSC 115290, also shown in Figure 7.17. Again using QUANTA/CHARMm, a superposition of low-energy conformers of these molecules led by inspection to the pharmacophore shown. Note that this pharmacophore includes a steric constraint in addition to the three features. The distance constraints were specified with a degree of precision that will unlikely ever be superseded.

A total of 267 compounds emerged from a virtual screen, of which 60 were chosen based on physical availability for testing, estimate solubility, and druglikeness. Another 59 molecules were randomly selected from the database and assayed. Of the 60 compounds in the hit list, 19 were active (inhibitory concentration 50% or $IC_{50} < 200$ μM), a 32% success rate. Some of the more interesting of those are shown in Figure 7.17. Of the 59 randomly selected compounds, 6 were active, indicating an enrichment ratio of 3.2; this enrichment was deemed statistically significant to $p < 0.01$. They make the interesting observation that some of those randomly selected 6 actives matched their pharmacophore as shown with interactive graphics, but did not appear among the 267 compounds in the hit list, suggesting weaknesses in the treatment of conformational flexibility in that search software. They describe some intriguing studies evaluating the multiple pharmacophore possibilities consistent with their SAR, some of which were automatically determined by a commercial, but unpublished, Pharmacophore Identification algorithm from Chem-X, which one may surmise is a variant of the MNMM algorithm.

This case study stands out in many respects, mainly for rigor:

- They demonstrate, using their random sampling of their database, that one can assess the enrichment ratio without testing the entire database. Furthermore, they use that same information to demonstrate that the entire process of VS is statistically significant.
- They also highlight the issue of *false negatives* — compounds in the database that are active but are not matched by the pharmacophore — and identified one component contributing to that, namely weaknesses in handling conformational flexibility.
- They investigated the implications of multiple candidate pharmacophores, one of the first VS studies to do so.
- Their use of automated pharmacophore discovery methods to determine the pharmacophore is among the first reports of such in the literature. It is unfortunate that those methods are unpublished.

7.7.5 Muscarinic M3 Antagonists (Astra)

Marriott et al. at Astra reported a VS success in lead discovery applied to the muscarinic M3 receptor [54]. The therapeutic potential for M3 antagonists is thought

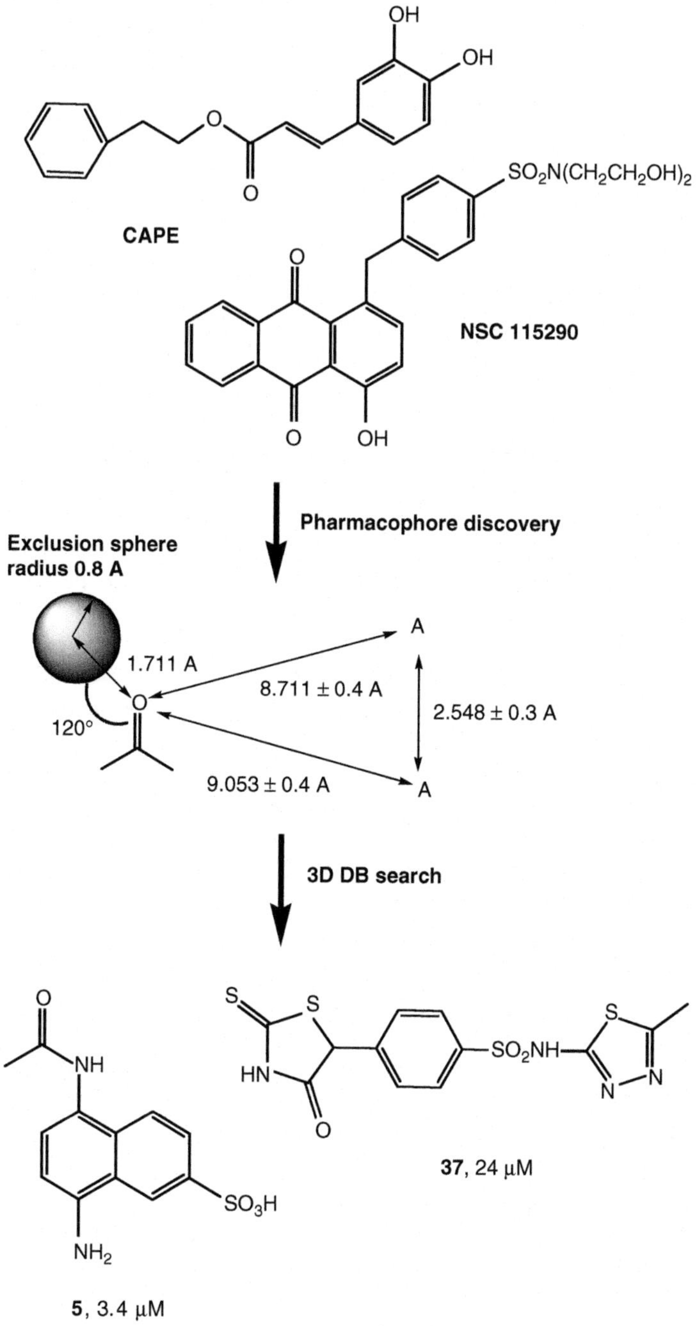

FIGURE 7.17 The HIV integrase dataset and pharmacophore of Nicklaus et al. and some active hits. (Used with permission from Nicklaus et al., *J. Med. Chem.* 40:920–929, 1997.)

to be in the treatment of irritable bowel syndrome. Their utility has now been shown for chronic obstructive pulmonary disease and urinary incontinence.

Their starting point was a series of molecules synthesized for a M1 agonist program, which also displayed M3 antagonism. Some of those structures are shown in Figure 7.18. Conformational searching was performed in SYBYL®, with a value of τ taken at 5 kcal/mol, resulting in 44 conformations of Compound 2. They were fortunate to have two rigid molecules in their data set, Compound 3 and Compound 4. The automated pharmacophore discovery procedure DISCO [55] was used, using Compound 3 as the reference conformation with 0.5 Å tolerances chosen arbitrarily and constraining DISCO's operation to 3 to 8 features. Five candidate pharmacophores emerged and 2 were selected based on visual inspection. These will be hereafter denoted Pharmacophore A and Pharmacophore B and are shown in Figure 7.18; they are both tetrads. Both were used in the VS.

Their 3D database was their corporate collection, augmented by a compilation of vendor's databases (Maybridge, Aldrich, Specs, etc.); the total size was not reported, but was likely in the range 500,000 to 1,000,000 compounds. Their corporate collection was likely in the range 100,000 to 200,000 compounds. The 3D structures were generated using CORINA, and the 3D search was performed using Tripos' UNITY system [56].

Pharmacophore A yielded 176 hits from the Astra compound collection; Pharmacophore B yielded 173 hits. The intersection of those two hit lists is 172 compounds, which suggests that the differences between Pharmacophore A and Pharmocophore B are minor. The union of the two hit lists yielded 177 compounds, of which 172 were available and sufficiently soluble for biological testing. The total number of actives was not reported, but Compound 5, Compound 6, and Compound 7 (Figure 7.18) were hits showing unusual activity, with pA2 values of 4.8, 5.5, and 6.7 respectively; the pK_B of Compund 7 was determined to be 6.7 (subμM).

The aspects to this case study that make it noteworthy are:

- They explicitly generate and evaluate multiple candidate pharmacophores and show that when used prospectively in a 3D database search that the differences in the model are minor. They used an automated pharmacophore discovery method with an atypically small value of = 5 kcal/mol. Detracting from their rigor was their arbitrary assignment of constraint tolerances.
- They show how they can combine a small amount of internally generated data with some competitor's data to generate useful pharmacophores, allowing them to discover interesting molecules that appear to be capable of being leads.
- The useful compounds all came from the proprietary corporate compound collection, not the much larger collection of vendor compounds.

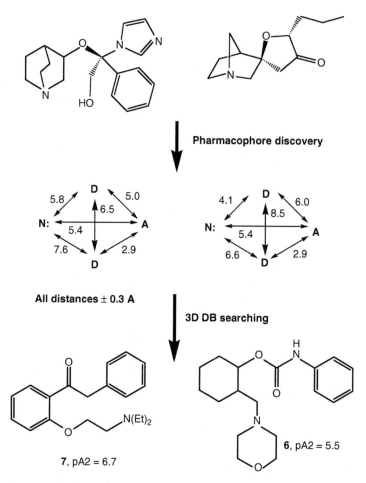

FIGURE 7.18 The initial M3 data set of Marriott et al., their pharmacophore and some active hits. Both hydrogen bond donors (denoted D) are projected points on the putative receptor; the hydrogen bond acceptor (denoted A) maps to an acceptor on the ligand, paired to the closer donor. (Used with permission from Marriott et al., *J. Med. Chem.* 42:3210–3216, 1999.)

7.7.6 α4β1 Antagonists (Biogen)

The antigen α4β1 (also known as late antigen, VLA-4) has been implicated in many inflammatory diseases. A group at Biogen, Singh et al. [57] successfully discovered novel nonpeptidic leads to this target via VS.

Their starting point was Compound 1, a 0.6 nM compound shown in Figure 7.19. Novel features were constructed in Catalyst and selected based on trends in the SAR; the two features were PUPA (4-[N'-(2-methylphenyl)ureido]phenylacetyl) and a carboxyl. The conformation of Compound 1 was based on the crystal stucture of the integrin binding region of vascular cell adhesion molecule-1 (VCAM-1), with systematic conformational searching of portions of the molecule undetermined by that x-ray structure using Discover CVFF. Features were positioned based on that conformation of Compound 1, with distances tolerances arbitrarily set to 2.0 Å. A 8624-compound subset of the ACD was created by searching for compounds with a free amine or nitro group. A virtual library was created by coupling in silico the carboxyl of the PUPA to the free amine or nitro. A multiconformation 3D database was built with Catalyst/Fast, up to 250 conformers/molecule = 15 kcal/mol. This 3D database was screened using Catalyst and resulted in 416 hits; after removal of peptides, they were left with 170 compounds. Of those, 12 were selected for reasons related to chemical synthesis. These 12 were coupled to the PUPA capping group and tested. All showed biological activity with the most potent, Compound 2, displaying 1.3 nM (Figure 7.19); activities ranged from 1.3 nM to 20 uM. Compound 2 was active in an animal model of allergic airways disease.

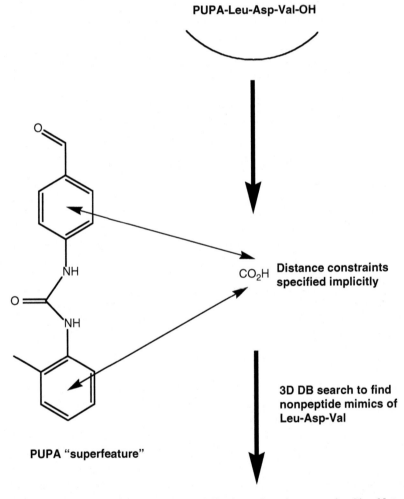

FIGURE 7.19 The VLA-4 pharmacophore of Singh et al. and some active hits. Note that there is some SAR among the hits. (Used with permission from Singh et al., *J. Med. Chem.* 45:2988–2993, 2002.)

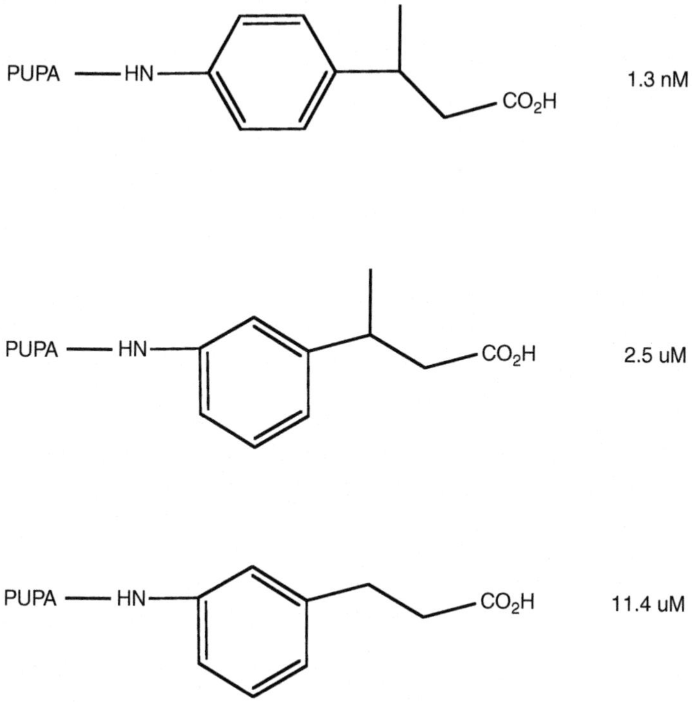

PUPA ——— HN ———⟨ ⟩ — CO₂H   1.3 nM

PUPA ——— HN ———⟨ ⟩ — CO₂H   2.5 uM

PUPA ——— HN ———⟨ ⟩ — CO₂H   11.4 uM

FIGURE 7.19 (continued)

Salient aspects of this case study are:

- This VS success rate of 100% is exceptional, as is the observation that these hits are also among the most potent compounds reported from a virtual screen. However, it should be kept in mind that the screen identified reagents for a virtual library, and the pharmacophore included a substructural element that was exceptionally large — over half the MW of the resulting hits. Also, a high degree of subjectivity may be inserted in the process of manually filtering the 170 hits into the 12 submitted for testing.
- This may be the first report in the literature of a success screening of a virtual library, as in this case the compounds in the hit list were first synthesized, and then tested. Screening a virtual library raises the bar for the confidence one must have in the quality of the pharmacophore, which may be the reason such a large substructural element was included.

7.7.7 ESTROGEN ERα RECEPTOR (ORGANON)

Mestres and Veeneman at Organon recently reported a remarkable success story against the estrogen receptor [45]. This receptor has been implicated in a variety of indications (e.g., breast cancer, osteoporosis).

They describe as latent hits compounds that emerge from a virtual screen and that are close, but not exactly the structure one needs to be active against the receptor. They began with a subset of their compound collection containing 134,000 compounds. This was trimmed further to 11,047 compounds based on "remov[ing] all compounds having a low probability of possessing the characteristics we were looking for in a nonsteroidal ERα agonist," using a dyad pharmacophore, constraining the number of rotatable bonds ≤ 5, and demanding that no steroids are present. A dyad pharmacophore constraining a 6-membered hydroxyl-containing-ring to lie within 9.2 to 12.6 Å from another –OH. The single-conformer 3D database was built using CORINA. The final virtual screen of the 11,047-compound database was performed, amazingly enough, using a variation of MIMIC, which performs an electrostatic-field-based match of the target conformation to each conformation in the database [19]. Figure 7.20 summarizes their results, where Compound 4 and Compound 5 were identified via the virtual screen, but showed activity EC_{50} (median effective concentration) > 10 μM. Compound 4 was transformed into Compound 6, a 4 nM compound; Compound 5 was transformed into Compound 7, Compound 8, and Compound 9 with activities 0.8 μM, 1.2 μM, and 0.08 μM, respectively. For such a virtual screen, it does not make sense to calculate success rates, enrichment ratios, etc.

A couple of aspects to this case study deserve attention:

- They present a novel way to think about how virtual chemistry must be performed on those compounds present in the database. Their notion of latent hits is a powerful one.
- They present a hierarchy of screens from practical issues to pharmacophores and finally to the use of a 3D similarity method. This succession of triage steps allows each step to be performed to a higher degree of refinement.

7.8 WHAT ARE THE COMPONENTS OF THE ART OF PHARMACOPHORE-BASED VIRTUAL SCREENING THAT ARE CRUCIAL FOR SUCCESS?

One of the first things that should be apparent is that success does not appear to hinge on the particular 3D database searching software used; successes have been enjoyed with a variety of different types of software. However, as Nicklaus et al. imply, the impact of poor treatment of conformational flexibility may be in the realm of what one does not see: the appearance of false negatives. Implicitly, the same is true about any weaknesses in handling chirality, though this is somewhat mitigated by the presence of numerous analogs of the same molecule in most corporate compound collections.

It is also clear that success arises frequently with unsophisticated pharmacophores constructed from small data sets. The most frequent occurrence is a mix of a small number of in-house compounds with a small number of compounds from the literature.

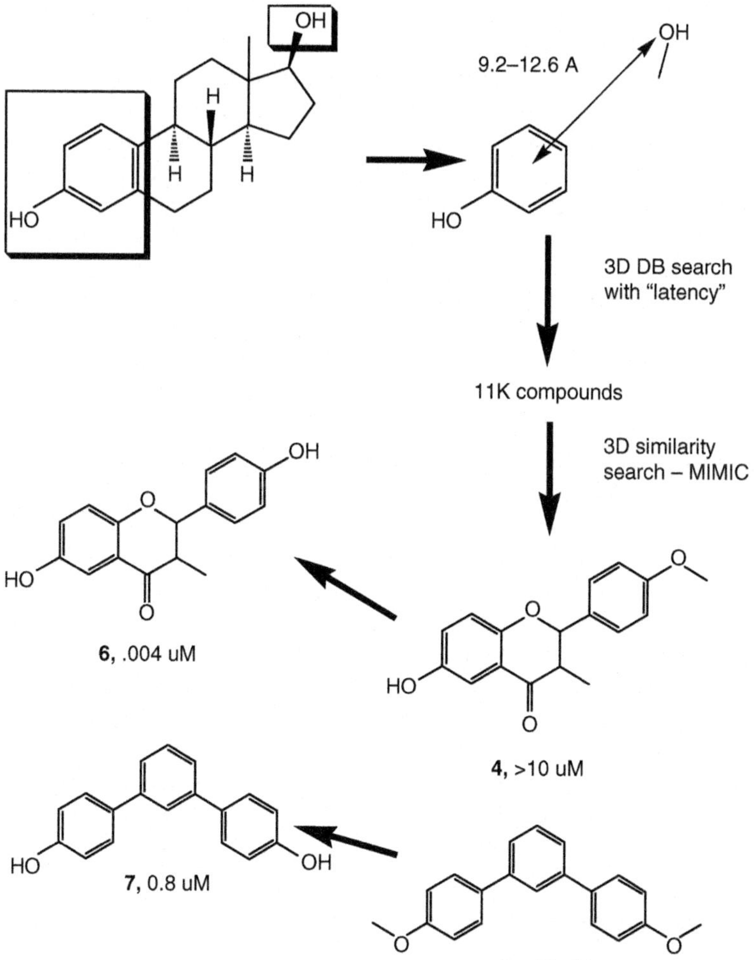

FIGURE 7.20 The estrogen receptor pharmacophore of Mestres and Veeneman and some latent hits and their conversion into actual hits. (Used with permission from Mestres and Veeneman, *J. Med. Chem.* 46:3441–3444, 2003.)

One may conclude that the key challenge is to discover a pharmacophore consistent with two structurally distinct scaffolds. In most cases, they were fortunate to have a potent, rock-rigid molecule upon which to base a pharmacophore. It appears that discovering such pharmacophores has not relied heavily on automated pharmacophore discovery tools. One might conclude that either one does not need sophisticated pharmacophores, or the currently available automated pharmacophore discovery tools are only marginally useful. It is more likely the latter. As researchers continue to push harder and especially as they shift more toward VS applied to lead optimization, the need for more sophisticated pharmacophores built from larger data sets and better pharmacophore discovery tools will be more apparent.

Note that in most of these case studies, the authors are improvising on the basic theme of pharmacophore-based VS. They display creative approaches toward adapting the existing commercial software to their needs. In many cases, filters were applied based on purely practical concerns to restrict the search space.

Successes have appeared over a wide range of biological targets: antibiotics, integrins, GPCRs, and nuclear hormone receptors. It is curious that no successes against ion channels or cytokine receptors have been reported.

It is difficult to gauge from these publications how much of a practical impact VS has had in advancing compounds into the clinic — the ultimate test of any technology for drug discovery research. Practical experience suggests, however, that our experience with ALADDIN on D1 is a rarity: that lead discovered by VS led to a clinical candidate. Why it has not had more of an impact may be more a comment on the sociology of the research environment and the ability of modelers to have an impact on the direction of discovery teams.

The final two case studies both have the common characteristic that the authors were not solely relying on the molecules in a database, but that virtual chemistry was performed on hits. VS of existing compound libraries, in large part, has been overtaken by the phenomenal advances of HTS in the 1990s. However, one capability available to computational screening not easily imitated by HTS is this screening of virtual libraries or identifying latent hits or whatever one wishes to call the process of computationally discovering molecules that are close to the desired goal.

Finally, the most important characteristic in common to all the case studies above is that those scientists knew what they were looking for; they had a clear view of the goal in mind. It is important to understand the biology and understand the needs of the discovery project. VS demands that one encode to the computer at an abstract (i.e., nonatomistic) level what the ultimate molecule should look like; if one cannot do that, the VS cannot succeed.

7.9 OPEN ISSUES AND FUTURE DIRECTIONS

The factors driving future innovations are:

- Influence of HTS — It is futile for VS to try to compete head-to-head with HTS. In those increasingly rare cases where an assay is not amenable to high throughput format, pure VS will still be useful. However, the introduction of virtual chemistry — either by screening virtual libraries, as in the Biogen case study, or by identifying latent hits as in the Organon case study — appears to provide VS with new opportunities for making an impact in drug discovery.

- Shift from lead discovery to lead optimization — This is intrinsically more difficult and more demanding on the methodology, but if one looks at how much time drug discovery teams spend in lead discovery vs. lead optimization, it is clear the needs are much stronger in the latter. Practical experience suggests that this will drive the improvement in the quality of the software, especially pharmacophore discovery. Naïve, simplistic pharmacophores may work in lead discovery where one is playing the odds, but in lead optimization, the expectations are higher — one must be right more often. Empirical evidence suggests that one of the most valuable contributions that one may reasonably expect in lead optimization is the ability to both define the limits of the playing field (i.e., describe the boundaries within which the chemists should limit their exploration) and to outline the terra incognita, regions of structural space the SAR has not yet explored [39].

- Integration of VS tools with tools for predicting ADME-tox properties — The recognition of the importance of modeling ADME-tox properties continues to grow. This must happen if the shift continues into virtual chemistry and lead optimization. For example, either the "Rule of 5" of Lipinski et al. [58] or the "Egan egg" model for predicting pharmacokinetic properties [59] is ideally suited as a computational postprocessing filter to a virtual screen.

Search speed is rarely an issue anymore; therefore, those aspects have been consciously suppressed in the case studies above. The early ALADDIN runs were overnight runs on a VAX (Virtual Address eXtension) machine; the NCI runs were week-long runs on an unspecified machine. Computers are now sufficiently fast that these issues have shrunk to insignificance. It still makes sense sometimes to implement a hierarchy of screens to adapt to different speed/subtlety tradeoffs, as in the Organon case study, where first a pharmacophore search was done, and then a more compute-intensive MIMIC match was performed. (The notion of VS with MIMIC would have seemed ludicrous in 1995.) Given the speeds of current computers, 3D database searching should be an interactive process (i.e., it should take only seconds, not dozens of minutes), though no such software exists; this could transform how this technology is used within the context of drug discovery.

7.10 CONCLUSIONS

The case studies listed above are just a sample of the successes that pharmacophore-based VS has achieved. There is an activation barrier to overcome — usually the construction of the 3D databases — but the validity of this method should now be beyond doubt.

The primary challenge for a computational chemist in pharmaceutical research, given any computational method, is to understand what its limits of applicability are and to distinguish between those situations which are appropriate for that method and which are not. Hopefully this chapter will assist the reader in understanding the limits of applicability of pharmacophore-based VS. When it comes to lead discovery, it usually requires vigilance on the part the modeler to recognize that a team is in the situation where they are ready to pursue a new lead. Modelers should be alert for opportunities to apply these methodologies to lead optimization. Finally, a modeler must be prepared to improvise as needed, while maintaining a clear understanding of the team's ultimate goals.

7.11 POSTSCRIPT

Circulation of an early draft of this chapter stimulated interesting discussions, primarily centering around the concept of a pharmacophore. There is clearly a malaise surrounding this word, exemplified by the brief discussion in Marriott et al. [54]:

> Like many words used in science, as in life generally "pharmacophore" has many meanings. Some use it to describe somewhat vague models of the environment within a ligand binding site. We take it to be something more specific and, possibly, more useful: an ensemble of interactive functional groups with a defined geometry.

Their definition is almost identical to that given at the outset of this chapter and represents the modern concept of a pharmacophore, consistent with the view originally articulated by Marshall et al. [2]. One of our aims with ALADDIN [3] was to explicitly provide a language "to objectively describe a receptor map hypothesis" (in that paper, receptor map and receptor mapping were used as synonyms for pharmacophore and pharmacophore discovery). The purpose of having an objective description is to eliminate vagueness and to allow a precise unambiguous description of that thing we wish to study. That this ALADDIN language for describing pharmacophores was so widely mimicked (MACCS-3D/ISIS, Catalyst, UNITY-3D, Chem-3DBS) is a testament to its utility. Furthermore, it is a common characteristic of all the case studies highlighted in this chapter that those authors have successfully achieved an objective description, which may have been a factor contributing to their success.

Fundamentally, a pharmacophore must correspond to something physically real and measurable. Measurement leads to the ultimate objective description, and the challenge in pharmacophore discovery is to infer a pharmacophore that parallels the details of ligand-receptor recognition when that later measurement is achieved. For example, the CNS pharmacophores shown throughout this chapter can now be directly related to the underlying physical reality of the corresponding GPCRs. Although we still are awaiting the first x-ray structure of a ligand bound to a drug-target GPCR, a steady succession of increasingly detailed biophysical data has revealed a consistent picture of those ligand–receptor interactions:

- Mutation data — Landmark studies by Strader et al. in the late 1980s [60] demonstrated that mutations on key residues of the 2-adrenergic receptor could profoundly influence ligand affinity. In particular, a crucial aspartate in putative transmembrane helix 3 (TM III) was implicated as the interaction site for the basic amine, and a series of aromatic residues in TM VI were implicated as the interaction site for the aromatic ring. These mutation data and the pharmacophore are in complete concordance: evidence that this pharmacophore indeed is objective and mirrors physical reality.
- Solvent-accessibility data — Javitch et al. [61] have studied the D2 dopamine receptor by mutating individual residues and determining which are solvent accessible, using a technique known as SCAM (substituted-

cysteine-accessibility method). These data complement the mutation data, allowing one to infer which residues are solvent accessible when certain ligands are bound.

- Homology models — GPCR homology models based on the x-ray structure of bacteriorhodopsin were introduced in the early 1990s by many groups (e.g., Hibert et al. [62]), but these models were problematic and inconsistent with knowledge from other sources. However, with the publication of the x-ray structure of rhodopsin in 2000 [63], the world of GPCR homology modeling was transformed as these models acquired real utility [64]. In particular, the SCAM data was now brought into perfect alignment and allowed detailed hypotheses on ligand–receptor interactions [65] also consistent with the pharmacophore inferences.

Although some might view the ultimate metric for pharmacophore-based VS to be maximizing the hit rate or the success ratio, this author's view is that the ultimate metric will be for those pharmacophores to mirror physical reality, and for 3D database searching thereby to function like a scientific instrument. For example, in the shape-enhanced pharmacophore shown in Figure 7.13, that picture of the oxazolidinone binding site may soon be subject to physical measurement, as those molecules target the ribosome, whose crystal structure was first determined in 2000; it will be interesting to compare Figure 7.13 with that binding site when it is determined and published. When these pharmacophores mirror physical reality, their greatest utility will come in the discovery of novel molecules with comparable biological properties. The case studies highlighted earlier are all examples of this.

The concept of a pharmacophore has had a curious evolution in the literature, which may lead to the confusion that surrounds it. Ehrlich is widely credited with coining the term, though the citation (Ehrlich, *Chem. Ber.* 42:17, 1909) usually given for that does not contain that term. Based on his coinage of a variety of terms ending in -phore, it is likely that he indeed did coin the term, however using it in a manner different than we now use. Even if Ehrlich did indeed coin the term in the early 1900s, the concept lay fallow until computational methods arose in the 1960s to allow their computation. In fact, it appears that the first pharmacophore was published in 1967 by Kier [66], from quantum mechanical studies on muscarinic ligands, though in that publication he calls it a "proposed receptor pattern." It appears that the modern concept of a pharmacophore was introduced shortly thereafter by Kier [67]; his introduction to that book chapter is a marvelously clear statement of the intellectual endeavor later amplified by Marshall and others using conformational analysis (italics added in the sentence introducing the term pharmacophore):

A useful concept to the scientist interested in drug phenomena in the body is that there is some substance present in tissue which possesses features enabling the tissue to interact with a drug molecule. Langley (*J. Physiol.*, 1:339, 1878) first proposed that there was some substance in tissue which made it capable of this interaction with a drug molecule, and the term "receptor" has since been generally used to describe this tissue feature. Considering all that has been written, receptors and drug-receptor interactions remain remarkably elusive to physical and chemical description. Indeed, not a great deal more is known about the nature of the receptor or its drug interaction than was known in the days of Langley.

The characterization of receptors by isolation and structural analysis of tissue components has not met with success. The view is widely held that the receptor is not an unalterable physical entity but a pattern of forces arrayed over the secondary and even tertiary structures of biopolymers. A conventional isolation and analysis could thus disrupt this pattern of forces, leaving only a biopolymer devoid of features complementary and susceptible to the drug molecule.

It has become necessary to attempt to characterize the nature of receptors indirectly, by characterizing the interacting drug molecule. This approach has centered around the elucidation of key atoms, groups, charged regions, and their spatial interrelation, all of which impart biological activity of a defined sort to a drug molecule. The approach is predicated on the hypothesis that the key features of a drug molecule are complementary (either upon initial interaction or sequentially upon subsequent engagement) to receptor features. It must be recognized that complementarity must be defined more broadly than within the rigid context of the "lock and key" connotation. This is necessary since it is conceivable that a primary engagement of a drug-molecule feature with a receptor feature may induce a change in the receptor which brings into juxtaposition additional drug and receptor features. Nevertheless, the key features of an active drug molecule must still be optimally positioned in the molecule to participate in the ultimate drug-receptor interaction in an efficacious manner, regardless of the sequence of events.

The medicinal chemist and the molecular pharmacologist have thus addressed themselves to the problem of defining the properties and positions of these essential drug-molecule features, and the term "pharmacophoric moiety" has been used to designate this receptor-specific pattern. An early approach was the dissection of large active drug molecules (from natural sources) into smaller synthetic version in which a variety of original features were retained. Classical examples are the dissection of atropine into smaller molecules possessing the bulky group and the aminoethanol "spasmophoric" moiety, and the dissection of cocaine into smaller molecules possessing the local-anesthetic moiety. With subsequent synthetic modification of these molecules and testing, the essential nature of certain atoms in specific relationships was deduced and some insight into the possible nature of the complementary receptor pattern was gained.

As an extension of this early work, a large amount of information has been accumulated on the effect of different chemical groups in different positions on the molecule. This has given rise to a host of empirical "structure–activity relationships" (SAR) for many drug classes. Using valence-bond reasoning (resonance theory) and SAR, medicinal chemists have reached conclusions on the nature of electronic charges at certain positions in the molecule.

The problem of relating SAR information to complementary receptor features and to the design of active synthetic analogs has been complicated by the fact that the conformation of many potent drug molecules is not readily apparent from inspection or chemical intuition. Single bonds (sigma bonds) have in the past been generally regarded as being "free to rotate," hence their conformations were not readily predictable. One attempt to circumvent this dilemma was to synthesize rigid molecules containing the presumed active sites in configurations in which interatomic distances could be nearly exactly calculated; comparative activity data were then used to select the configuration which had the optimum key interatomic distances. Out of all of these efforts have emerged some receptor hypotheses and some successful rationale for the design of new and useful drug molecules.

Some of the confusion surrounding the concept of a pharmacophore comes from the numerous computational methods that appeared in the 1990s that use the term pharmacophore, but do not aspire to determining an objective picture that mirrors the underlying physical reality, but rather are mathematical constructs (that nonetheless may have value in VS). For example, multiple researchers [68,69] adopted the approach of computing a molecular fingerprint, as is commonly done in 2D similarity methods, but rather than using a series of bits to record the presence/absence of a substructure as in 2D similarity, they use a series of bits to record the presence/absence of various 3D arrangements of features on a molecule. These bits do not represent pharmacophores in the sense used above; rather, they are spatial arrangements of features that may or may not be essential for biological activity. Also, even though pharmacophores may be used to create overlays, they are distinct from methodologies that use field-based similarity metric to perform overlays of molecules [19,70]. There is an important physical distinction between pharmacophores as physical objects mirroring the physical reality of ligand–receptor interactions and these field-based similarity overlay methodologies: the latter assume that each patch of ligand–receptor interaction contributes equally to the overall binding affinity. An often-overlooked wrinkle on the concept of a pharmacophore is it can only work because there are privileged ligand–receptor interactions, ones which when disrupted cause profound changes in affinity. Additivity is a pervasive assumption in computational biophysics [71]. Although most attempts to model ligand–protein interactions explicitly assume additivity (e.g., most scoring functions explicitly sum up distinct individual contributions), a pharmacophore is the simplest model of ligand–protein interactions that does not assume additivity (i.e., it acknowledges that some interactions are more important than others).

ACKNOWLEDGMENTS

The author has benefited from interactions with dozens of scientists over the years. This includes former colleagues at Abbott: Yvonne Martin, Mark Bures, Cathy May; former colleagues at BioCAD who with me codeveloped the database search component of Catalyst: Scott Kahn, Shneor Berezin, Siu-Ling Ku; former colleagues at Upjohn, later Pharmacia & Upjohn, still later Pharmacia: Gerry Maggiora, Mic Lajiness, Jim Blinn, Scott Larsen, and Dick Nugent; the many geographically dispersed early users of ALADDIN; the users of Catalyst during my days at BioCAD, especially the groups at Roche/Basel, including Klaus Gubernator and Manfred Kansy, and the group at Lilly, including Jim Wikel; colleagues at Vertex, especially Mark Murcko, Craig Marhefka, and Paul Charifson, all of whom reviewed the initial draft of this chapter and made many valuable suggestions. Paul in particular first articulated to me the role of the shift from lead discovery to lead optimization.

REFERENCES

[1]   I.D. Kuntz, J.M. Blaney, S.J. Oatley, R. Langridge, and T.E. Ferrin, A geometric approach to macromolecule-ligand interactions, *J. Mol. Biol.* 161:269–288, 1982.

[2]   G.R. Marshall, C.D. Barry, H.E. Bosshard, R.A. Dammkoehler, and D.A. Dunn, The conformational parameter in drug design: the active analog approach. In E.C. Olson and R.E. Christofersen, Eds., *Computer-Assisted Drug Design,* Washington, DC: American Chemical Society, pp. 205–226, 1979.

[3]   J.H. van Drie, D. Weininger, and Y.C. Martin, ALADDIN: an integrated tool for computer-assisted molecular design and pharmacophore recognition from geometric, steric, and substructure searching of three-dimensional molecular structures, *J. Computer-Aided Mol. Des.* 3:225–251, 1989.

[4]   S.E. Jakes, N. Watts, P. Willett, D. Bawden, and J.D. Fisher, Pharmacophoric pattern matching in files of 3D chemical structures: evaluation of search performance, *J. Mol. Graph.* 5:41–48, 1987.

[5]   Y.C. Martin, E.B. Danaher, C.S. May, and D. Weininger, MENTHOR, a database system for the storage and retrieval of three-dimensional molecular structures and associated data searchable by substructural, biologic, physical, or geometric properties, *J. Computer-Aided Mol. Des.* 2:15–29, 1988.

[6]   R.S. Pearlman, the University of Texas — Austin and www.optive.com.

[7]   Http://Www.Netsci.Org/Science/Cheminform/Feature06.Html.

[8]   Accelrys, www.accelrys.com.

[9]   P.A. Greenidge, B. Carlsson, L.G. Bladh, and M. Gillner, Pharmacophores incorporating numerous excluded volumes defined by x-ray crystallographic structure in three-dimensional database searching: application to the thyroid hormone receptor, *J. Med. Chem.* 41:2503–2512, 1998.

[10]   E.J. Lloyd and P.R. Andrews, A common structural model for central nervous system drugs and their receptors, *J. Med. Chem.* 29:453–462, 1986.

[11]   D. Mayer, C.B. Naylor, I. Motoc, and G.R. Marshall, A unique geometry of the active site of angiotensin-converting enzyme consistent with structure-activity studies, *J. Computer-Aided Mol. Des.* 1:3–16, 1987. It is interesting to note that this was the first paper published in this journal.

[12]   R.P. Sheridan, A. Rusinko, III, R. Nilakantan, and R. Venkataraghavan, Searching for pharmacophores in large coordinate data bases and its use in drug design, *PNAS* 86:8165–8169, 1989.

[13]   P. Seeman, M. Watanabe, D. Grigoriadis, J.L. Tedesco, S.R. George, U. Svensson, J.L. Nilsson, and J.L. Neumeyer, Dopamine D2 receptor binding sites for agonists: a tetrahedral model, *Mol. Pharm.* 28:391–399, 1985. In addition to being a first-rate scientific contribution, this is a wonderfully nostalgic paper from the precomputer days when models were constructed physically, photographs taken, and drawings made abstracted from those photographs.

[14]   J.H. van Drie, "Shrink-wrap" surfaces: a new method for incorporating shape into pharmacophoric 3D database searching, *J. Chem. Inf. and Comp. Sci.* 37:38–42, 1997.

[15]   J.H. van Drie, 3D databases on the desk of the medicinal chemist. In J. Gasteiger, Ed., *Software Entwicklung in der Chemie-10,* Frankfurt: Springer-Verlag, 1996. Duplicate information available also on http://www2.chemie.uni-erlangen.de/external/cic/tagungen/workshop95/vandrie/index.html.

[16]   T.E. Moock, D.R. Henry, A.G. Ozbabak, and M. Alamgir, Conformational searching in ISIS/3D databases, *J. Chem. Inf. Comp. Sci.* 34:184–189, 1994.

[17] T.J. Hurst, Flexible 3D searching: the directed tweak technique, *J. Chem. Inf. Comp. Sci.* 34: 190–196, 1994.

[18] M.A. Johnson and G.M. Maggiora, Eds., *Concepts and Applications of Molecular Similarity*, New York: Wiley, 1990.

[19] J. Mestres, D.C. Rohrer, and G.M. Maggiora, MIMIC: a molecular-field matching program — exploiting applicability of molecular similarity approaches, *J. Comp. Chem.* 18:934–954, 1997.

[20] D. Weininger, SMILES: 1. Introduction and encoding rules, *J. Chem. Inf. Comp. Sci.* 28:31, 1988.

[21] Details on the MOL format may be obtained from Molecular Design Ltd., www.mdli.com.

[22] The NCI database is available from the National Cancer Institute: http://dtp.nci.nih.gov.

[23] The Comprehensive Medicinal Chemistry database is available from MDL: www.mdli.com.

[24] The MDL Drug Data Report database is available from MDL: www.mdli.com.

[25] CORINA was developed in the lab of J. Gasteiger, University of Erlangen and is available from www.mol-net.de.

[26] W.P. Walters and M.A. Murcko, Library filtering systems and prediction of drug-like properties. In H.-J. Böhm and G. Schneider, Eds., *Virtual Screening for Bioactive Molecules*, Weinheim: Wiley-VCH, pp. 15–32, 2000.

[27] S.K. Kearsley, D.J. Underwood, R.P. Sheridan, and M.D. Miller, Flexibases: a way to enhance the use of molecular docking methods, *J. Computer-Aided Mol. Des.* 8:565–582, 1994.

[28] E.M. Ricketts, J. Bradshaw, M. Hann, F. Hayes, N. Tanna, and D.M. Ricketts, Comparison of conformations of small molecule structures from the Protein Data Bank with those generated by Concord, Cobra, ChemDBS-3D and Convertor and those extracted from the Cambridge Structural Database, *J. Chem. Inf. Comp. Sci.* 33:905–925, 1993.

[29] G. Chang, W. Guida, and W. Still, An internal coordinate Monte-Carlo method for searching conformational space, *J. Amer. Chem. Soc.* 111:4379–4386, 1989. See also MacroModel 8.0 reference manual, available from Schrodinger, Inc.

[30] Ph.D. thesis of A. Rusinko, 1988 at the University of Texas — Austin. Available from University Microfilms, Ann Arbor, MI.

[31] G.M. Crippen, *Distance Geometry and Conformational Calculations,* New York: John Wiley and Sons, 1981.

[32] A. Smellie, S.L. Teig, and P. Towbin, Poling: promoting conformational variation, *J. Comp. Chem.* 16:171–187, 1995.

[33] OMEGA is an extension of earlier work on Skizmo (unpublished work, described at the 1999 Spring ACS meeting. W.P. Walters, M.T. Stahl, J.J. Corkery, and M.A. Murcko, Skizmo: a new program for rapid conformational analysis and structure-based design of combinatorial libraries); it is available from www.eyesopen.com.

[34] Confort was developed by R.S. Pearlman et al. at the University of Texas — Austin and is available from www.optive.com.

[35] J.H. van Drie, An inequality for 3D database searching and its use in evaluating the treatment of conformational flexibility, *J. Computer-Aided Mol. Des.* 10:623–630, 1996.

[36] J.H. van Drie, Pharmacophore discovery — lessons learned, *Curr. Pharm. Des.* 9:1649–1664, 2003.

[37] J.H. van Drie, Pharmacophore discovery — a critical review, In P. Bultinck, H. de Winter, J. Tollenaere, and W. Langenaeker, Eds., *Computational Medicinal Chemistry,* New York: Marcel Dekker, 2003.

[38]   J.H. van Drie, Strategies for the determination of pharmacophoric 3D database queries, *J. Computer-Aided Mol. Des.* 11:39–52, 1997.

[39]   J.H. van Drie and R.A. Nugent, Addressing the challenges of combinatorial chemistry: 3D databases, pharmacophore recognition and beyond, *SAR and QSAR in Env. Res.* 9:1–21, 1998.

[40]   S.J. Brickner, D.K. Hutchinson, M.R. Barbachyn, P.R. Manninen, D.A. Ulanowicz, S.A. Garmon, K.C. Grega, S.K. Hendges, D.S. Toops, C.W. Ford, and G.E. Zurenko, Synthesis and antibacterial activity of U-100592 and U-100766, two oxazolidinone antibacterial agents for the potential treatment of multidrug-resistant gram-positive bacterial infections, *J. Med. Chem.* 39:673–679, 1996.

[41]   O. Roche, P. Schneider, J. Zuegge, W. Guba, M. Kansy, A. Alanine, K. Bleicher, F. Danel, E.M. Gutknecht, M. Rogers-Evans, W. Neidhart, H. Stalder, M. Dillon, E. Sjogren, N. Fotouhi, P. Gillespie, R. Goodnow, W. Harris, P. Jones, M. Taniguchi, S. Tsujii, W. von der Saal, G. Zimmermann, and G. Schneider, Development of a virtual screening method for identification of "frequent hitters" in compound libraries, *J. Med. Chem.* 45:137–142, 2002.

[42]   S.L. McGovern, B.T. Helfand, B. Feng, and B.K. Shoichet, A specific mechanism of nonspecific inhibition, *J. Med. Chem.* 46:4265–4272, 2003.

[43]   M.S. Lajiness and J.H. van Drie, unpublished results.

[44]   Y.C. Martin, E.B. Danaher, C.S. May, D. Weininger, and J.H. van Drie, Strategies in drug design based on 3D-structures of ligands, *Prog. Clin. Biol. Res.* 291:177–181, 1989.

[45]   J. Mestres and G.H. Veeneman, Identification of "latent hits" in compound screening collections, *J. Med. Chem.* 46:3441–3444, 2003.

[46]   Y.C. Martin, 3D database searching in drug design, *J. Med. Chem.* 35:2145–2154, 1992.

[47]   J.W. Kebabian and D.B. Calne, Multiple receptors for dopamine, *Nature* 277:93–96, 1979.

[48]   At that time, the Med Chem software was available from the Hansch lab at Pomona College. By 1989, Weininger had started a company to distribute that; ALADDIN was made commercially available through Daylight: www.daylight.com.

[49]   (A) M.P. DeNinno, R. Schoenleber, K.E. Asin, R. MacKenzie, and J.W. Kebabian, (1R,3S)-1-(aminomethyl)-3,4-dihydro-5,6-dihydroxy-3-phenyl-1H-2-benzopyran: a potent and selective D1 agonist, *J. Med. Chem.* 33:2948–2950, 1990; (B) J.W. Kebabian, C. Briggs, D.R. Britton, K. Asin, M. DeNinno, R.G. MacKenzie, J.F. McKelvy, and R. Schoenleber, A68930: a potent and specific agonist for the D-1 dopamine receptor, *Am. J. Hypertens.* 3:40S–42S, 1990.

[50]   G.D. Hartman, M.S. Egbertson, W. Halczenko, W.L. Laswell, M.E. Duggan, R.L. Smith, A.M. Naylor, P.D. Manno, R.J. Lynch, and G. Zhang et al., Non-peptide fibrinogen receptor antagonists. 1. Discovery and design of exosite inhibitors, *J. Med. Chem.* 35:4640–4642, 1992.

[51]   J. Samanen, F. Ali, T. Romoff, R. Calvo, E. Sorenson, J. Vasko, B. Storer, D. Berry, D. Bennett, and M. Strohsacker et al., Development of a small RGD peptide fibrinogen receptor antagonist with potent antiaggregatory activity *in vitro*, *J. Med. Chem.* 34:3114–3125, 1991.

[52]   S. Wang, D.W. Zaharevitz, R. Sharma, V.E. Marquez, N.E. Lewin, L. Du, P.M. Blumberg, and G.W. Milne, The discovery of novel, structurally diverse protein kinase C agonists through computer 3D-database pharmacophore search: molecular modeling studies, *J. Med. Chem.* 37:4479–4489, 1994.

[53] M.C. Nicklaus, N. Neamati, H. Hong, A. Mazumder, S. Sunder, J. Chen, G.W. Milne, and Y. Pommier, HIV-1 integrase pharmacophore: discovery of inhibitors through three-dimensional database searching, *J. Med. Chem.* 40:920–929, 1997.

[54] D.P. Marriott, I.G. Dougall, P. Meghani, Y.J. Liu, and D.R. Flower, Lead generation using pharmacophore mapping and three-dimensional database searching: application to muscarinic M(3) receptor antagonists, *J. Med. Chem.* 42:3210–3216, 1999.

[55] Y.C. Martin, M.G. Bures, E.A. Danaher, J. DeLazzer, I. Lico, and P.A. Pavlik, A fast new approach to pharmacophore mapping and its application to dopaminergic and benzodiazepine agonists, *J. Computer-Aided Mol. Des.* 7:83–102, 1993.

[56] UNITY version 2.4 is software available from Tripos Associates, www.tripos.com.

[57] J. Singh, H. van Vlijmen, Y. Liao, W.C. Lee, M. Cornebise, M. Harris, I.H. Shu, A. Gill, J.H. Cuervo, W.M. Abraham, and S.P. Adams, Identification of potent and novel alpha4beta1 antagonists using *in silico* screening, *J. Med. Chem.* 45:2988–2993, 2002.

[58] C.A. Lipinski, F. Lombardo, B.W. Dominy, and P.J. Feeney, Experimental and computational approaches to estimate solubility and permeability in drug discovery and development settings, *Adv. Drug Delivery Rev.* 23:3–25, 1997.

[59] W.J. Egan, K.M. Merz, and J.J. Baldwin, Prediction of drug absorption using multivariate statistics, *J. Med. Chem.* 43:3867–3877, 2000.

[60] C.D. Strader, I.S. Sigal, R.B. Register, M.R. Candelore, E. Rands, and R.A. Dixon, Identification of residues required for ligand binding to the beta-adrenergic receptor, *PNAS* 84:4384–4388, 1987.

[61] J.A. Javitch, J.A. Ballesteros, J. Chen, V. Chiappa, and M.M. Simpson, Electrostatic and aromatic microdomains within the binding-site crevice of the D2 receptor: contributions of the second membrane-spanning segment, *Biochem.* 38:7961–7968, 1999.

[62] M.F. Hibert, S. Trumpp-Kallmeyer, J. Hoflack, and A. Bruinvels, This is not a G protein-coupled receptor, *TIPS* 14:7–12, 1993.

[63] K. Palczewski, T. Kumasaka, T. Hori, C.A. Behnke, H. Motoshima, B.A. Fox, I. Le Trong, D.C. Teller, T. Okada, R.E. Stenkamp, M. Yamamoto, and M. Miyano, Crystal structure of rhodopsin: a G protein-coupled receptor, *Science* 289:739–745, 2000.

[64] T.A. Berkhout, F.E. Blaney, A.M. Bridges, D.G. Cooper, I.T. Forbes, A.D. Gribble, P.H. Groot, A. Hardy, R.J. Ife, R. Kaur, K.E. Moores, H. Shillito, J. Willetts, and J. Witherington, CCR2: characterization of the antagonist binding site from a combined receptor modeling/mutagenesis approach, *J. Med. Chem.* 46:4070–4086, 2003.

[65] J.A. Ballesteros, L. Shi, and J.A. Javitch, Structural mimicry in G protein-coupled receptors: implications of the high-resolution structure of rhodopsin for structure-function analysis of rhodopsin-like receptors, *Mol. Pharm.* 60:1–19, 2001.

[66] L.B. Kier, Molecular orbital calculation of preferred conformations of acetylcholine, muscarine, and muscarone, *Mol. Pharm.* 3:487–494, 1967.

[67] L.B. Kier, Receptor mapping using molecular orbital theory. In J.F. Danielli, J.F. Moran, and D.J. Triggle, Eds., *Fundamental Concepts in Drug-Receptor Interactions,* London: Academic Press, 1969.

[68] J.S. Mason, A.C. Good, and E.J. Martin, 3-D pharmacophores in drug discovery, *Curr. Pharm. Des.* 7:567–597, 2001.

[69] E.K. Bradley, P. Beroza, J.E. Penzotti, P.D. Grootenhuis, D.C. Spellmeyer, and J.L. Miller, A rapid computational method for lead evolution: description and application to alpha(1)-adrenergic antagonists, *J. Med. Chem.* 43:2770–2774, 2000.

[70] G. Jones, P. Willett, and R.C. Glen, A genetic algorithm for flexible molecular overlay and pharmacophore elucidation, *J. Computer-Aided Mol. Des.* 9:532–549, 1995.

[71] K.A. Dill, Additivity principles in biochemistry, *J. Biol. Chem.* 272:701–704, 1997.

8 Using Pharmacophore Multiplet Fingerprints for Virtual High Throughput Screening

Robert D. Clark, Peter C. Fox, and Edmond J. Abrahamian

8.1 INTRODUCTION

The term *pharmacophore* was originally coined by Paul Ehrlich to describe the particular substructure in a drug molecule responsible for its biological activity, but the concept has since evolved to mean the particular geometric disposition of groups in space through which a ligand binds to its target protein [1]. Each such interacting group constitutes a *pharmacophoric feature*. It has proven enormously useful to move away from defining ligand features solely in terms of their narrowly defined chemical nature (for example, amides or nitriles), but there is a limit to how effective such generalizations can be. In particular, features can be characterized as falling into one of a relative handful of distinct *feature types*, based on the complementary functionalities found in protein binding sites, limited for the most part, to those presented by the 20 naturally occurring amino acids. Widely used feature types fall into 3 classes:

- Polar — Hydrogen bond donors (HDOs) and hydrogen bond acceptors (HACs)
- Ionic — Positive and negative centers
- Nonpolar — Aromatic centers and hydrophobes

Pharmacophoric features are generalizations across atom types, so a HAC may be either a carbonyl oxygen or a pyridine-type nitrogen. Similarly, a HDO may be either a basic amino group or the OH group in an alcohol. They are, however, more fundamental than many of the functional groups commonly recognized in medicinal chemistry. Hence, an individual substructure can, and often does, present more than one feature, and the features presented may be of different types.

A primary amide, for example, is composed of both a HAC (the carbonyl group) and a HDO (the NH_2 group). A benzylic substituent is aromatic as well as hydrophobic, whereas a *t*-butyl group is hydrophobic but not aromatic. The most commonly encountered positive center encountered in drug molecules is a protonated tertiary amine, which is also a HDO. The most commonly encountered negative center is a carboxylate group, which bears two HAC atoms. Indeed, hydrogen bonds (H-bonds) often greatly reinforce ionic interactions where the geometry is favorable. The strong interaction often seen between ligand carboxylates and the guanidinium group of a binding site arginine is a particularly important example of this.

In many cases, H-bonding features are better represented as vectors than as atoms. This can be accomplished by placing an associated *site point* at the position in space where the complementary protein feature would lie given a particular geometry of the ligand. Acceptor site points are associated with donor atoms, and donor site points are associated with acceptor atoms. Polar heteroatoms bearing multiple hydrogen atoms (or lone pairs), therefore, have multiple associated acceptor (or donor) sites. An additional level of complication arises when the molecular orbitals involved are radially symmetrical (e.g., sulfonyl oxygens as acceptor atoms). In such cases, the corresponding site point is actually a torus centered along the axis of the corresponding σ bond.

It is important to bear in mind that the pharmacophoric features in a ligand molecule represent potential interaction points. Many may not actually be directly involved in binding to a particular protein. Indeed, placing a polar feature into a nonpolar region of the binding pocket will generally reduce binding affinity, and placing an ionic feature into such a region may abolish binding altogether. The same is true of the protein: it may present features for which no complementary feature exists in a particular ligand. Such considerations often make it difficult to tell which features are actually involved in the binding of a particular ligand and which are required for binding to a particular protein. Consider the two compounds shown in Figure 8.1, for example. Each presents a distinct array of pharamacophoric features, yet they compete for binding to the target, protoporphyrinogen IX oxidase [2,3].

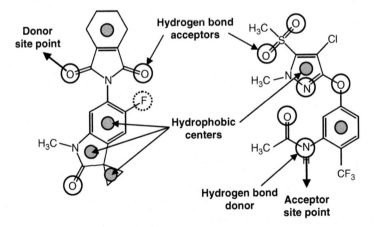

FIGURE 8.1 2D structures for two inhibitors of protoporphyrinogen IX oxidase with their pharmacophoric features indicated.

8.2 OVERVIEW

Pharmacophore models are important for understanding ligand binding, but their use in three-dimensional (3D) searching probably represents their most important contribution to modern drug and agrochemical discovery and development. Once a characteristic pharmacophore query has been formulated for a given target protein, a structural database of potential ligands can be scanned for molecules that fit it and are therefore good candidates for assay. Good ligands are often more or less flexible, however, and different conformations generally present different pharmacophoric patterns. Important hits will be missed in the search if such ligand flexibility is not properly accounted for.

One approach to this problem has been to populate 3D databases with many conformers for each molecule in the database, then to run rigid pharmacophore searches against all of the conformers. This presumes that all biologically relevant conformations can be in the database, which is not an unreasonable assumption for small molecules with only a few torsionally constrained rotatable bonds, but starts to break down for larger structures containing many rotatable bonds. It is particularly problematic where bonds between sp^2 and sp^3 hybridized atoms are involved, because such bonds typically have broad, shallow torsional energy profiles. The directed tweak technique was developed to address this concern, thereby enabling fully flexible 3D searching [4].

To run such a pharmacophore search and get useful results, one must know the correct pharmacophore query to use. Appropriate queries can, in some cases, be deduced from the structure of ligand–protein complexes. More typically, however, the query is an hypothesis based on the binding properties of a set of known ligands combined with some kind of conformational analysis. This is often difficult to do properly because, as noted above, each substructure can present multiple features and all pharmacophoric elements need not be matched by all ligands. An adequate discussion of the scope and limitations of the various programs developed for carrying out this task is an interesting subject [5], but is outside the scope of the present work.

Beginning late in 2000, Tripos, Inc., and Novo Nordisk S/A embarked on a collaborative research and development project to determine whether an alternative similarity searching strategy was feasible. Rather than try to match a rigid query to a single conformer of a target, the full pharmacophoric pattern found for each ligand across a range of conformers was decomposed into its constituent two-, three-, or four-feature multiplets. Molecules were then compared by comparing their multiplet signatures (multiplet fingerprints, or *tuplets*), the expectation being that ligands that can bind to the same target are likely to have similar fingerprints.

Pharmacophore triplet fingerprints — derived from trios of features together with the pairwise distances separating them — were originally developed primarily for measuring diversity rather than for the similarity applications envisioned for this project [6]. They were initially commercialized as part of the ChemX suite of programs [7] and later in the PDT module in SYBYL [8]. Such fingerprints subsequently saw a great deal of use in compound acquisition and library design programs, where the focus was generally on identifying compounds that presented individual multiplets not found in compounds already in a corporate collection [7,9–11].

Four-feature pharmacophores (quartets) eventually saw use in the design of focused libraries [12–14], but generally as filters applied to make sure that particular quartets appeared in the products. Overall pharmacophoric similarity has never been widely used to drive database searching, in part because the memory requirements were too great and because processing speeds were too slow.

8.3 CONCEPTUAL ADVANCES

The ChemX software was no longer commercially available when scientists working at Tripos and at Novo Nordisk began development of the Tuplet Toolkit[1]. Preliminary descriptions of the methodology involved were presented at several National American Chemical Society (ACS) meetings [15,16]. A more complete description of the methodology, including its application to corporate database screening for estrogen antagonists, has been published recently as well [17]. Our program differs from other implementations in several key respects.

8.3.1 BITSETS VS. BITMAPS

Full multiplet bitsets are never generated, manipulated, or stored. Instead, each bitset is represented by compressed form known as a bitmap that is created by replacing long strings of 0s (or 1s) with an indicator of how long that string of 0s is. Hence a *bitmap* of

$$0,1000;1,1;0,200;1,2;0,100;\dots$$

represents a *bitset* of 1000 0s followed by a single 1, followed by 200 more 0s, then by 2 1s and 100 0s, and so on. Multiplet fingerprints are characteristically rather sparse, so this compression scheme typically yields a considerable savings in storage space and working memory. Moreover, bitmaps can be manipulated much more efficiently than can the corresponding bitsets, especially for the large, sparse fingerprints generated from pharmacophore quartets and related multiplets. This is true even for Boolean operations such as ANDing and ORing, which are computationally efficient in other contexts.

8.3.2 MULTIPLET MAPPING

Each multiplet must map to one specific bit in the fingerprint, and it must always be possible to recover the identity of the multiplet that led to a specified bit being set in a fingerprint. In the Tuplet Toolkit, each triplet is given a unique index by first sorting its edges in decreasing order of bin index, and then assigning the vertex feature types in the order ABC where B is the feature type of the vertex lying at the intersection of the longest and shortest edges; A is the vertex joined to B by the longest edge; and C is the vertex joined to B by the shortest edge. Quartets are generally indexed by first identifying a base triangle that includes the longest and shortest edges, then appending edge bin indices in the order in which they connect the "left out" vertex to the three making up the base triangle. One possible quartet is then specified by:

8 7 2 5 3 2 Hy Da Da Aa

where 8 7 2 Hy Da Da constitute a base triangle composed of a hydrophobic center (Hy) and two donor atoms (Da), with an acceptor atom (Aa) at the fourth vertex. The other three faces of the quartet's tetrahedron would then be triplets definable by:

8 5 3 Hy Da Aa

7 5 2 Hy Da Aa

3 2 2 Da Aa Da

Ambiguities between vertex assignments are resolved by putting the degenerate types in the order in which their definitions appear in the configuration file. The ways in which other, more complex ambiguities are resolved is described elsewhere [17]. Encoding multiplets in this way consolidates the geometrically impossible multiplets into long stretches of zeros that can then simply be compressed out.

The bit positions that would otherwise map back to impossible multiplets can also be used to encode the chirality of quartets. The indexing described above is used if the fourth vertex lies to the "right" of the base triangle, whereas the edge indexing of the base triangle is inverted if the fourth vertex lies to the "left." Hence the left-handed version of the quartet described above would be:

2 7 8 2 3 5 Da Da Hy Aa

Note that the vertex indices follow the edges during inversion. Conversely, the inverted order of the initial edge indices immediately identifies this as a left-handed quartet.

8.3.3 ACCOMMODATING A RANGE

Besides facilitating compression, this coding scheme is readily extensible to accommodate a broad range of multiplet species. The fourth vertex in a quartet can be assigned to a specific substructure previously identified as important for binding (e.g., a privileged substructure [13]), such as a hydroxamic acid or sulfonamide group involved in binding to Zn metalloproteins. It can also be assigned to a site point associated with one of the features making up the base triangle, thereby forming an augmented triplet. Fingerprints based on such multiplets are a more effective way of capturing the information carried by site points than is incorporating them directly into the triplet definitions.

8.3.4 BITMAP SIZE

Druglike molecules usually bear a fair number of pharmacophoric features, so the number of bits set in a tuplet generally increases as the dimensionality of the multiplet increases: there are more quartets than triplets and more triplets than doublets. The number increases more slowly, however, than does the number of possible multiplets, which is what dictates the length of the corresponding bitset. The net result is that the corresponding fingerprints grow progressively more sparse and the growth in bitmap size can be handled without having to set artificial limits on the number of distance bins allowed. The indices are another matter, however; even a relatively modest feature set and binning scheme will generate indices greater than 2^{32}, which necessitates the use of 64-bit (or higher) executables. Hence some mechanism had to be found by which a tuplet generation and manipulation program compiled in a 64-bit architecture could communicate with database manipulation code compiled in a 32-bit environment. This was accomplished by generating an intermediary American Standard Code for Information Interchange (ASCII) data stream formatted in Extensible Markup Language (XML). The XML hierarchy consists of tags for a structure's SLN-ID (an index into the UNITY database), each conformer, a feature type, properties, and XYZ coordinates. The XYZ tag, then, sets off a list of x, y, and z coordinates for feature of the corresponding type found in each conformer of each SYBYL line notation (SLN) [17].

8.3.5 STOCHASTIC COSINE

Earlier tools for generating and manipulating multiplet fingerprints were predicated on the assumption that complete conformational sampling could be achieved by starting from 3D structures created using a rule-based system, then applying large torsional increments to each bond. The approach had the added virtue of being nominally deterministic. As noted above, however, it is likely to be adequate only for bonds between pairs of sp^3 atoms. We chose to make the means of evaluating similarity independent of the way in which the conformers were generated. Hence a new similarity measure, the *Stochastic Cosine*, was developed that is applicable when the conformational sampling method used is stochastic in nature:

$$C*(a,b) = \frac{E(|a \cap b|)}{\sqrt{E(|a \cap a'|) \times E(|b \cap b'|)}}$$

Here, a and a' correspond to tuplets summed across different samples from one conformational population, whereas b and b' represent tuplets for samples drawn from a different population (e.g., for a different molecule). The vertical bars indicate application of the cardinality operator, and E indicates that an expectation is being taken. For determinate cases, a and a' are identical, as are b and b', so the expression reduces to the well-known cosine similarity coefficient.

As is discussed in more detail below, generating a tuplet hypothesis involves truncating a fingerprint rather severely, which can distort symmetrical similarity coefficients like the Stochastic Cosine. It is often more appropriate when a represents a tuplet hypothesis to use the Stochastic Asymmetric Similarity defined by:

$$A*(a,b) = \frac{E(|a \cap b|)}{E(|a \cap a'|)}$$

In part to support such analyses, each tuplet file is actually made up of four separate bitmaps: the intersection across all conformers, the union across all conformers, and the unions across each of two subsets obtained by random assignment of the conformers.

8.4 APPLICATION

The example described here is based on a filtered subset of 31,639 pharmacologically active compounds taken from the World Drug Index (WDI) [18]. Of these, 42 included the keywords "angiotensin-2 antagonist" in the MECHANISM field; these were used as known actives and served as the training set for generating the hypothesis. Because the same procedure is used to generate multiplet fingerprints for the training set as for the database being searched, protocols involved in doing so are outlined first, followed by procedures for characterizing the training set, generating a hypothesis, and searching the full 31.6 K WDI database.

8.4.1 PREPARING THE DATASET

1. Structures need to be entered into a UNITY database using the dbimport utility. They are generally brought in as hit lists of structures defined in SLN [19], although other file formats can be accommodated directly or by conversion from SMILES, SD, or MOL2 formats using dbtranslate.
2. Every structure needs to have 3D coordinates associated with it. These may already be available from the corresponding corporate or commercial UNITY database or they may need to be generated specifically for this application. Several alternative methods for generating 3D structures from 2D ones are available in the latter case:
   a. CONCORD is a rule-based system developed by Pearlman et al. (University of Texas — Austin) for generating biologically relevant, low-energy conformers from 2D structures. It is available exclusively from Tripos, Inc. [8].
   b. CORINA was developed at the University of Erlangen by Gasteiger et al. and is available from Molecular Networks GmbH [20]. It combines rules and energy minimization to generate 3D structures from 2D input.
   c. OMEGA is available from OpenEye Scientific Software [21]. It utilizes a systematic search method based on the Clean force field.
3. Pharmacophore multiplet analysis can be carried out using a single conformer for each molecule of interest, but the similarities obtained in that

case may reflect structural commonalities as much pharmacophoric ones [17]. The ensembles of conformers from which multiplet fingerprints are more usually created can be generated by using the CORINA or OMEGA programs cited above, but several other methods are available:

a. Confort is the companion tool to CONCORD and was also developed at the University of Texas. It generates conformational ensembles from input 2D or 3D structures. It identifies the global energy minimum with respect to the Tripos force field, along with all local minima within some specified energy range. Alternatively, it can be set to return a diverse set of low-energy conformers.

b. Catalyst is available from Accelrys [22], a subsidiary of Pharmacopeia. The program takes 3D structures as input and generates a multiconformer database of diverse conformers by running a poling algorithm. This involves running a series of torsional minimizations wherein parts of the energy surface near conformations identified in earlier iterations are distorted, thereby driving subsequent minimizations away from them [23]. The arrays of conformers produced are intended for use in 3D searching within Catalyst, but, presumably, can be exported to UNITY for use with the Tuplet Toolkit.

c. The Tuplet Toolkit itself can generate conformers from input 3D structures on the fly. It does so by assigning a random torsion to each rotatable bond and then relaxing the molecule using the directed tweak algorithm [4] to relieve steric clashes. The internal conformational generator will be called upon whenever more conformations are called for in the tuplets configuration file (see below) than are available in the input database. If a hitlist is supplied as input, a temporary UNITY database will be created in the background, and any conformers above and beyond the ones supplied in that hitlist will be generated on the fly.

8.4.2 SPECIFYING TUPLET GENERATION PARAMETERS

The selection of the multiplet styles generated is determined by the contents of the tuplets configuration file. If no file path is supplied at the command line, tuplets.cfg will be used. If there is no local copy of a file by that name, Tuplet Tools will use the one supplied with UNITY. Critical information supplied by the entries in this file include:

* *The number of conformers to use* — 100 conformers by default. Conformational space is more thoroughly sampled at higher conformer counts, but the association between multiplets from each single conformation tends to be lost. This default represents a compromise level at which the similarity between randomly chosen pairs of molecules stays relatively low and the similarity between pharmacologically related ones is relatively high. Initial experiments indicate that looking at more conformers tends to oversaturate the multiplet fingerprints and reduce discrimination.

- *The kinds of pharmacophoric features to allow as vertices* — The default set of feature types consists of DONOR_ATOM, ACCEPTOR_ATOM, HYDROPHOBIC, NEGATIVE_CENTER, and POSITIVE_N. These are drawn from feature definitions found in the user-editable sln3d_macros.def file (the local one if such exists, otherwise the one found at $TA_3DB/tables. This is also where privileged substructures (e.g., POSITIVE_N) and augmentation points (e.g., DONOR_SITE and ACCEPTOR_SITE) are specified. These can also be defined in-line using SLN, as can the main feature types. Aromatic feature types are not included at present because of the difficulty in formulating a precise definition capable of capturing those cases (e.g., oxopyrimidine lactams) that are not hydrophobic.
- *How multiplet edges are to be binned* — Results are not strongly dependent on the exact binning scheme used, but a preliminary survey of multiplet frequencies found within and between pharmacological classes indicates that the following binning scheme (in angstroms) generally works well for triplets:

0–1.75	1.75–3.0	3.0–4.0	4.0–5.0	5.0–6.0
6.0–7.0	7.0–8.0	8.0–8.75	8.75–9.75	9.75–10.75
10.75–11.75	11.75–13.0	13.0–15.0	15.0–	

Bin ranges most appropriate for working with quartets may differ somewhat. Note that the bin ranges do not have to be contiguous.
- *How multiplet edges are to be weighted for creating hypotheses* — These values are unity by default and were not changed for the analyses described here. Flexible 3D searches carried out against databases of druglike compounds using model queries indicate that HDOs and HACs are similar to hydrophobic centers in their intrinsic discriminating power, but that ionic features carry greater weight in that context. Studies are currently under way to determine whether this is also true for tuplets.

Triplet fingerprints generated with the above parameter set will be 275,674 b in length (3 vertices of 5 types and 3 edges of 14 possible length implies $5^3 \times 14^3$ distinct combinations, including the geometrically impossible ones). Quartet fingerprints, defined as they are by 4 vertices and 6 edges, will be 3,016,755,625 b in length ($5^4 \times 14^6$). The dimensionality itself is specified on the command line or from the SYBYL GUI (graphical user interface), not in the configuration file.

Note that comparisons between multiplet fingerprints are unlikely to be meaningful if they were generated using different configuration files.

8.4.3 CREATING MULTIPLET FINGERPRINTS

1. Tuplet generation can be initiated directly from the UNIX command line using the dbmktuplets command[2] or via the Tuplet Tools dialog available within SYBYL.

2. Each structure in the database is parsed using the `mkstream` executable to determine the coordinates for all features of any type identified in the configuration file. This is done for every conformation found in the database up through the specified conformer count, with additional random conformations generated on the fly as needed to make up any shortfall. Results are written out to a XML file for subsequent tuplet generation. This file can be retained by setting the appropriate command flag, but is normally deleted immediately after use. Calculation of XML feature streams for the full 31.6 K data set took 11.1 central processing unit (CPU) hours on a single 600 MHz SGI R14000 processor in a Origin 300 Server or 1.26 sec/structure. Most of this time was spent in creating 99 conformers (or as many as possible) for each molecule.

3. The XML stream is parsed by a 64 b executable that takes the bin boundary definitions from the configuration file and does the combinatorial calculations required to generate the corresponding bitmaps. Calculation of triplet fingerprints from the feature stream for 100 conformers from 31.6 K compounds required 1.08 hours on the machine described above (0.12 sec/structure).

4. The tuplet bitmaps are stored within their own directory as binary files indexed to the `SLN_ID` of the compound in the original UNITY database or serially for data sets input as hitlists.

$$\frac{1354626kB}{31639cpds} = 42.81\frac{kB}{cpd}$$

By default, each of the files contains four bitmaps for each compound (see above). For triplets, the bitmaps for the full database required 1.29 GB of storage, an average of 42.8 kB per compound. Given that each file actually represents four bitmaps, the compression ratio obtained is:

$$42.81\frac{kB}{cpd} \times \frac{1cpd}{4Fprints} \times \frac{1024Bytes}{kB} \times \frac{8bits}{Byte} = 87685\frac{bits}{Fprint}$$

$$Compression = \frac{275674bits\,/\,Fprint}{87685bits\,/\,Fprint} > 3{:}1$$

Much greater compressions are achievable for pharmacophore quartets.

8.4.4 EXAMINATION OF TUPLETS FOR A CLASS OF ACTIVE COMPOUNDS

1. The same procedure described above for the full database was applied to the 42 compounds from the training set. The structures for these compounds are given in Figure 8.2 to Figure 8.4, where they are split out by

structural class into 30 biphenyltetrazoles (Figure 8.2); 5 sulfonimides and the acidic sulfonanilide GR-138950-C (Figure 8.3); and 5 benz- and pyridino-imidazoles, along with one flavone (Figure 8.4). Each structure is annotated with its Tanimoto and Stochastic Cosine self-similarity coefficients; the latter is set off in parentheses in each case.

2. The tuplets created from the training set database were created with a different random number seed from those created for the same compounds in the full database. Hence the conformers used were not identical in the two cases, causing differences in the overall fingerprints that could potentially affect the ability of a compound to retrieve itself in a similarity search. As expected, the net result is that the Tanimoto similarity values obtained are less than unity – considerably so in some cases. When the stochastic effect is taken into account by using the Stochastic cosine, the self-similarities come out, as expected, much closer to 1. The relationship between Tanimoto and Stochastic Cosine self-similarities is illustrated graphically in Figure 8.5.

8.4.5 Clustering Actives By Pharmacophoric Similarity

1. Having a set of high throughput screening (HTS) hits or some other set of active compounds in hand, the next task is to characterize the relationships among compounds within the descriptor space defined by pharmacophore mulitplet fingerprints. This was done by running `tuplet_compare` across the training set 42 times, taking a different compound as reference each time and producing a distance matrix derived from the Stochastic Cosine similarities between molecules.

2. The distance matrix so obtained was brought into a molecular spreadsheet (MSS) within SYBYL for subsequent analysis with tools from the quantitative structure–activity relationship (QSAR) module. Hierarchical clustering was then carried out using the complete linkage method, with no weighting or scaling of the values. The dendrogram obtained is shown in Figure 8.6.

3. That the clustering seen within the training set follows the groupings set out in Figure 8.2 through Figure 8.4 is actually somewhat surprising, given the overlaps in 2D structure. Interestingly, however, benzimidazole TAK-536 clusters with the sulfonimides rather than with the fused-ring imidazoles, whereas L-158809 clusters with fused-ring imidazoles rather than with the biphenyl tetrazoles (Figure 8.6). Note that PD-098059 (2'-amino-3'-methoxyflavone) is a singleton pharmacophorically as well as structurally. These clustering results indicate that the structural subclasses carry over into the pharmacophoric realm and suggest that it might be profitable to consider creating separate hypotheses for each class.

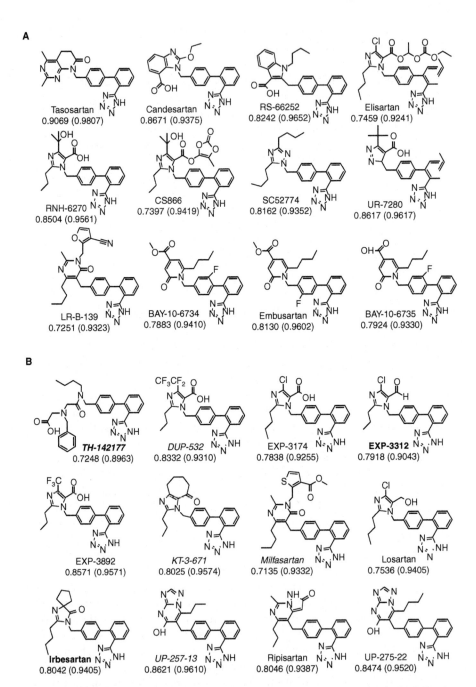

FIGURE 8.2 Biphenyl tetrazole angiotensin II receptor antagonists included in the full training set. Names set off in **boldface** type were used to create the *diverse* hypothesis as well, whereas names set off in *italics* were used to create the *tetrazoles* hypothesis. Appended values indicate the Tanimoto and Stochastic Cosine self-similarities for each compound, with the latter set off in parentheses.

c

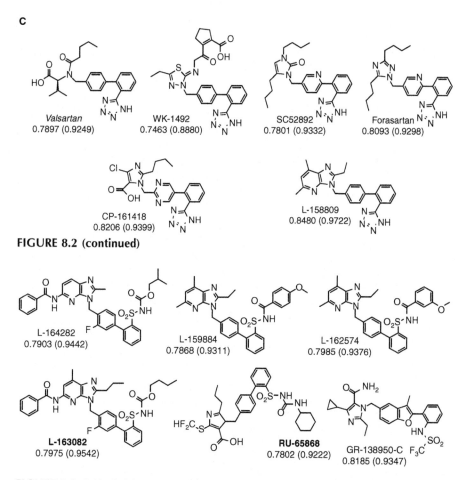

Valsartan
0.7897 (0.9249)

WK-1492
0.7463 (0.8880)

SC52892
0.7801 (0.9332)

Forasartan
0.8093 (0.9298)

CP-161418
0.8206 (0.9399)

L-158809
0.8480 (0.9722)

FIGURE 8.2 (continued)

L-164282
0.7903 (0.9442)

L-159884
0.7868 (0.9311)

L-162574
0.7985 (0.9376)

L-163082
0.7975 (0.9542)

RU-65868
0.7802 (0.9222)

GR-138950-C
0.8185 (0.9347)

FIGURE 8.3 Sulfonimide and sulfonanilide angiotensin II receptor antagonists included in the full training set. Names set off in **boldface** type were used to create the diverse hypothesis. Appended values indicate the Tanimoto and Stochastic Cosine self-similarities for each compound, with the latter set off in parentheses.

8.4.6 Creating Tuplet Hypotheses and Screening a Database

1. The process of creating a tuplet hypothesis begins with the creation of a count vector, each element of which represents the number of bitmaps in which the corresponding bit was set to 1 (actually, this is a count *map*, because it, too, is compressed).

2. Some multiplets are intrinsically more discriminating than others. In particular, large multiplets are more discriminating than small ones are. Hence, the count vector is converted into a scores vector by multiplying the count for each bit by two weights: one obtained by summing the weights for the contributing feature types, and the other by summing the weights for the bins to which each edge was assigned. For the analyses

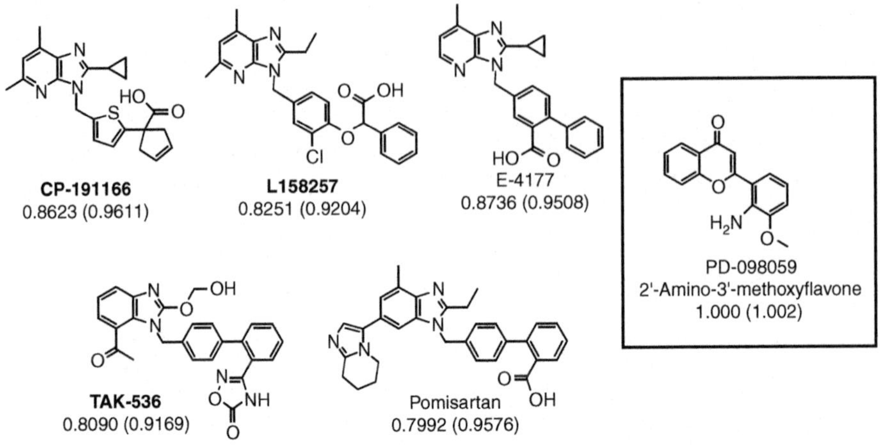

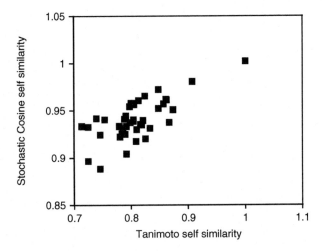

FIGURE 8.4 Benz- and pyridinoimidazole angiotensin II receptor antagonists included in the full training set. The sole flavone antagonist is also shown. Names set off in **boldface** type were used to create the diverse hypothesis as well. Appended values indicate the Tanimoto and Stochastic Cosine self-similarities for each compound, with the latter set off in parentheses.

FIGURE 8.5 Tanimoto self-similarities of training set compounds plotted as a function of Stochastic Cosine self-similarity.

done here, all feature types received a weight of 1. The weights used for the distance bins were the default values, which are roughly proportional to their upper boundaries [17].

3. The bits corresponding to the k highest-scoring bits in the scores vector are then set in the hypothesis. The value of k specified at the command line or through the GUI then determines how far into the data the hypoth-

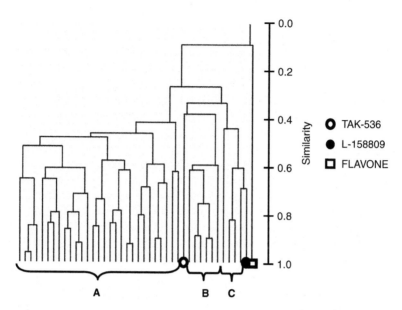

FIGURE 8.6 Hierarchical clustering based on Stochastic Cosine similarities between pharmacophore triplets obtained from training set antagonists. The side-scale indicates the degree of similarity at various levels in the dendrogram. (A) Biphenyl tetrazoles. (B) Sulfonimides and sulfonanilide. (C) Benz- and pyridinoimidazoles.

esis is to reach (i.e., the depth of detail included in the model). The default value used here — 100 — is a reasonably good starting point, but it will often prove necessary to examine an array of hypothesis depths.

4. Based on the information garnered from the clustering study described above, 3 different hypotheses were created. The first, rather obvious approach was to use all 42 compounds in the training set. This will be referred to as the *full* query.

5. A second, more specific *tetrazoles* hypothesis was created using only the 10 biphenyl tetrazoles whose names are highlighted in *italics* in Figure 8.2. This was done to see how the structural diversity of the training set would affect the recovery of active compounds from the full database.

6. A training set composed of the 10 compounds highlighted in **boldface** in Figure 8.2 through Figure 8.4 was used to create a third, more broadly based *diverse* hypothesis. Results from searches using this hypothesis were expected to provide useful information about the expected efficiency with which actives not included in the training set could be recovered from the full database.

7. Figure 8.7 illustrates the results from using each of the three hypotheses to screen the full 31.6 K database. The compounds were sorted in ascending order of Stochastic Asymmetric similarity to the hypothesis, and the number of known actives recovered at each level was plotted. The number of compounds recovered from the training set are indicated by the filled

circles in the 3 plots shown. The tetrazoles hypothesis gave somewhat better recovery early in the search (i.e., among the 10 compounds most similar to the query) and the diverse hypothesis performed somewhat less well. Nonetheless, all three showed excellent enrichments (310- to 420-fold) at the 50% recovery mark.

8. Recovery of a second class of compounds is indicated by the heavy solid lines in Figure 8.7. This class is composed of the training set plus the 150 compounds in the full dataset that are marked as being angiotensin antagonists, but not as being angiotensin II antagonists *per se*. For the full hypothesis, this line tracks the light weight line that indicates perfect recovery through the first 10 compounds. Recovery is slightly inferior for this class using the tetrazoles hypothesis and 2- to 3-fold worse using the diverse query.

9. Recovery of a third class of compound is indicated by the broken lines in Figure 8.7. This group included all compounds marked as angiotensin antagonists, but was extended to include those compounds for which no *mechanism* was indicated in the database but which, by visual inspection, were nonetheless likely to be angiotensin II antagonsists. In some cases, the lack of annotation is clearly an oversight. In others, the compound in question may well not be active but is one that most medicinal chemists would still want to see assayed before allowing it to be set aside.

    The striking thing about the curves for this class of compounds is how disproportionately superior the *diverse* hypothesis is at recovering them than at recovering those labeled as "angiotensin antagonists." The take-home message would seem to be that a narrowly defined training set may identify actives more efficiently, but that a more broadly based hypothesis is likely to capture a more diverse crop of inhibitors. How generally true that conclusion may be remains to be seen, but the tools needed to address the question are clearly now in hand.

10. Although there are significant differences between the three sets of recovery curves shown in Figure 8.7, these differences are relatively small. This, taken together with the good recovery of "external actives," indicates that the system is probably not overtraining and is reasonably robust. Indeed, plots of similarity ranks obtained from the different hypotheses against each other are remarkably linear. Interestingly, the outliers are all analogs of captopril, which are classed as both angiotensin antagonists and as angiotensin converting enzyme inhibitors. This observation suggests that suitably constructed tuplet hypotheses may prove useful in identifying *selective* chemistries as well as active ones.

ACKNOWLEDGMENTS

Lars Nærum, Henning Thøgersen, and Inge Thøger Christensen (Novo Nordisk A/S) made several key contributions to the development of the software described here. Malcolm Cline, Essam Metwally, David Lowis, and David Baker contributed data for this chapter or have been involved in the "productization" of the program described here.

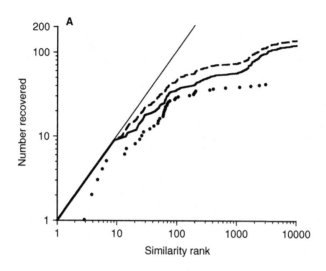

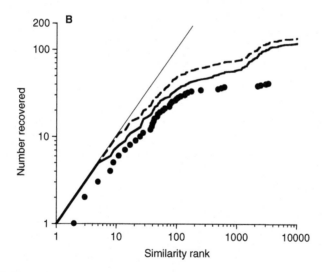

FIGURE 8.7 Pharmacophore triplet search recovery curves for compounds marked as "angio-tensin-2" antagonists (●) in the WDI; angiotensin antagonists (——), and probable angio-tensin antagonists (- - -). The line defining perfect recovery is also shown (——). (**A**) Results for a tuplet search using the full hypothesis. (**B**) Results for the tetrazoles hypothesis. (**C**) Results for the diverse hypothesis.

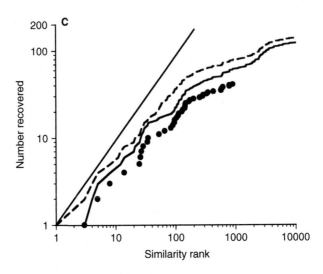

FIGURE 8.7 (continued)

NOTES

1. Tuplet Toolkit is the name tentatively assigned to the software package described here. It is being prepared for inclusion in SYBYL 7.0, which is scheduled for release late in 2003 or early in 2004.
2. One can get a list of all command line flags and arguments for any UNITY command by entering the command followed by -#.

REFERENCES

[1] P. Gund, Evolution of the pharmacophore concept in pharmaceutical research. In O. Güner, Ed., *Pharmacophore Perception, Development, and Use in Drug Design*, La Jolla, CA: International University Line, pp. 3–11, 2000.

[2] R.D. Clark, Synthesis and QSAR of herbicidal 3-pyrazolyl α,α,α-trifluorotolyl ethers, *J. Agricult. Food Chem.* 44:3643–3652, 1996.

[3] U.B. Nandihalli and S.O. Duke, Structure-activity relationships of protoporphyinogen oxidase inhibiting herbicides. In S.O. Duke and C.A. Rebeiz, Eds., *Porphyric Pesticides: Chemistry, Toxicology, and Pharmaceutical Applications*, Washington, D.C.: American Chemical Society, Symposium Series 559, pp. 133–146, 1994.

[4] T. Hurst, Flexible 3D searching: the directed tweak technique, *J. Chem. Inf. Comput. Sci.* 34:90–196, 1994.

[5] Y. Patel, V.J. Gillet, G. Bravi, and A.R. Leach, A comparison of the pharmacophore identification programs: Catalyst, DISCO and GASP, *J. Computer-Aided Mol. Des.* 16:653–681, 2002.

[6] A.C. Good and I.D. Kuntz, Investigating the extension of pairwise distance pharmacophore measures to triplet-based descriptors, *J. Computer-Aided Mol. Des.* 9:373–379, 1995.

[7]   S.J. Cato, Exploring phrarmacophores with Chem-X, In O. Güner, Ed., *Pharmacophore Perception, Development, and Use in Drug Design*, La Jolla, CA: International University Line, pp. 107–125, 2000.

[8]   Tripos, Inc.: http://www.tripos.com.

[9]   H. Matter and T. Pötter, Comparing 3D pharmacophore triplets and 2D fingerprints for selecting diverse compound subsets, *J. Chem. Inf. Comput. Sci.* 39:1211–1225, 1999.

[10]   M.J. McGregor and S.M. Muska, Pharmacophore fingerprinting: 1. Application to QSAR and focused library design, *J. Chem. Inf. Comput. Sci.* 39:569–574, 1999.

[11]   M.J. McGregor and S.M. Muskal, Pharmacophore fingerprinting: 2. Application to primary library design, *J. Chem. Inf. Comput. Sci.* 40:117–125, 2000.

[12]   J.E. Eksterowicz, E. Evensen, C. Lemmen, G.P. Brady, J.K. Lanctot, E.K. Bradley, E. Saiah, L.A. Robinson, P.D.J. Grootenhuis, and J.M. Blaney, Coupling structure-based design with combinatorial chemistry: application of active site derived pharmacophores with informative library design, *J. Mol. Graphics Modelling* 20:469–477, 2002.

[13]   J.S. Mason, I. Morize, P.R. Menard, D.L. Cheney, C. Hulme, and R.F. Labaudiniere, New 4-point pharmacophore method for molecular similarity and diversity applications: overview of the method and applications, including a novel approach to the design of combinatorial libraries containing privileged substructures, *J. Med. Chem.* 42:3251–3264, 1000.

[14]   J.S. Mason and B.R. Beno, Library design using BCUT chemistry-space descriptors and multiple four-point pharmacophore fingerprints: simultaneous optimization and structure-based diversity, *J. Mol. Graphics Modelling* 18:438–451, 2000.

[15]   P. Fox, L. Nærum, E. Abrahamian, H. Thøgersen, R. Clark, and T. Heritage, An efficient bitmap container package for very high-dimensional fingerprints, 223rd ACS National Meeting, New Orleans, 2003, COMP 377.

[16]   P. Fox, E. Abrahamian, R.D. Clark, I.T. Christensen, and H. Thøgersen, Fully flexible pharmacophore multiplet bitmaps as molecular descriptors: implementation and applications, 225th ACS National Meeting, New Orleans, 2003, COMP 377.

[17]   E. Abrahamian, P.C. Fox, L. Nærum, I.T. Christensen, H. Thøgersen, and R.D. Clark, Efficient generation, storage and manipulation of fully flexible pharmacophore multiplets and their use in 3-D similarity searching, *J. Chem. Inf. Comput. Sci.* 43:458–468, 2003.

[18]   Thomsom Derwent: http://thomsonderwent.com/worlddrugindex.

[19]   S. Ash, M.A. Cline, R.W. Homer, T. Hurst, and G.B. Smith, SYBYL line notation (SLN): a versatile language for chemical structure representation, *J. Chem. Inf. Comput. Sci.* 37:71–79, 1997.

[20]   Molecular Networks GmbH: http://www.molecular-networks.de.

[21]   OpenEye Scientific Software: http://www.eyesopen.com.

[22]   Accelrys, Inc.: http://www.accelrys.com.

[23]   A. Smellie, S.L. Teig, and P. Towbin, Poling: promoting conformational variation, *J. Comput. Chem.* 16:171–187, 1995.

Part IV

Important Considerations Impacting Molecular Docking

9 Potential Functions for Virtual Screening and Ligand Binding Calculations: Some Theoretical Considerations

Kim A. Sharp

9.1 INTRODUCTION

Virtual screening (VS) of compounds for possible drug leads requires identifying the relatively few candidates, out of perhaps many thousands, which can bind with significant affinity (typically 100 μM or better) to a target of a known structure. An enrichment factor (EF) can be defined as [1]

$$EF = \frac{a/n}{A/N} \tag{9.1}$$

where a is the number of active compounds in the n top-ranked compounds of a total database of N compounds of which A are active. Successful screening implies EF >> 1. This in turn requires the identification of the best ligand conformation/position/orientation (pose) in the target binding site, that is, the solution of the docking problem. Solution of the docking problem in turn requires the ability to accurately calculate the binding affinity of a given pose (at least relative to another pose), which is the solution of the binding problem. Clearly VS is a formidable task that must ultimately depend on the solution of the binding problem at some level of accuracy. The binding affinity defined here as the association constant $K_a = [PL]/[P][L]$ (or often in the literature by the dissociation constant $K_d = 1/K_a$), or equivalently, the binding free energy $\Delta G_{bind} = -kT\ln(K_a)$, is an equilibrium thermodynamic property that is connected to the underlying molecular and atomic detail structure of the target, ligand, and solvent through the laws of statistical mechanics. The view taken in this chapter is that any method that successfully scores or ranks candidates or

229

poses must be doing it by emulating, however approximately, the true binding free energy. This must be so whether one calls the ranking method a scoring function, a filter function, a screening potential, and even if it was derived with the explicit lesser goal of "not trying to calculate a real binding free energy." As they say, "if it walks like a duck and talks like a duck...." Referring to the enrichment factor, Equation 9.1, even if EF >> 1 is achieved in an initial screen simply by dropping out a large number of compounds that are difficult or impossible to fit into the binding site (i.e., a shape filter type function), the criteria used to do this must emulate, in a crude way and over the whole range of compounds, the difference in actual binding free energy between the worst compounds and the others.

The goal of this chapter is to present a theoretician's perspective on the necessary ingredients that must go into a real binding affinity or free energy calculation. It is recognized that the extreme computational demands of screening large numbers of compounds currently make the more realistic methods of binding free energy calculation impractical in early stages of screening. However, two factors must be borne in mind: Computers get faster and algorithms get smarter. It is impossible to predict what shortcuts or clever tricks may be developed to mimic binding free energy components, providing one knows the physical underpinnings of those components. For example, it is now routine to replace the complex solvent rearrangements that give rise to the hydrophobic effect with a simple surface area term. This is both quick and successful because the physical origin, magnitude, and functional form of this free energy contribution are well known.

The goal of this chapter is not to provide an extensive review of different screening, scoring, or binding potential functions. Excellent reviews have recently been published [1–4]. The goal is to provide some concise but hopefully practical principles to enable the virtual screeners to parse either their own or published screening potentials with a view to improving them. One aim is to help the researcher decide whether a screening potential's deficiencies arise because the potential is unphysical, has an inadequate functional form, or because the parameters in it need to be improved. The practical consequences can be significant: In the first two cases, no amount of parameterization with training and test sets will improve the screening potential significantly. The effort would be better spent retooling the potential function.

The outline of the chapter is as follows. First, a short review of the statistical mechanics of binding is presented with a discussion of some practical consequences. Next, some general issues are discussed that are common to any method of scoring. Finally, some issues specific to particular types of scoring/binding potentials are discussed.

9.2 TYPES OF SCORING/BINDING POTENTIALS

For the purpose of this chapter, screening/scoring/binding potential functions will be divided into three types (with some overlap) — physics-based, statistical-based, and empirical. These types are defined below and discussed in more detail in the appropriate sections below.

Physics-based potentials describe the interactions of atoms, groups and molecules through the fundamental forces recognized by physical chemists — covalent bonding, van der Waals (vdW), electrostatic forces, etc. They are usually described by what is termed a Hamiltonian, force field, or potential function, plus a prescription (e.g., molecular dynamics, Monte Carlo (MC) sampling, finite difference methods) for extracting equilibrium or dynamic quantities. A distinguishing feature or goal of physics-based potentials is their transferability. The required input parameters are obtained, perhaps entirely, from physical measurements other than binding affinities and they may be applied to many problems other than binding. This gives them a high degree of statistical and parameter value independence from binding data. Thus to the extent that they reproduce a set of binding data, they may be taken to model the real physical interactions that determine affinity. Examples of physics-based potentials include the AMBER [5], CHARMm [6], and Poisson-Boltzmann [7] descriptions of molecule/solvent systems.

Statistical-based (or knowledge-based) potentials were first applied to protein folding and later applied to binding [8,9]. These extract effective atomic or group interaction potentials as a function of distance from known protein–ligand complex structures. In principle, data from both protein stability and binding can be used in obtaining these potentials, although current versions designed for binding use only binding data in their parameterization.

Empirical potentials are derived by picking some functional form for the potential (using physics-based knowledge if possible) that relates structural features of the binding complex (e.g., number of contacts of type X) to the binding affinity. The coefficients of the individual terms are then parameterized on as large a set of experimental binding data and in the most statistically powerful and valid way permissible. An early, widely used example is the potential function for Ludi [10].

9.3 BASIC THEORY OF ABSOLUTE AND RELATIVE BINDING AFFINITY

9.3.1 ABSOLUTE BINDING AFFINITY

Detailed treatments of the theory of binding have recently been published [11,12] to which the reader is referred for more details. Here a condensed version of the treatment of Luo and Sharp is given [12]. Consider a binding reaction between a ligand L and its target protein P governed by the equilibrium $L + P \Leftrightarrow LP$. The evaluation of equilibrium properties, including the binding affinity, requires ensemble averaging over all the degrees of freedom of the system. If there are N_s solvent atoms, N_l ligand atoms and N_p target protein atoms, the total number of degrees of freedom is $3(N_s + N_l + N_p) - 6$. It is helpful to divide these degrees of freedom into three types, based on their role in binding, and the fact that one might use different methods of treating each type: The $3N_s$ solvent degrees of freedom, $3N_p - 6$ protein internal coordinates, $3N_l - 6$ ligand internal coordinates, and six coordinates (three translational, three rotational) that describe the relative position and orientation of the ligand with respect to the protein. The relative position and orientation degrees of freedom are unrestricted when the protein and ligand are unbound (technically the position is restricted only by the volume of the solution) and become restricted when binding occurs. In contrast, the internal coordinates may change upon binding, but their range is restricted in both bound and unbound states by the dictates of the molecular structures. The internal and translation/rotation degrees of freedom involve one or two molecules, respectively, and the solvent degrees of freedom will involve many molecules. These differences guide the strategies used in their averaging.

It can be shown that a simple, but general and rigorous expression for the association constant $K_a = [LP]/[L][P]$ is given by

$$K_a = \frac{1}{8\pi^2} \int H(\mathbf{r},\Omega) e^{-\omega(\mathbf{r},\Omega)/kT} d\mathbf{r} d\Omega \tag{9.2}$$

where k is the Boltzmann constant, T is the absolute temperature, $\mathbf{r}$ and Ω specify the position and orientation of A with respect to B, with three degrees of freedom each respectively. $\omega(\mathbf{r},\Omega)$ is the potential of mean force (PMF) between L and P (i.e., the thermodynamic work or free energy for bringing L from far away to a position/orientation $\mathbf{r},\Omega$ with respect to P). The PMF ω completely describes the interaction between L and P. It depends explicitly on just position and orientation, so this means that the solvent and internal molecular coordinate contributions have been averaged over and are thus included implicitly. The treatment of the solvent terms through a PMF provides a sound physical justification for the use of implicit solvation treatments in scoring functions [13].

$H(\mathbf{r},\Omega)$ is a binary function that is 1 for configurations of P/L where they form a complex, and 0 otherwise. $H(\mathbf{r},\Omega)$ simply serves to ensure that the integral of the Boltzmann factor $e^{-\omega/kT}$ is evaluated only over the position/orientation phase volume that constitutes the complex. In spite of its simplicity, evaluation of Equation 9.2 in actual binding reactions is a formidable task because of the enormous number of protein, ligand, and solvent atom configurations that can contribute to $\omega(\mathbf{r},\Omega)$.

To illustrate the most important consequences of Equation 9.2 for binding, consider a simplified case, where the interaction depends only on the position of L with respect to P. Further, consider the interaction to be identical in the three spatial directions x, y, and z and described by a square well PMF with depth E (mean energy) and width σ (Figure 9.1A).

For this case, Equation 9.2 can be evaluated analytically as

$$K_a = \sigma^3 e^{E/kT} \tag{9.3}$$

This expression has two terms, one that depends on the exponential of the well depth, the other that depends on the width or, more precisely, the volume of the binding well. It is customary to define an absolute free energy of binding as

$$\Delta G_{bind} = -kT \ln(K_a) \tag{9.4}$$

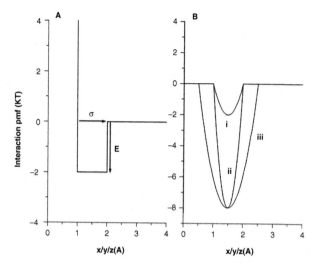

FIGURE 9.1 Model binding interaction PMFs. Abscissa represents position in x, y, or z directions. A: A square well potential of width σ and depth E. B: Truncated three-dimensional harmonic potentials with depths of 2kT (potential i), 8kT (Potential ii and Potential iii), respectively. Potential i and Potential iii have identical curvature. Potential i and Potential ii have identical width binding regions.

with K_a expressed in units of 1/M, which is how much of the experimental affinity data is published. For this simple example this yields

$$\Delta G_{bind} = -E - kT\ln \sigma^3 \tag{9.5}$$

where, simplifying somewhat, the second term is the residual translational entropy retained by A when bound to B, or vice versa (the terms *cratic* or *mixing entropy* should be avoided as entirely too misleading, controversial, and unenlightening).

From this analysis we may draw the following conclusions:

- Determination of the binding constant requires a definition of the complex (i.e., specification of the configurations where $H(\mathbf{r},\Omega) = 1$). In the square well example, we naturally took the region of the square well. In practice, the choice may not be so obvious. However, it must be made, explicitly or implicitly, to correspond, in some sense, to that measured by the techniques used to obtain the experimental binding affinities.
- K_a has units of volume or length³. The concentration units sets the conversion factor, or unit of length for the Boltzmann integration over x/y/z, or vice versa. For example, if K_a is in units of l/mol the unit of length would be 11.8 Å (obtained from $(1000cm^3/N_a)^{1/3}$, where N_a is Avogadro's number).
- Determination of K_a requires not just the evaluation of the depth of the minimum in the binding well (using say the crystal structure of the complex), but determination of the volume of the binding well or the

residual entropy. In more realistic cases, Equation 9.2 tells us this requires a six-dimensional Boltzmann weighted phase volume integral over three position and three orientation coordinates, not simply the width of the potential well in the simple example.

- As a corollary, a single determination of the binding interaction energy ω does not yield the binding affinity; even if it is a real free energy and evaluated at the minimum of ω.
- From Equation 9.4 the numerical value of ΔG_{bind} depends on the units used for K_a. Conventionally, this is described as the effect of the *standard state* on the free energy: It costs less to take the ligand out of a higher concentration solution and put it in the binding site, because the free ligand has a lower translational or liberational entropy at higher concentration (e.g., [14]). Thus ΔG_{bind} will have a more negative (and the net translational entropy loss a less negative) numerical value if K_a is expressed in units of M rather than say the lower concentration units of μM. However, the binding affinity (the constant K_a) is unchanged by choice of units. The common use of 1M as the standard state is purely convention.
- The difficulty in calculating binding affinities has nothing to do with the fact that ΔG_{bind} as given by Equation 9.4 varies with the units used for K_a, nor can it be solved by choice of the right standard state. As Equation 9.2, Equation 9.3, and Figure 9.1A demonstrate, to obtain K_a the binding phase volume (colloquially, the residual rotational and translational (R/T) entropy due to motion of A with respect to B in the complex) must be evaluated. This depends on the details of the shape and atomic interactions of the binding partners.
- It is misleading to view the R/T entropy loss upon binding as due to either loss of 6 degrees of freedom (3T + 3R), or the conversion of two particles (L + P) to one particle (PL), because formally in Equation 9.2 or any equivalent treatment, no degrees of freedom are lost: The number of atoms is constant. Furthermore, L and P are always considered as individual molecules, albeit in close proximity, because L can always move somewhat relative to P (i.e., the residual R/T entropy is never zero).
- As a consequence, attempts to extract the translational entropy loss upon binding by measuring the change in affinity upon covalently attaching L to P with a flexible linker (i.e., preconverting two particles into one) are misdirected and should not be relied upon to provide useful estimates for binding calculations, see discussion in [12,15]. The linking experiments modify the entropy of the unbound state and thus do not get at the real unknown quantity: the residual translational entropy in the bound state.
- Thus determination of the binding affinity requires either the explicit or implicit integration of the ligand–target interaction free energy over relative positions and orientations around the interaction potential minimum or minima that correspond to the complexed state (see [12,16–21] for several recent approaches) or some *a priori* knowledge of the amount of residual R/T entropy, perhaps from closely related complexes.

As a practical matter, screening and docking involve evaluation of multiple poses, but usually information from just the best one is used. However poses close in position and orientation to the best in effect already sample, albeit coarsely, the integrand in Equation 9.2 required for the binding calculation. In principle, this information could be used, especially if supplemented by finer sampling, to evaluate the binding affinity integral and make the scoring more physically realistic. It should be emphasized that to be valid, the sampling in position and orientation should either be done uniformly, and then Boltzmann weighted, or sampled with a Boltzmann distribution using for example the Metropolis MC method.

9.3.2 RELATIVE BINDING AFFINITY

A common response to the aforementioned difficulty in determining the absolute affinity is to restrict calculations to the affinity of one compound relative to another (closely related?) reference compound. Because VS only requires one to get the right ranking of compounds to work, this is eminently sensible if it solves the problem. The argument goes as follows: The residual R/T entropy (or equivalently, the binding phase volume) is difficult and expensive to calculate because many binding poses must be evaluated. However, if two compounds are similar or bind similarly, these entropy terms are probably similar and thus subtract out when differences in affinity are measured or calculated. In terms of the simple square well binding example, if the two compounds had identical width wells, their different affinities would be due solely to different well depths. The well depth is presumably easier to calculate, because it requires the evaluation of one pose (the minimum energy one).

To examine this issue further, consider the somewhat more realistic binding interaction potentials in Figure 9.1B, consisting of three truncated harmonic potential wells denoted i, ii, and iii. Well ii and Well iii are both four times as deep as Well i. Well ii was chosen to have the same width binding region as Well i. Well iii was chosen to have the same curvature as Well i. Again taking the natural definition of the complex as the region of the well, the binding affinity can be evaluated exactly using Equation 9.2. The results are shown in Table 9.1.

TABLE 9.1
Relative Binding Affinities in a Truncated Three-Dimensional Harmonic Potential Model

Potential (See Figure 9.1)	i	ii	iii
Depth E(kT)	-2	-8	-8
Well width at top σ(Å)	1	1	2
Curvature (kT/Å^2)	16	64	16
Relative affinity, actual[a]	1	68	546
Relative affinity, expected[b]	1	403	403

[a] Exact result using Equation 9.2. Expressed relative to potential i: K_a/K_a^i.
[b] From the exponential of the difference in well depths: $e^{-(E-Ei)}$.

Clearly for neither of the deeper wells is the change in affinity determined solely by the difference in well depths. For Well ii, as it becomes deeper, it becomes narrower at the bottom, so the ligand becomes more confined, leading to a smaller residual entropy, and a less than expected increase in affinity: An example of partial entropy–enthalpy compensation (although the well depth is not truly enthalpic because the interaction is a governed by a PMF with entropic contributions). For Potential iii, because it has the same curvature, it becomes wider when deeper, leading to a larger residual entropy and a greater than expected affinity increase. The point of this example is to show that even with a simple binding well shape, it is not so easy to change the depth and keep the residual entropy constant. It can be judged how much less likely this is to happen in binding reactions with interaction potentials of more complex shape and perhaps multiple minima. Strictly, one cannot even rely on some kind of compensation between binding enthalpy and translation entropy as a guide to the *direction* of the effect: Potential iii shows anticompensation relative to Potential i, a phenomenon seen in binding and discussed by Gallichio et al. [22]. The phenomenon of anticompensation discussed in this reference merits wider attention from scoring function developers. In summary, avoiding absolute binding affinity calculations — doing just relative binding calculations — is no magic bullet for the residual R/T entropy loss problem.

Again, from practical standpoint, we may conclude that even for relative binding, evaluation of the binding phase volume by explicit pose sampling or some implicit means would significantly improve relative binding affinity calculations. Although this may not be computationally practical for large scale screening at this time, it should be factored into the strategy for future improvements.

9.4 SCREENING POTENTIAL OR FREE ENERGY CALCULATION?

Referring back to the definition of enrichment, Equation 9.2, in principle a method that just ranked a series of compounds correctly, but got the actual binding free energies quite wrong would suffice to produce a good enrichment factor EF in large scale screening. In statistical terms, such a potential would have a rank correlation coefficient of $R^2 = 1$. The question that is considered here is whether it might be easier to design such a scoring potential than compute a real binding free energy. Consider the following scenario: Such a scoring function is devised, which on a test set produces a perfect correlation with experimental binding data, but whose magnitudes are off. Figure 9.2 illustrates two such cases where the error is a factor of two:

1. Line i — The potential underestimates ΔG_{bind}.
2. Line ii — The potential overestimates ΔG_{bind}.

In the first case, because the potential is underestimating differences between compounds, it is missing some of the interaction. We can assume with confidence than when let loose on a wider range of compounds outside the parameterization set, the missing parts will be the ones needed to distinguish between some com-

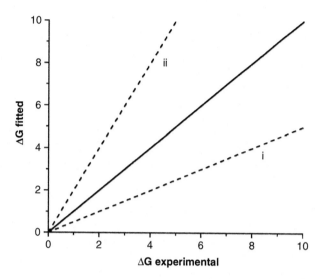

FIGURE 9.2 Chematic result of scoring potential fitting, with smaller or large slopes (dashed Line i and Line ii, respectively). Line of unit slope (calculated = experiment) is solid.

pounds, and the potential will fail on these compunds. The second case is more curious: Somehow our potential is more discriminating than nature herself. This can only be so because it is overweighting some interactions that are prevalent in the test set. Clearly this potential will run into transferability problems. This analysis simply reiterates the point in the introduction that calculation, or at least emulation, of an actual binding free energy is the only meaningful guiding strategy for scoring function development.

It may be objected that one can simply scale the potential function by a constant factor, 1/slope. First, the true measure of the potential is how well it does on a validation set (i.e., its free R^2). Uniform scaling of the training set is not guaranteed to improve the slope or correlation on the validation set. Second, in the over- or underestimate of slope, there is presumably useful information about the relative importance of the different terms in the potential function that would be missed by uniform scaling. Achieving a slope of one by adjustment of individual terms is more physically meaningful. The major point is that an overemphasis on a good correlation coefficient and neglect of the actual numerical values, particularly with empirical potentials which depend on a large amount of parameterization, may be counterproductive.

Interestingly, statistical potentials for binding give a slope much greater than unity for ΔG_{bind} calculated vs. experiment (e.g., [8]). A possible reason for this is discussed below in the section on statistical potentials, as an example of where this kind of discrepancy is telling us something important about the function.

As a final point on this topic, empirical potentials are typically developed with the following general strategy:

1. Select a functional form.
2. Train on a test set of complexes with known affinities.
3. Validate (i.e., compute the free R^2 or root mean squared (rms) error) on a disjoint set of compounds.
4. Publish, patent, or perish.

In reality, the first attempt may produce disappointing results, so one does a postmortem, modifies the potential function, and repeats the entire cycle. This meta-iteration, guided by improvements in the free R may be repeated. Unhappily, the R is now no longer free. It will underestimate, sometimes severely, the true error. This cycle may not be evident from the final published description. The anecdotal observation from many labs that "using other published functions on our compounds" often gives a poorer result suggests that this must be factored into the assessment of scoring functions. Again, this problem is exacerbated by overemphasis on the goodness of fit rather than binding free energy fundamentals.

9.5 SEPARABILITY OF BINDING FREE ENERGY TERMS

A crucial issue for the development of practical screening functions is whether one can identify and separately quantify different contributions to ΔG_{bind}. If it is not valid to make such a separation, then it calls into question the usual divide and conquer approach to potential development, both in terms of decomposing measured ΔGs for parameterization and in the construction of the function itself out of different terms, each of which might be modeled in a different way for the best trade-off in speed and accuracy. To frame the question more formally, consider the physics-based, statistical mechanics perspective. Here one starts with a Hamiltonian or potential energy function, U, which can be expressed as a sum of terms (electrostatic, vdW, etc.)

$$U = \sum U_j \qquad (9.6)$$

(the issue of a many-body polarizable force field is beyond the scope of this discussion). Equilibrium quantities are then obtained by Boltzmann weighting by U and averaging over the atomic configurations (ensemble averaging). For example, the mean energy (or enthalpy, the tiny pressure-volume [PV] term will be ignored here) can be decomposed into terms corresponding to each term in the potential function U, because

$$H = E \ \ =< U >=< \sum_j U_j >= \sum_j < U_j >= \sum_j E_j \qquad (9.7)$$

where $<>$ indicates the ensemble averaging. The crucial question arises: Can one decompose the free energy, G, into corresponding terms as

$$G = \sum_j G_j \qquad (9.8)$$

Because the free energy is obtained from the log of the partition function, which is a sum of nonlinear Boltzmann terms, $e^{-U_i/kT}$, and because of other factors (i.e., the path dependence of components obtained from some free energy perturbation calculations), the separability of free energy terms has been called into question [23,24]. It can however be shown that a valid decomposition like Equation 9.8 can be made if one defines the components correctly in terms of temperature derivatives of the energy terms [25]

$$G = \sum_j \left(\sum_{n=0}^{\infty} \frac{(-\beta)^n}{(n+1)!} \frac{\partial^n <U_j>_\beta}{\partial \beta^n} \right) = \sum_j G_j \qquad (9.9)$$

where $\beta = 1/kT$. Equation 9.9 thus provides formal justification for the almost universal practice of decomposing the binding free energy into terms, often each modeled in different ways. Moreover, because $G = H - TS$, Equation 9.7 and Equation 9.9 together imply that the entropy may also be decomposed into a sum of corresponding terms. Although Equation 9.9 does not give a recipe for any particular case, it does however provide important reassurance for the overall divide and conquer strategy. Because each of the subcomponent free energies itself is a sum of terms (the second summation in Equation 9.9), it also provides a way to get successively better approximations by including first and second higher terms. It is suggested that this may be a way to improve linear interaction energy methods for binding calculations, discussed below.

A final but important point about decomposability must be stressed. Decomposability of G, H (enthalpy), and S (entropy) into components corresponding to the terms used in the potential function U is quite different from the independence of the individual terms. For example if van der Waals (vdW) and electrostatic terms are both present in U, as U_{vdw} and U_{elec}, then one can identify and interpret, in physical terms, their corresponding terms in the free energy: G_{vdw} and G_{elec}. However if one repeated the calculation with say the electrostatics switched off ($U_{elec} = 0$), then the resulting vdW term in the free energy, G_{vdw} ($U_{elec} = 0$) would in general not be the same: the components G_{vdw} and G_{elec} are not independent of each other or the other terms [24–26]. So for example G_{vdw} ($U_{elec}=0$) could not be taken as a reliable estimate of the G_{vdw}. Exactly the same considerations apply to the enthalpy and entropy terms. There may be situations where individual G/S/H components are independent, but this has to be demonstrated first. This is an important point to bear in mind when considering a tiered screening strategy, in which one might put in the cheapest interactions first, screen, and then add the more expensive terms later when the number of compounds has been winnowed down.

9.6 WHAT TO PUT INTO FREE ENERGY-BASED SCORING FUNCTIONS

To paraphrase, the functions should be made as simple as possible, but no simpler. In modern physics-based potential functions (e.g., all atom force fields) aside from the covalent bonded terms there are usually just two others, the vdW and electrostatic nonbonded terms. For example, a separate H-bond term was long since subsumed into electrostatics, as was a salt-bridge term. Numerous simulations and comparisons with experiment have testified to the reliability of this short list of fundamental forces. It hardly needs mentioning that solvent, either explicit or implicit, must be included in binding calculations of any sophistication. The Poisson-Boltzmann- (PB) and generalized Born- (GB) based implicit solvent methods retain the same simplicity as the modern potential functions, with an electrostatic term that treats polar and charged interactions, and an apolar term that subsumes vdW and hydrophobic contributions. Nevertheless, the PB and GB methods successfully model other thermodynamics properties, notably pKa shifts and solvation free energies that are measured independently of binding affinities, so this is strong evidence that these two terms are also sufficiently inclusive of the important physics, hence their increasing popularity for treating solvation in screening and binding potentials.

When considering what terms to include, the case of linear interaction energy (LIE) methods is instructive. These use atomic force fields, and in their initial implementation, also contained just vdW and electrostatic terms [27]. Interestingly, an extra solute cavity term was later added [28], but its magnitude was found, on further parameterization to shrink to insignificance [29], again suggesting the completeness, in a functional sense, of the current terms. Given this, the main efforts, for all these physics-based methods are currently directed at improved sampling and better force field parameters like atomic charges. As a corollary, when developing empirical scoring potentials, one should look hard at extra terms beyond the aforementioned ones: Are they giving improvement simply by increasing the number of adjustable parameters rather than by improving the physical representation?

A perennial problem in potential function calculations is whether to include hydrogens. From a computational standpoint, clearly one would rather not: They are often not available in the structure, building them in involves some modeling, and they significantly increase the number of atoms in the calculation. Thus they add both uncertainty and computational cost. It has been amply demonstrated that for all atom physics-based methods, inclusion of at least the polar Hs (ones attached to N, O, S, and P), is required to get the right behavior. Similarly for the widely used PB- and GB-implicit solvent models. Clearly at least some hydrogens must be included if the binding involves changes in protonation state. Moreover, the protonation states of the relevant ionizable groups must be determined, often requiring an expensive interacting site pKa calculation.

For statistical and empirical potentials, the picture is murkier, because parameterization may put the effect of the hydrogens in implicitly. Statistical potentials work quite well without hydrogens [8,9], probably because the major effect of H-bonding is included through the parameterization of acceptor and donor group types. Our experience in adding hydrogens to the protein–ligand complexes used by

Muegge and Martin [8] and reparameterizing their scoring functions yielded improvement in some cases, worse results in others, with a slight worsening of overall performance [30], which we attributed to uncertainty in H placement outweighing the few specific interactions (e.g., carboxyl-protonated carboxyl) in which specific H representation matters.

9.7 LINEAR INTERACTION ENERGY METHODS

The most rigorous way to calculate a binding free energy is through an explicit atom simulation with a physics-based potential using, for example, free energy perturbation or thermodynamic integration method [31] or direct sampling [32]. However, these methods are hideously expensive for drug screening. The LIE methods represent a powerful shortcut [27], which can be further combined with implicit solvent treatments [33]. Another linear method used for calculating binding free energies is based on the quasiharmonic model [12]. Although still too expensive for initial screening, these methods can be used at later stages of the process. Their use of both the mean and fluctuation in energy terms to approximate the full free energy again relies on the validity of free energy separation discussed above. In effect, these methods correspond to using the zeroth and first terms in the full free energy component expression in Equation 9.9. Parameterization of the LIE methods have demonstrated that for best results the electrostatic and vdW (and nonpolar) terms must be separately scaled [33,34]. This may be a consequence of the missing higher terms in Equation 9.9, which suggests a possible route for improvement.

9.8 STATISTICAL POTENTIALS

Statistical, or knowledge-based, potentials for binding have been successfully developed and show promise for screening because they combine speed, reasonable accuracy, and a fairly objective, physicslike functional form and parameterization [8,9]. The basic concepts are drawn from liquid state theory. One analyzes the distribution of distances betweens pairs of atom or group types in available complex structures to derive the potentials. If the probability of Group i and Group j being at Separation r has the distribution $g_{ij}(r)$ then a statistical potential for $i–j$ interaction is defined as

$$\Delta\omega_{ij}(r) = -kT \ln\left(\frac{g_{ij}(r)}{g(r)}\right) \tag{9.10}$$

where $g(r)$ is a normalization factor, which varies in detail between implementations. The score W of a complex is then obtained by summing the statistical potentials for each ligand i–protein j pair

$$\Delta W = \sum_{i}^{ligand} \sum_{j}^{protein} \Delta\omega_{ij}(r_{ij}) \qquad (9.11)$$

The Δs in Equation 9.10 and Equation 9.11 indicate that the potentials and the resulting score are relative quantities.

Like many successful methods for binding, the initial implementation works well, but successive improvements become harder to achieve. This may be a question of the physicality of the function or the quality of the parameters. If the distribution of i and j were obtained with a full ensemble averaging over the positions of all other groups or atoms (as happens in a liquid) then a true potential of mean force would result. This is clearly not the case for a sample of distances from a survey of crystal structures and has been carefully considered by the developers. Nevertheless, as the sample of complexes gets larger and more diverse, it may well mimic this overall averaging.

A second, and in some ways more fundamental, issue may be responsible for the bottoming out of the improvement process and may point the way to go. Consider the extraction of a PMF using the example of hard sphere liquid, which is well understood. The Hamiltonian or potential function between two hard spheres is shown in Figure 9.3A. It is zero until the spheres overlap, where it becomes infinite.

The exact radial distribution (or local relative density) function for a fluid of hard spheres, $g(r)$, can be calculated and is shown in Figure 9.3B. There is a maximum at the contact (nearest neighbor) separation, and others at one and two diameters separation where two spheres are separated by another one or two, respectively. The log of $g(r)$ yields the PMF $\omega(r)$ between two spheres, Figure 9.3C. This

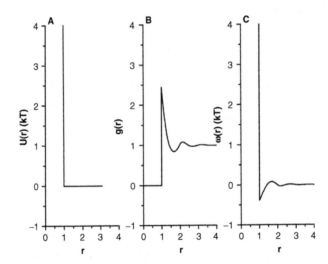

FIGURE 9.3 A: Pair potential or Hamiltonian for hard spheres. Distance is in units of sphere diameter. B: Radial distribution or local density function, $g(r)$, for fluid of hard spheres at 50% volume fraction. Distribution is normalized to bulk sphere density. C: PMF between two spheres, obtain as $\omega(r) = -kT \ln(g(r))$.

has a minimum at the contact separation and another at the sphere separated distance. Thus an effective attraction has appeared between spheres that have no intrinsic attractive term. Now imagine that we simply have $g(r)$ and are trying to discover what the potential is. If we use $g(r)$ to get $\omega(r)$, and then subsequently use this as a pairwise potential to sum over a configuration of spheres, it would give the wrong answer: for one, it would overestimate the cohesiveness. An exactly analogous problem is encountered in using protein–ligand complex derived $g(r)$ to obtain $\omega(r)$s, which are then used in pairwise summing to score other complexes. A symptom may well be the slope of $\gg 1$ seen in the statistical PMF score plotted against the measured $\log K_a$ [8]. In a formal sense, $\omega(r)$ is a convolution of $U(r)$ over configuration space and to recover the latter, we need to deconvolute. There is no analytical way to do this. However, an iterative adjustment of $U(r)$ can bring the calculated $g(r)$ closer to experiment. This is the manner in which, say, atomic water pair potential functions are derived so that the model liquid has close to the experimental $g(r)$. It is suggested that some version of this procedure is needed to further improve statistical potentials and put them on a firmer theoretical basis.

9.9 INTERNAL CONFORMATIONAL CHANGES

As discussed above, the total degrees of freedom of the protein–ligand–solvent system may be classified into those from solvent and from protein–ligand relative R/T motion, both discussed above, and lastly from internal conformational degrees of freedom of ligand and protein. To introduce discussion of the latter, the specific case where an unbound ligand may have multiple conformations is considered, and then the analysis is generalized.

9.9.1 EFFECT OF MULTIPLE UNBOUND LIGAND CONFORMATIONS

Many, perhaps most compounds in a screening database can adopt more than one conformation in the unbound state. Some consequences and possibilities of multiple ligand conformations for binding calculations are worth considering. To illustrate the issue, consider a ligand with just two unbound conformations, denoted L and l. (Generalization to multiple conformations is straightforward.) The binding of this ligand to a protein P is described by the following thermodynamic cycle:

$$
\begin{array}{ccc}
 & K_L & \\
L + P & \Leftrightarrow & LP \\
\Uparrow & & \Uparrow \\
K_{Ll} & & K_L K_{Ll}/K_l \qquad (9.12)\\
\Downarrow & & \Downarrow \\
l + P & \Leftrightarrow & lP \\
 & K_l &
\end{array}
$$

where $K_{Ll} = [L]/[l]$ is the ligand conformation equilibrium constant, K_L and K_l are the binding affinities of the L and l conformations, respectively. The conformational equilibrium of the bound state ligand is given by closure of the thermodynamic cycle. What matters for screening and drug candidacy is the net binding affinity of L in all its conformations, which is just

$$K_{net} = \frac{[LP] + [lP]}{([L] + [l])[P]} = K_L f_L + K_l f_l \approx K_L f_L \qquad (9.13)$$

where $f_L = K_{Ll}/(1 + K_{Ll})$ and $f_l = 1/(1 + K_{Ll})$ are the fractions of unbound ligand in conformations L and l, respectively. It is often the case that the binding site will fit only one conformation of the ligand, say L, thus the affinity for the l conformation is much lower due to steric clash ($K_l \ll K_L$ leading to the second equality in Equation 9.13). This means that the fraction of L in the bound state is close to 100% (i.e., there is induced fit). However, the net affinity is just $K_L f_L$. This leads to the following practical considerations:

Induced fit, sometimes thought to be an advantage for binding is a mixed blessing: if the ligand is rigidly constrained in the wrong conformation, things would be worse than with induced fit. More commonly, though, the right conformation is accessible but the net affinity has to be scaled by its fraction in the unbound state. If l now stands for all the wrong conformations of a ligand with a modest number of degrees of internal freedom, then f_L may easily be < 1% (because of the combinatoric explosion in conformations) dropping a presumably good score for L by two orders of magnitude or more. The fact that the ligand is induced to adopt the right conformation in the site does not matter, because it is the unbound fraction of L that is relevant. In typical screening strategies, all conformations that have no strain or internal steric clash, or are below some generous energy cutoff, are considered in screening and docking. This is done to avoid false negatives, but neglect of the unbound population effect will result in unnecessary false positives.

On the positive side, recent high throughput screening (HTS) protocols now use pregenerated libraries with multiple conformations for each compound [35,36]. The libraries are usually generated once and reused for many different target screens. This suggests that an extra investment in realistically modeling the relative energies, and hence the relative populations of all the conformations of a ligand, is well worth the extra time. Admittedly, the number of compounds or conformations will be large, but the cost will be amortized over multiple applications. Conformations with low populations could be flagged as such, either to correctly scale down their score or simply eliminate them from the initial screening to reduce computation. Only if no decent candidates emerged would it be worth screening them.

9.9.2 GENERAL EFFECT OF INTERNAL COORDINATE CHANGES

If we write the effect of the ligand conformational equilibrium on the net binding free energy by taking the logarithm of both sides of Equation 9.13 we find

$$\Delta G_{net} = -kT \ln(K_L) - kT \ln(f_L)$$
$$= \Delta G_L - f_l kT \ln(K_{Ll}) + \{-f_L kT \ln(f_L) - f_l kT \ln(f_l)\}$$

(9.14)

where, in the second equality we can identify the first term as the binding free energy of the ligand in the "correct" conformation, the second term as the thermodynamic work necessary to convert the fraction of unbound ligand in the incorrect conformation to the correct conformation before binding, and the third term, delimited by {}, as the entropy cost of "collapsing" the ensemble of unbound conformations into the "single" correct conformation.

This expression is quite general, and it should be noted that l and L may refer in fact to two sets of conformations, rather than two single conformations, as long as one can specify: an effective equilibrium constant between the two sets, K_{Ll}, and a net binding constant for the high affinity set, K_L. Similarly, Equation 9.14 can be generalized to include the effects of protein conformations, by adding a third dimension to scheme 9.12, resulting in a thermodynamic cube with equilibria between low and high affinity states of the protein and ligand, denoted P/p and L/l respectively, and their bound forms. For the analogous simple case where the protein also has two distinct sets of conformations — one with low affinity, one with affinity, the generalization of Equation 9.14 is

$$\Delta G_{net} = \Delta G_{LP} - f_l kT \ln(K_{Ll}) + \{-f_L kT \ln(f_L) - f_l kT \ln(f_l)\}$$
$$- f_p kT \ln(K_{Pp}) + \{-f_P kT \ln(f_P) - f_p kT \ln(f_p)\}$$

(9.15)

where K_{Pp} is the effective equilibrium constant between the high and low affinity protein conformations, and $f_P = K_{Pp}/(1 + K_{Pp})$, $f_p = 1/(1 + K_{Pp})$ are the fractions of high and low affinity protein conformations, respectively.

For the special case where a ligand has n_l iso-energetic conformations, of which only one can bind, Equation 9.14 reduces to $\Delta G_{net} = \Delta G_L + kT\ln(n_l)$, that is, the penalty is simply the logarithm of the number of conformations. This, and a similar result for the case of iso-energetic protein conformations, is the basis of the widely used *rotamer* counting method of accounting for protein and ligand conformational changes upon binding. The assumption behind the rotamer counting method is clear: All conformations are counted equally.

Several insights may be drawn from this analysis:

- An obvious (although not easy) way to improve the rotamer counting method is to evaluate the relative energy, and Boltzmann weight, of the different rotamers.
- Rather than try to explicitly count and weight all the rotamers in the L/l or P/p classes, one could use the effective equilibrium constants K_{Ll} and K_{Pp} between high and low affinity sets, if these single quantities could be obtained from some other kind of calculation or experimental data.

- The most rigorous, but most expensive way to account for the internal degrees of freedom is to sample internal conformational space continuously and Boltzmann weight. Conformations with all ranges of affinity are thereby represented. Examples of this approach range from the more expensive free energy perturbation/thermodynamic integration methods [37,38] to the the less expensive LIE and quasiharmonic methods that treat fluctuations as Gaussian [12,29,33] and hybrid methods that combine molecular dynamics and implicit solvent models [19,20,39,40].

9.10 CONCLUSIONS

In this chapter, I have tried to survey the major theoretical issues involved in the realistic calculation of binding affinities. Two aspects have been stressed. First, that ultimate progress must be based on evaluation of real free energies, which requires attention to the physicality and consistency of the underlying models. Second, that this free energy requires the explicit or implicit sampling of three types of degrees of freedom — solvent, internal, and inter-ligand–protein (R/T). Although the best methods we have are usually too expensive for initial screening and are typically applied later when the number of candidates has been winnowed to a dozen or less, they provide the form and behavior that quicker, more approximate initial screening methods must emulate; therefore, they are useful in a more strategic sense in these early stages. I have also tried to identify aspects of existing methods where theoretical improvement, rather than better parameterization, needs to be made.

ACKNOWLEDGMENT

Support from the National Science Foundation grant MCB02-35440 is gratefully acknowledged.

REFERENCES

[1]   N. Brooijmans and I.D. Kuntz, Molecular recognition and docking algorithms, *Ann. Rev. Biophys. Biomol. Struct.* 32:335–373, 2003.

[2]   B.K. Shoichet and I.D. Kuntz, Predicting the structure of protein complexes: a step in the right direction, *Chem. & Biol.* 3:151–156, 1996.

[3]   M.R. Reddy, M.D. Erion, and A. Agarwal, Free energy calculations: use and limitations in predicting ligand binding affinities, *Rev. Comput. Chem.* 16:217–304, 2000.

[4]   H. Gohlke and G. Klebe, Approaches to the description and prediction of the binding affinity of small-molecule ligands to macromolecular receptors, *Angew. Chemie. Int. Ed.* 41:2644–2676, 2002.

[5]   W.D. Cornell, P. Cieplak, C.I. Bayly, I.R. Gould, K.M. Merz, D.M. Ferguson, D.C. Spellmeyer, T. Fox, J.W. Caldwell, and P.A. Kollman, A second generation force field for the simulation of proteins, nucleic acids, and organic molecules, *J. Am. Chem. Soc.* 117:5179–5197, 1995.

[6]   A.D. MacKerell, D. Bashford, M. Bellott, R.L. Dunbrack, M.J. Field, S. Fischer, J. Gao, H. Guo, S. Ha, D. Joseph, L. Kuchnir, K. Kuczera, F.T.K. Lau, C. Mattos, S. Michnick, T. Ngo, D.T. Nguyen, B. Prodhom, B. Roux, M. Schlenkrich, J.C. Smith, R. Stote, J. Straub, J. Wiorkiewicz-Kuczera, and M. Karplus, Self-consistent parameterization of biomolecules for molecular modeling and condensed phase simulations, *FASEB Journal* 6:A143, 1992.

[7]   K. Sharp and B. Honig, Electrostatic interactions in macromolecules: theory and applications, *Ann. Rev. Biophys. Biophys. Chem.* 19:301–332, 1990.

[8]   I. Muegge and Y.C. Martin, A general and fast scoring function for protein–ligand interactions: a simplified potential approach, *J. Med. Chem.* 42:791–804, 1999.

[9]   H. Gohlke, M. Hendlich, and G. Klebe, Knowledge-based scoring function to predict protein–ligand interactions, *J. Mol. Biol.* 295:337–356, 2000.

[10]   H.J. Bohm, The development of a simple empirical scoring function to estimate the binding constant for a protein–ligand complex of known three-dimensional structure, *J. Computer-Aided Mol. Des.* 8:243–256, 1994.

[11]   M.K. Gilson, J.A. Given, B.L. Bush, and J.A. McCammon, The statistical-thermodynamic basis for computation of binding affinities: a critical review, *Biophys. J.* 72:1047–1069, 1997.

[12]   H. Luo and K.A. Sharp, On the calculation of absolute binding free energies, *PNAS* 99:10399–10404, 2002.

[13]   B.K. Shoichet, A.R. Leach, and I.D. Kuntz, Ligand solvation in molecular docking, *Proteins: Structure, Function, and Genetics* 34:4–16, 1999.

[14]   A. Ben-Naim, Standard thermodynamics of transfer: uses and misuses, *J. Phys. Chem.* 82:792–803, 1978.

[15]   M. Karplus and J. Janin, Comment on: the entropy cost of protein association, *Prot. Eng.* 12:185–186, 1999.

[16]   J. Hermans and S. Shankar, The free energy of xenon binding to myoglobin from molecular dynamics simulation, *Israel. J. Chem.* 27:225–227, 1986.

[17]   P. Cieplak and P.A. Kollman, Calculation of the free energy of association of nucleic acid bases in vacuo and water solution, *J. Am. Chem. Soc.* 110:3734–3739, 1988.

[18]   B. Roux, M. Nina, R. Pomes, and J.C. Smith, Thermodynamic stability of water molecules in the bacteriorhodopsin proton channel: a molecular dynamics free energy perturbation study, *Biophys. J.* 71:670–681, 1996.

[19]   B. Kuhn and P.A. Kollman, A ligand that is predicted to bind better to avidin than biotin: insights from computational fluorine scanning, *J. Am. Chem. Soc.* 122:3909–3916, 2000.

[20]   V. Tsui and D.A. Case, Calculations of the absolute free energies of binding between RNA and metal ions using molecular dynamics simulations and continuum electrostatics, *J. Phys. Chem.* B105:11314–11325, 2001.

[21]   L. David, R. Luo, and M.K. Gilson, Ligand-receptor docking with the mining minima optimizer, *J. Computer-Aided Mol. Des.* 15:157–171, 2001.

[22]   E. Gallicchio, M.M. Kubo, and R.M. Levy, Entropy-enthalpy compensation in solvation and ligand binding revisited, *JACS* 120:4526-4527, 1998.

[23]   A.E. Mark and W.F. van Gunsteren, Decomposition of the free energy of a system in terms of specific interactions, *J. Mol. Biol.* 240:167–176, 1994.

[24]   P.E. Smith and W.F. van Gunsteren, When are free energy components meaningful? *J. Phys. Chem.* 98:13735–13740, 1994.

[25]   G.P. Brady, A. Szabo, and K.A. Sharp, On the decomposition of free energies, *J. Mol. Biol.* 263:123–125, 1996.

[26]   S. Boresch and M. Karplus, The meaning of component analysis: decomposition of the free energy in terms of specific interactions, *J. Mol. Biol.* 254:801–807, 1995.

[27] J. Aqvist, C. Medina, and J.-E. Samuelsson, A new method for predicting binding affinity in computer aided drug design, *Protein Eng.* 7:385–391, 1994.

[28] N.A. McDonald, H.A. Carlson, and W.L. Jorgensen, Free energies of solvation in chloroform and water from a linear response approach, *J. Phys. Org. Chem.* 10:563–576, 1997.

[29] T. Hansson, J. Marelius, and J. Aqvist, Ligand binding affinity prediction by linear interaction energy methods, *J. Computer-Aided Mol. Des.* 12:27–35, 1998.

[30] V. Heredia, Y. Huyen, and K.A. Sharp, unpublished results.

[31] J.A. McCammon, Computer aided molecular design, *Science* 238:486, 1987.

[32] M.S. Head, J.A. Given, and M.K. Gilson, Mining minima: direct computation of conformational free energy, *J. Phys. Chem.* A101:1609–1618, 1997.

[33] R. Zhou, R.A. Friesner, A. Ghosh, R.C. Rizzo, W.L Jorgensen, and R.M. Levy, New linear interaction method for binding affinity calculations using a continuum solvent model, *J. Phys. Chem.* B105:10388–10397, 2001.

[34] J. Marelius, T. Hansson, and J. Aqvist, Calculation of ligand binding free energies from molecular dynamics simulations, *Int. J. Quantum Chem.* 69:77–88, 1998.

[35] D.M. Lorber and B.K. Shoichet, Flexible ligand docking using conformational ensembles, *Protein Sci.* 7:938–950, 1998.

[36] D. Joseph-McCarthy, B.E. Thomas, M. Belmarsh, D. Moustakas, and J.C. Alvarez, Pharmacophore-based molecular docking to account for ligand flexibility, *Proteins* 51:172–188, 2003.

[37] P. Bash, C. Singh, F. Brown, R. Langridge, and P. Kollman, Calculation of the relative change in binding free energy of a protein-inhibitor complex, *Science* 235:574–576, 1987.

[38] T. Lybrand, J.A. McCammon, and G. Wipf, Theoretical calculation of relative binding affinity in host-guest system, *Proc. Natl. Acad. Sci.* 83:833, 1986.

[39] R. Luo and M.K. Gilson, Synthetic adenine receptors: direct calculation of binding affinity and entropy, *JACS* 122:2934–2937, 2000.

[40] G. Archontis, T. Simonson, and M. Karplus, Binding free energies and free energy components from molecular dynamics and Poisson–Boltzmann calculations: application to amino acid recognition by aspartyl-tRNA synthetase, *J. Mol. Biol.* 306:307–327, 2001.

10 Solvation-Based Scoring for High Throughput Docking

Thomas S. Rush III, Eric S. Manas, Gregory J. Tawa, and Juan C. Alvarez

10.1 INTRODUCTION AND SCOPE OF THE PROBLEM

Solvent effects play a fundamental role in mediating the affinity of small molecules for protein binding sites. For instance, it is well known that water has the ability to severely dampen intermolecular electrostatic interactions, to reduce affinities by solvating polar or charged groups, and to contribute to binding through the hydrophobic effect. Given the significance of this role, it is essential that these effects be considered when computationally evaluating protein–ligand complexes. However, because these effects are notoriously difficult and often time-consuming to evaluate, they are frequently neglected when computational speed is an issue; this is especially true for high throughput docking (HTD) studies.

In a typical HTD exercise, one poses (orients) and scores a database of molecules within a protein binding site, hoping that the molecules with the most favorable scores will have a greater likelihood of binding to the protein than a molecule chosen at random. When this occurs, the top X% of the HTD hit list is said to be enriched with active molecules. During this computational exercise, many poses are generated, and it is the job of the scoring function to discern the most likely pose(s) from this often sizeable set. Basic scoring functions typically use measures of shape complementarity between the ligand and protein, or the direct calculation of molecule–molecule interactions to evaluate a given pose. Due to the sheer number of scoring events in a typical HTD study, most scoring functions trade accuracy for speed, and it is often the case that speed is improved by neglecting solvent effects. For example, in physics-based scoring functions, it is common for only simple gas-phase Coulombic and van der Waals (vdW) potentials to be considered (e.g., DOCK [1–3]). In some cases, a distance-dependent dielectric model is used to crudely estimate solvent effects (e.g., AMBER [4,5]). On the other hand, empirically based scoring functions exist that are parameterized with experimental data and are often presumed to include some of the effects of solvation. Other types of scoring functions also exist (some of which are discussed in more detail later), and in general, indirectly approximate the effects of solvation or ignore it completely. (For more detailed information on docking and scoring functions in general, see [6].)

Nevertheless, any scoring function intended to rank the more potent molecules higher on the HTD hit list than less potent ones must, at some level, be a measure of the experimentally observed protein–ligand binding free energy, in which the solvent clearly plays an important role. This does not imply that the scoring function should be able to predict an absolute or even a relative binding free energy within a reasonable confidence interval. However, it does suggest that the scoring function and the binding free energy should be *correlated*, even if the correlation is a weak one. Toward this end, we discuss in this chapter the current theories behind determining the free energies of binding between small molecules and proteins and how available HTD scoring functions relate to these theories. We then discuss some of the newer, cutting-edge efforts to develop scoring methods that will correlate better with actual binding data by incorporating solvent effects, including some work of our own.

10.2 EVALUATING THE FREE ENERGY OF BINDING BETWEEN SMALL MOLECULES AND PROTEINS

An expression for the free energy (ΔG) of the protein (P)–ligand (L) binding reaction

$$P(aq) + L(aq) \rightarrow PL(aq)$$

can be derived by equating the solution chemical potentials of the reactants and products and written in terms of configurational integrals (Z) involving P, L, PL, and the solvent S [7]

$$\Delta G_{PL}^{\circ} = -RT \ln\left(\frac{C^{\circ}}{8\pi^2} \frac{\sigma_P \sigma_L}{\sigma_{PL}} \frac{Z_{N,PL} Z_{N,0}}{Z_{N,P} Z_{N,L}} \right) + P^{\circ} \Delta \bar{V}_{PL}, \qquad (10.1A)$$

where:

$$Z_{N,PL} = \int I(\zeta_L) J_{\zeta_L} e^{-\beta U\left(\mathbf{r}_P, \mathbf{r}_L, \zeta_L, \mathbf{r}_s\right)} d\mathbf{r}_P d\mathbf{r}_L d\zeta_L d\mathbf{r}_S \qquad (10.1B)$$

$$Z_{N,X} = \int e^{-\beta U(\mathbf{r}_X, \mathbf{r}_s)} d\mathbf{r}_X d\mathbf{r}_S \quad (X = P, L) \qquad (10.1C)$$

$$Z_{N,0} = \int e^{-\beta U(\mathbf{r}_s)} d\mathbf{r}_S \qquad (10.1D)$$

In the above equations, C° and P° represent the standard state concentration and pressure, respectively, σ_X is the symmetry number of species X (which are P, L, and PL), $\beta = (RT)^{-1}$, and $\Delta \bar{V}_{PL} = \bar{V}_{PL} - (\bar{V}_P + \bar{V}_L)$, where $\bar{V}_X$ is the equilibrium volume change induced when one molecule of solute X is added to a large number N of solvent molecules. This term is typically small in a biological assay at ambient pressure and will therefore be dropped in subsequent equations. The configurational integrals are performed over solute and solvent internal coordinates $\mathbf{r}_i$ (i = P, L, PL, and/or S), and the relative coordinates ζ_L of the ligand with respect to the protein. J_ζ is the Jacobian determinant for the rotation of the ligand with respect to the protein, which is used because the coordinates ζ_L are considered internal coordinates of the complex. $U(\mathbf{r}_i)$ is the energy of the system with respect to the species i. Finally, $I(\zeta_L)$ is a step function that equals 1 when P and L form a complex (i.e., when PL exists) and 0 otherwise.

The difficulty in evaluating Equation 10.1A lies in the calculation of the configurational integrals in Equation 10.1B to Equation 10.1D. Integration over solvent configurations is often approximated by utilizing an implicit, or continuum solvation model. This is done by rewriting Equation 10.1A to Equation 10.1D as [7]

$$\Delta G_{PL}^{\circ} = -RT \ln \left(\frac{C^{\circ}}{8\pi^2} \frac{\sigma_P \sigma_L}{\sigma_{PL}} \frac{Z_{PL}}{Z_P Z_L} \right)$$
(10.2A)

$$Z_{PL} = \int I(\zeta_L) J_{\zeta_L} e^{-\beta \left[U(\mathbf{r}_P, \mathbf{r}_L, \zeta_L) + W(\mathbf{r}_P, \mathbf{r}_L, \zeta_L) \right]} d\mathbf{r}_P \, d\mathbf{r}_L \, d\zeta_L$$
(10.2B)

$$Z_X = \int e^{-\beta[U(\mathbf{r}_X) + W(\mathbf{r}_X)]} d\mathbf{r}_X \quad (X = P, L)$$
(10.2C)

where

$$W(\mathbf{r}_x) = -RT \ln \left[\frac{\int e^{-\beta \Delta U(\mathbf{r}_x, \mathbf{r}_s)} e^{-\beta U(\mathbf{r}_s)} d\mathbf{r}_s}{\int e^{-\beta U(\mathbf{r}_s)} d\mathbf{r}_s} \right]$$
(10.2D)

$W(\mathbf{r}_x)$ is the nonpressure–volume work of transferring a solute in conformation $\mathbf{r}_x$ from gas phase to solvent. As the ratio in brackets represents an average over solvent coordinates of the solute–solvent interaction, $\Delta U(\mathbf{r}_x, \mathbf{r}_s)$, Equation 10.2D is naturally approximated by using a continuum dielectric model to calculate solvation energies of P, L, and PL.

Various levels of approximation have been applied to calculate the configurational integrals necessary for determining absolute and relative binding affinities within the context of the above formalism [8–13]. These methods typically make use of extensive sampling techniques like Monte Carlo, molecular dynamics, and systematic sampling. For the relatively simple case of a ligand binding in a single multidimensional harmonic potential well, the binding free energy expression reduces to [11].

$$\Delta G_{PL}^{\circ} = \omega_{min} - kT \ln \left[\frac{\sqrt{8\pi^3} \delta_x \delta_y \delta_z}{V_{ref}} \right] - kT \ln \left[\frac{\delta_\Omega^3}{\sqrt{6^3}\pi} \right] - kT \ln \left[\sqrt{\frac{\delta_{P(bnd)}^2 \| \delta_{L(bnd)}^2 \|}{\delta_{P(free)}^2 \| \delta_{L(free)}^2 \|}} \right]$$

(10.3)

In the above equation, ω_{min} is the depth of the well, representing all solvent-averaged protein–ligand interactions and internal coordinate strain of the ligand and protein relative to their respective global minima in solution. The well depth is often calculated in terms of the following thermodynamic cycle

$$
\begin{array}{ccccc}
 & & \omega_{min} & & \\
P_{solv} & + & L_{solv} & \longrightarrow & PL_{solv} \\
\big\downarrow -W^{P}_{solv} & & \big\downarrow -W^{L}_{solv} & & \big\uparrow W^{PL}_{solv} \\
 & & \Delta U^{gas}_{bind} & & \\
P_{gas} & + & L_{gas} & \longrightarrow & PL_{gas}
\end{array}
\tag{10.4}
$$

The second and third terms in Equation 10.3 contain the quantities $\delta_{x,y,z,\Omega}$, which are the root mean square (rms) fluctuations in position (x,y,z) and orientation (Ω) of the ligand with respect to the protein, and V_{ref}, which is the standard state reference volume ($C° = 1M$ corresponds to 1660 Å³/molecule). These two terms yield the positional and orientational entropic penalties upon binding. These quantities are often considered to be invariant for druglike ligand binding reactions [14–16], although this is not necessarily always the case [7,11,17,18]. The final term represents the entropic penalty for restricting protein and ligand internal coordinate degrees of freedom upon binding and involves determinants of the internal coordinate fluctuation variance–covariance matrix [$\delta_i\delta_j$] for the protein and ligand in both the bound and free states [11,19]. In the quasiharmonic method, which implicitly includes some degree of anharmonicity, these fluctuation terms are estimated by performing extensive sampling with a simulation method like molecular dynamics [11]. Generalization to multiple binding wells can also be done [7–10,13,20]. Alternatively, normal mode analysis has also been applied to calculate these entropic contributions [18,21].

10.3 SCORING FUNCTIONS FOR HIGH THROUGHPUT DOCKING

When attempting to estimate the binding free energy (i.e., score) of a molecule within the context of HTD, there are two main challenges to overcome that are highlighted by the above discussion: namely, sampling and evaluation of protein–ligand interactions. Clearly, extensive molecular-mechanics-based sampling methods such as molecular dynamics are too slow to accurately evaluate fluctuations about the minimum for even a single well per ligand (forget about the protein fluctuations). This difficulty is compounded by the fact that one does not know the true binding mode (or modes) for any given molecule, so all reasonable modes must be sampled and evaluated. In HTD, this sampling of binding modes is usually achieved using stochastic (genetic algorithms [GAs] [22]), systematic (FRED [23,24], Glide [25]), or heuristic (DOCK [1–3], PhDOCK [26,27]) searching methods. Ligand internal coordinate sampling is often performed by docking ensembles of low-energy conformations [26–28] or by on-the-fly conformer generation techniques (DOCK's anchor and grow, Glide, while protein internal coordinate sampling is ignored). In principle, positional and orientational sampling could be done finely enough to map out the energy landscape of the binding site, such that either converged fluctuations such as those in Equation 10.3 can be obtained; or in the case of a uniform sampling algorithm such as in FRED, the integrals over ligand position and orientation in Equation 10.2 can be evaluated. However, in practice, one typically considers and evaluates only a relatively small set of low energy conformations and their poses, and only the single conformation or pose able to achieve the most favorable interaction with the protein is assumed to contribute to binding. Fluctuations in ligand position and orientation within the binding pocket are typically neglected. In addition, changes in internal coordinate motion upon binding (for this one binding mode) are either ignored or, at best, a constant penalty of $kT \ln N$ is applied to each rotatable bond, representing a crude estimate of the entropic penalty associated with restricting N rotamers down to a single rotamer upon ligand binding.

Regardless of the chosen level or method of sampling, we contend that one generally cannot identify the correct (or at least energetically accessible) binding modes without a reliable estimate of the protein–ligand interactions in the presence of solvent. However, for the sake of speed, many scoring functions in existence neglect or roughly approximate the effects of solvent and thus have generally met with mixed success. Until now, this was certainly a necessity if one wished to evaluate databases of any reasonable size (e.g., 100,000 molecules or more).

In what follows, we provide a brief overview of methods previously developed to evaluate protein–ligand interactions during HTD. In general, these methods can be divided into three categories, and we discuss them in this context: force-field-based, empirically based, and knowledge-based schemes. We also summarize recent work on HTD studies that utilize continuum solvent models, which appear to be able to capture the essential thermodynamics of solvent-mediated binding within the context of force-field-based scoring methods.

10.3.1 FORCE-FIELD-BASED SCHEMES

Force-field-based scoring functions are typically easy and fast to evaluate, and it is for this reason we believe they appeared so early on in the development of HTD. In general, they consist of the nonbonded terms from an established molecular-mechanics force field (e.g., AMBER, CHARMm), and evaluate gas-phase interactions between the protein and the ligand. Two well-known HTD programs that utilize this type of scoring function are DOCK [1–3] and GOLD [22,29].

DOCK's scoring function [30,31] utilizes the electrostatic and vdW terms (and the corresponding parameters) from the AMBER potential

$$E_{\text{interaction}} = \sum_{ij} \frac{A_{ij}}{R_{ij}^{12}} - \frac{B_{ij}}{R_{ij}^{6}} + 332.0 \frac{Q_i Q_j}{\varepsilon\left(R_{ij}\right) R_{ij}} \tag{10.5}$$

Here, the first term is a 12-6 Leonard–Jones potential for the vdW interaction, and the second is a Coulombic interaction energy to evaluate electrostatics. It is purely a gas-phase evaluation, which, for practical purposes, is evaluated using a grid encompassing the protein active site. Although the authors chose to use the non-bonded terms from AMBER, any molecular-mechanics force field could have been used in a similar manner. An approximate way to account for the effects of solvation using this type of scoring function is to adjust the value of the dielectric constant (ε) for the Coulombic interaction, or turn it into a distance dependent function.

In a recent implementation [31], DOCK and its energy-based function were evaluated using a panel of 15 protein–ligand crystal structure test cases. For all test cases, at least one docked ligand pose is generated within 2 Å rms of the crystal structure geometry of the ligand. DOCK was also recently used to test the feasibility of accounting for solvation by using only ligand-based solvation terms [32] and through a generalized-Born/surface area (GB/SA) approach [33]; we discuss these in more detail in Section 10.3.4.

GOLD [22,29] utilizes a GA to find optimal ligand conformations for a given protein target and thus evaluates poses with a fitness function. The fitness function is force-field-based and includes a directional hydrogen (H)-bonding term, a soft vdW term, and an internal energy term

$$\Delta G_{bind} = E_{H-bond} + E_{complex} + E_{internal} \tag{10.6}$$

The interesting features of this function are the additional H-bonding term, the indirect consideration of desolvation through the H-bonding term, and the evaluation of internal energies. To approximate the effects of solvation, the authors add on donor-water and acceptor-water energy terms that approximate the costs of desolvating each to form the H-bond. The GOLD algorithm and the above potential were tested on a data set of 100 crystal structure complexes [22]. When used to dock the ligands back into the binding sites, it was found that GOLD identified the proper experimental binding for 71 out of the 100 complexes.

Another notable, soft molecular-mechanics scoring function is PLP (Piecewise Linear Potential), created by Gehlhaar et al. [34]. This potential is given by

$$E_{total} = E_{H-bond} + E_{repulsion} + E_{contact} + E_{internal} \tag{10.7}$$

However, only 4 atom types are considered in this evaluation, and the interaction potential between each of the atom types (categorized as H-bond, repulsion, or contact) is defined by a linear-based representation of a nonbonded potential. For example, the H-bonding potential goes from +20.0 at d = 0.0 Å to 0.0 at d = 2.3 Å. Then, from 2.3 Å to 2.6 Å, the potential linearly decreases to 2.0, where it remains flat until d = 3.1 Å. Finally, the score raises linearly to 0.0 at a distance of 3.4 Å. The internal energy of the ligand is given by the following torsional potential

$$E_{internal} = A\left[1 - \cos(n\phi - \phi_0)\right] \tag{10.8}$$

where A = 3.0, n = 3.0, $\phi_0 = \pi$ for sp^3–sp^3 bonds; A = 1.5, n = 6, and $\phi_0 = 0$ for sp^3–sp^2 bonds; and torsional angles about sp^2–sp^2 bonds are held fixed.

10.3.2 EMPIRICALLY-BASED SCHEMES

In the empirically-based schemes, the protein–ligand binding affinity is decomposed into a series of terms believed to be important to the binding free energy, and each term is multiplied by a weighting factor determined by a mathematical fit to experimental data. One of the earlier and probably most well-known empirical scoring functions was that derived by Bohm (Ludi [35]). In this case, the binding free energy is given by

$$\Delta G_{bind} = \Delta G_0 + \Delta G_{H-bond} \sum_{H-bond} f(\Delta R, \Delta\alpha) + \Delta G_{ionic} \sum_{ionic} f(\Delta R, \Delta\alpha)$$
$$+ \Delta G_{lipo} \mid A_{Lipo} \mid + \Delta G_{ROT} N_{ROT} \tag{10.9}$$

where $\Delta G_{H\text{-}bond}$ is the contribution due to an ideal neutral H-bond; ΔG_{ionic} is the contribution due to unperturbed ionic interaction; and $f(\Delta R, \Delta\alpha)$ is a penalty function that accounts for deviations from ideal H-bond geometry. This penalty function is given by

$$f(\Delta R, \Delta\alpha) = f1(\Delta R) + f2(\Delta\alpha) \tag{10.10}$$

where

$$f1(\Delta R) = \begin{cases} 1 & \Delta R \leq 0.2\,\mathring{A} \\ 1 \quad -(\Delta R - 0.2)/0.4 & 0.2 < \Delta R \leq 0.6\,\mathring{A} \\ 0 & \Delta R > 0.6\,\mathring{A} \end{cases} \tag{10.10A}$$

and

$$f2(\Delta\alpha) = \begin{cases} 1 & \Delta R \leq 30^{o} \\ 1 \quad -(\Delta\alpha - 30)/50 & 0.2 < \Delta R \leq 80^{o} \\ 0 & \Delta R > 80^{o} \end{cases} \tag{10.10B}$$

ΔR is the deviation from ideal H-bond length (1.9 Å), and $\Delta\alpha$ is deviation from ideal H-bond angle (180°). ΔG_{lipo} is the contribution due to lipophilic interactions; $\mid A_{lipo} \mid$ is the lipophilic contact surface between the ligand and the protein; ΔG_{rot} is the contribution due to freezing out of rotational degrees of freedom upon binding; N_{rot} is the number of ligand rotors; and ΔG_0 is everything else (e.g., loss in translational entropy). This function was originally calibrated by fitting known dissociation constants of 45 protein–ligand complexes and was found to be able to reproduce the binding constants of the dataset with a standard deviation of 1.9 kcal/mol [35].

A notable extension of Bohm's function is the one implemented in the docking program FlexX [36]. Here, the lipophilic interaction term is replaced by

$$\Delta G_{lipo} \mid A_{Lipo} \mid = \Delta G_{aromatic} \sum_{aromatic} f(\Delta R, \Delta \alpha) + \Delta G_{lipo} \sum_{lipo} f(\Delta R) \qquad (10.11)$$

where the distinction is made between protein–ligand aromatic contacts and all other lipophilic contacts. The functions $F(\Delta R, \Delta \alpha)$ and $F(\Delta \alpha)$ are similar in form to those described for H-bonding in the LUDI potential, however the determinants for ideal bond distance and angle are modified to reflect aromatic and lipophilic interactions (e.g., the ideal distance between two lipophilic atoms is given by the sum of their vdW radii). The coefficients in this scoring function were originally calibrated using 19 protein–ligand complexes [36]. The experimentally observed binding mode of the ligands in this set was reproduced with 0.5 to 1.2 Å root mean square deviation (rmsd) when utilizing the FlexX docking engine.

Examples of other functions that generally hold true to this form, but that are somewhat different in their weights or geometric constraints, include ChemScore [37], Fresno [38], Score [39], and the scoring function of Hammerhead [40,41]. Relevant to the discussion in this chapter is the fact that the latter three functions include additional terms to specifically account for the effects of solvation. In the case of Fresno

$$\Delta G_{bind} = K + \alpha(HB) + \beta(LIPO) + \gamma(ROT) + \delta(BP) + \gamma(DESOLV) \quad (10.12)$$

Here, the two terms added to directly account for the effects of solvation include the buried polar atom term (BP) and the ligand desolvation term (DESOLV). The buried polar term is given by

$$BP = \sum_{L,P} f(r_{L,P}) + \sum_{P,L} f(r_{p,L}) \qquad (10.13)$$

where the functions $f(r)$ are similar in form to LUDI's $F(\Delta R)$. $f(r_{L,p})$ describes burial of ligand polar atoms by lipophilic protein atoms, and the term $f(r_{p,L})$ describes burial of protein polar atoms by lipophilic atoms of the ligand. The desolvation term is given by

$$DESOLV = -\Delta G_{reac(L)}^{o} \qquad (10.14)$$

where $\Delta G_{reac(L)}^{o}$ is the electrostatic solvation free energy determined by solving the linear form of the Poisson–Boltzmann (PB) equation using a finite-difference method. The coefficients in the Fresno function were originally calibrated using 2 training sets — 1 involving 5 HLA-A*0201-peptide structures, the other involving 37 H-2K^k-peptide structures [38]. For both training sets, a good cross-validated fit to experimental binding free energies was obtained with predictive errors of 3 to 3.5 kJ/mol.

On the other hand, Score [39] uses the form:

$$\Delta G_{bind} = K_0 + K_{vdw} + K_{metal} + K_{hbond} + K_{desolvation} + K_{deformation} \qquad (10.15)$$

where $K_{desolvation}$ is the contribution due to ligand and protein desolvation and is given by the term

$$K_{HM} = \sum_i F_i HM_i \qquad (10.16)$$

where HM_i is a hydrophobic matching term set to 1 if a ligand atom i is hydrophobic and placed in a hydrophobic environment, otherwise it is set to 0. F_i is the hydrophobicity scale of atom i. The F_i values were determined by assuming that the logP of a molecule could be expressed as a sum of weighted atomic contributions the weights being determined by regression analysis of a large set of compounds. In the end, the overall scoring function reproduced the binding free energies of the training set with a cross-validated deviation of 6.3 kJ/mol.

Last, we would like to mention here the scoring function found in AutoDock [42,43], which is a function that combines both force-field-like and empirically based attributes. This scoring function has the form

$$\Delta G_{bind} = \Delta G_{VDW} \sum_{ij} \left(\frac{A_{ij}}{R_{ij}^{12}} - \frac{B_{ij}}{R_{ij}^{6}} \right)$$

$$+ \Delta G_{HBOND} \sum_{ij} E(t) \left(\frac{C_{ij}}{R_{ij}^{12}} - \frac{D_{ij}}{R_{ij}^{6}} \right)$$

$$+ \Delta G_{elec} \sum_{ij} \frac{Q_i Q_i}{\varepsilon(R_{ij}) R_{ij}} \qquad (10.17)$$

$$+ \Delta G_{TOR} N_{TOR}$$

$$+ \Delta G_{SOL} \sum_{ij} \left(S_i V_j + S_j V_i \right) e^{\frac{-R_{ij}^2}{2\sigma}}$$

The first three terms of this equation are similar to ones found in molecular-mechanics force fields for vdW, H-bonds, and electrostatics, but in this instance, they are weighted by empirical weighting factors (the ΔGs). The last term is the protein–ligand desolvation penalty. Here, S_i and S_j are the atomic solvation parameters for ligand atom i and protein atom j, respectively, V_j is the volume of protein atom j in the vicinity of ligand atom i, and V_i is the volume of ligand atom i in the vicinity of protein. The whole term is scaled by an exponential function of the distance between the atom pair. The empirical free energy function was calibrated using a set of 30 known protein–ligand complexes with known binding constants [43]. The authors reproduced the experimental binding free energies with a residual standard error of 2.2 kcal/mol.

10.3.3   KNOWLEDGE-BASED SCHEMES

In knowledge-based schemes, one derives a potential by analyzing experimentally determined protein–ligand crystal structure complexes [44]. In this analysis, a set of ligand and protein fragment types is identified, a list of all of the pairwise distances between the fragment types across all crystal structure complexes is compiled, the probability of occurrence of particular interfragment separations for each pair of fragment types (g_{ij}) is calculated, then a distance-dependent two-body fragment potential describing the interaction of the pair is written, i.e.,

$$\Delta W_{ij}(R) = -\ln\left(\frac{g_{ij}(R)}{g(R)}\right) \tag{10.18}$$

Here $g_{ij}(R)$ is the probability distribution of fragment pair i,j being R distance apart and $g(R)$ is a reference distribution selected such that $\Delta W_{ij}(R)$ approaches 0 as R approaches infinity. By summing over the protein–ligand two-body fragment potentials one obtains the total binding free energy

$$\Delta G_{bind} = \sum_{i,j} \Delta W_{ij} \tag{10.19}$$

The recent knowledge-based scoring function DrugScore [45] was derived by converting data from crystal structures of 1376 protein–ligand complexes of the Protein Data Bank (PDB) [46] into distance-dependent pair potentials. The method was first used in second tier fashion to successfully rerank docking results generated by FlexX using a test set of 158 PDB complexes. The method was subsequently used in first tier fashion as the cost function during docking optimization. This approach was tested on a set of 41 PDB complexes using the AutoDock engine. Results were shown to be successful in many cases where DrugScore reranking of already docked ligand conformations does not yield satisfactory results.

Similarly, Muegge and Martin [47] derived a PMF (potential of mean force) score for pairs of 16 protein-atom and 34 ligand-atom types. They used 697 protein–ligand crystal structure complexes from the PDB to derive the distance-dependent PMF functions. Best results were obtained for a set of 16 serine protease complexes where the correlation between calculated PMF scores and experimental log(K_i) values was 0.86.

In the BLEEP (Biomolecular Ligand Energy Evaluation Protocol) method derived by Mitchell et al. [48,49], over 58 atom types were defined and 351 crystal structure complexes from the PDB were used to construct 1711 distance-dependent pair potentials. The pair potentials were used to calculate binding free energies for a set of 90 protein–ligand complexes. The correlation coefficient of the calculated binding energies with experiment was 0.74.

An interesting related technique is that of Ishchenko et al. [50] in a method called SMoG (Small Molecule Growth). Again, one identifies a set of ligand and protein fragment types (σ_p, σ_l), however the scoring function is based on two-body contact potentials instead of distance-dependent pair potentials. The protein–ligand binding free energy is given as a sum over protein–ligand contact potentials

$$\Delta G_{bind} = \sum_{pl} g(\sigma_p \sigma_l) \Delta_{pl} \tag{10.20}$$

The parameter $\Delta_{pl} = 1$ if a protein atom p of type σ_p is within 5 Å of a ligand atom l of type σ_l and 0 otherwise. The function $g(\sigma_p, \sigma_l)$ is a two-body contact potential

$$g(\sigma_p \sigma_l) = -\ln \frac{p(\sigma_p \sigma_l)}{p_{ref}} \tag{10.21}$$

where $p(\sigma_p, \sigma_l)$ is the probability that protein atom p of type σ_p is within 5 Å of ligand atom l of type σ_l. This probability comes from the measured frequency of contacts between atoms of type σ_p and σ_l in a training database. P_{ref} is a reference probability. The authors believe that by using the 5 Å contact sphere, they are implicitly taking into account solvent entropy. This is because the 5 Å distance is approximately equal to the sum of the radii of the first water coordination shell around the atoms under consideration, and therefore, when the 2 atoms come into contact, the shell waters must be released into free solvent. A recent implementation of SMoG utilized 725 protein–ligand crystal structure complexes, 13 protein-atom types and 13 ligand-atom types to determine the two-body contact potentials, $g(\sigma_p, \sigma_l)$ [50]. Binding free energy calculations were performed on a test set of 119 protein–ligand complexes, and the authors found an overall correlation coefficient of 0.435.

10.3.4 CONTINUUM-BASED SCHEMES

An intermediate but practical approach to address solvation effects is to treat the solvent as a continuum dielectric medium [51,52]. The simplest implementation of continuum electrostatics in a HTD scoring function was described in a publication by Shoichet et al. in 1999 [32]. Here, the authors chose to streamline the evaluation of

$$\Delta G_{bind}^{solv} = \Delta G_{bind}^{gas} - \Delta G_{solv}^{P} - \Delta G_{solv}^{L} + \Delta G_{solv}^{PL} \qquad (10.22)$$

($\approx \omega_{min}$; as described in the thermodynamic cycle of Equation 10.4) with the use of

$$\Delta G_{bind}^{solv} \approx \left(\Delta E_{vdW}^{gas} + \Delta E_{elec}^{gas} \right) - \Delta G_{solv}^{L} \qquad (10.23)$$

Here, continuum electrostatics is used to evaluate the ligand solvation term. This function thus assumes that the ligand is completely desolvated upon binding, and that every ligand desolvates the protein equally. Although a simple use of continuum electrostatics, the function is designed for speed and is straightforwardly implemented within the DOCK program. The authors start with the DOCK energy function and add separate electrostatic and nonpolar corrections to ligand solvation (on-the-fly) as determined by the program HYDREN [53]. Nonetheless, this simple implementation had a profound effect on the ranking of the knowns and the size and charge of other ligands populating the top of the hit list [32].

In the same year, two other publications appeared with more rigorous treatments of solvation using continuum methods — one by Kuntz and coworkers [33], and the other by Caflisch and coworkers [54]. The paper by Kuntz and coworkers described an approximate GB/SA model to score DOCK generated poses [33]. Here, Equation 10.22 is approximated by

$$\Delta G_{bind}^{solv} \approx \left[\Delta E_{elec}^{gas} + \Delta E_{vdW}^{gas} \right] - \Delta G_{solv}^{P} - \Delta G_{solv}^{L} + \Delta G_{solv}^{PL} \qquad (10.24)$$

and evaluated using

$$\Delta G_{bind}^{solv} \approx \left[\Delta E_{elec}^{gas} + \Delta E_{vdW}^{gas} \right] + \Delta\Delta G_{GB} + \Delta\Delta G_{SA} \qquad (10.25A)$$

where

$$\Delta\Delta G_{GB} = \Delta G_{GB}^{PL} - \left(\Delta G_{GB}^{P} + \Delta G_{GB}^{L} \right) \qquad (10.25B)$$

and

$$\Delta\Delta G_{SA} = \Delta G_{SA}^{PL} - \left(\Delta G_{SA}^{P} + \Delta G_{SA}^{L} \right) \qquad (10.25C)$$

The surface area terms of Equation 10.25B involve contributions from both total and nonpolar surface areas to account for vdW and cavity contributions to solvation, respectively. True to the DOCK methodology, the authors precalculate the protein contributions to electrostatic screening and store them on a grid surrounding the active site. This includes both the solvation energy term for the protein alone and the effect of the protein atoms on the Born radii of atoms involved in complex formation. Although a force-field-based function, the authors used variable parameters to scale the vdW and the two surface area terms. These parameters were explored/optimized using dihydrofolate reductase (DHFR) and trypsin test cases. In the end, ranking and pose selection were found to be much improved over calculations performed using the standard DOCK energy function.

Lastly, the paper by Caflisch and coworkers described a GB-based scoring function for their fragment docking program SEED (Solvation Energy for Exhaustive Docking) [54]. Here, the solvent-based scoring function is of the form of Equation 10.24, but is implemented using

$$\Delta G_{bind}^{solv} = E_{vdW,gas}^{PL} + E_{elec,solv}^{PL} + E_{desolv}^{P} + E_{desolv}^{L} \qquad (10.26)$$

Thus the intermolecular vdW energy is summed with three terms that account for the effects of solvation:

1. A screened intermolecular electrostatic interaction (GB approximation)
2. A ligand desolvation energy (GB approximation)
3. A protein desolvation energy

This latter term is calculated in an *ad hoc* fashion using a grid surrounding the binding site to save time. In general, their method relies on a fast preprocessing step that captures the effect of the ligand displacing the first shell of water molecules within the active site. This method was validated by evaluating more than 2500 complexes of small molecules with thrombin and HIV-1 (human immunodeficiency virus) protease, in that docked fragments reproduced known interaction patterns for the complexes.

10.4 FULL SOLVATION-BASED SCORING FOR HIGH THROUGHPUT DOCKING

The scoring functions discussed above aim to approximate the important contributions to the free energy of binding in a manner consistent with the demands of HTD. Most of the terms added to these functions to address the effects of solvation are included to capture the qualitative effects in an easily implemented atom-based fashion (i.e., weight down pairwise Coulombic interactions, penalize buried polar groups, reward buried hydrophobic interactions). It has only been recently that researchers have used more rigorous, physics-based approaches to capture the effects of solvent in a HTD scoring function (i.e., continuum electrostatics). However, to speed up the calculations, these more rigorous approaches still utilize an approximate continuum electrostatics method (GB) and take algorithmic shortcuts. With the availability of faster methods to calculate PB-based electrostatics — a more rigorous continuum electrostatics method (upon which GB has been parameterized) — we felt that we were ready to take the next step toward the development of a full solvation-based HTD scoring function. The impetus and benefits for doing so will now be discussed.

10.4.1 The Benefits of Full Solvation-Based Scoring

The practice of neglecting solvation when scoring docked poses is supported by successful structure-based design efforts that demonstrate good predictivity using only gas-phase protein–ligand interaction terms. One such example is found in a study performed by Holloway et al. on HIV-1 protease ligands, in which the sum of gas-phase electrostatic and vdW terms had a good linear correlation (cross-validated R^2 as high as 0.76) with measured enzyme inhibition [55]. However, we point out that the emphasis in this study was on optimizing potency within a congeneric series of ligands. In contrast, in a HTD experiment, there is typically no relationship between randomly chosen members of a HTD database. In this case, any assumptions regarding the constant contribution of solvation or solute entropy are expected to be invalid. Furthermore, the linear trends observed by Holloway et al. have slopes far from unity (~ 0.29 to 0.4) if $-RT \ln(IC_{50})$ is plotted against the calculated interaction energy. This suggests that the raw scoring function is actually overestimating some contribution to the binding free energy, which is corrected for by linear regression in the case of the HIV-1 protease inhibitors. In a HTD study, the actives are unknown, so linear regression to correct the raw score cannot be performed. Even if the actives were known, they are not necessarily members of a congeneric series; therefore, the potencies might not follow a simple linear correlation with a gas-phase interaction energy. Therefore, we feel that a scoring function for HTD needs to take into account all of the contributions to the binding free energy, including solvent. This will in all likelihood lead to a tradeoff between a considerably lower R^2 value when plotting measured vs. calculated binding affinity, and a slope closer to unity. However, such a tradeoff will be worthwhile if arbitrary ligand scaffolds with a broad range of binding affinities are to be treated adequately.

We find that a continuum solvent model is capable of mitigating the overestimation of favorable electrostatic interactions seen for the HIV-1 protease example. This is not surprising. The electrostatic contribution to desolvation is expected to be nearly always unfavorable. For buried interactions, the desolvation penalty typically outweighs Coulombic interactions and descreening effects [56–64]. For solvent exposed interactions where pure desolvation is not an issue, solvent screening is expected to strongly reduce the strength of H-bonds.

As an exercise to demonstrate this behavior, we docked and scored (using Poisson–Boltzmann surface area [PB/SA]) a series of ligands from Holloway et al. to the x-ray crystal structure of L-700,417 complexed with HIV-1 protease (4PHV [65]; see Figure 10.1 and Figure 10.2). We chose this structure specifically because it is well resolved (2.10 Å), and because the temperature factors for solvent exposed residues arginine-8 (Arg8) and aspartic acid-29 (Asp29) in the S1 and S1' pockets are low, apparently due to formation of a salt bridge. As reported in Holloway et al., molecular-mechanics overpredicts the potency of L-700,417 by nearly three orders of magnitude, which the authors hint might be due to solvent exposure of the H-bond between one of the indanol groups and Asp29. Furthermore, although the overall correlation between predicted and observed potency is pretty good, there are subseries that we find can form solvent exposed H-bonds with Arg8, for which gas-phase electrostatics incorrectly predicts improvements in potency.

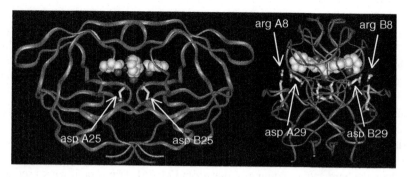

FIGURE 10.1 The x-ray crystal structure of L-700,417 bound to HIV-1 protease. The protein structure is shown as a ribbon, and the small molecule atoms are represented as vdW spheres. The two views (left and right) are rotated by 90°.

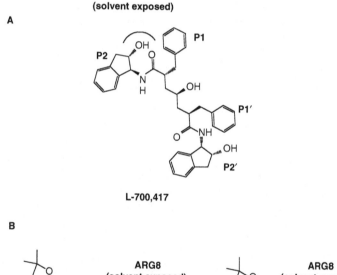

FIGURE 10.2 (A) The structure of L-700,417 and (B) the subset of analogs studied and discussed in the text.

Based on the binding mode of L-700,417, 32 analogs were built into the binding site and minimized using the MMFF94 (Merck Molecular Force Field) force field [66–70] in SYBYL [71]. Care was taken to make sure that none of the analogs exhibited poor potency due to intramolecular strain or steric clashes with the protein, as evaluated by a molecular-mechanics force field after minimization. Electrostatic (Coulombic + solvation) energies were calculated using ZAP [72]. In this method, solutions to the Poisson equation are obtained using an exponentially switched atomic Gaussian function to represent the dielectric boundary, such that the dielectric constant varies smoothly from $\varepsilon = 2$ for the molecular region to $\varepsilon = 80$ for the solvent [73]. Integration over the Gaussian function for each atom is parameterized to reproduce the hard sphere volume of that atom [74–76]. Tripos radii were used here to calculate the atomic volumes. Atomic charges (taken from MMFF94 for this work) are distributed on a grid with 0.5 Å spacing by quadratic inverse interpolation. Electrostatic solvation energies ΔG_{el} are then obtained by summing the product of charge $\times$ potential over all atoms and subtracting out the self-energy and Coulombic (i.e., uniform dielectric) terms. The apolar contribution to desolvation is calculated using $\Delta\Delta G_{ap} = \gamma\,\Delta A$, where ΔA is the total loss of solvent exposed surface area of the protein and ligand upon forming a complex (also calculated using ZAP). The quantity γ ($= 47$ cal/mol/Å^2) was chosen such that $\Delta\Delta G_{ap}$ represents the difference (complex vs. protein + ligand) in transfer energy from a low dielectric environment (such as an alkane solvent or binding site in a protein) with $\varepsilon = 2$, to a water with $\varepsilon = 80$ [77–79]. Thus we used

$$\Delta G_{bind}^{solv} = \Delta E_{elec}^{gas} + \Delta\Delta G_{solv} + \Delta\Delta G_{ap}$$

As one can see in Figure 10.3, the sum of electrostatic + loss area contribution correlates well with the observed potency, although the overall correlation is comparable to what was seen with molecular-mechanics. This is not surprising, as we find that the majority of the correlation comes from the loss area term. However, one noteworthy point is that L-700,417 itself is no longer an outlier, presumably a result of taking solvent screening into account. In Holloway et al., L-700,417 was predicted to have an IC$_{50}$ of 0.0009 nM, whereas the subsequently determined IC$_{50}$ was found to be 0.69 nM, a 767-fold disparity corresponding to an error of about 3.9 kcal/mol in the binding free energy. As mentioned above, the authors point out one possible reason for this discrepancy is that an additional H-bond to the active site may be overemphasized in their gas-phase molecular-mechanics calculation. In contrast, when solvation-based scoring is used, the predicted binding affinity of L-700,417 falls within only a few tenths of a kcal/mol of the value estimated from the experimental IC$_{50}$. We also point out that the slope of calculated vs. experimental binding is considerably closer to unity when using solvation-based scoring, suggesting that the overcounting that occurs with gas-phase-only electrostatics has been compensated for to some extent. Neglect of solute entropy may provide an explanation for any remaining deviation of the slope from unity.

Perhaps even more interesting is the way ZAP is able to predict the structure–activity relationship (SAR) seen in the S1′ pocket, for groups that are able to

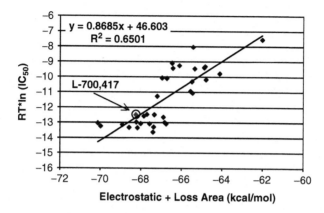

FIGURE 10.3 A plot of the observed potency for the HIV-1 protease ligands vs. the calculated Electrostatic + Loss Area derived score as described in text.

approach Arg8 closely enough to form Coulombic interactions. Figure 10.4 shows

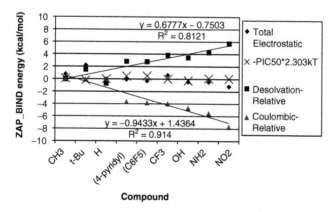

FIGURE 10.4 A plot of the total electrostatic energy and its components vs. the PIC_{50} for 9 of the HIV-1 protease ligands from Figure 10.2B.

the total electrostatic binding energy, as well as its Coulombic and desolvation components (relative to the smallest value), for a series of analogs, ranked in order of increasing magnitude of the Coulombic interaction. Also shown are the potencies of the analogs relative to $R_1 = H$. Clearly, as the Coulombic interaction improves, the desolvation term becomes more unfavorable, demonstrating that Poisson electrostatics is properly handling the solvent exposed interaction. Considering that gas-phase electrostatics by itself would predict the nitro group to have a more favorable interaction with Arg8 by almost 8 kcal/mol (>700,000-fold difference in potency), the mitigating effects of ZAP would clearly be beneficial in a HTD milieu.

10.4.2 A FULL SOLVATION-BASED SCORING FUNCTION FOR HIGH THROUGHPUT DOCKING

The benefits of using a fast PB-based program like ZAP to more rigorously evaluate protein–ligand complexes, along with our continual frustration from not being able to identify suitable leads during HTD studies, led us to begin developing a new PB-based scoring function for HTD. An excellent example of the persistent issues we face while performing HTD studies is exemplified by our experience with a controlled DHFR test case (described in detail in Chapter 13 written by Joseph-McCarthy et al.). Here, we screened a small, seeded database consisting of 2012 molecules against the "3dfr" (PDB code) DHFR structure [46,80]. This small database consists of 2000 random compounds from the Available Chemicals Directory (ACD) [81] and 12 molecules known to bind to the *Lactobacillus casei* DHFR. Ten of the known active compounds have been previously crystallized with a DHFR protein, and we refer to them by their PDB codes: 3dfr/methotrexate, 7dfr/folic acid, 1rf7/dihydrofolic acid, 1dyh/5-deazafolate, 1dyj/5,10-dideazatetrahydrofolate, 1bzf/trimetrexate, 1jol/folinic acid, 1dis/brodimoprim-4,6-dicarboxylate, 1dyr/trimethoprim and 1boz/N6-(2,5-dimethoxy-benzyl)-N6-methyl-pyrido[2,3-D]pyrimidine-2,4,6-triamine. The two without publicly available structures are piritrexim and brodimprim. Each molecule was conformationally expanded using OMEGA [82] with a 5 kcal strain energy cutoff and a 1.0 Å rmsd cutoff to generate unique structures. For docking, we used our own PhDOCK HTD program [26] with the DOCK energy function (described in Section 10.3) [1–3]. In general, PhDOCK rigidly docks molecules by matching subsets of molecule pharmacophores with subsets of user-generated, active site pharmacophore site points. The 67 sitepoints used in this exercise were generated by an automated procedure [27] and contained subsets consistent with each of the 12 known active molecules.

The results of this PhDOCK run are shown in Figure 10.5. PhDOCK, when combined with the gas-phase, AMBER-based energy function (with solvent effects crudely incorporated by using a distance-dependent dielectric) finds 7 out of the 12 known active compounds within the top 2.5% of the database (50 molecules), and the other 5 fall below the top 10% of the HTD hit list (200 molecules). However, upon closer examination, we were quite surprised to see that many of the poses chosen as the best scoring pose were in fact false positives (i.e., the core ring system had greater than 2 Å rmsd from the crystal structure pose; for brodimporim and piritrexim, we used 1dyr and 1bzf as representative poses, respectively). An example of a false positive (Docked 1) is shown in Figure 10.6. Once all false positives of this type were removed (all molecules marked with an X in Figure 10.5), the enrichment rate dropped dramatically. Now, only 4 out of the 12 molecules show up in the top 2.5% of the list. Figure 10.6 highlights the fact that even though PhDOCK sampled a reasonable pose (Docked 2), the scoring function preferred a completely different, and presumably irrelevant, pose for the molecule (Docked 1). Also observed were false negatives, such as the one shown in Figure 10.7. Here, the docked pose had a rmsd of 0.4137 Å from that of the crystal structure, but was ranked 352 out of 2012. Finally, within the set of random molecules, we found numerous examples of formally charged and large molecular weight molecules with

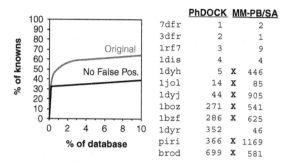

	PhDOCK	MM-PB/SA
7dfr	1	2
3dfr	2	1
1rf7	3	9
1dis	4	4
1dyh	5 X	446
1jol	14 X	85
1dyj	44 X	905
1boz	271 X	541
1bzf	286 X	625
1dyr	352	46
piri	366 X	1169
brod	699 X	581

FIGURE 10.5 Enrichment plots and associated data for the PhDOCK-based DHFR HTD study. See text for details.

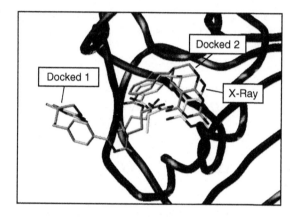

FIGURE 10.6 HTD poses and x-ray structure for 5,10-dideazatetrahydrofolate (1dyj). Docking 1 is the best scoring pose from the HTD run. Docking 2 is the HTD pose with the smallest rmsd to the crystal structure. X-ray denotes the conformation of the molecule as found in the PDB file 1dyj.

scores high enough to displace the known active compounds from the top of the list. As previously described, similar results were observed for DHFR and 2 other test systems by Shoichet et al. using DOCK [32].

With the release of a faster new algorithm (ZAP) to calculate PB-based electrostatics [72], we felt it prudent to try our own hand at evaluating ΔG_{bind}^{solv} ($=\omega_{min}$) in the thermodynamic cycle shown in Equation 10.4 within the context of our HTD experiments [83]. Our hope was that we would see further hit list enrichments by eliminating the false positives and negatives of the types described above and appropriately accounting for the full effects of solvation on the entire system; no approximations are made to speed up the evaluation of the effects of solvent on either the protein or protein–ligand complex. We thus evaluate Equation 10.22 using [83]

$$\Delta G_{bind}^{solv} \approx \left[\Delta E_{elec}^{gas} + \Delta E_{vdW}^{gas} \right] + \Delta\Delta G_{PB} + \Delta\Delta G_{SA} \qquad (10.27)$$

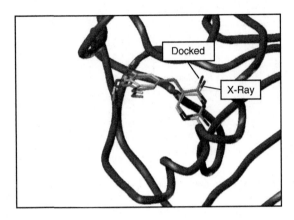

fIGURE 10.7 HTD and X-ray pose for Trimethoprim (1dyr). Docked is the best scoring pose from the HTD run. X-ray denotes the conformation of the molecule as found in the PDB file 1dyr.

where

$$\Delta\Delta G_{PB} = \Delta G_{PB}^{PL} - \left(\Delta G_{PB}^{P} + \Delta G_{PB}^{L}\right) \qquad (10.27A)$$

and

$$\Delta\Delta G_{SA} = \Delta G_{SA}^{PL} - \left(\Delta G_{SA}^{P} + \Delta G_{SA}^{L}\right) \qquad (10.27B)$$

Here, we evaluate all solvation terms explicitly by the way of PB/surface area calculations (PB/SA) and the gas-phase contributions to binding via all-atom molecular-mechanics calculations. Specifically, in this exercise, we evaluated and summed three quantities:

1. $\Delta E_{elec}^{gas} + \Delta\Delta G_{PB}$ via ZAP [72]

2. $\Delta\Delta G_{SA}$ = Loss Area × 6 cal/mol/Å²; loss area also output by ZAP

3. ΔE_{vdW}^{gas} via a molecular-mechanics evaluation in SYBYL [71]

This has all been straightforwardly implemented in a C shell script that spawns both an SPL (SYBYL Programming Language) script for a quick minimization and E_{vdW} evaluation within SYBYL and then a `zap_bind` calculation of the resulting minimized complex. These three quantities are subsequently summed and used to keep a rank order of the poses being evaluated. All in all, this implementation of Equation 10.22 is making the standard DOCK assumptions, assuming the ligand internal coordinate energy changes are equal and assuming that each of the conformers in our database have relatively equal strain energies.

We decided to test the above scoring function (which we refer to as "MM-PB/SA" in the figures) by rescoring the poses generated by PhDOCK in the aforementioned DHFR docking exercise. The results are depicted in Figure 10.5. Quite noticeable is both the agreement with the true hits at the top of the list and the disparity in the ranking of the false positives and negatives in the remainder of the list. Thus, as we anticipated, this PB-based scoring function thought favorably of the X-ray-like, PhDOCK-generated poses; penalized the false positives (such as 1dyj; Figure 10.6); and recovered false negatives (such as 1dyr; Figure 10.7). We have since witnessed the same type of results when rescoring hits from other docking algorithms (e.g., FRED [84]), using other approximate scoring functions (e.g., PLP [34], ChemScore [85], ScreenScore [86]).

10.5 IMPLEMENTING SOLVATION-BASED SCORING IN A TIERED HIGH THROUGHPUT DOCKING SCHEME

Although we found this PB-based scoring function to be quite useful as a rescoring tool, our intention was to use it to identify true (or at least plausible) hits *during* the HTD run. Thus we would avoid populating our hit lists with unrealistic poses or molecules. However, even though ZAP afforded us better timing relative to other PB-based methods, it still generally takes us on the order of 1 to 3 minutes/pose/processor to obtain a full score; this includes both a minimization step (currently our limiting factor) and a PB calculation step. With well over a billion poses to be evaluated in a typical HTD study, it is needless to say that on-the-fly evaluations of each and every pose in this manner is not feasible. Our solution was to implement the evaluations in a tiered fashion.

Tiered HTD schemes are nothing new and can be found in commercial programs such as FRED [84] and Glide [25]. DOCK also uses a coarse tiered approach in that it only allows poses that pass its bump filter to be scored by either the Contact or Energy scoring functions. Although currently unpractical in a HTD sense, we have also seen a skillful implementation of tiered docking in a recent paper by Wang et al., which includes a much more rigorous evaluation of the free energy of binding [21]. Here the authors used DOCK to identify five plausible binding modes for a HIV reverse transcriptase (RT) inhibitor with an unknown binding mode, then dynamics simulations coupled with molecular-mechanics plus PB/SA scoring to identify the correct binding mode, as later identified by researchers at DuPont Pharmaceuticals. In addition to the terms of Equation 10.22 (implemented like Equation 10.27), this demanding scoring procedure also included entropic penalties as evaluated by a normal mode calculation, as well as energetic strains of both the protein and ligand. In a similar manner, we settled on the tiered docking scheme depicted in Figure 10.8 [83]. This scheme carries with it the philosophy that fast, crude, and approximate methods can rapidly and reliably eliminate noncompetent poses or molecules, and that more sophisticated, time-consuming calculations can then be applied to identify likely binders. The only requirement here is that the first stage filter must effectively remove the nonsense while maintaining a manageable set of realistic poses for further evaluation. A little bit of experimentation with our DHFR test case defined the details of our current scheme.

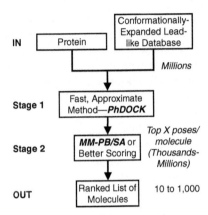

FIGURE 10.8 A tiered scheme for HTD studies with PB-based or better scoring functions.

We first set out to see how many poses per molecule should be kept and which DOCK scoring function should be used to capture the X-ray pose for false positives, such as seen for 1dyj in our DHFR test run (Figure 10.6). At first, we saved increasing numbers of poses for 1dyj as evaluated by DOCK's Energy function. In this particular example, even after saving the top 100 Energy ranked poses (a number currently impractical for our typical HTD studies), we still did not capture a pose resembling the x-ray pose (see Figure 10.9A). However, when we switched to DOCK's Contact function, all the poses in the top 10 were consistent with the x-ray determined binding mode, with Pose 3 having the smallest rmsd (see Figure 10.9B).

We then reran the DHFR HTD study by saving the top 10 Contact poses per molecule (Stage 1) and evaluating each of them with our PB-based scoring function (Stage 2) to produce a final hit list populated with the best scoring binding mode per molecule. The results of this exercise are shown in Figure 10.10 and Figure 10.11. Here we obtained a good score for the right pose of 10 of the 12 knowns, with 8 of them falling within the top 0.8% of the list. This was a dramatic improvement to what we originally found when using DOCK with the Energy score alone.

Upon further investigation, we determined that the negative results for 1jol and piritrexim were due to the fact that a correct pose for each of the molecules was not passed on for further evaluation by the first stage filter. Although further improvements to Stage 1 scoring could potentially resolve this, at this point in time, we accept the loss of a few knowns for the positive identification of 66% of them in the top 1% of the database. If one was to consider hit rates, in this circumstance, we have a 40% hit rate (8/20) out of a possible overall 60% hit rate (12/20) assuming the other 8 random molecules are inactive.

Although quite pleased with these results, we could not help but wonder how much the electrostatic component influences the score and rank of the molecules. After all, these scores are generally dominated by the contributions from the vdW energy and the surface area term. So given a set of molecules that make comparable vdW interactions with the protein and bury relatively the same amount of surface area, can the zap_bind energies differentiate the binders from the nonbinders? This question was recently answered by an inspired experiment performed by the

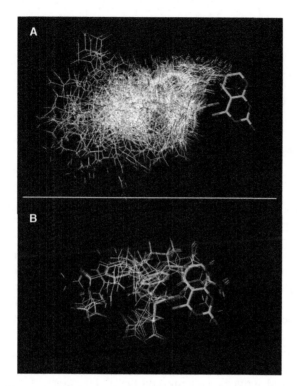

FIGURE 10.9 The top PhDOCK poses for 5,10-dideazatetrahydrofolate (1dyj). A: The top 100 Energy poses of 1dyj overlaid with the x-ray conformation (rendered in capped sticks). B: The top 10 Contact poses of 1dyj overlaid with the x-ray conformation (rendered in capped sticks).

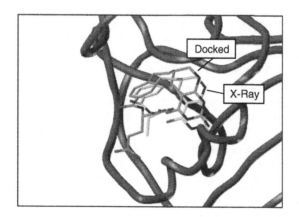

FIGURE 10.10 MM-PB/SA-based HTD pose and x-ray structure for 5,10-dideazatetrahydrofolate (1dyj). Docked is the best scoring pose from the HTD run. X-ray denotes the conformation of the molecule as found in the PDB file 1dyj.

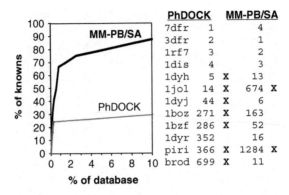

	PhDOCK		MM-PB/SA	
7dfr	1		4	
3dfr	2		1	
1rf7	3		2	
1dis	4		3	
1dyh	5	X	13	
1jol	14	X	674	X
1dyj	44	X	6	
1boz	271	X	163	
1bzf	286	X	52	
1dyr	352		16	
piri	366	X	1284	X
brod	699	X	11	

FIGURE 10.11 Enrichment plots and associated data for the PhDOCK and MM-PB/SA-based DHFR HTD studies. See text for details.

developers at OpenEye Scientific Software [87,88]. First, they generated a set of 1320 methotrexatelike molecules (i.e., docking decoys) by making reasonable atom substitutions in the structure of methotrexate as given in the 3dfr crystal structure. This not only included enumerating all possible protonation states and tautomers of the parent molecule, but also changing atom types throughout the core of the molecule. After appropriately charging the new molecules, they then calculated the electrostatic contribution to the binding energy using ZAP for each molecule in the presence of the protein, also as given in the 3dfr PDB file. In essence, this experiment holds the vdW and surface area terms constant in our scoring function and varies only the electrostatic contributions to binding. In the end, they found that PB electrostatics not only captured the gross electrostatic effects of the system (e.g., correctly identified the appropriate charge and protonation states of methotrexate and its close analogs), but also reproduced the SAR of known analogs with subtly different heterocycles — a desirable trait for any scoring function.

10.6 CONCLUSIONS AND FUTURE DIRECTIONS

With the rapid accumulation of crystal structures and the increasing ability to produce reliable homology models, HTD studies are progressively becoming a valuable tool in the drug discovery arsenal. If one had the ability to consistently identify a small set of molecules likely to bind to a given target by computational means, one could considerably reduce the costs associated with lead discovery. However, by all accounts, we currently lack the ability to completely solve the HTD problem because we have not yet identified a truly universal method for producing enriched hit lists from any given database. Two of the most obvious shortcomings in our current schemes are the insufficient accounting of both solvent effects and entropy. Until recently, the lack of rigor was necessary to keep the time associated with a screen to within a reasonable time period.

With the recent advances in computer hardware and algorithm efficiencies, however, we now have the ability to more accurately treat physical binding phenomena. We strongly feel that methods described in the latter half of this chapter capitalize on these technological advances and have provided us with a means to more reproducibly identify bioactive molecules and their binding modes. In particular, the ability to quickly evaluate the full effects of solvation on the poses passed down from a simple shape-based filter has made a world of difference in the quality of the results we observe.

In the end, this development is only one step along the path to solving the HTD problem. We fully appreciate that we still do not take into account induced strain, protein flexibility, tautomer and protonation states of both the ligand and protein side chains, the possible involvement of water molecules in the binding event, and solute internal coordinate entropy. Nevertheless, we feel that our procedure is a suitable compromise between efficiency and rigorousness at this current point in time. We will continue to streamline it and evolve with new technology.

ACKNOWLEDGMENTS

The authors would like to thank their colleagues Diane Joseph-McCarthy and Iain McFadyen for their work on the DHFR test system, Anthony Nicholls and Roger Sayle from OpenEye Scientific Software for their development of and experimentations with ZAP, and Kim Sharp and Katharine Holloway for useful intellectual discussions.

REFERENCES

[1] E.C. Meng, D.A. Gschwend, J.M. Blaney, and I.D. Kuntz, Orientational sampling and rigid-body minimization in molecular docking, *Proteins* 17(3):266–278, 1993.
[2] I.D. Kuntz, J.M. Blaney, S.J. Oatley, R. Langridge, and T.E. Ferrin, A geometric approach to macromolecule-ligand interactions, *J. Mol. Biol.* 161(2):269–288, 1982.
[3] E.C. Meng, B.K. Shoichet, and I.D. Kuntz, Automated docking with grid-based energy evaluation, *J. Comput. Chem.* 13(4):505–524, 1992.
[4] W.D. Cornell, P. Cieplak, C.I. Bayly, I.R. Gould, K.M. Merz, Jr., D.M. Ferguson, D.C. Spellmeyer, T. Fox, J.W. Caldwell, and P.A. Kollman, A second generation force field for the simulation of proteins, nucleic acids, and organic molecules [Erratum for *J. Am. Chem. Soc.* 117:5179–5197, 1995], *J. Am. Chem. Soc.* 118(9):2309, 1996.
[5] W.D. Cornell, P. Cieplak, C.I. Bayly, I.R. Gould, K.M. Merz, Jr., D.M. Ferguson, D.C. Spellmeyer, T. Fox, J.W. Caldwell, and P.A. Kollman, A second generation force field for the simulation of proteins, nucleic acids, and organic molecules, *J. Am. Chem. Soc.* 117(19):5179–5197, 1995.
[6] I. Muegge and M. Rarey, Small molecule docking and scoring. In K.B. Lipkowitz and D.B. Boyd, Eds., *Reviews in Computational Chemistry,* New York: John Wiley and Sons, pp. 1–60, 2001.
[7] M.K. Gilson, J.A. Given, B.L. Bush, and J.A. McCammon, The statistical-thermodynamic basis for computation of binding affinities: a critical review, *Biophys. J.* 72(3):1047–1069, 1997.
[8] M. Vieth, A. Kolinski, and J. Skolnick, A simple technique to estimate partition functions and equilibrium constant from Monte Carlo simulations, *J. Chem. Phys.* 102(15):6189–6193, 1995.

[9]   L. David, R. Luo, and M.K. Gilson, Ligand-receptor docking with the mining minima optimizer, *J. Computer-Aided Mol. Des.* 15(2):157–171, 2001.

[10] V. Kairys and M.K. Gilson, Enhanced docking with the mining minima optimizer: acceleration and side-chain flexibility, *J. Comput. Chem.* 23(16):1656–1670, 2002.

[11] H. Luo and K. Sharp, On the calculation of absolute macromolecular binding free energies, *Proc. Natl. Acad. Sci. USA* 99(16):10399–10404, 2002.

[12] J. Wang and E.O. Purisima, Analysis of thermodynamic determinants in helix propensities of nonpolar amino acids through a novel free energy calculation, *J. Am. Chem. Soc.* 118(5):995–1001, 1996.

[13] J. Wang, Z. Szewczuk, S.Y. Yue, Y. Tsuda, Y. Konishi, and E.O. Purisima, Calculation of relative binding free energies and configurational entropies: a structural and thermodynamic analysis of the nature of non-polar binding of thrombin inhibitors based on hirudin55-65, *J. Mol. Biol.* 253(3):473–492, 1995.

[14] A. Ajay and M.A. Murcko, Computational methods to predict binding free energy in ligand-receptor complexes, *J. Med. Chem.* 38(26):4953–4967, 1995.

[15] P.R. Andrews, D.J. Craik, and J.L. Martin, Functional group contributions to drug-receptor interactions, *J. Med. Chem.* 27(12):1648–1657, 1984.

[16] M.I. Page and W.P. Jencks, Entropic contributions to rate accelerations in enzymic and intramolecular reactions and the chelate effect, *Proc. Natl. Acad. Sci. USA* 68(8):1678–1683, 1971.

[17] B. Tidor and M. Karplus, The contribution of vibrational entropy to molecular association: the dimerization of insulin, *J. Mol. Biol.* 238(3):405–414, 1994.

[18] S.M. Schwarzl, T.B. Tschopp, J.C. Smith, and S. Fischer, Can the calculation of ligand binding free energies be improved with continuum solvent electrostatics and an ideal-gas entropy correction? *J. Comput. Chem.* 23(12):1143–1149, 2002.

[19] I. Andricioaei and M. Karplus, On the calculation of entropy from covariance matrices of the atomic fluctuations, *J. Chem. Phys.* 115(14):6289–6292, 2001.

[20] M.K. Gilson, J.A. Given, and M.S. Head, A new class of models for computing receptor-ligand binding affinities, *Chem. & Biol.* 4(2):87–92, 1997.

[21] J. Wang, P. Morin, W. Wang, and P.A. Kollman, Use of MM-PBSA in reproducing the binding free energies to HIV-1 RT of TIBO derivatives and predicting the binding mode to HIV-1 RT of efavirenz by docking and MM-PBSA, *J. Am. Chem. Soc.* 123(22):5221–5230, 2001.

[22] G. Jones, P. Willett, R.C. Glen, A.R. Leach, and R. Taylor, Development and validation of a genetic algorithm for flexible docking, *J. Mol. Biol.* 267(3):727–748, 1997.

[23] FRED, OpenEye Scientific Software: www.eyesopen.com.

[24] M.R. McGann, H.R. Almond, A. Nicholls, J.A. Grant, and F.K. Brown, Gaussian docking functions, *Biopolymers* 68(1):76–90, 2003.

[25] Glide, Schrödinger, LLC: www.schrodinger.com.

[26] D. Joseph-McCarthy, B.E. Thomas, M. Belmarsh, D. Moustakas, and J.C. Alvarez, Pharmacophore-based molecular docking to account for ligand flexibility, *Proteins* 51(2):172–188, 2003.

[27] D. Joseph-McCarthy and J.C. Alvarez, Automated generation of MCSS-derived pharmacophoric DOCK site points for searching multiconformation databases, *Proteins* 51(2):189–202, 2003.

[28] D.M. Lorber and B.K. Shoichet, Flexible ligand docking using conformational ensembles, *Protein Sci.* 7(4):938–950, 1998.

[29] G. Jones, P. Willett, and R.C. Glen, A genetic algorithm for flexible molecular overlay and pharmacophore elucidation, *J. Computer-Aided Mol. Des.* 9(6):532–549, 1995.

[30] T.J. Ewing, S. Makino, A.G. Skillman, and I.D. Kuntz, DOCK 4.0: search strategies for automated molecular docking of flexible molecule databases, *J. Computer-Aided Mol. Des.* 15(5):411–428, 2001.

[31] T.J.A. Ewing and I.D. Kuntz, Critical evaluation of search algorithms for automated molecular docking and database screening, *J. Comput. Chem.* 181175–181189, 1997.

[32] B.K. Shoichet, A.R. Leach, and I.D. Kuntz, Ligand solvation in molecular docking, *Proteins* 34(1):4–16, 1999.

[33] X. Zou, Y. Sun, and I.D. Kuntz, Inclusion of solvation in ligand binding free energy calculations using the generalized-born model, *J. Am. Chem. Soc.* 1218033–1218043, 1999.

[34] D.K. Gehlhaar, G.M. Verkhivker, P.A. Rejto, C.J. Sherman, D.B. Fogel, L.J. Fogel, and S.T. Freer, Molecular recognition of the inhibitor AG-1343 by HIV-1 protease: conformationally flexible docking by evolutionary programming, *Chem. & Biol.* 2(5):317–324, 1995.

[35] H.J. Bohm, The development of a simple empirical scoring function to estimate the binding constant for a protein–ligand complex of known three-dimensional structure, *J. Computer-Aided Mol. Des.* 8(3):243–256, 1994.

[36] M. Rarey, B. Kramer, T. Lengauer, and G. Klebe, A fast flexible docking method using an incremental construction algorithm, *J. Mol. Biol.* 261(3):470–489, 1996.

[37] D. Eldridge, C.W. Murray, T.R. Auton, G.V. Paolini, and R.P. Mee, Empirical scoring functions: 1. The development of a fast empirical scoring function to estimate the binding affinity of ligands in receptor complexes, *J. Computer-Aided Mol. Des.* 11(5):425–445, 1997.

[38] D. Rognan, S.L. Lauemoller, A. Holm, S. Buus, and V. Tschinke, Predicting binding affinities of protein ligands from three-dimensional models: application to peptide binding to class I major histocompatibility proteins, *J. Med. Chem.* 42(22):4650–4658, 1999.

[39] R. Wang, L. Liu, L. Lai, and Y. Tang, SCORE: a new empirical method for estimating the binding affinity of a protein–ligand complex, *J. Mol. Model.* 4(12):379–394, 1998.

[40] W. Welch, J. Ruppert, and A.N. Jain, Hammerhead: fast, fully automated docking of flexible ligands to protein binding sites, *Chem. & Biol.* 3(6):449–462, 1996.

[41] A.N. Jain, Scoring noncovalent protein–ligand interactions: a continuous differentiable function tuned to compute binding affinities, *J. Computer-Aided Mol. Des.* 10(5):427–440, 1996.

[42] G.M. Morris, D.S. Goodsell, R. Huey, and A.J. Olson, Distributed automated docking of flexible ligands to proteins: parallel applications of AutoDock 2.4, *J. Computer-Aided Mol. Des.* 10(4):293–304, 1996.

[43] G.M. Morris, D.S. Goodsell, R.S. Halliday, R. Huey, W.E. Hart, R.K. Belew, and A.J. Olson, Automated docking using a Lamarckian genetic algorithm and an empirical binding free energy function, *J. Comput. Chem.* 191639–191662, 1998.

[44] H. Gohlke and G. Klebe, Statistical potentials and scoring functions applied to protein–ligand binding, *Curr. Opin. Struct. Biol.* 11(2):231–235, 2001.

[45] H. Gohlke, M. Hendlich, and G. Klebe, Knowledge-based scoring function to predict protein–ligand interactions, *J. Mol. Biol.* 295(2):337–356, 2000.

[46] H.M. Berman, J. Westbrook, Z. Feng, G. Gilliland, T.N. Bhat, H. Weissig, I.N. Shindyalov, and P.E. Bourne, The Protein Data Bank, *Nucleic Acids Res.* 28(1):235–242, 2000.

[47] I. Muegge and Y.C. Martin, A general and fast scoring function for protein–ligand interactions: a simplified potential approach, *J. Med. Chem.* 42(5):791–804, 1999.

[48]   J.B.O. Mitchell, R.A. Laskowski, A. Alex, and J.M. Thornton, BLEEP — potential of mean force describing protein–ligand interactions: I. Generating potential, *J. Comput. Chem.* 201165–201176, 1999.

[49]   J.B.O. Mitchell, R.A. Laskowski, A. Alex, M.J. Forster, and J.M. Thornton, BLEEP — potential of mean force describing protein–ligand interactions: II. Calculation of binding energies and comparison with experimental data, *J. Comput. Chem.* 201177–201185, 1999.

[50]   A.V. Ishchenko and E.I. Shakhnovich, SMall Molecule Growth 2001 (SMoG2001): an improved knowledge-based scoring function for protein–ligand interactions, *J. Med. Chem.* 45(13):2770–2780, 2002.

[51]   M.K. Gilson, K. Sharp, and B. Honig, Calculating the electrostatic potential of molecules in solution: method and error assessment, *J. Comput. Chem.* 9327–9335, 1988.

[52]   B. Honig and A. Nicholls, Classical electrostatics in biology and chemistry, *Science* 268(5214):1144–1149, 1995.

[53]   A.A. Rashin and K. Namboodiri, A simple method for the calculation of hydration enthalpies of polar molecules with arbitrary shapes, *J. Phys. Chem.* 916003–916012, 1987.

[54]   N. Majeux, M. Scarsi, J. Apostolakis, C. Ehrhardt, and A. Caflisch, Exhaustive docking of molecular fragments with electrostatic solvation, *Proteins* 37(1):88–105, 1999.

[55]   M.K. Holloway, J.M. Wai, T.A. Halgren, P.M. Fitzgerald, J.P. Vacca, B.D. Dorsey, R.B. Levin, W.J. Thompson, L.J. Chen, and S.J. deSolms, A priori prediction of activity for HIV-1 protease inhibitors employing energy minimization in the active site [Erratum appears in *J. Med. Chem.* 39(11):2280, 1996], *J. Med. Chem.* 38(2):305–317, 1995.

[56]   E. Kangas and B. Tidor, Electrostatic specificity in molecular ligand design, *J. Chem. Phys.* 112(20):9120–9131, 2000.

[57]   L.P. Lee and B. Tidor, Optimization of electrostatic binding free energy, *J. Chem. Phys.* 106(21):8681–8690, 1997.

[58]   E. Kangas and B. Tidor, Optimizing electrostatic affinity in ligand-receptor binding — theory, computation, and ligand properties, *J. Chem. Phys.* 109(17):7522–7545, 1998.

[59]   E. Kangas and B. Tidor, Electrostatic complementarity at ligand binding sites: application to chorismate mutase, *J. Phys. Chem.* B105(4):880–888, 2001.

[60]   J. Shen and J. Wendoloski, Electrostatic binding energy calculation using the finite difference solution to the linearized Poisson–Boltzmann equation — assessment of its accuracy, *J. Comput. Chem.* 17(3):350–357, 1996.

[61]   J. Novotny, R.E. Bruccoleri, M. Davis, and K.A. Sharp, Empirical free energy calculations: a blind test and further improvements to the method, *J. Mol. Biol.* 268(2):401–411, 1997.

[62]   V.K. Misra, J.L. Hecht, K.A. Sharp, R.A. Friedman, and B. Honig, Salt effects on protein-DNA interactions. The lambda cI repressor and EcoRI endonuclease, *J. Mol. Biol.* 238(2):264–280, 1994.

[63]   K.A. Sharp, Electrostatic interactions in hirudin-thrombin binding, *Biophys. Chem.* 61(1):37–49, 1996.

[64]   J. Novotny and K. Sharp, Electrostatic fields in antibodies and antibody/antigen complexes, *Prog. Biophys. & Mol. Biol.* 58(3):203–224, 1992.

[65]   R. Bone, J.P. Vacca, P.S. Anderson, and M.K. Holloway, X-ray crystal structure of the HIV protease complex with L-700,417, an inhibitor with pseudo C2 symmetry, *J. Am. Chem. Soc.* 113(24):9382–9384, 1991.

[66]   T.A. Halgren, Merck molecular force field: 1. Basis, form, scope, parameterization, and performance of MMFF94, *J. Comput. Chem.* 17(5–6):490–519, 1996.

[67] T.A. Halgren, Merck molecular force field: 2. MMFF94 van der Waals and electrostatic parameters for intermolecular interactions, *J. Comput. Chem.* 17(5–6):520–552, 1996.

[68] T.A. Halgren, Merck molecular force field: 3. Molecular geometries and vibrational frequencies for MMFF94, *J. Comput. Chem.* 17(5–6):553–586, 1996.

[69] T.A. Halgren, Merck molecular force field: 5. Extension of MMFF94 using experimental data, additional computational data, and empirical rules, *J. Comp. Chem.* 17(5–6):616–641, 1996.

[70] T.A. Halgren and R.B. Nachbar, Merck molecular force field: 4. Conformational energies and geometries for MMFF94, *J. Comput. Chem.* 17(5–6):587–615, 1996.

[71] SYBYL 6.9, Tripos, Inc.: www.tripos.com.

[72] ZAP, OpenEye Scientific Software, www.eyesopen.com.

[73] J.A. Grant, B.T. Pickup, and A. Nicholls, A smooth permittivity function for Poisson-Boltzmann solvation methods, *J. Comput. Chem.* 22(6):608–640, 2001.

[74] J.A. Grant and B.T. Pickup, Gaussian shape methods, *Comput. Sim. Biomol. Sys.* 3150:176, 1997.

[75] J.A. Grant, M.A. Gallard, and B.T. Pickup, A fast method of molecular shape comparison: a simple application of a Gaussian description of molecular shape, *J. Comput. Chem.* 17(14):1653–1666, 1996.

[76] J.A. Grant and B.T. Pickup, A Gaussian description of molecular shape [Erratum for Vol. 99, 1995], *J. Phys. Chem.* 100(6):2456, 1996.

[77] D. Sitkoff, K.A. Sharp, and B. Honig, Correlating solvation free energies and surface tensions of hydrocarbon solutes, *Biophys. Chem.* 51(2–3):397–403, [discussion 404–399], 1994.

[78] D. Sitkoff, N. Bental, and B. Honig, Calculation of alkane to water solvation free energies using continuum solvent models, *J. Phys. Chem.* 100(7):2744–2752, 1996.

[79] K.A. Sharp, A. Nicholls, R.F. Fine, and B. Honig, Reconciling the magnitude of the microscopic and macroscopic hydrophobic effects, *Science* 252106–252109, 1991.

[80] J.T. Bolin, D.J. Filman, D.A. Matthews, R.C. Hamlin, and J. Kraut, Crystal structures of *Escherichia coli* and *Lactobacillus casei* dihydrofolate reductase refined at 1.7 A resolution: I. General features and binding of methotrexate, *J. Biol. Chem.* 257(22):13650–13662, 1982.

[81] MDL Information Systems: www.mdli.com.

[82] OMEGA, OpenEye Scientific Software: www.eyesopen.com.

[83] T.S. Rush III and J.C. Alvarez, manuscript in preparation, 2003.

[84] FRED, OpenEye Scientific Software: www.eyesopen.com.

[85] C.A. Baxter, C.W. Murray, D.E. Clark, D.R. Westhead, and M.D. Eldridge. Flexible docking using Tabu search and an empirical estimate of binding affinity, *Proteins* 33(3):367–382, 1998.

[86] M. Stahl and M. Rarey, Detailed analysis of scoring functions for virtual screening, *J. Med. Chem.* 44(7):1035–1042, 2001.

[87] R. Sayle and A. Nicholls, manuscript in preparation.

[88] WABE, OpenEye Scientific Software: www.eyesopen.com.

11 Classification of Ligand–Receptor Complexes Based on Receptor Binding Site Characteristics

Marguerita S.L. Lim-Wilby, Teresa A. Lyons, Michael Dooley, Anne Goupil-Lamy, Sunil Patel, Christoph Schneider, Remy Hoffmann, Hugues-Olivier Bertrand, and Osman F. Güner

11.1 INTRODUCTION

The efforts of Kuntz and his group in the early 1980s [1] are generally considered the beginning of docking of ligands to proteins. Much has been published since in the field of ligand–protein docking. The present abundance of commercially available docking tools and number of publications on docking are testimony to the increasing success of docking and scoring tools for drug design applications. Recent reviews cover these various topics in docking and scoring: comparison of docking tools and their methodologies [2], test sets for docking validations [3], docking as a virtual screening (VS) tool [4–6], docking in context with other screening methods [7], and the application of scoring functions to predicting binding affinity [8,9]. The fact that many recently released drugs (e.g., Crixivan® [10,11], Viracept® [12], Trusopt® [13]) were designed by structure-based design techniques has increased the importance of the use of target receptor structures in drug discovery. Of particular interest in the last decade has been the application of docking methods in direct analogy to high throughput screening (HTS), that is, virtual HTS.

The purpose of this chapter is to discuss approaches to docking and scoring based on the requirements of the binding site and to provide a novel classification of drug–receptor interactions based on characteristics of the binding site. The Kuntz group [1] used the concept of matching putative ligand atoms to surface spheres in contact with the receptor. Such a definition for a ligand-binding site within a macromolecule acknowledges the basic assumption that ligands bind to receptors based on complementary chemical interactions at the interface of the complex. As first used by Kuntz et al., we refer to *docking* as the computational simulation of ligands binding to receptors. During the process of docking, the placement of ligands in a myriad of poses in the binding site of a receptor necessitates the objective evaluation of these poses, because it is currently not feasible to save all ligand poses. Ideally, this objective evaluation of the interaction energy of the ligand for the receptor represents an estimate of the affinity of ligand for the receptor. Such estimates of affinity are usually derived from trained scoring functions with differing degrees of accuracy.

In the *in vitro* world of molecular pharmacology, automation has led to advances in assay technologies, from being able to perform chromogenic or fluorogenic assays in 1 cm path length cuvettes to plate assays with Microtiter™ 96-well plates, HTS with 3456-well plates (Aurora Instruments, LLC) and now ultra-HTS with microfluidic chips. In the molecular modeling laboratory, advances in computational methods and reducing costs of processing speed, bandwidth, and data storage are now propelling docking methods into the high throughput realm. Automation and parallelization of docking and scoring methods will require that these methods become increasingly accurate, with reproducible protocols for obtaining the required results.

11.2 CONSIDERATIONS IN BINDING SITE DEFINITIONS

In vitro assays typically proceed with collisions between receptor and ligand molecules, either in solution or on membranelike surfaces, with the aid of Brownian motion and long range electrostatic interactions, which when productive, lead to binding and a resultant signal usually mediated by substrate cleavage. In real time, millions of such collisions per second would result in a maximal signal from several thousand interacting ligand–receptor pairs per cubic centimeter volume. Computationally, nonproductive collisions between ligand and receptor can be just as costly to calculate as favorable interactions between receptor and good-binding ligand conformations. A typical well-folded target protein of about 300-residues, for example the structure of CDK2 in pdb1his.ent from the Protein Data Bank (PDB) [14], has a total exposed van der Waals (vdW) surface area of more than 14,000 Å^2, compared to less than 1700 Å^2 when a binding site of only 16 residues is defined (Figure 11.1).

A reduction in the Cartesian search problem is the first step for most docking algorithms. This is accomplished by allowing the user to define a binding site. To define a binding site, one needs to understand the receptor's structure and the likely direction of a drug discovery strategy so that biochemical assays are consistent with the target site of candidate ligands as modeled. Determining the target binding site is not always straightforward. The simplest means of defining a binding site is to use the location of known bound ligands, which is available only when multiple structures of the receptor — or a close homolog — in complex with several ligands from diverse classes have been experimentally determined. If cognate complexes are not available, as is the case for modeled receptors or an apoprotein structure, one can use various algorithms for locating binding sites. The most fundamental of

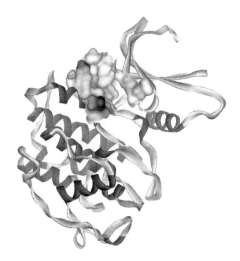

FIGURE 11.1 Solvent-accessible surface area of ligand binding site of CDK2. The receptor from the pdb1h1s complex [14] is depicted as a ribbon colored by secondary structure and its surface is colored by electrostatic charges.

these is a flood-filling algorithm, which works well for binding sites that are cavities [15] or that take advantage of cavities for aggregation of spheres in potential site points [16,17]. Many binding sites do not include cavities, so many other methods have been devised for prioritizing the location of a binding site [18].

Once a binding site has been identified, it can be tested by investigation of goodness of fit with known ligands. Models can be generated by hand or a docking algorithm can be applied. Methodologies for defining a binding site depend on the requirements of the particular docking application (Table 11.1).

Whether a binding site is defined as a list of residues, box, sphere, surface, or shape, the aim is to restrict the search space for the initial placement of a ligand to begin the docking process. One exception is Affinity [19], which allows the binding site to be flexible and requires the user to manually place the ligand in the binding site; these are the two main reasons precluding it from use as a high throughput docking method.

11.3 CONCEPTUAL ADVANCES IN USING BINDING SITE INFORMATION

The use of information availed by ligands with known binding modes at known binding sites is illuminating to the computational chemist. Such knowledge can be deduced from a variety of experimental methods, including mutagenesis results, evolutionary trace, and most importantly for docking, ligand–protein contacts at specific receptor sites. Restraining ligand poses by the imposition of selective interactions to specified receptor sites can greatly aid computational methods in reproducing known binding modes [26]. An interesting advance of late is the application

TABLE 11.1
Comparison of Binding Site Definition Methodologies for Various Commercially Available Docking Applications

Application	Definition of Binding Site
Affinity [19]	List of receptor residues.
AutoDock [20]	Grid map surrounding region of interest or entire protein.
DOCK [1]	Surface spheres.
FlexX [21]	List of receptor residues.
FRED [22]	Box around region of interest.
Glide [23]	Enclosing box for containing all ligand atoms and bounding box for positions of ligand centers.
GOLD [24]	Either a spherical volume encompassing all site atoms or a list of binding site atoms. Plus, a cavity detection algorithm to further define concave solvent-accessible surfaces.
LigandFit [25]	Either the volume occupied by a ligand or a cavity detection algorithm. Plus, user editing of site points.

of consensus selection of poses from different docking applications, namely Cons-Dock [27].

Another development is the use of softened potentials, as with the Piecewise Linear Potential (PLP) docking function that was found to smoothen the energy landscape when used with two different docking algorithms [28]. Originally devised as a docking function, the published method was implemented as a scoring function in the C2.LigandFit toolkit [29] and later integrated as an alternative docking function that has led to some improvement in LigandFit results as well [30]. Most notably, PLP1's softened potential allows close ligand–protein contacts where necessary for a native dock for the complexes, for example, 1a9u.pdb [14] (Table 11.2).

As was mentioned in the previous section, the definition of a binding site reduces the Cartesian search problem to only the region of interest. However, if a binding site is truly large and there are multiple subsites that small ligands can select from the total search area, one is still faced with a fairly challenging problem. This can be further exacerbated by a difficult binding site topology where several local minima can exist. Parameters that address the ability to adjust conformational trials cannot overcome a special search problem. For this particular set of binding sites, site partitioning [30] is particularly useful. Partitioning a binding site increases the number of starting positions in the placement of ligands significantly smaller than the defined binding site, such that all subsites can be effectively explored.

A concept that we would like to introduce is the consideration of selecting and adjusting docking parameters for specific classes of binding sites. Binding sites on macromolecules can be for a range of ligand types, from the large, as in protein–protein interactions and protein-nucleic acid complexes, to the smallest substrates, cofactors, allosteric factors, agonists, and antagonists. It has been proposed that binding sites for small molecule ligands can be categorized into six main classes [31], namely:

TABLE 11.2
RMS Accuracy of Docking for Diverse Ligand–Protein Complexes

Receptor	PDB code	Site Volume (Å3)¶	Ligand Fit (25)§	FlexX (33)#	FlexX (27)#	GOLD (34)	GOLD (27)	Dock (27)	Cons Dock (27)
Class I Small, Well-defined									
FADH oxidoreductase	1PBD	60		0.33	(1.61)	good	1.33	0.44	0.73
Streptavidin	1STP	99	0.29	0.65	(0.98)	good	0.51	0.62	0.51
Lysozyme	186L	120	0.71						
Thymidine kinase	1KIM	200	0.32						
Cytochrome P-450	1PHG	250	0.57	4.74		close			
Cholesterol oxidase	1COY	260		(1.06)	1.04		0.70	0.65	0.66
Class II Large, Well-defined									
Renin	1RNE	520		12.24	(18.86)	close	5.62	15.48	5.73
HIV protease	1AAQ	500			1.56	wrong	2.21	13.65	1.76
Class III Open Binding Site, with Well-defined Subsites									
Acetylcholinesterase	1ACL	330	1.95						
Phospholipase A2	5P2P	400		1.00		close			
Dihydrofolate reductase -NADPH	4DFR	630	1.24	(1.40)	1.03	good	1.13	5.37	0.65
Thrombin	1ETR	670	3.48	(7.24)	2.93	errors	5.72	5.11	4.29
Collagenase	1HFC	750		(2.51)	2.42	close	2.75	10.73	2.50
Elastase	1ELA	760	2.17	9.71					
Penicillopepsin	1APT	725	8.35	1.89	(6.23)	close	8.26	9.03	6.57
Hydroxysteroid dehydrogenase	1HDC	1100		(11.74)	13.58	errors	9.59	2.85	9.66
Class IV Large Binding Site, with No Well-defined Subsites									
Ribonuclease T1	6RNT	350		4.79		close			
Acetyltransferase	3CLA	420		6.42		wrong			
Ribonuclease A	1ROB	440		7.70	(8.55)	errors	1.17	1.04	0.91
Ribonuclease Ms	1RDS	720		4.89		errors			
Endothiapepsin	1EED	740		(9.78)	1.42	wrong	12.13	11.89	1.31
MAP kinase p38	1A9U	840	8.08	10.10	(10.33)	wrong	8.50	1.89	1.19
Class V Shallow Sites, with No Well-defined Subsites									
Cyclodextrin glycosyl transferase	1CDG	610							
Class VI Ill-defined Sites/Hinge Regions									
Adenylate kinase	2AK3	590							

The LigandFit results from Venkatachalam et al. [25] were obtained with the first version of LigandFit in Cerius2 4.6 and without consideration of the best LigScore2, which was not available at the time.

¶ Site volumes were determined by defining the site as a grid at 0.5 Å resolution in DS LigandFit, using criteria as discussed in Section 11.4. Site volumes are calculated as the number of site points multiplied by the volume of a site point (0.125 Å³).

Where rms results are available in multiple publications for FlexX, the best result is considered, and the unused data points are in parentheses.

1. Class I — Small, well-defined (smaller than 300 Å³)
2. Class II — Large, well-defined (larger than 300 Å³)
3. Class III — Open binding site, with well-defined subsites
4. Class IV — Large and open binding site, with no well-defined subsites
5. Class V — Shallow and superficial binding site
6. Class VI — Ill-defined binding site at hinge region

One can further clarify and define Class I through Class VI using examples of ligand–receptor pairs as shown in Table 11.2. Note that Table 11.2 lists published results for various docking software. It is likely that these results do not necessarily reflect the current results that can be obtained with these programs that are consistently being improved and enhanced. Our intention is not to compare results from different programs, but rather to provide historical published data for the sake of completeness.

11.3.1 CLASS I — SMALL, WELL-DEFINED BINDING SITES

A Class I site is well-defined by receptor atoms in that most of the binding interactions are contained within a cavity that is enclosed by the receptor. The site volumes are typically more than 60 Å³ and less than 300 Å³. Cavity volume includes locations for potential water molecules — as assigned by a crystallographer — that may or may not be displaced by ligand binding. A small, well-defined site is exemplified (Figure 11.2) by cholesterol oxidase from the 1COY.pdb complex [14].

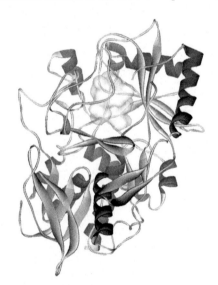

FIGURE 11.2 Small, Well-Defined Binding Site
The binding site for cholesterol oxidase from the pdb1coy complex [14] was found by LigandFit's cavity search and is represented as a solid transparent surface. The receptor is depicted as a ribbon colored by secondary structure.

Docking to Class I receptors is usually not problematic. The only point in question here is the selection of the best docking and scoring functions such that the best root mean square (rms) or reproduction of binding mode as found in the highest ranking pose, such that only three to five poses are necessary to be saved. This is especially important if the receptor is to be a target for virtual HTS.

11.3.2 CLASS II — LARGE, WELL-DEFINED BINDING SITES

These sites are still well-defined by receptor atoms, but typically one or two ends may be exposed to solvent. More importantly, the volume enclosed by receptor is more than 300 Å^3; there are no observed examples greater than 800 Å^3 in Class II receptors that bind small molecule ligands, although a thorough survey of published complexes has not been performed. Archetypical representatives for this class are the well-known aspartyl protease of human immunodeficiency virus (HIV) and the mammalian aspartyl protease, renin. An HIV-1 aspartyl protease-inhibitor complex from the pdb1aaq [14] is shown in Figure 11.3.

Docking to Class II receptors can be challenging, due to two main reasons:

1. The large number of ligand torsions leading to a large conformational search space, as ligands for this type of receptor are typically large and can have many rotatable bonds.
2. Ligands that can bind to this type of receptor do not necessarily utilize all of the subsites and hence can be significantly smaller that the receptor site.

The challenge then becomes orientation and initial placement of the ligand to the correct binding location.

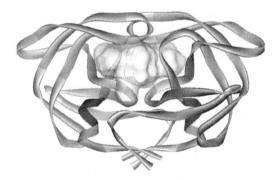

FIGURE 11.3 Class II — Large, Well-Defined Binding Site
The binding site for HIV-1 aspartyl protease from the pdb1aaq complex [14] was found by LigandFit's cavity search and is represented as a solid transparent surface. The receptor is depicted as a ribbon colored by secondary structure.

11.3.3 CLASS III — OPEN BINDING SITE, WITH WELL-DEFINED SUBSITES

A Class III open binding site is one that is mostly solvent exposed, but still having subsites that present well-defined binding pockets to small ligands or portions of larger ligands. Open binding sites are more typical for the serine, cysteinyl, and metalloproteases. Well-known examples of well-defined subsites are S2 of the cysteine protease papain, S1 of the trypsin proteases, and S2 of thrombin, also a trypsinlike enzyme, as shown in Figure 11.4.

11.3.4 CLASS IV — LARGE BINDING SITE, WITH NO WELL-DEFINED SUBSITES

This class of binding site is extremely well populated by many synthases and transferases, as well as protein kinases, for which binding sites can be variable in size. The characteristics of binding sites in this class are that:

* More than two sides of the binding site are exposed to solvent
* Less than half the binding site is typically occupied
* Regions of flexibility exist in one or both sides defining the site

An example for thymidylate synthase is shown in Figure 11.5.

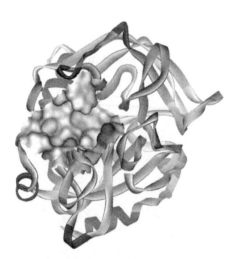

FIGURE 11.4 Class III — Open Binding Site, with Well-Defined Subsites
An example of this class of site is represented as the solvent-accessible surface of residues in contact with the binding site of thrombin from the pdb1etr complex [14]. The binding site was defined by expansion of the volume occupied by the bound ligand from the complex; the receptor is depicted as a ribbon colored by secondary structure.

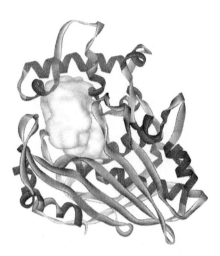

FIGURE 11.5 Class IV — Large Binding Site, with No Well-Defined Subsites
The binding site for thymidylate synthase from the pdb1tdb complex [14] was defined by extensive expansion of the volume occupied by the bound ligand and is represented as a solid transparent surface. The receptor is depicted as a ribbon colored by secondary structure.

11.3.5 CLASS V — SHALLOW AND SUPERFICIAL BINDING SITE

A Class V site is shallow and superficial (i.e., like a dent on the receptor surface). Such binding sites can be variable in size and shape. This class of binding sites is not commonly addressed as drug design targets. If a Class V site can be defined by a bound ligand, or model of a potentially bound ligand, then it should be treated as either a Class I or Class II site, where the best protocol to be used depends on the actual volumes occupied by bound ligands. A graphical definition of a Class V site is shown in Figure 11.6.

11.3.6 CLASS VI — ILL-DEFINED BINDING SITE AT HINGE REGION

Class VI is a difficult target for docking and, due to the tremendous flexibility found in such receptors, docking methods applied to these targets are nonideal. In Figure 11.7, adenylate kinase, pdb2ak3 [14] with a large binding site folding around the hinge region is displayed. Despite the large and flexible binding site, the ligand occupies only a small portion of the site.

Class VI type binding sites are the most difficult cases for docking due to additional complications being introduced by relatively large receptor flexibility around the hinge region. Use of site partitioning is essential in such cases, as well as use of the region dictated by any known bound ligand to direct the docking to the correct subsite.

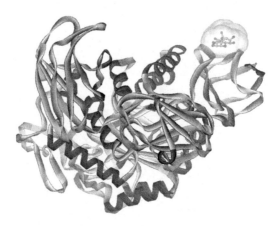

fIGURE 11.6 Class V — Shallow and Superficial Binding Site
The binding site for cyclodextrin glycosyl transferase from the pdb1cdg complex [14] was defined by expansion of the volume occupied by the bound ligand and is represented as a solid transparent surface. The bound ligand is displayed as a ball-and-stick model visible within the binding site. The receptor is depicted as a ribbon colored by secondary structure.

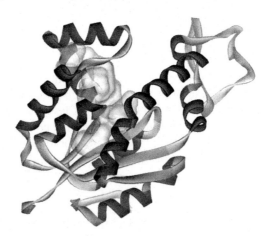

FIGURE 11.7 Class VI — Ill-Defined Binding Site at Hinge Region
The binding site for adenylate kinase from the pdb1ak3 complex [14] as defined by the default cavity search tool in LigandFit is represented as a solid transparent surface. The receptor is depicted as a ribbon colored by secondary structure.

11.4 PROTOCOLS FOR ADDRESSING BINDING SITES ACCORDING TO CLASSES

Use of default parameters for docking applications will not necessarily suit all types of ligands and all classes of binding sites, and certainly not for all desired applications, regardless of whether one is docking to explore binding modes or docking as part of a VS workflow. In an ideal world, where computer processor power would be unlimited, ligand libraries not too large, and receptors highly amenable to docking, it would be easy to recommend the use of the most accurate settings available. However, compromises are often made in the real world, where time constraints and a fixed number of computational servers dictate the number of ligands that can be processed as well as the accuracy of the results.

The protocols presented here — resulting from our limited access to docking applications — are necessarily tailored to LigandFit, although the general concepts should be transferable to other methodologies in docking. For all ligand–receptor pairs presented in Table 11.2, the procedures outlined below were applied. These recommendations based on different classes of receptor active site are also summarized in Table 11.3.

11.4.1 Check that There Are No Obvious Problems with Ligand–Receptor Interactions

Example problems may arise from:

1. Close contacts between ligand and receptor within vdW radii unless hydrogen bonding is involved.
2. Multiple poses for ligand in complex (for example, the alternative poses for bound ligand in pdb1RDS [14] in Table 11.1).

11.4.2 Check X-ray Pose of Ligand

1. Copy ligand from ligand–receptor complex into separate modeling area.
2. Check that all atoms are present and correctly identified.
3. Save x-ray pose for rms calculations.

11.4.3 Prepare Ligand for Native Docking

1. Copy x-ray pose into a separate modeling area.
2. Correct bond orders.
3. Add hydrogens (LigandFit uses force fields — parameterized with hydrogens enumerated — to determine atom-types for atomic radii and partial charges).
4. Check that ligand charges and tautomeric form (if any) is compatible with binding to receptor.
5. Idealize bonds and angles to simulate ligand from a two-dimensional–three-dimensional (2D-3D) converter.
6. Save newly cleaned ligand for docking.

11.4.4 Prepare Receptor Structure for Docking

1. Remove ligand from ligand–receptor complex.
2. Consider if any metal ions and cofactors are missing from correct representation of receptor, especially in vicinity of binding site.

TABLE 11.3
Summary of Recommendations for Different Classes of Binding Sites

Binding Site Class and Characteristics	Site Definition	Site Partitioning	Docking Parameters
Class I — Small, well-defined	Cavity search	Not necessary	Dock Energy for docking function. Okay to use fixed number of trials of 5000.
Class II — Large, well-defined	Cavity search	Use (Rounddown of) [Site volume/Ligand Volume]	If close contacts exist, use PLP1 docking function. Use variable number of trials per torsion, as shown in Table 11. 4. Refine poses with 200 to 500 steps of molecular-mechanics-based minimization for accurate scoring.
Class III — Large, with well-defined subsites	Cavity search with selective expansion, or ligand-defined with site expansion	Use (Rounddown of) [Site volume/Ligand Volume]. Increase rms threshold, for comparison of ligand shape to site shape, from default 2.0 to 10.0 or higher if site topology is not smooth.	If close contacts, use PLP1, otherwise use Dock Energy. Use variable number of trials per torsion, as shown in Table 11.4.
Class IV — Large, with no well-defined subsites	Ligand-defined with expansion and editing of site	Use (Rounddown of) [Site volume/Ligand Volume]	If close contacts exist, use PLP1 docking function. Use variable number of trials per torsion, as shown in Table 11.4. In addition, increase Default Maximum Number of Trials per Torsion to be the same as Maximum Number of Trials per Torsion. Refine poses with 200 to 500 steps of molecular-mechanics-based minimization for accurate scoring.
Class V — Shallow and superficial binding site	Ligand-defined with expansion and editing of site	Depends on area of binding site to be covered, typically not necessary	For sites of up to 300 Å³, use Class I parameters. For sites of more than 300 Å³, use Class II parameters.
Class VI — Ill-defined binding site at hinge region	Ligand-defined with extensive expansion and editing of site	Use (Rounddown of) [Site volume/Ligand Volume]	Use PLP1 docking function. Results will be highly approximate and ranking by score may be highly inaccurate.

3. Check that temperature factors for receptor are reasonable; B-factors should be less than 40.

4. Remove waters of crystallization unless absolutely necessary. (For results in Table 11.1, waters were retained only for the 3CLA complex when docking with LigandFit.) If waters of crystallization are to be included because they are known to help define binding modes that would otherwise not be found, then care should be taken to optimize the locations of the added hydrogens to be consistent with the binding modes intended.

5. Correct bond orders by inspecting each residue in sequence along the backbone. If time is limited or docking is only a preliminary test, inspection of the binding site for these potential problems is usually sufficient:
   a. Truncated regions where no density is observed for known regions of structure.
   b. Tautomeric forms of histidyl imidazole rings.
   c. Residues that have been either phosphorylated or glycosylated.

6. Add hydrogens (LigandFit uses force fields — parameterized with hydrogens enumerated — to determine atom-types for atomic radii and partial charges).

7. Charge receptor according to conditions under which binding affinity (K_i) is experimentally determined; this mostly affects histidyl imidazole rings, but occasionally may affect active site cysteinyl thiols, depending on the pH of the site's microenvironment.

11.4.5 DEFINE BINDING SITE

If one is performing a native dock, it is possible to inspect the environment presented by the receptor to the ligand. In the absence of a native ligand, various binding site detection or prediction methods can be used and the spheres or points resulting from such methods can be used as dummy atoms for the purposes of defining a ligand. Depending on the ligand and receptor contacts, one of these procedures below can usually apply:

1. Ligand is fully or mostly (70%, as judged by eye) surrounded by receptor (Class I — small, well-defined site and Class II — large, well-defined site). Use cavity search with default opening size (5 Å). With this method, ligand atoms should be then fully contained by volume described by site points. If ligand atoms protrude from envelope of site points, or if a larger site is desired, try a larger opening size. Up to 9 Å usually is sufficient. When larger cavity opening sizes are used, it may be necessary to delete site points that may be considered as spurious subsites if site partitioning is later applied.

2. Ligand is only partially enclosed by receptor and receptor subsites that are fully occupied by parts of ligand exist (Class III — open binding site, with well-defined subsites). Use cavity search followed by Expand site function until ligand atoms are covered. If necessary, partial expansion of site may be accomplished by selection of site points near area requiring coverage followed by Expand command.

3. Ligand is only partially enclosed by receptor and no receptor subsites exist (Class IV — large binding site, with no well-defined subsites). Use ligand-defined volume as site, followed by either selective expansion or Expand with no selected site points to increase coverage in all directions where not occupied by receptor. An example of a site that is created beginning with a ligand-defined volume followed by several expansions and editing is shown for 1TDB in Figure 11.8.

4. Ligand is not enclosed by receptor at all and binding pocket is shallow for cavity search to be effective (Class V — shallow and superficial binding site). A Class V site is shallow and superficial. Such binding sites can be extremely variable in size and shape. If a Class V site can be defined by a bound ligand or model of a potentially bound ligand, then it should be treated as either a Class I or Class II, depending on the volume occupied by the bound ligand. A graphical definition of a Class IV site is shown in Figure 11.6.

5. Ligand is not enclosed by receptor at all and ligand makes contact with receptor at a site where cavity search reveals a ring around hinge region of a receptor (Class VI — ill-defined binding site at hinge region). Class VI is an extremely difficult target for docking and, due to the tremendous flexibility found in such receptors, docking methods applied to these targets are still nonideal.

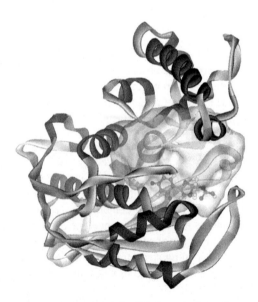

FIGURE 11.8 Binding site definition for thymidylate synthase from pdb1tdb [14], with a Class IV binding site. The receptor, depicted as a ribbon, partially encloses the binding site, shown as a transparent surface. The ligand, 5-fluoro-2′-deoxyuridine-5′-monophosphate, is shown in ball-and-stick representation within the binding site that it helped to define.

11.4.6 SELECT LEVEL OF SITE PARTITIONING

For native docks, if the volume of the ligand — in the pose to be reproduced — is more than half the site volume, the Cartesian search problem is not an issue. For sites in other classes, especially if one is docking unknown ligands that may be substantially smaller than the site, the following rule-of-thumb can be used for selecting the level of site partitioning:

Site Partitioning Level = (Rounddown of) [Site volume/Ligand Volume]

where ligand volume is calculated on the smallest ligand of interest.

11.4.7 FIND BEST DOCKING PARAMETERS

A rigid dock of the native conformation may be desirable as a quick means to investigate which docking parameters are useful from this list below. After these parameters are found, an unbiased conformation of the ligand with idealized bond lengths, angles, and dihedrals may be used to flexibly fit the ligand back into the partner receptor from the original complex.

The adjustable docking parameters for LigandFit are listed below, with the optimal settings depending on the various factors.

1. Docking function for energy grid calculation — A docking function supplies the means to prioritize poses during the docking process, as well as to drive the docking to produce good poses.
    a. DockScore for ligand pose-receptor surface interactions parameterized with Dreiding force field and Gasteiger charges — This is the default setting, which works well for most Class I and many Class II and Class III sites. The Dreiding force field allows almost all types of ligands to be parameterized without additional parameter estimation and thus is a suitable general purpose force field for VS of ligand libraries.
    b. DockScore for ligand pose-receptor surface interactions parameterized with Consistent Force Field (CFF) — CFF is a Class II force field, which has been demonstrated to work well with druglike molecules. CFF2000 with parameter equivalence is able to type 95% of the molecules in Derwent's World Drug Index [34].
    c. PLP1-based DockScore — The docking function using PLP1 provides a useful means of softening the likely repulsions arising from close contacts that may be required by ligands in tight complexes (Class II and Class III) or where receptor flexibility — not accounted for by LigandFit in the docking process — would lead to contacts that may otherwise be considered repulsive (Class IV and Class VI).
2. Force field selection — Regardless of the docking function, ligand internal energy is calculated. Conformations that produce intramolecular clashes in ligand being docked will be deleted.

11.4.8 Number of Conformational Trials per Ligand

1. Rigid docking — Docking may be performed in a rigid fashion on a library of ligands with enumerated conformations, for example with FRED (22) or LigandFit (25). In rigid docking, no conformation search trials are applied; instead, another conformation search engine is used to enumerate ligand conformations into a database that is then searched. Rigid docking also has applications in testing the docking function, checking for ligand–receptor clashes, and investigating the number of steps to be applied during rigid-body minimization phases of docking.

2. General issues in flexible docking — Most docking applications include at least one form of conformational searching, using either genetic algorithms, fragment-based construction, simulated annealing, or Monte Carlo (MC)-based torsion scrambling. The major consideration in flexible docking is that more flexible ligands require more conformational searching than less flexible ligands. Large and flexible ligands with more than 25 torsions are not recommended for inclusion in a ligand library for VS, as such molecules would generally not be useful hits.

   Even if the extent of torsions is only 0 to 25, one still has a spectrum of flexibility in ligands that need to be effectively and efficiently addressed. Using a single number of conformational trials for all ligands in a library would put one at risk of oversampling rigid ligands, while not sufficiently searching conformational space for the more flexible ones. Rather, the number of trials to be applied to any one ligand should be determined by its own flexibility, which can be considered in torsion bins or ranges in a look-up table. Two examples for LigandFit are presented in Table 11.4. The choice of whether to use the greater or fewer number of trials depends on the computational resources available to the user and the thoroughness of the search required. Thus, for larger sites (Class II, Class III, and Class IV), we recommend that a greater number of MC trials be applied, as in the second column of Table 11.4 — Maximum Number of MC Trials for Larger Sites, compared to the default (third column, Table 11.4). Means to shorten the time spent on each ligand along with reasonable default values have been supplied (last column, Table 11.4). However, if the site volume warrants a more thorough search, we recommend that the values in the fourth column of Table 11.4 — Number of MC Trials Allowed without Update to Save List for Larger Sites be used.

3. Conformational searching applied to different classes of binding sites:
   a. In general, Class I binding sites cannot accommodate large ligands, so conformational searching is not an issue when using VS for small, well-defined binding sites.
   b. Class II binding sites tend to require larger ligands to effectively occupy the majority of the site, as is typically found for known high-affinity inhibitors of these targets. Thus conformational searching is one of the most significant issues here, as hits are generally found among the

TABLE 11.4
Number of Monte Carlo Trials Recommended for LigandFit According to Ligand Torsions

Number of Ligand Torsions	Maximum Number of MC Trials for Larger Sites	Default Maximum Number of MC Trials	Number of MC Trials Allowed without Update to Save List for Larger Sites	Default Number of MC Trials Allowed without Update to Save List
0–2	500	500	120	120
3–4	1200	1200	300	300
5–6	1500	1500	700	350
7–10	2000	2000	1000	500
11–14	3000	3000	1500	750
15–19	4500	3000	1500	750
20–25	6000	3000	1500	750

more flexible molecules. In addition, for Class II sites, some pose optimization with molecular-mechanics-based minimization, which is necessary for more accurate scoring of saved poses.

c. Class III through Class VI binding sites can accommodate ligands from small to large, so torsion binning as in Table 11.4 may prove to be most helpful.

11.4.9 OPTION TO PERFORM MINIMIZATION WITH MOLECULAR MECHANICS ON SAVED POSES

Docking functions help to prioritize poses for saving. The analysis of these poses cannot be done manually for a typical library in VS, so scoring functions are called upon to further select ligand poses, as discussed in the next section. Most scoring functions are trained against a set of heavily screened high-quality x-ray structures of ligand–protein complexes, not docked ligand poses. Thus, an intermediate polishing or refining step using energy minimization of ligand poses can further optimize ligand–protein interactions such that the poses will score accurately with the selected scoring function. For ligand poses in most Class I sites, ligand minimization is not required; but for Class II and Class III sites where larger ligands can bind, scores are generally much better after only about 200 to 500 steps of molecular-mechanics-based ligand minimization.

11.4.10 SELECT SCORING FUNCTIONS

An objective scoring function should be used to score and rank ligand poses and, ultimately, ligands from a library that is being screened. A general purpose scoring function is ideally one that is trained against diverse ligands and diverse receptors for an initial robust screen. If sufficient data on related complexes are available, a customized and presumably more accurate scoring function can then be obtained. Methods to train, select, and customize scoring functions are outside the scope of this chapter.

11.5 CONCLUSIONS

It is important to be able to classify binding sites for the purposes of appreciating the extent of the search problems inherent in docking to diverse receptors. Options and selections for any algorithm in docking need to be set appropriately for the requirements of the model biological system being studied. Understanding the binding site being investigated in a drug discovery program will also help to set expectations in terms of the potential quality of the results as well as the meaning of the ligand poses that may be returned by the docking tool being applied.

This chapter proposes a classification of binding sites in six classes to help computational chemists assess the magnitude of the docking problem. It is well known for many flexible receptors that binding sites can change quite dramatically in size and loop conformation when diverse ligands are bound. Thus, it is entirely possible that different ligand–receptor complexes may present binding sites that would be in different classes. Nevertheless, it is still useful to be able to classify a binding site so that the extent of problems in a docking study can be systematically considered. Practitioner guidelines and recommended practices are provided to help handle the challenges posed by each class of binding sites. The binding sites increase in complexity from Class I to Class VI as listed in Table 11.3. Commensurate with the difficulty of accurately docking to the binding site, more options to improve prediction of known binding modes are recommended to tackle difficult cases. These options include editing of site definition, exploration of docking functions, use of site partitioning, increasing the number of MC trials, and refining ligand poses with molecular-mechanics-based in situ ligand minimization. As development of tools for structure-based design is an area of active research, it is hoped that developers of new algorithms in this area continually address challenging areas requiring new solutions.

The suggestions provided in this chapter are simple practical guidelines but not absolute requirements for the computational scientist working in structure-based design. Many of these guidelines will need to be adjusted depending on the problem at hand; however, providing a good starting point will expedite the practitioner's development of expertise in using these tools. Ultimately, discovery is the realm of the scientist and not the software. A scientist's own intuition and experience are necessary for the successful discovery and design of new bioactive compounds using computational tools and guidelines as proposed here.

ACKNOWLEDGMENTS

We gratefully acknowledge the work by Amit Kulkarni, Hongwei Huang, Luke Fisher, Nazy Khosrovani, Shikha Varma, and Swati Puri in testing and validating LigandFit. Full details of LigandFit applied to complexes listed in this work are being prepared for publication [35]. We are wholly indebted to our developers: Marvin Waldman for his vision and for reviewing this chapter, Jeff Jiang and Venkat Venkatachalam for algorithmic improvements to LigandFit, Andre Krammer for developing LigScore2, and Daniel McDonald for the tools in DS Modelling 1.2 — SBD with which the data and figures were generated.

REFERENCES

[1]   I.D. Kuntz, J.M. Blaney, S.J. Oatley, R. Langridge, and T.E. Ferrin, A geometric approach to macromolecule-ligand interactions, *J. Mol. Biol.* 161:269–288, 1982.

[2]   R.D. Taylor, P.J. Jewsbury, and J.W. Essex, A review of protein-small molecule docking methods, *J. Computer-Aided Mol. Des.* 16:151–166, 2002.

[3]   J.W.M. Nissink, C. Murray, M. Hartshorn, M.L. Verdonk, J.C. Cole, and R. Taylor, A new test set for validating predictions of protein–ligand interaction, *Proteins* 49:457–471, 2002.

[4]   B. Waszkowycz, T.D.J. Perkins, R.A. Sykes, and J. Li, Large-scale virtual screening for discovering leads in the postgenomic era, *IBM Systems J.* 40:360–376, 2001.

[5]   C. Bissantz, G. Folkers, and D. Rognan, Protein-based virtual screening of chemical databases: 1. Evaluation of different docking/scoring combinations, *J. Med. Chem.* 43:4759–4767, 2000.

[6]   C. Bissantz, P. Bernard, M. Hibert, and D. Rognan, Protein-based virtual screening of chemical databases: II. Are homology models of G-protein coupled receptors suitable targets? *Proteins: Structure, Function, and Genetics* 50:5–25, 2003.

[7]   J.F. Blake and E.R. Laird, Recent advances in virtual screening. In A.M. Doherty, Ed., *Annual Reports in Medicinal Chemistry* 38:305–314, 2003.

[8]   R. Wang, Y. Lu, and S. Wang, Comparative evaluation of 11 scoring functions for molecular docking, *J. Med. Chem.* 46:2287–2303, 2003.

[9]   H. Gohlke and G. Klebe, Approaches to the description and prediction of the binding affinity of small-molecule ligands to macromolecular receptors, *Angew. Chem. Int. Ed.* 41:2644–2676, 2002.

[10]   B.D. Dorsey, R.B. Levin, S.L. McDaniel, J.P. Vacca, J.P. Guare, P.L. Darke, J.A. Zugay, E.A. Emini, W.A. Schleif, J.C. Quintero, J.H. Lin, I.-W. Chen, M.K. Holloway, P.M.D. Fitzgerald, M.G. Axel, D. Ostovic, P.S. Anderson, and J.R. Huff, L-735,524: the design of a potent and orally bioavailable HIV protease inhibitor, *J. Med. Chem.* 37:3443–3451, 1994.

[11]   M.K. Holloway, et al. In C.H. Reynolds, et al., Eds., *Computer-Aided Molecular Design*, ACS Symp. Series 589:36–50, 1995.

[12]   S.W. Kaldor, V.J. Kalish, J.F. Davies, II, B.V. Shetty, J.E. Fritz, K. Appelt, J.A. Burgess, K.M. Campanale, N.Y. Chirgadze, D.K. Clawson, B.A. Dressman, S.D. Hatch, D.A. Khalil, M.B. Kosa, P.P. Lubbehusen, M.A. Muesing, A.K. Patick, S.H. Reich, K.S. Su, and J.H. Tatlock, Viracept (nelfinavir mesylate, AG1343): a potent, orally bioavailable inhibitor of HIV-1 protease, *J. Med. Chem.* 40:3979–3985, 1997.

[13] J. Greer, J.W. Erickson, J.J. Baldwin, and M.D. Varney, Application of the three-dimensional structures of protein target molecules in structure-based drug design, *J. Med. Chem.* 37:1035–1054; 1994.

[14] H.M. Berman, J. Westbrook, Z. Feng, G. Gilliland, T.N. Bhat, H. Weissig, I.N. Shindy-alov, and P.E. Bourne, The Protein Data Bank, *Nucleic Acids Res.* 28:235–242, 2000.

[15] C.M.W. Ho and G.R. Marshall, Cavity search: an algorithm for the isolation and display of cavity-like binding regions, *J. Computer-Aided Mol. Des.* 4:337–354, 1990.

[16] G.P. Brady, Jr. and P.F.W. Stouten, Fast prediction and visualization of protein binding pockets with PASS, *J. Computer-Aided Mol. Des.* 14:383–401, 2000.

[17] J. Liang, H. Edelsbrunner, and C. Woodward, Anatomy of protein pockets and cavities: measurement of binding site geometry and implications for ligand design, *Protein Science* 71884–71897, 1998.

[18] R.C. Willis, Surveying the binding site, *Mod. Drug Discov.* 5:28–34, 2002.

[19] B.A. Luty, Z.R. Wasserman, P.F.W. Stouten, C.N. Hodge, M. Zacharias, and J.A. McCammon, A molecular mechanics/grid method for evaluation of ligand-receptor interactions, *J. Comput. Chem.* 16:454–464, 1995.

[20] D.S. Goodsell and A.J. Olson, Automated docking of substrates to proteins by simulated annealing, *Proteins: Structure, Function, and Genetics* 8:195–202, 1990.

[21] M. Rarey, B. Kramer, T. Lengauer, and G. Klebe, A fast flexible docking method using an incremental construction algorithm, *J. Mol. Biol.* 261:470–489, 1996.

[22] M. McGann, H. Almond, A. Nicholls, J.A. Grant, and F.K. Brown, Gaussian docking functions, *Biopolymers* 68:76–90, 2003.

[23] R. Friesner, J. Banks, H. Beard, T.A. Halgren, T.F. Hendrickson, J. Klicic, D. Mainz, R.B. Murphy, J. Perry, and P. Shenkin, Flexible ligand docking with GLIDE: evaluation of structure prediction performance and utility for virtual ligand screening, *Abstracts of Papers American Chemical Society* 223:COMP 8, 2002.

[24] G. Jones, P. Willett, and R.C. Glen, Molecular recognition of receptor sites using a genetic algorithm with a description of desolvation, *J. Mol. Biol.* 245, 43–53, 1995.

[25] C.M. Venkatachalam, X. Jiang, T. Oldfield, and M. Waldman, LigandFit: a novel method for the shape-directed rapid docking of ligands to protein active sites, *J. Mol. Graphics Modelling* 21:289–307, 2003.

[26] S. Hindle, M. Rarey, C. Buning, and T. Lengauer, Flexible docking under pharmaco-phore type constraints, *JCAMD* 16:129–149, 2002.

[27] N. Paul and D. Rognan, ConsDock: a new program for the consensus analysis of protein–ligand interactions, *Proteins: Structure, Function, and Genetics* 47:521–533, 2002.

[28] G.M. Verkhiver, D. Bouzida, D.K. Gehlhaar, P.A. Rejto, S. Arthurs, A.B. Colson, S.T. Freer, V. Larson, B.A. Luty, T. Marrone, and P.W. Rose, Deciphering common failures in molecular docking of ligand–protein complexes, *J. Computer-Aided Mol. Des.* 14:731–751, 2000.

[29] Cerius2, version 4.7, Accelrys Software, Inc., 2001: www.accelrys.com.

[30] M. Lim-Wilby, J. Jiang, M. Waldman, and C.M. Venkatachalam, Virtual high through-put screening using LigandFit as an accurate and fast tool for docking, scoring, and ranking. In Wendy Warr & Associates, *Chemical Information and Computation 2002*, 224th ACS National Meeting and Exposition, Boston, August 18–22, 2002, pp. 87–91.

[31] T.A. Lyons, M. Dooley, A.-G. Lamy, S. Patel, C. Schneider, R. Hoffmann, H.-O. Bertrand, and M. Lim-Wilby, Docking of diverse ligands to diverse binding sites: six degrees of application. In Wendy Warr & Associates, *Chemical Information and Computation 2002*, 224th ACS National Meeting and Exposition, Boston, August 18–22, 2002, pp. 57–60.

[32] B. Kramer, M. Rarey, and T. Lengauer, Evaluation of the FlexX incremental construction algorithm for protein–ligand docking, *Proteins: Structure, Function, and Genetics* 37:228–241, 1999.

[33] G. Jones, P. Willett, R.C. Glen, A.R. Leach, and R. Taylor, Development and validation of a genetic algorithm for flexible docking, *J. Mol. Biol.* 267, 727–748, 1997.

[34] S. Szalma, X. Ni, and K. Olszewszki, personal communication.

[35] M. Lim-Wilby, H.-O. Bertrand, M. Dooley, A. Goupil-Lamy, T. Lyons, and S. Patel, manuscript in preparation. All LigandFit and LigandScore work was performed in the Discovery Studio® Modelling 1.2 — SBD environment using DS LigandFit and DS LigandScore.

Part V

Docking Strategies and Algorithms

12 A Practical Guide to DOCK 5

Demetri T. Moustakas, Scott C.H. Pegg, and Irwin D. Kuntz

12.1 INTRODUCTION

The docking problem is the search for the most energetically favorable binding pose of a ligand to a receptor, often specialized to a small molecule ligand and a macromolecular receptor. A docking algorithm should correctly predict the docked structure of a ligand–receptor complex, as well as calculate a binding energy that can be used to rank order different molecules according to their affinity for a receptor.

The DOCK algorithm implements a solution to the docking problem. It uses molecular shape descriptors to position a ligand molecule into a macromolecular receptor and evaluates these poses to generate predicted binding modes for a ligand–receptor complex.

The original DOCK program [1,2] implemented rigid body docking, which coupled with molecular graphics systems, allowed users to generate binding mode predictions of ligands. The subsequent versions of the DOCK program have implemented molecular-mechanics force field scoring (DOCK 3.0 [3]), energy minimization (DOCK 3.5 [4,5,6]), and ligand conformational flexibility (DOCK 4.0 [7,8]).

The most recent version, DOCK 5.0, has been developed in a new C++ codebase to maximize the portability and modularity of the DOCK algorithm. Each major component of the DOCK algorithm has been implemented as a class with a documented interface, allowing these DOCK functions to be modified or replaced easily. DOCK 5 features solvation scoring, rigid docking clustering analysis, new ligand conformational search methods, new minimization methods, and includes support for parallel computing using the Message Passing Interface (MPI) standard.

This chapter describes the steps needed to prepare a receptor structure and a library of small molecules for docking. There is a description of the DOCK algorithm, followed by a discussion of the optimal DOCK parameters and the theory behind them. Finally, we consider the methods used to evaluate a docking method's success. The primary focus of this chapter is the DOCK parameters that influence the performance of the algorithm and how to maximize the accuracy of DOCK results. This chapter does not describe all of the DOCK parameters, but is rather intended as a practical guide. For a complete description of DOCK parameters, please consult the *DOCK User's Manual*.

12.2 TARGET PREPARATION

12.2.1 X-ray Structures

Using DOCK requires a target structure, typically one generated from x-ray crystallography. These files are usually stored in the Brookhaven Protein Data Bank (PDB) format [9]. Many PDB files contain coordinates for water molecules. These should be removed, with the exception of water molecules that are known to remain in the binding site. Often not all of the residues of the target structure are observed via x-ray crystallography due to their being disordered in the crystal itself. If residues within or near the binding site are missing, they need to be inserted. Building these residues can be done via SYBYL [10] or other similar programs.

Once the binding site has all of its residues represented and has been cleared of nonessential waters, hydrogen atoms should be added. Again, this can be performed easily by third-party molecular modeling programs such as SYBYL. In some cases it is important to have histidine residues in specific protonation states, requiring careful assignment of the PDB residue type and addition of hydrogens.

To create the grids used by DOCK, the PDB file needs to be converted into MOL2 format. If a DOCK scoring function that includes electrostatics is going to be used, charges need to be added to the atoms in the MOL2 file. Multiple methods for calculating these charges are available via packages such as SYBYL and MOE (Molecular Operating Environment) [11].

12.2.2 Structures from Other Sources

Target structures derived from sources other than x-ray crystallography (for example, nuclear magnetic resonance or homology modeling) can also be used with DOCK. When preparing such targets for docking, one should generally follow the guidelines listed above, resulting in a MOL2 formatted file representing the target structure with hydrogens and electrostatic charges.

The results of using DOCK with a structure of greater positional variance, such as those generated by homology modeling, logically contain an increased uncertainty. In our own investigations on the use of homology models, we found that ligand score rankings generated by docking to a model began to differ considerably from the rankings generated by docking a crystal structure of the same protein as models moved beyond 2 Ås root-mean-square deviation (rmsd) [12]. However, even models with higher rmsds generated rankings with positive correlations to the crystal rankings, providing a rough screen for general ligand size and shape [12]. At least one method utilizing DOCK has been developed to combine multiple structures of an identical target protein in an effort to improve docking results [13].

12.2.3 Generating Matching Points

The DOCK program guides its orientational search by matching the heavy atoms of the ligand to matching points that lie within the target binding site. For more information on the orientational search, refer to Section 12.5. One can create these matching points in any manner (for example, using the heavy atoms of a substrate analog bound in the crystal structure), but we suggest using the SPHGEN program provided in the DOCK distribution. This program generates matching points by filling the binding site with spheres of varying radii (specified by the user, but typically between 1.4 and 2.5 Ås). The center of each sphere becomes a matching point.

The SPHGEN program will typically generate more spheres than needed as matching points. Having too many matching points will result in marked slowing of the orientational search, yet having too few points will result in undersampling. Depending on the size of the active site, anywhere from 20 to 40 matching points will provide good coverage in the orientational search. Reducing a large number of spheres into a smaller set typically requires converting the SPHGEN output file (.sph) into a PDB file using the SHOW_SPHERES utility (provided in the DOCK distribution) and displaying it visually along with the target structure. Spheres that appear to lie impractically far from (or close to) the binding site can be removed.

12.2.4 GRID CONSTRUCTION

To make the scoring step of docking faster, DOCK employs grids based on the target MOL2 file. Each point on the grid represents the score contribution from all of the target atoms; therefore computing the score for a ligand atom at a grid point is equivalent to summing the scores between the ligand atom and all of the receptor atoms. When scoring a point that does not lie directly on a grid point, the target score contribution is determined via linear interpolation of the eight points surrounding it. For more information about score calculations, refer to Section 12.4.

The first step in grid construction is to determine the grid size. This is done by defining a rectangular box containing the target's binding site of interest. The box should enclose all of the residues of the binding site and all of the spheres (matching points). The program SHOWBOX, a utility within the DOCK distribution, builds the enclosing box and provides a PDB-format output file that can be used to view the box boundaries.

The next step is to generate the grids using the GRID program provided in the DOCK distribution. The GRID program uses the bounding box and the receptor's MOL2 description to produce the scoring grids. The resolution of the grids is user-defined. Because the grids are loaded into memory, it is preferable to have lower (coarser) resolution grid files, but not ones so coarse as to lose important scoring information. The overall size of the grid depends on its dimensions (determined by the bounding box) and its resolution, expressed as the distance between adjacent grid points. (It is important to note that the number of grid points, and thus the amount of memory required to store the grid, will grow as the cube of the dimensions.) Resolutions between 0.3 Ås and 0.15 Ås have proven most effective in practice. One may choose to generate a grid for the energy score function (these grids have a .nrg file extension) or a steric bump grid (a .bmp file extension).

12.3 LIGAND PREPARATION

12.3.1 CONFORMATION

DOCK requires that ligands be in MOL2 format, with complete three-dimensional (3D) coordinates, even when performing flexible docking. Generating 3D coordinates from lower dimensional representations (e.g., SMILES [14], MDL® [15]) can be done using third-party software such as SYBYL (via the CONCORD [16] module) or OMEGA [17].

Typically, one generates a single 3D conformer per ligand molecule. When performing rigid docking, it is preferable that this single conformer be similar to that of the true binding conformation, which is typically a low-energy conformation (as calculated in simple gas-phase models). The CONCORD program outputs a single, low-energy conformer, and OMEGA outputs many conformers, ranked by energy. When performing flexible docking, the ligand conformation matters only for the nonrotatable bonds.

An alternative to flexible docking is to generate many conformers of each ligand and rigidly dock each of them to the target separately. This is commonly referred to as the flexibase approach [18]. Building a flexibase library of a ligand can be done using the OMEGA program.

12.3.2 ELECTROSTATICS

If a score function that uses electrostatic contributions (such as the default DOCK energy score) is to be used, each ligand needs to have proper electrostatic charges assigned to each atom. The charge model used matters little when using only the DOCK energy score, because this score is almost completely dominated by the van der Waals (vdW) contributions. However, when using a scoring function that employs a solvation correction, such as the generalized-Born (GB) [19,20] or Poisson–Boltzmann (PB) [21] models, differences in the charges on the ligand can lead to significant variations in scoring. Which charge model provides the best performance is still a matter of debate and an active research area in docking. In practice, when charging a small number of molecules, we prefer to use the RESP (Restrained ElectroStatic Potential) charge model [22]. However, due to the high computational expense of calculating these charges, the AM1-BCC [23] or AMSOL [24] charge models are recommended if a large number of molecules require charging.

12.4 DOCKING

Once the receptor and ligands have been prepared for docking as described previously in this chapter, DOCK can be run. These sections of this chapter will describe the DOCK algorithm, its optimum parameter choices, and the principles underlying these choices.

To characterize and better understand the behavior of DOCK [8], we have focused on the problem of recreating crystallographic ligand poses. DOCK is considered to succeed when it predicts a ligand pose that falls within 2.0 Å of the crystallographic structure. Such testing methodology is discussed further in Section

12.8. The default values of the parameters discussed in this section are derived from experiments applying DOCK to a version of the GOLD test set [25] of protein–ligand crystal structure complexes. Docking to nucleic acid targets is similar to docking to protein targets, with one notable exception. The highly charged nature of nucleic acids causes a greater contribution of electrostatics to the binding energy. As a result, it is important to use scoring functions that incorporate a more sophisticated treatment of the effects of a high dielectric polar solvent upon the electrostatic component of the binding energy.

The DOCK algorithm can be thought of as a number of sampling algorithms that probe a complex multidimensional function — the nonbonded AMBER force field binding energy between a small molecule ligand and a macromolecular receptor — to find its minimum energy values. A schematic overview of the DOCK algorithm shown in Figure 12.1 illustrates this point. Accordingly, the following sections are organized to first describe the details of the scoring functions and optimizers, followed by a description of the sampling methods and a discussion of proper choices of parameters.

12.4.1 DOCK SCORING FUNCTIONS

DOCK has several scoring functions, each of which is best suited for different applications. DOCK allows the user to select different scoring functions to be used during rigid docking, flexible docking, and library ranking. They are *bump filtering, contact scoring, energy scoring,* and *solvation scoring* (either GB surface area (GB/SA) or PB surface area (PB/SA)). The scoring functions provide a range of resolutions (accuracies) and speeds. In general the scoring functions trade off resolution and accuracy for speed, so it is important to use the low and high accuracy scoring functions properly, to avoid highly inaccurate results or excessively long calculations. The proper scoring function choices will be discussed in Section 12.5 and Section 12.6.

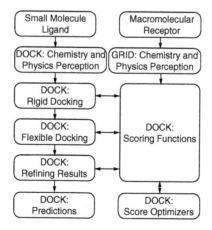

FIGURE 12.1 Schematic illustrating the interactions between different modules in the DOCK algorithm.

To speed up scoring function evaluations, DOCK can use potential energy grids generated by the GRID program (as described in Section 12.2.4 "Grid Construction" in this chapter). When GRID is used, some of the parameters described in the paragraphs below will be used during grid creation. When GRID is not used, and scoring is done using continuous scoring (all ligand and receptor atom pairs are evaluated), all scoring function parameters will be requested by DOCK.

The DOCK scoring functions are all derived from the AMBER force field [26,27]. When DOCK reads a ligand from a MOL2 database, it has only atomic coordinates, bond information, and partial charges. DOCK then assigns all the ligand atoms the appropriate SYBYL atom type [10] based on bonding patterns, as defined in the file vdw.defn, whose path is specified with the vdw_defn_file parameter. For each ligand and receptor atom, a vdW radius and well depth are assigned based on atom type. Each of the scoring functions can then calculate an all-atom nonbonded interaction energy by summing contributions from all ligand–receptor atom pairs. For all the scoring functions described below, L and R represent the ligand and receptor atoms, while D_{LR} represents the distance between L and R.

12.4.1.1 Bump Filtering

The bump filter is a Boolean scoring function that calculates the number of vdW clashes between the ligand and the protein and eliminates ligand orientations that exceed a user-defined threshold, set with the bump max parameter. A vdW clash occurs when $D_{LR} < (L_{radius} + R_{radius})$ x bump_overlap, where bump_overlap is a user-defined parameter between 0 and 1 that defines how much overlap between the ligand and receptor atom spheres is tolerated. The bump filter requires the user to calculate a bump grid using the GRID accessory program (as described in Section 12.2). The prefix to the bump grid is passed to the bump filter in DOCK as the bump_grid_prefix parameter. Bump filtering is especially useful when a complex has unusually short contacts such as those involving metal ligation or partial covalent bonds, when structures are of low resolution, or when macromolecular ligands are being docked.

12.4.1.2 Contact Scoring

The contact score [3] is an all-atom integer scoring function that is a step function approximation of the Lennard-Jones vdW function (Figure 12.2). The Lennard-Jones function is discretized into three regions defined by the value of D_{LR}. The repulsive region is defined as $0 < D_{LR} < (L_{radius} + R_{radius})$ x contact_clash_overlap, where the contact_clash_overlap parameter (default value = 0.75) defines how much overlap between the ligand atom and receptor atom volumes constitutes a clash. In the repulsive region, the ligand–receptor atom pair contributes a positive value set by the contact_clash_penalty parameter (default value = 50) to the score. The attractive region is defined as $(L_{radius} + R_{radius})$ x contact_clash_overlap $< D_{LR}$ < contact_cutoff_distance, where the contact_cutoff_distance parameter (default value = 4.5 Å) determines the maximum distance for a favorable vdW interaction. In the attractive region, a ligand–receptor atom pair contributes a value

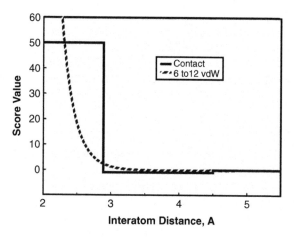

FIGURE 12.2 Comparison of the contact score with a 6 to 12 Lennard–Jones potential.

of 1 to the score. The noncontact region is defined as $D_{LR} >$ contact_
cutoff_distance. In this region, a ligand–receptor atom pair contributes 0 to the
score. Contact scoring provides a useful way to examine shape matching of ligand
and receptor. Some of the configurations that score well with contact scoring would
not score well with a full force field treatment.

The contact score returns an integer approximation of the Lennard-Jones vdW
function and, as such, does not have formal units assigned to its value.

12.4.1.3 Energy Scoring

The *energy score* [3] is a derivative of the AMBER force field [27]. It is an all atom
scoring function that considers all nonbonded interactions between ligand and receptor atoms.

The DOCK energy function, which contains two terms is shown in Figure
12.3. The first is a Lennard-Jones vdW term, which is parameterized using the
AMBER 94 force field parameters. The A_{LR} and B_{LR} parameters are derived using
the vdW radius and well depth for each atom. The parameters a and b are the
Lennard-Jones attractive and repulsive exponents and are set using the
attractive_exponent and repulsive_exponent parameters (default values
= 6, 12). The second term is a Coulombic electrostatic term with a distance
dependant dielectric scaling term. The partial charges q_L and q_R are determined in
the predocking steps as outlined in Section 12.2 and Section 12.3. The dielectric
coefficient S is set using the dielectric_factor parameter (default value = 4).
DOCK users can also scale the vdW or the Coulombic terms by using the
vdw_scale and es_scale parameters.

$$E = \sum_{L}^{Lig} \sum_{R}^{Rec} \left(\frac{A_{LR}}{D_{LR}^a} - \frac{B_{LR}}{D_{LR}^b} + C \times \frac{q_L q_R}{S \times D_{LR}} \right)$$

FIGURE 12.3 Functional form of the DOCK energy score equation.

The energy score returns what is essentially the enthalpic component of the binding free energy and as such can be considered to have units of kcal/mol. However, due to the lack of an entropic term, as well as the lack of a free energy of desolvation term, the energy score will have differences from experimentally determined binding free energies. These errors limit the ability of the energy score to resolve differences between ligands with similar energy scores.

12.4.1.4 Solvation Scoring

A large contribution to the binding free energy of a ligand to a receptor comes from solvation effects, specifically the energy of desolvating the ligand and the receptor site. There are many approaches taken to approximate the effects of solvent on ligand binding, ranging from a distance-dependent dieletric in the electrostatic term of the energy score function [28,29,30] to complex models based on the explicit treatment of solvent molecules [31,32].

DOCK 5 implements two implicit solvent solvation scoring functions: one that uses the GB/SA model as implemented by Zou [33]; one that uses the PB/SA model as implemented by Nicholls [17,34]. Although the specific details of both models differ, they both begin by calculating the DOCK energy score, using a dielectric scaling factor of 1, which treats the ligand–receptor complex as if it was in pure vacuum. Both the GB and PB functions compute a correction for the electrostatic term based on the Reaction Field Energy of the system. This correction accounts for the effects of the high dielectric value of the solvent, and the lower dielectric value of the solute, upon the electrostatic energy of the complex. Finally, a solvent accessible surface area (SASA) term is computed to account for the free energy cost of desolvating polar atoms upon complex formation. These terms are combined with the appropriate coefficients to yield an approximate free energy of binding.

The solvation score returns the enthalpic component of the binding free energy, plus a free energy of desolvation term. Additionally, solvation scoring computes the electrostatic component of the enthalpy more accurately than the DOCK energy score does. As such, the units are given in kcal/mol. This scoring function does not include the entropy of binding, which will result in differences between the solvation score of a molecule and an experimentally determined binding free energy, though the errors should generally be smaller than those of the energy score.

As mentioned above in Section 12.3, the electrostatic charges on the ligand can cause the results when using GB or PB to vary considerably, due to the increased contributions of charged interactions. Determining the ligand charge model, which produces the best results under each solvation scoring scheme, is still an active area of docking research.

12.4.2 DOCK Score Optimization

The DOCK energy landscape of a typical small molecule/receptor complex can be quite complex. There are many different minima with a wide range of energies and spatial positions. Due to the rugged nature of these energy landscapes, DOCK does not use a global minimizer, but rather uses a combination of global sampling and robust local minimization to find the global minimum position.

This motivates the choice of the simplex minimizer in DOCK. The simplex minimizer is a fairly robust local minimization algorithm that generates an initial random set of molecule configurations (the simplex) about a starting structure. The simplex algorithm then performs a set of deterministic geometric steps to move the ensemble of molecule configurations toward convergence at the local minimum. Simplex minimization does not require analytical derivatives of the scoring functions, and it scales well to the dimensionality typically encountered in small molecule docking [35].

The simplex minimizer contains a number of parameters. A single cycle of simplex minimization begins with the simplex creation, where a series of random molecule configurations are generated that lie within the bounds set by the trans_step, rot_step, and tors_step parameters. The trans_step parameter is the maximum translational step size in angstroms. The rot_step parameter sets the maximum rigid body rotation step size in units of (radians/). The tors_step parameter sets the maximum rotatable bond torsion angle step size, in degrees. Once the simplex is created, simplex minimization steps are taken to find lower energy positions for the starting molecule configuration. A maximum of max_iterations steps are taken, though the minimization can converge and complete before this point. Convergence is reached if the score difference between the highest scoring molecule configuration and the lowest scoring molecule configuration in the simplex is less than the score_converge parameter. If the max_cycles parameter is greater than one, then upon completing one cycle of simplex minimization, a new cycle is initiated until the limit of max_cycles is reached or until the simplex cycles converge. Simplex cycle convergence occurs when the difference between the final score of one minimization cycle and the final score of the next minimization cycle is less than the cycle_converge parameter.

As with the scoring functions, DOCK allows users to apply different minimization parameters to minimization during rigid docking, flexible docking, and library ranking. The proper minimization parameters will be discussed in Section 12.5 and Section 12.6.

12.5 RIGID DOCKING

We define rigid docking as the process of fitting a rigid small molecule ligand to a rigid protein receptor. The current version of DOCK does not directly consider receptor flexibility, therefore rigid or flexible docking refers to whether DOCK searches the conformational space of the ligand. Although this chapter deals with small molecule/protein docking, rigid docking has also been applied to macromolecular systems (protein/protein, nucleic acid/protein, and small molecule/nucleic acid) [36]. Most applications of macromolecular docking require a full molecular-mechanics postdocking optimization [37] that we will not expand on in this chapter.

Rigid docking is used during the docking of both rigid and flexible ligands. For rigid ligands (either ligands with no rotatable bonds or flexible ligands that are conformationally expanded into a number of discrete conformers), the goal of rigid docking is to generate and evaluate many possible ligand poses within the protein receptor to identify the most favorable pose. For flexible ligands, rigid docking is used to generate a set of poses of a rigid portion of the molecule from which to base the conformational search (see Section 12.6 for more details). In either case, it is important to understand the process of sampling rigid orientations to ensure that adequate sampling is being performed.

Rigid docking can be thought of as two processes, which will be described in detail in Section 12.5.1, Section 12.5.2, and Section 12.5.3, which discuss generating, scoring, and optimizing orientations, respectively. In brief, a set of orientations of the ligand is first generated within the receptor. For each orientation, the intermolecular energy between the protein and the ligand is calculated using a molecular-mechanics force field. The orientation can then be locally optimized to obtain a score that represents the nearest local energy minimum. All of the local minima are compared, and the lowest energy minimum is selected as the global minimum.

12.5.1 GENERATION OF ORIENTATIONS

The purpose of generating orientations during rigid docking is to heuristically identify a set of ligand configurations that have favorable vdW interactions with the target protein. These ligand orientations are likely to be in or near the major minima on the binding energy landscape. DOCK uses receptor spheres (described in Section 12.2) to identify approximately favorable positions for ligand atoms in the receptor, which greatly reduces the search space of possible orientations. This is important because there is a large space of potential ligand orientations within the receptor, as there are 6 rigid degrees of freedom (3 vectors of translation and 3 euler angles for rotation). The orientation search over this space must be fine grained enough to sample at least the major features of the interaction energy landscape, which anecdotally has important features with a rotational resolution of 2 to 3°, and a translational resolution of 0.1 to 0.2 Å [38,39]. To exhaustively generate orientations of a ligand in a protein target site, with a rotational resolution of 3°, over 1.7 million (120^3) rotations need to be sampled at each ligand translational position (the position of the ligand's center of mass). To sample a 10 x 10 x 5 Å box at a 0.2 Å translational resolution, 62,500 translational positions need to be sampled, resulting in a total of 108 x 10^9 orientations to evaluate. This extremely large number of orientations is why DOCK does not use an exhaustive orientation generation method and instead relies on the spheres generated by SPHGEN to restrict the generated orientations to a relevant area of orientational space.

As described in Section 12.2, SPHGEN places spheres in the concave regions of the receptor's solvent accessible surface. These sphere positions represent an estimation of energetically favorable ligand atom positions. By sampling orientations that place ligand atom centers on the sphere centers, it is possible to focus the orientation search on a much smaller region of orientational space that contains most of the favorable orientations.

The first step in generating orientations of a ligand to receptor spheres is to generate a set of matches between ligand atom centers (centers) and receptor spheres (spheres). This is analogous to the isomorphous subgraph matching problem [40], if the spheres are considered to be a completely connected, undirected graph, then the ligand atoms are likewise considered to be a completely connected, undirected graph. The graph edge lengths are simply the Euclidian distances between the atom (and sphere) centers. A node is defined as a pairing of one center with one sphere. A match is defined as a set of nodes where the intersphere distances and the intercenter distances are equivalent (to within a user-specified tolerance). Starting with one node, a match is extended by adding nodes that satisfy the distance constraints until there are no valid nodes left to add. In this manner, DOCK will exhaustively generate a set of nondegenerate matches. Matches are sorted by size and are retrieved individually starting with the largest matches. First DOCK performs a least-squares fitting between the center coordinates and the sphere coordinates that produces a transformation matrix (translation vector + a 3×3 rotation matrix) that minimizes the error between the sphere/center match. Each transformation matrix is applied to the original ligand molecule position, which generates the orientation for that particular match.

There are a number of parameters that control matching in DOCK. These are `distance_tolerance`, `distance_minimium`, `minimum_nodes`, `maximum_nodes`, and `maximum_orientations`. The `distance_tolerance` is the maximum error (in angstroms) between any pair of nodes in a match (the error is the difference between the sphere–sphere and center–center distances). The size of the `distance_tolerance`, with a default value of 0.25 Å, determines how similar the ligand atom positions need to be to the sphere positions to be considered a match; increasing the distance tolerance will allow more matches to be produced. The `distance_minimum` is the smallest distance between two nodes (the sphere–sphere distance) to be included in the match. This is used to prevent matches that are dominated by small clusters of spheres and instead encourage matches that represent more of the overall shape of the molecule. The `minimum_nodes` parameter defines the fewest number of nodes that need to be included in a match for it to be considered valid. This parameter defaults to 3 nodes, because with fewer than 3 matching points, there is not a unique spatial transformation. This parameter can be increased when docking large molecules to select matches that better represent the shape of the molecule. The `maximum_nodes` parameter defines the maximum number of nodes that will be added to a match before the search is truncated. The default value for this parameter is 10 because the search can become time-consuming if it is allowed to proceed too far. Searches that terminate when the match reaches this size are considered a success, and the match is stored despite the fact that it is not fully grown (there still are nodes that could be added). This will allow some degenerate matches (matches that produce almost the same transformation matrix); however, this is preferable to undersampling these orientations. The `maximum_orientations` parameter is used to limit the number of orientations generated. Once the matches are computed and sorted, orientations will be generated until the `maximum_orientations` limit has been reached. These parameters can be manually set, and DOCK will generate as many orientations as the parameters allow, which in some cases may be fewer than the `maximum_orientations` parameter, resulting in uneven sampling (some molecules in the database may sample more orientations than others). To solve this problem, DOCK has one additional parameter, `automated_matching`, which automatically sets the matching parameters to the default values and generates orientations. If there are fewer matches than the maximum_orientations parameter allows, automated_matching will increase the distance_tolerance until a sufficient number of matches have been produced.

Chemical matching is a feature that allows users to apply chemical labels (such as H-bond donor, H-bond acceptor, and hydrophobe) to both ligand atoms and receptor spheres and define rules for which chemical labels can be paired together. When matches are generated, only atom/sphere pairs that satisfy the chemical matching rules, as well as satisfying the distance constraints described above, are added to the match. This method reduces the orientational space that is explored; however, because the orientations that are explored have some degree of chemical complementarity to the receptor, the orientational space explored is likely more relevant. Chemical matching is enabled by setting the chemical_match parameter to Yes. The receptor sphere labeling takes place during target preparation and is described in more detail in the *DOCK User's Manual*. The chemical labels for the ligand atoms are defined in the chem.defn file, whose path is set using the chemical_definition_file parameter. The syntax of this file is described in detail in the *DOCK User's Manual*. The chemical matching rules are defined in the chem._match.tbl file, and its path is specified with the chemical_match_file parameter. The syntax of this file is described in detail in the *DOCK User's Manual*.

Critical point matching is a feature that allows users to apply critical labels to receptor spheres and group these critical spheres into clusters. If the critical_points parameter is set to Yes, then DOCK will only generate matches that include at least one critical sphere from each critical cluster. This feature reduces that orientational space being sampled with the assumption that the critically labeled spheres populate an interesting region of orientational space.

When rigidly docking small molecules, it is appropriate to enable automated_matching and set the maximum_orientations parameter to 5000. Generating fewer than 500 orientations results in a marked decrease in rigid docking success rates and generating more than 5000 orientations tends to have no effect upon rigid docking success, though it certainly increases the computational expense. Generating a large number of orientations (approximately 50,000) results in a decrease in the rigid docking success rate, likely due to the identification of more orientations that lead to scoring failures, which are discussed further in Section 12.8.

12.5.2 SCORING LIGAND ORIENTATIONS

During rigid docking, the primary purpose of a scoring function is to rank order the orientations. Due to the potentially large number of orientations, it is advantageous to use the fastest scoring function that will yield accurate results. Section 12.4 describes scoring functions ranging from least to most accurate, where the higher accuracy functions tend to have more terms than the lower accuracy scoring functions and are more computationally expensive. By better understanding the nature of rigid docking, it is possible to choose the most appropriate scoring function for any application.

12.5.2.1 Bump Filtering

The bump filter eliminates ligand orientations that fit tightly into the receptor, and will therefore have an unfavorable score. These tight fitting orientations are undesirable for two reasons. First, during optimization their scores may not improve much. Second, even if they can be optimized to an energetically favorable position, this position is likely to be entropically unfavorable (because it could only be sampled through the use of minimization). Using the bump filter therefore speeds up the DOCK algorithm by conserving the use of the minimizer on ligand orientations that are likely to have poor scores (either high scores or low scores that result from a favorable enthalpic component at the expense of an unfavorable entropic component). The bump filter becomes particularly important when a large number of orientations are generated (oversampling). It has been demonstrated [41] that oversampling orientations can lead to an increase in scoring failures, as more poorly accessible (entropically unfavorable) orientations are generated. Using the bump filter with a low bump threshold (max_bumps = 2) should mitigate this effect.

12.5.2.2 Contact Scoring

The contact score is fast to compute, because it is not necessary to calculate the exponential terms in the Lennard-Jones equation. The speed of this method needs to be balanced against the low detail of the vdW energy surface that this method provides. In general, contact scoring is not appropriate for small molecule docking, because the shape of a small molecule requires a fair amount of detail to describe.

For this reason, the contact score is useful for problems such as macromolecular docking or pharmacophore docking [42], where an approximate shape-fit of the ligand and receptor are desired. In these cases, it is likely that even the best initial ligand orientations will have a number of vdW clashes with the receptor, due to the large size of the ligand with respect to the receptor. The contact score bounds the positive contributions of the clashes, which allows this scoring function to have a much smaller dynamic range of possible values than the Lennard-Jones vdW function. This smaller dynamic range facilitates smoother optimization that tends to relieve the unfavorable vdW clashes, without undoing the favorable vdW interactions.

Although contact scoring has some advantages during scoring, optimization, and ranking of orientations of certain docking systems, its units are abstract and most certainly not an accurate measure of the binding energy. Therefore, if it is used during rigid docking, it will be important to rescore with a more accurate scoring function during library ranking, as described in Section 12.7.

12.5.2.3 Energy Scoring

The energy score requires more computational time than the contact score; however, it is still rather fast, making the energy score the most widely used of the DOCK scoring functions. The energies of most of the orientations in a typical docking are dominated by their vdW terms. The scoring function in rigid docking essentially is used to discard many orientations that have a poor shape fit to the protein's vdW surface. Because the energy score fully describes the vdW energy surface as well as the electrostatic energy surface (albeit with a simplistic treatment of the solvent dielectric effects), it should be used during rigid docking of most small molecule/protein target systems.

However, there are a number of cases when extra care should be taken. Any systems that have atoms that might have nonstandard vdW parameters (e.g., proteins complexed with metals) should be carefully examined to ensure that DOCK can assign a SYBYL atom type to the atoms in question, and when it does, that the parameters are reasonable (with regard to values from the literature). Any systems that are likely to have energies dominated by electrostatics (such as deoxyribonucleic acid [DNA] or ribonucleic acid [RNA] targets) should be carefully examined as well. In electrostatically dominated systems, it is important to run controls, such as docking known ligands to the target, to ensure that the DOCK energy score can succeed on the controls. In these cases it may be necessary to use a solvation scoring function to calculate a more accurate approximation of the electrostatic energy.

12.5.2.4   Solvation Scoring

Solvation scoring functions such as GB/SA and PB/SA are much slower than the DOCK energy score. This is because the electrostatic solvation correction terms cannot be precomputed onto a potential grid. Because rigid docking generates many orientations to evaluate, and the dominant component of the energy at this point of docking is the vdW term, solvation scoring is not advised for use during rigid docking of typical small molecule/protein target systems.

As described earlier, systems that are dominated by electrostatic interactions, such as nucleic acid targets, may benefit from solvation scoring during rigid docking. In these cases, it is advisable to be aware of the average time required to dock a molecule to ensure that a full library docking is feasible. It may be necessary to reduce the orientational sampling or the amount of minimization to achieve a reasonable level of performance.

12.5.3   Optimizing Ligand Orientations

During rigid docking the minimizer serves two purposes. The first is to correct for errors that occur during the matching and orienting routines. There are two sources of error while orienting. First, the sphere positions that DOCK matches to are only approximations of favorable vdW potential positions. Second, once a match is generated, the orienting function uses a least squares fitting procedure to generate the orientation. For both of these reasons, the resulting orientations may require slight optimization to alleviate any small bumps between the ligand and the receptor. The scores of these slightly optimized orientations are a much more meaningful representation of the binding energy landscape.

The other purpose for the minimizer during rigid docking is to effectively increase the orientational sampling by increasing the radius of minimization about each generated orientation. As the radius of minimization is increased, the probability of a set of orientations identifying all the important minima (including the global minimum) increases. The one caveat regarding the use of the minimizer is that it should not be treated as the primary method of sampling. Due to the rugged nature of the energy landscape, it is far too inefficient to run DOCK with few orientations and a large amount of minimization.

During rigid docking of small molecules using the DOCK energy score, it is recommended to use the following simplex minimization parameters: max_iterations = 500, max _cycles = 1, trans_step = 1.0, rot_step = 0.1, tors_step = 10.0, score_converge = 0.1, and cycle_converge = 1.0. One important detail to note is that we encourage the use of fully flexible minimization (minimizing both the rigid and torsional degrees of freedom in the ligand) during rigid docking, as we find it tends to increase the radius of minimization and subsequently increases rigid docking success rates.

12.6 FLEXIBLE DOCKING

DOCK uses the Anchor and Grow algorithm [8] to search over ligand conformational space in the context of the receptor. Anchor and Grow is an incremental construction algorithm that generates conformations of the ligand on the fly within the receptor. This differs from flexible database docking methods that explicitly generate multiple conformations for each ligand in a database and then use rigid docking to evaluate each conformer independently.

The steps taken during Anchor and Grow are ligand anchor perception, rigid docking of the anchor, and flexible growth. In brief, a rigid substructure of the ligand is used to generate a number of possible orientations of the ligand in the receptor. From these orientations, a conformational search is performed over the flexible portions of the ligand. For a schematic of the Anchor and Grow algorithm, see Figure 12.4. Because the conformational search occurs within the receptor, it is possible to sample conformational space more finely than explicit conformation generation methods.

12.6.1 LIGAND ANCHOR PERCEPTION

Anchor and Grow first identifies all the rotatable bonds in a molecule using a set of rotatable bond definitions based on the types of the atoms flanking each bond. The rotatable bonds are defined in the file flex.defn, whose path is specified with the flex_defn_file parameter. Next it decomposes a ligand into a set of rigid atom segments connected by rotatable bonds and identifies the largest segment as the *anchor*. The remaining segments are assigned to flexible *layers* that radiate away from the anchor. This is illustrated below in Figure 12.5. The anchor perception is controlled by two parameters. If the multiple_anchors parameter is set to No, then the largest rigid segment is selected as the anchor. If multiple anchors are selected, then the segments are ranked according to size (the number of heavy atoms), and starting with the largest segment, are added to the anchor list until a segment sized smaller than the min_anchor_size parameter is reached. If no segment is larger than the min_anchor_size parameter, the largest segment is selected as the only anchor. The rest of the flexible docking algorithm needs to be carried out once

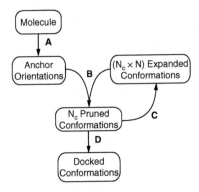

FIGURE 12.4 Schematic illustrating the steps of the Anchor and Grow algorithm. A: The molecule's anchor is rigidly docked, and the orientations are added to the conformer list. B: The conformer list is pruned by the rank/rmsd measure. C: Each conformer in the list is conformationally expanded by one flexible layer. D: When all the layers have been added, the algorithm exits.

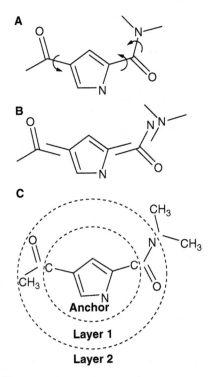

FIGURE 12.5 Illustration of rigid anchor and flexible layer perception. A: All rotatable bonds are identified. B: Molecule is divided into overlapping rigid segments. C: The largest segment is selected as the anchor, and all nonoverlapping rigid segments are assigned to layers.

for each anchor, therefore it is advisable to not use multiple anchors when docking a large library. If the library molecules warrant using more than one anchor (multiple large rigid substructures), then make sure that the min_anchor_size is set large enough (at least 10 to 15) to prevent a large number of anchors being selected.

12.6.2 RIGID DOCKING OF THE ANCHOR

For each anchor identified, rigid docking routines are used to position the anchor in the receptor, and a sorted list of all generated anchor orientations are returned. For the flexible growth step of this algorithm to have any chance for success, the rigid docking of the anchor must generate good orientations of the anchor that score well enough to be near the top of the orientation list. Exactly how close to the top of the list depends on the flexible growth parameters that will be discussed in the next section. The important point is that oversampling orientations incurs the risk of displacing the good anchor orientations with many bad or decoy orientations, so that no good orientations of the anchor reach the flexible growth stage. Furthermore, there is another reason to be wary of oversampling anchor orientations. The energy landscape of a ligand anchor is simplified with respect to the energy landscape for the entire ligand, because many of the ligand atoms are not present in the anchor. Therefore, less sampling is required to identify the major features of the landscape.

As a result of these considerations, it is appropriate to use automated_matching with maximum_orientations = 500 to 1000. The optimal minimization parameters are max_iterations = 200 with trans_step = 0.1, rot_step = 0.1, and tors_step = 10.0 degrees.

The bump filter should be used during anchor docking to exclude orientations that position the anchor tightly into small receptor spaces, as these orientations likely will not be able to accommodate any flexible growth. Max_bumps should be set to 0 or 1 to require orientations to be more accessible and, therefore, more likely to accommodate flexible growth by having room for the layers to be placed.

At the least, the DOCK energy scoring function should be used during anchor docking. The contact score is far too simplistic to accurately position small fragments like ligand anchors. In highly polar receptors, such as nucleic acid targets, it may be necessary to use solvation scoring for the rigid docking of the anchor, because an accurate ordering of anchor orientations is required for successful flexible growth.

12.6.3 FLEXIBLE GROWTH

The flexible growth routine searches through the conformations of the layer segments, essentially growing the flexible portions of the molecule onto the best scoring anchor orientations.

The first step of flexible growth is to prune the list of orientations to remove the worst scoring and redundant orientations from the list. The pruning is done with a greedy clustering algorithm that compares the first orientation in the ranked list to each of the subsequent orientations. The rmsd between the two orientations is calculated, and if the (rank of the subsequent orientation)/rmsd exceeds the parameter N_c, then the subsequent orientation is removed from the list. This is repeated for all remaining orientations, until all pairs of orientations have been compared. The remaining orientations are copied into an empty list of molecules, the conformer list.

The next step in flexible growth is the expansion of the conformer list. For each molecule in the list, the next flexible layer of atoms is added, and all conformations of the rotatable bonds are generated based on a library of torsion angles for each rotatable bond type defined in flex.defn. The torsion angles are stored in the flex_drive.tbl file, whose path is specified with the flex_drive_file parameter. Each new conformation is scored and optimized, and then added to the list. If the use_clash_penalty parameter is set to Yes, then an internal vdW clash check is performed. If any of the atoms of one segment clash with atoms from another segment, the conformation is discarded. A clash occurs when the distance between two atoms is less than the sum of their vdW radii scaled by the clash_overlap parameter. After all molecules in the list have been conformationally expanded one flexible layer, the list is sorted and pruned according to the procedure described above. This is repeated until all of the flexible layers have been added, resulting in a set of full conformations.

12.6.3.1 Conformational Sampling

The conformational sampling during flexible growth is controlled by both the N_c parameter and the granularity of the torsion library. The optimum value of N_c during flexible growth is between 25 and 50. Setting N_c to 100 improves the anchor docking success rate; however, it results in extremely slow flexible growth because many more conformers are expanded at each step in the growth. For this reason, we are implementing, in a future DOCK update, separate N_c values for anchor pruning and for flexible growth. Decreasing the granularity of the torsion library may also improve flexible docking results, but it results in increased computation times. As a result, it is important to selectively edit the torsion angle values and to sample more finely certain bonds that are important to the ligands being docked without increasing the number of torsion steps for all of the rotatable bonds.

The clash check is a useful tool for avoiding the generation of unrealistic self-clashing conformations. It speeds up the flexible growth phase by removing clashing conformers and reducing the list size. However, if the clash_overlap parameter is set too high, it may consider conformers that have compact conformations, but not necessarily substantial vdW overlap, to be clashed. This can result in early elimination of too many conformers, and the failure of flexible growth to generate any full conformations. As a result, the clash_overlap parameter should optimally be set to 0.25. Anchor and Grow can also compute an intramolecular energy score, where the energy score is computed between all atoms of different segments. This score will similarly eliminate self-clashing conformations; however, it is more computationally expensive to calculate. Therefore, it is recommended for use on systems with flexible, large ligands, such as small peptides. The intramolecular score is enabled by setting the use_internal_energy parameter to Yes. The internal_energy_att_exp, internal_energy_rep_exp, and internal_energy_dielectric parameters should be set to 6, 12, and 4.0, respectively.

12.6.3.2 Scoring and Optimization

Energy scoring should be used during flexible growth, due to the large number of scoring function evaluations that are performed, making solvation scoring computationally infeasible. This is acceptable because the flexible growth tends to be dominated by the vdW component of the energy score. The contact score yields too simple an energy landscape to effectively guide the conformational search.

Optimization during flexible growth primarily compensates for the coarse-grained nature of the torsion library. As a result, a small amount of minimization is required to adjust torsion angles to local minimum positions, without moving the partially grown ligand much. The optimum parameters are max_iterations = 10 to 25, trans_step = 0.1, rot_step = 0.1, and tors_step = 10.0.

12.7 REFINEMENT OF RESULTS

12.7.1 CLUSTERING

Due to the uncertainty inherent in the approximations in the scoring functions, it is prudent to consider a range of top scoring DOCK predictions. There can be dominant minimum energy positions that are redundantly reported, resulting in many nearly identical orientations with slight differences in their scores. These redundant orientations can overwhelm the ranked list of all the results, reducing their spatial diversity. Clustering provides a useful tool for pruning redundant results from this top scoring list.

As with nearly all clustering problems, choosing cluster boundaries and cluster representatives can be considered as much an art as a science [43]. Most schemes used for clustering docked molecules are simple and use rmsd as their distance metric. A common technique is to use a simple greedy scheme, where the list of docked orientations is first ranked by score. The best scoring orientation is then chosen, and all other orientations within 1.5 to 2.0 Å of it are grouped into a single cluster. This step is repeated with the best scoring unassigned orientation until all orientations have been assigned to a cluster. A few of the lowest scoring members in each cluster are used as representatives of possible binding modes.

Clustering the orientations found by rigid docking can be used to provide a rough description of the nature of the energy landscape for a particular receptor–ligand pair. The recognition of multiple binding modes of the ligand is aided by the identification of significant low scoring clusters. The size of each cluster provides a rough estimate of the width of the energy well in that particular region of orientation/conformation space.

Due to the rmsd-based pruning steps during the flexible growth, clustering the conformations produced by flexible docking is generally not necessary because they are already spatially diverse.

12.7.2 SOLVATION RESCORING

When docking a single molecule to predict the binding mode, the DOCK energy score is generally sufficient for identifying the correct global minimum energy pose.

This is because all of the orientations/conformations of the single molecule have the same charges and the same number of atoms; therefore the DOCK energy scores are directly comparable. However, when comparing the docked poses of different molecules, the differences in charges and the number of atoms can result in a misleading comparison as the effects of solvation for both molecules are not equivalent. Rescoring the docked poses with a solvation scoring function helps alleviate this problem [44]. As mentioned above, DOCK implements both GB/SA and PB/SA solvation scoring functions. DOCK users employing the GB rescoring methods have reported increased abilities to correctly rank ligand binding affinities [33,45].

12.8 EVALUATION OF DOCKING RESULTS

12.8.1 REPRODUCTION OF CRYSTAL STRUCTURES

There are many possible applications and measures of success reported in the literature for docking methods. A common test of docking algorithms is to attempt to reproduce the conformation and orientation of ligands as observed in x-ray crystal structures [46,47,48]. The reproduction of crystal structures is a measure of docking success, because it is the logical prerequisite to correctly ranking libraries of compounds. Typically, a threshold of 2.0 Å rmsd between the docked structure and the crystal structure is used [38,39].

The performance of DOCK 5 was evaluated using a modified version of the GOLD test set that had been prepared for docking [25]. From an initial set of 224 crystallographically determined protein–ligand complex structures, covalent complexes, metal coordinating complexes, and structurally uncertain complexes were removed, leaving 203 complexes in our test set. For the purposes of these tests, a success was evaluated by measuring the heavy atom rmsd between the best scoring DOCK prediction and the crystallographic ligand coordinates. DOCK 5 was first evaluated for its ability to rigidly dock the ligand, using its crystallographic conformation, to the receptor. DOCK 5 was then evaluated for its ability to flexibly dock the ligand, using only the standard library torsion angles, to the receptor. DOCK 5 achieved success rates of 81% for rigid docking and 43% for flexible docking. For reasons of simplicity, only the top scoring docked pose for each complex was examined; in reality, it would be appropriate to consider all top scoring poses within an energy window to account for scoring function error. For more details of these ongoing experiments, please refer to the DOCK 5 paper [41].

12.8.2 TYPES OF FAILURES

There are two major classes of failures that occur when any docking program fails to reproduce the pose (the orientation and conformation) of a ligand from a crystal structure. The first type of failure, termed *sampling failures*, occur when the algorithm fails to sample any ligand poses near the crystal pose. These failures generally can be overcome by simply increasing the sampling of the orientation/conformation space. The amount of sampling required to identify the correct values may often be impractical, however.

The second type of failure, a scoring failure, occurs when the algorithm samples a ligand pose that scores more favorably than the crystal pose. These are generally considered to occur due to deficiencies in the scoring function, and no amount of increased sampling or clever searching can recover from this faulty energy landscape. Scoring failures can sometimes be the result of a crystal structure that was not built carefully, which is not as uncommon, given that the precise positioning of the ligand may not have been a priority to those building the crystal model. Some of these structures contain clashes between hydrogens of the ligand and the receptor, such that the crystallographic ligand pose is truly not the lowest energy pose in the landscape. Thus, it is important when attempting to regenerate a crystallographic structure that the crystal pose itself is examined closely first and then perhaps allowed to minimize any clashes which may exist in the structure as built by the crystallographers. Despite such careful preparations, scoring failures still occur and emphasize the need for improved scoring functions.

12.8.3 Ranking of Docked Molecules

Reproducing the position of a ligand from a crystal structure is only a partial test of docking. One of the major uses of docking is in the ranking of ligands in order of their relative binding affinities. Although being able to recognize the correct positioning of a single ligand can be useful, the ability to perform such tests correctly does not necessarily mean that ligand rankings will also be correct.

The most statistically informative method for evaluating a docking program's ability to correctly rank ligands is calculating the correlation between correct (i.e., experimentally validated) rankings and those generated by the docking program. Unfortunately, comprehensive, validated datasets are extremely rare, especially those in which the assay conditions were uniform and the ligands and targets diverse. At this time, there exists no such publicly available dataset where each ligand was tested against each receptor target in a uniform manner. Such a dataset would be of particular interest given the promiscuity of molecules predicted by DOCK when a single library is docked against multiple diverse targets [49].

Lacking this comprehensive test set, the most common approach is to consider *enrichment*, the ability of the docking program to put ligands that are known to bind the target near the top of a score ranked list. Although less statistically rigorous, these tests have shown that DOCK is able to significantly enrich rankings [8].

12.9 DOCK RESOURCES

For more general information about DOCK, including availability, please visit http://dock.compbio.ucsf.edu. For more detailed information about running DOCK, including a complete explanation of all parameters, please refer to the *DOCK User's Manual,* which is also available on the DOCK Web site.

ACKNOWLEDGMENTS

The authors wish to thank Natasja Brooijmans for her work adapting the GOLD test set for DOCK use, Rob Rizzo for his work investigating small molecule charge models, and Terry Downing for her work investigating the structural basis of scoring failures.

This work was supported by the Institute of General Medical Sciences, National Institutes of Health.

REFERENCES

[1]   I.D. Kuntz, J.M. Blaney, S.J. Oatley, R. Langridge, and T.E. Ferrin, A geometric approach to macromolecule-ligand interactions, *J. Mol. Biol.* 161:269–288 1982.

[2]   R.L. DesJarlais, R.P. Sheridan, G.L. Seibel, J.S. Dixon, I.D. Kuntz, and R. Venkataraghavan, Using shape complementarity as an initial screen in designing ligands for a receptor binding site of known three-dimensional structure, *J. Med. Chem.* 31:722–729, 1988.

[3]   E.C. Meng, B.K. Shoichet, and I.D. Kuntz, Automated docking with grid-based energy evaluation, *J. Comput. Chem.* 13:505–524, 1992.

[4]   E.C. Meng, D.A. Gschwend, J.M. Blaney, and I.D. Kuntz, Orientational sampling and rigid-body minimization in molecular docking, *Proteins* 17:266–278, 1993.

[5]   B.K. Shoichet and I.D. Kuntz, Matching chemistry and shape in molecular docking, *Protein Eng.* 6:223–232, 1993.

[6]   D.A. Gschwend and I.D. Kuntz, Orientational sampling and rigid-body minimization in molecular docking revisited: on-the-fly optimization and degeneracy removal, *J. Computer-Aided Mol. Des.* 10:123–132, 1996.

[7]   T.J. Ewing and I.D. Kuntz, Critical evaluation of search algorithms for automated molecular docking and database screening, *J. Comput. Chem.* 18:1175–1189, 1997.

[8]   T.J. Ewing, S. Makino, A.G. Skillman, and I.D. Kuntz, DOCK 4.0: search strategies for automated molecular docking of flexible molecule databases, *J. Computer-Aided Mol. Des.* 5:411–428, 2001.

[9]   H.M. Berman, J. Westbrook, Z. Feng, G. Gilliand, T.N. Bhat, H. Weissig, I.N. Shindyalov, and P.E. Bourne, The protein data bank, *Nuc. Acids Res.* 28:235–242, 2000.

[10]   Tripos, Inc.: www.tripos.com.

[11]   Chemical Computing Group: www.chemcomp.com.

[12]   S.C.H. Pegg, Remote homology and drug design, PhD dissertation, University of California — San Francisco, 2001.

[13]   R.M. Knegtel, I.D. Kuntz, and C.M. Oshiro, Molecular docking to ensembles of protein structures, *J. Mol. Biol.* 266:424–440, 1997.

[14]   D. Weininger, SMILES 1. Introduction and encoding rules, *J. Chem. Inf. Comput. Sci.* 28:31, 1988.

[15]   MDL Information Systems, Inc.: www.mdl.com.

[16]   A. Rusinko, J.M. Skell, R. Balducci, C.M. McGarity, and R.S. Pearlman, CONCORD, a program for the rapid generation of high quality approximate 3-dimensional molecular structures, The University of Texas — Austin and Tripos Associates in St. Louis, MO, 1988.

[17]   OpenEye Scientific Software: www.eyesopen.com.

[18] S.K. Kearsley, D.J. Underwood, R.P. Sheridan, and M.D. Miller, Flexibases: A way to enhance the use of molecular docking methods, *J. Computer-Aided Mol. Des.* 8:565–582, 1994.

[19] W.C. Still, A. Tempczyk, R.C. Hawley, and T. Hendrickson, Semianalytical treatment of solvation for molecular mechanics and dynamics, *J. Am. Chem. Soc.* 112:6127–6129, 1990.

[20] G.D. Hawkins, C.J. Cramer, and D.G. Truhlar, Parameterized models of aqueous free energies of solvation based on pairwise descreening of solute atomic charges from a dielectric medium, *J. Phys. Chem.* 100:19824–19839, 1996.

[21] D. Sitkoff, K.A. Sharp, and B. Honig, Accurate calculation of hydration free-energies using macroscopic solvent models, *J. Phys. Chem.* 98:1978–1988, 1994.

[22] C.I. Bayly, P. Cieplak, W.D. Cornell, and P.A. Kollman, A well-behaved electrostatic potential based method using charge restraints for determining atom-centered charges: the RESP model, *J. Phys. Chem.* 97:10269, 1993.

[23] A. Jakalian, D.B. Jack, and C.I. Bayly, Fast, efficient generation of high-quality atomic charges. AM1-BCC model: II. Parameterization and validation, *J. Comput. Chem.* 23:1623–1641, 2002.

[24] J.B. Li, T.H. Zhu, C.J. Cramer, and D.G. Truhlar, New class IV charge model for extracting accurate partial charges from wave functions, *J. Phys. Chem.* 102:1820–1831, 1998.

[25] J.W. Nissink, C. Murray, M. Hartshorn, M.L. Verdonk, J.C. Cole, and R. Taylor, A new test set for validating predictions of protein–ligand interaction, *Proteins* 49:457–471, 2002.

[26] D.A. Case, D.A. Pearlman, J.W. Caldwell, T.E. Cheatham, III, J. Wang, W.S. Ross, C.L. Simmerling, T.A. Darden, K.M. Merz, R.V. Stanton, A.L. Cheng, J.J. Vincent, M. Crowley, V. Tsui, H. Gohlke, R.J. Radmer, Y. Duan, J. Pitera, I. Massova, G.L. Seibel, U.C. Singh, P.K. Weiner, and P.A. Kollman, AMBER7, University of California — San Francisco.

[27] W.D. Cornell, P. Cieplak, C.I. Bayly, I.R. Gould, K.M. Merz, Jr., D.M. Ferguson, D.C. Spellmeyer, T. Fox, J.W. Caldwell, and P.A. Kollman, A second generation force field for the simulation of proteins and nucleic acids, *J. Am. Chem. Soc.* 117:5179–5197, 1995.

[28] R.W. Pickersgill, A rapid method of calculating charge-charge interaction energies in proteins, *Prot. Eng.* 2:247–248, 1988.

[29] E.L. Mehler and T. Solmajer, Electrostatic effects in proteins: comparison of dielectric and charge models, *Prot. Eng.* 4:903–910, 1991.

[30] T. Solmajer and E.L. Mehler, Electrostatic screening in molecular dynamics simulations, *Prot. Eng.* 4:911–917, 1991.

[31] D.L. Beveridge and F.M. DiCapua, Free energy via molecular simulation: applications to chemical and biomolecular systems, *Ann. Rev. Biophys. Chem.* 18:431–492, 1989.

[32] P.A. Kollman, Free energy calculations: applications to chemical and biochemical phenomena, *Chem. Rev.* 93:2395–2417, 1993.

[33] X. Zou, Y. Sun, and I.D. Kuntz, Inclusion of solvation in ligand binding free energy calculations using the generalized-Born model, *J. Am. Chem. Soc.* 121:8033–8043, 1999.

[34] J.A. Grant, B.T. Pickup, and J.A. Nicholls, A smooth permittivity function for Poisson-Boltzmann solvation methods, *J. Comput. Chem.* 22:608–640, 2001.

[35] R. Fletcher, *Practical Methods of Optimization*, New York: Interscience, 1960.

[36] Z. Du, K.E. Lind, and T.L. James, Structure of TAR RNA complexed with a Tat-TAR interaction nanomolar inhibitor that was identified by computational screening, *Chem. & Biol.* 9:707–712, 2002.

[37] J. Wang, P. Morin, W. Wang, and P.A. Kollman, Use of MM-PBSA in reproducing the binding free energies to HIV-1 RT of TIBO derivatives and predicting the binding mode to HIV-1 RT of efavirenz by docking and MM-PBSA, *J. Am. Chem. Soc.* 123(22):5221–5230, 2001.

[38] Y.P. Pang, E. Perola, K. Xu, and F.G. Prendergast, EUDOC: a computer program for identification of drug interaction sites in macromolecules and drug leads from chemical databases, *J. Comput. Chem.* 22:1750–1771, 2001.

[39] M. McGann, H. Almond, A. Nicholls, J.A. Grant, and F. Brown, Gaussian docking functions, *Biopolymers* 68:76–90, 2003.

[40] F.S. Kuhl, G.M. Crippen, and D.K. Friesen, A combinatorial algorithm for calculating ligand binding, *J. Comput. Chem.* 5:24–34, 1984.

[41] D.T. Moustakas, N. Broojimans, P.T. Downing, S. Pegg, R. Rizzo, and I.D. Kuntz, Development and evaluation of an object-oriented docking program: DOCK 5, manuscript in preparation.

[42] D. Joseph-McCarthy, B.E. Thomas, M. Belmarsh, D. Moustakas, and J.C. Alvarez, Pharmacophore-based molecular docking to account for ligand flexibility, *Proteins* 51(2):172–188, 2001.

[43] L. Kaufman and P.J. Rousseeuw, *Finding Groups in Data: an Introduction to Cluster Analysis*, New York: John Wiley & Sons, 1990.

[44] B.Q. Wei, W.A. Baase, L.H. Weaver, B.W. Matthews, and B.K. Shoichet, A model binding site for testing scoring functions in molecular docking, *J. Mol. Biol.* 322(2):339–355, 2002.

[45] B. Shoichet, A.R. Leach, and I.D. Kuntz, Ligand solvation in molecular docking, *Proteins* 34:4–16, 1999.

[46] G. Jones, P. Willett, R.C. Glen, A.R.L. Leach, and R. Taylor, Development and validation of a genetic algorithm for flexible docking, *J. Mol. Biol.* 267:727–748, 1997.

[47] R. Wang, Y. Lu, and S. Wang, Comparative evaluation of 11 scoring functions for molecular docking, *J. Med. Chem.* 46(12):2287–2303, 2003.

[48] S. Ha, R. Andreani, A. Robbins, and I. Muegge, Evaluation of docking/scoring approaches: a comparative study based on MMP3 inhibitors, *J. Computer-Aided Mol. Des.* 14(5):435–448, 2000.

[49] R.T. Koehler and H.O. Villar, Statistical relationships among docking scores for different protein binding sites, *J. Computer-Aided Mol. Des.* 14(1):23–37, 2000.

13 Pharmacophore-Based Molecular Docking: A Practical Guide

Diane Joseph-McCarthy, Iain J. McFadyen,
Jinming Zou, Gary Walker, and Juan C. Alvarez

13.1 INTRODUCTION

Structure-based virtual screening (VS) is often time-intensive. To be predictive, docking of small molecules to a target structure requires adequate orientational and conformational sampling as well as accurate scoring within the target binding site. Our pharmacophore-based docking method, PhDock, whereby conformers of the same or different molecules sharing a common pharmacophore are overlaid into ensembles and simultaneously docked using their largest three-dimensional (3D) pharmacophore, allows for efficient sampling of the ligand conformations and orientations in the target structure. Theoretical pharmacophore points, which represent hot spots in the target binding site, more preferentially orient the candidate ligands in productive modes. The combination of simultaneously orienting all the ligands within an ensemble combined with the more efficient sampling dramatically improves search times.

The overall drug discovery process can be made more efficient through the use of computational methods for screening. VS of large 3D molecular databases enables the identification of novel small molecule drug leads for biologically relevant targets. High throughput molecular docking has emerged as a complement to experimental high throughput screening.

VS can either be ligand-based or target-based. Ligand-based VS methods include pharmacophore searches (e.g., [1–4]) and shape-based searches (e.g., ROCS [5]), where the pharmacophore or shape is determined by the structure of a ligand or series of ligands active against the target. Target-based VS can be performed once a 3D structure of the target is available, either from x-ray crystallography, nuclear magnetic resonance spectroscopy, or high quality homology modeling. In general, molecular docking involves searching a database for compounds that fit into the binding site of the target structure in terms of shape complementarity and chemical matching. To be a viable technique for screening commercial and corporate databases as well as virtual libraries of several hundred thousand molecules, search times on the order of seconds rather than minutes per molecule are desired. Furthermore, given that the conformation a ligand adopts in solution is not necessarily the conformation it will adopt when bound to the target, conformational flexibility of the database ligands must be considered. Once a virtual screen is completed, candidate compounds are selected and experimentally assayed at a relatively high concentration to look for weak binders with novel scaffolds.

In this chapter, we describe multiple issues and practicalities that are necessary to consider when carrying out a high throughput docking experiment. We will present some specific solutions as well as our attempts to automate different portions of the process. Finally, the results of sample cases in cross-docking and enrichment studies will be discussed. Although the overall process we describe herein is tailored to our in-house algorithm, PhDock, the concepts are important and most of the solutions presented apply to any docking methodology.

13.2 MOLECULAR DOCKING

13.2.1 LIGAND FLEXIBILITY

Ligand flexibility can be incorporated either by flexing the ligand as it is incrementally built into the binding site (e.g., [6–13]) or by docking rigid, precomputed conformers from conformationally expanded databases (e.g., [14–17]). Methods that treat ligand flexibility on the fly, suffer from a redundant sampling of torsions. That is, the conformers must be generated at every anchor position; starting with the first bond extending from the anchor, all of its allowed rotations must be sampled, and then the same must be done for the second bond, etc. Incremental construction methods additionally suffer from sensitivity to initial anchor positioning. That is, small variations in the initial placement of the anchor, particularly when the anchor is in a terminal portion of the molecule, have dramatic effects on the positioning of atoms distant to the anchor. Often the anchor represents only a small fraction of the total atoms in the molecule and therefore much information is lost in guiding the docking process. Finally, greedy algorithms require that each sequential torsion placement be optimal, as a result the most optimal overall molecule position may not be found. The bioactive conformer, particularly for a novel nonoptimized ligand, may require that one or more torsions not be fully optimal to allow for the best interaction of the entire small molecule with the target. Small ligand strain energies are often necessary to obtain the best fit. In a greedy algorithm, often these conformers would be discarded as the flexible docking progressed, never allowing the correct pose to be found, and in most the final internal strain energy of the ligand is never evaluated. Because exhaustive searching using incremental construction is not feasible due to time constraints, the final result is a loss in conformational and orientational sampling, particularly with molecules containing multiple rotatable bonds. In contrast, when docking rigid precomputed conformers (flexibases [14]), the cost for generating the conformers is incurred once, during the setup of the database, thus eliminating the redundancy in torsional sampling. The resulting flexibase can then be screened against any target. Furthermore, because conformer generation is done infrequently, the best, most time-intensive algorithms and parameters can be applied, and the resulting conformers can be carefully evaluated in terms of their internal strain energies. Because each conformer is evaluated as a complete molecule, not as an anchor or fragment, all the information in the molecule is used to guide the docking process. Due to the linear increase in search time with the number of conformers, however, the docking of individual conformers from a flexibase can be prohibitive [18]. Docking ensembles of conformers instead of individual conformers can dramatically speed up the search time allowing for the screening of databases of several hundred thousands of compounds [15,16]. Ensembles generated on anchor fragments, however, do not overcome the issues of anchor placement sensitivity.

13.2.2 TARGET FLEXIBILITY

A number of research groups are developing docking methods that allow for some degree of target flexibility. This has largely been limited to a few side chain motions or water displacements (e.g., SLIDE — Screening for Ligands by Induced-fit Docking, Efficiently [19]). Others try to account for protein flexibility by using multiple scoring grids representing multiple protein structures or a scoring grid with values averaged over or derived from a number of protein structures (e.g., AutoDock [10]).

13.2.3 PHDOCK AND MCSS2SPTS

All docking algorithms employ a heuristic for rapidly positioning the database molecules in the binding site. The program DOCK [20] uses a heuristic that matches ligand atoms to predefined site points in the binding site of the target structure. Our pharmacophore-based docking method, PhDock [16], is implemented in DOCK 4.0 [21] and allows for pharmacophore-based docking of ensembles of precomputed conformers (Figure 13.1). In a PhDock database, precomputed conformers of the same or different molecules are overlaid based on their largest 3D pharmacophore. The pharmacophore points (and not all of the ligand atoms) are matched to the predefined site points to determine the allowed orientations for the ensemble in the binding site. Each individual conformer within the ensemble is scored in the binding site and the best scoring conformer (or a set number of best scoring conformers) for each molecule is saved.

Docking pharmacophoric ensembles provides an efficient strategy for rapidly docking large databases of molecules. Computing the conformers during the database generation step circumvents problems with redundant rotations of bonds during flexible docking and overcomes the greed algorithm problem because all conformers within a reasonable internal energy cutoff are considered. Anchor sensitivity as is found in "Anchor and Grow" methods and docking ensembles overlaid on the largest rigid fragment are no longer a problem, because the pharmacophore features generally span the entire extent of the small molecule. Using the pharmacophore points for the matching greatly reduces the combinatorial problem associated with determining the orientations, and docking the ensemble eliminates the need to calculate possible orientations for each individual conformer. Although the overall sampling is reduced, greatly reducing the search time, when matching to chemically labeled site points, the relevant sampling is enhanced. Because the ligand pharmacophore defines the parts of the molecule most likely to interact with the target,

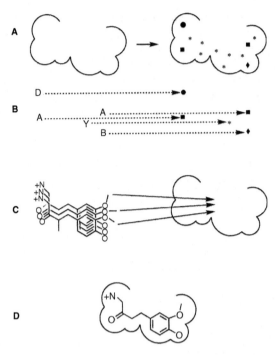

FIGURE 13.1 Schematic of PhDock methodology. In the first step (A), the target binding site is filled with chemically labeled site points. Then distances between pairs of atoms in a database ligand pharmacophore are matched to distances between pairs of pharmacophoric site points (B). Partial pharmacophore matches are allowed. A transformation matrix is calculated for each allowed orientation of the database pharmacophore and all conformers in that ensemble are docked into the binding site (C). Finally, the fit of each individual conformer is scored (D).

pharmacophore matching ensures the most productive orientations are still sampled (see Section 13.6.3 for an example). Thus, PhDock allows for efficient sampling of the ligand conformations and orientations in the target structure.

Theoretical pharmacophore points that represent hot spots in the target binding site can be used as the matching site points to orient the candidate ligands more rapidly. More specifically, chemically labeled site points can be generated in an automated fashion using the script MCSS2SPTS [22]. MCSS2SPTS employs the program MCSS (Multiple Copy Simultaneous Search) [23–25] to determine the target-based theoretical pharmacophores. MCSS functional group maps (or preferred binding sites) are calculated for nine different groups that span ionizable, polar, and hydrophobic atom types. Chemically labeled site points (labeled appropriately according to their pharmacophoric type) are automatically extracted from selected low energy functional-group minima and clustered together. Chemically labeling the site points can significantly reduce the search time by restricting the search space to areas relevant to the target, thereby reducing the combinatorial problem.

13.3 PHDOCK DATABASE GENERATION

The preparation of a conformationally expanded, 3D ligand database for rigid molecular docking involves generating a set of structures for each ligand that aims to accurately represent all of the low energy, accessible conformers for that ligand. Although ligands are unlikely to be found in high energy conformations when bound to a protein, they may adopt any number of low energy conformations. A sufficient number of low energy conformers must be generated to account adequately for the ligand flexibility. The preparation of a PhDock database involves generating realistic conformers, determining a 3D pharmacophore for each database molecule, and finally overlaying conformers, from the same and different molecules, by their largest 3D pharmacophore to form database ensembles. Databases of pharmacophoric ensembles are generated in a fully automated manner as described below and in Figure 13.2.

13.3.1 PHYSICAL PROPERTY FILTERING

Before generating conformers for a large molecular database, such as a corporate collection or external vendor library, the database should first be filtered to remove nonleadlike or at least nondruglike compounds [26,27]. Several studies have

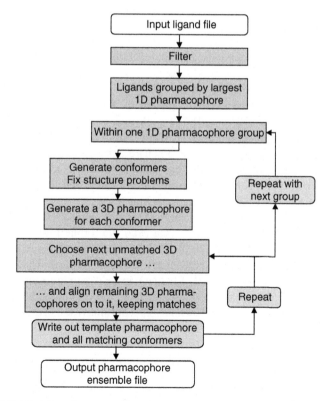

FIGURE 13.2 Workflow for generation of a pharmacophoric ensemble database.

indicated that molecular complexity (including molecular weight, lipophilicity, etc.) increases as a lead is fully optimized into a drug [28,29]. Because the goal of VS is to identify new lead compounds for a given target, it is desirable to eliminate nonleadlike compounds up front. The sequential filter we use for this purpose removes undesirable structures (those containing known reactive functionalities, inorganic elements, etc.) and retains only those compounds with molecular weight (MW) ≥ 180 and ≤ 400, ionizable groups ≤ 3, number of rings ≥ 1, logD at pH 7.4 ≤ 4.0, number of chiral centers < 2 and rotors ≤ 8 (nonhydrogen). The limits on rotors and chiral centers stem partly from consideration of leadlike properties and partly from the need to limit the overall size of the resulting database. Limiting the number of rotors, and thereby capping the entropic cost for the binding of a database ligand, is also desirable because many scoring functions lack consideration of this entropic term.

13.3.2 CONFORMER GENERATION

We use OMEGA [30] to generate systematically, via a torsion-driving beam search, all conformers within a specified internal energy cutoff (using the van der Waals (vdW) component of a modified Dreiding force field [31]) for each ligand. OMEGA has been shown to generate bioactive conformers rapidly [32,33]. A torsion library file lists allowed torsional angle values for the search. The defaults are $\pm 60°$, and $180°$ for sp³-sp³ bonds, $0°$, $\pm 30°$, $\pm 150°$, and $180°$ for sp²-sp² bonds, and $0°$ plus all $30°$ increments for sp²-sp³ bonds. The torsional library was customized to improve sampling. For example, for the bond type *=*–*=*, torsion values of $\pm 10°$, $\pm 40°$, $\pm 150°$, and $180°$ are allowed. For an sp³-sp³ carbon–carbon bond in an unbranched alkyl chain, specified with the SMARTS pattern *[CX4D2][CX4D2]*, torsion values of $\pm 30°$, $\pm 60°$, $\pm 90°$, $\pm 150°$, and $180°$ are allowed. Our customized torsion library lists 46 specific torsion bond types. With an upper limit on the number of conformers for any molecule of 10,000, a root-mean-square deviation (rmsd) tolerance of 1.0 Å to eliminate redundant conformers, and an internal energy cutoff of 5 kcal/mol, we found that conformer generation typically takes 450 to 550 seconds per thousand conformers on a SGI R12000® Octane® machine.

For automated generation of our large databases, input is in SMILES format and conformers are output in structure data (SD) file format [34]. Using SMILES input takes advantage of built-in 1D/2D to 3D conversion routines in OMEGA that produce molecules with idealized geometry (including rings). Using 3D ligand coordinate input (PDB, MOL2, or 3D SD files), instead, would result in conformers that retain all bond lengths and bond angles from the original structure. We have found the latter can yield poor results when nonideal values exist in crystal structures.

13.3.3 3D PHARMACOPHORE GENERATION AND OVERLAYS

The recognized pharmacophore types are hydrogen bond acceptors, hydrogen bond donors, duals that can act as both acceptors and donors, and ring centroids [16].

More specifically, nitrogen with an unconjugated lone pair, and oxygen or sulfur atoms with lone pairs and no attached hydrogens are considered acceptors; those with attached hydrogens and no lone pairs are considered donors; those with lone pairs and hydrogens are duals. Centroids are derived from five- and six-membered rings. We chose not to include a nonaromatic hydrophobe type, because it was difficult to define (e.g., should only alkyl length chains of three or more be considered, on what atom should the pharmacophore point be centered for an isopropyl or propyl group) and it would reduce the degeneracy of the overlays and therefore diminish the possible gains in speed. Furthermore, most molecules of interest are likely to be oriented productively with the target using at least four of the other features listed above.

For PhDock, each conformer that is generated has an associated 1D and 3D pharmacophore. For preparation of our PhDock databases, the filtered molecular database is first sorted by its 1D pharmacophore. A particular molecule might have, for example, one acceptor, two donors, one dual, and two centroids as its fingerprint. Typically many database ligands will share the same 1D pharmacophore; therefore, partitioning a large database in this manner allows for efficient parallelization of the overlay steps described below. The spatial arrangement of all pharmacophoric features in the molecule determines the largest 3D pharmacophore for that conformer. The full 3D pharmacophore is extracted from each conformer and output as a SD file.

Conformers and pharmacophores are then overlaid to generate ensembles (Figure 13.2). That is, conformers of the same and different molecules are overlaid based on their largest 3D pharmacophore. Each 1D pharmacophore partition is processed separately for efficiency. For a given partition, the first 3D pharmacophore is selected as a template, and that pharmacophore and its associated conformer is written to the ensemble database file. An alignment algorithm then attempts to overlay onto the template all other 3D pharmacophores from the file. Pharmacophores that match the template (with a 1.0 Å tolerance for individual pharmacophore features) are removed, and the corresponding conformers are written to the ensemble database file. This process is repeated, selecting the next remaining pharmacophore on the list as the new template, until there are no unprocessed pharmacophores left in the file. The final database contains multiple ensembles, each consisting of a pharmacophore (defined in MOL2 format), followed by one or more matching conformers (also in MOL2 format). Because the largest 3D pharmacophore is used to define the ensembles, each conformer appears in only one ensemble. Partial pharmacophore matches are allowed later during the docking steps.

Conformationally expanded databases (pharmacophoric ensembles or otherwise) of large numbers of molecules contain a vast amount of data, which can pose a challenge in terms of storage when using the typical MOL2 format. Therefore, a binary format was developed for PhDock databases. It is capable of significant data compression, typically requiring only 20% of the space required for the equivalent MOL2 file. It also has an advantage in input/output (I/O) operations due to faster binary reads, reducing search times by up to 30%.

13.3.4 DATABASE STATISTICS

The Available Chemicals Directory (ACD), version 2001.1, which contains 301,125 compounds, [34] was sequentially filtered as described above. The filter for undesirables removes 37%, MW removes another 20%, ionizables < 0.1%, rings 3%, logD 13%, chiral centers 5%, and rotors 3% of the database molecules. Our final PhDOCK ACD database contains approximately 39% of the compounds in the original MDL ACD database (116,204 out of 301,125 compounds). The resulting 116,204 compounds were conformationally expanded and arranged into ensembles.

The final PhDOCK database (14.0 Gb in our binary format) contains 16 million conformers, 1.7 million unique pharmacophores, and has an average of 138 conformers per molecule. The distribution of the number of rotors peaks at 4, with more than 22,000 database molecules having 4 rotatable bonds, and only about 4000 molecules containing 8 rotors. However, molecules with 4 rotatable bonds have an average of about 32 conformers, and molecules with 8 rotatable bonds average about 1200 conformers. Thus, although the molecules containing 8 rotors represent less than 4% of the molecules in the database, they represent greater than 40% of the conformers. This data illustrates the need to restrict rotors to keep the overall size of the database manageable as well as adequately cover conformational space. An analysis of the database reveals that the average number of pharmacophore points per pharmacophore definition is 9.6, and the average number of conformers per pharmacophore is 8.7. Of the pharmacophores, 30% have only one conformer and 20% have two. It is likely that the molecules with 8 rotors significantly contribute to this lack of degeneracy as they tend to be the molecules for which it is most difficult to generate overlays; given that they contain the largest number of pharmacophoric centers, any rotatable bond that affects the position of a pharmacophoric center will result in a new 3D pharmacophore. Eliminating molecules with 8 rotors would decrease the overall size of the database by 40% and likely increase the pharmacophoric degeneracy and therefore docking efficiency. Currently, however, we have chosen to retain these molecules so as to not miss potential ligands.

13.4 THE DOCKING PROCESS

13.4.1 PROTEIN PREPARATION

The first step in a virtual screen is preparing the protein structure. In general, hydrogens need to be added, ligands and solvent removed, chain breaks and termini appropriately capped, tautomers checked, etc. Processing the protein structure into a state suitable for docking can be carried out in any number of molecular modeling programs. We present the steps necessary utilizing the Molecular Operating Environment (MOE), version 2003.02 [36]. We have chosen MOE because of the presence of various utilities and the ability to script the process using their Scientific Vector Language (SVL). A flowchart for our Automated Preparation of Proteins (APOP) script is shown in Figure 13.3.

Once the protein or protein–ligand complex is read into MOE, atom types and bond orders are automatically corrected by the program. Chain breaks are detected

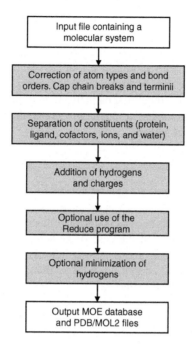

```
┌─────────────────────────┐
│  Input file containing a │
│     molecular system     │
└─────────────────────────┘
            │
            ▼
┌─────────────────────────┐
│ Correction of atom types │
│ and bond orders. Cap     │
│ chain breaks and terminii│
└─────────────────────────┘
            │
            ▼
┌─────────────────────────┐
│ Separation of constituents│
│ (protein, ligand,        │
│ cofactors, ions, and water)│
└─────────────────────────┘
            │
            ▼
┌─────────────────────────┐
│   Addition of hydrogens  │
│        and charges       │
└─────────────────────────┘
            │
            ▼
┌─────────────────────────┐
│    Optional use of the   │
│      Reduce program      │
└─────────────────────────┘
            │
            ▼
┌─────────────────────────┐
│  Optional minimization of│
│        hydrogens         │
└─────────────────────────┘
            │
            ▼
┌─────────────────────────┐
│    Output MOE database   │
│    and PDB/MOL2 files    │
└─────────────────────────┘
```

FIGURE 13.3 Schematic of the APOP script.

by comparing SEQRES records (at the beginning of a PDB file) with the sequence found in the protein structure; chain breaks are capped with ACE/NME, and true chain termini with NH_3^+/COO^-, respectively, at N- and C-ends. Any ligand, cofactor, ion, or water residues are identified. Residues are defined as protein if they belong to a chain (based on the chain identifier) of length 16 or more. Water is defined as residues named WAT, HOH, or T3P. A list of heterogroups in the Protein Data Bank (PDB) is maintained online in a HET dictionary [37]. Approximately 100 cofactor residue names were selected from those heterogroups appearing in 10 or more PDB structures; the resulting list is used to define cofactor. Ions are defined as molecules containing 6 or less heavy atoms. Ligands are defined as all molecules with < 16 residues remaining after cofactors, ions, and waters have been identified. Next hydrogens and MMFF94s (MMFF — Merck Molecular Force Field) charges [35] are added to all residues (protein, cofactor, and ligand).

At this point, before optimizing hydrogens, when manually preparing a protein structure, residues near the active site are visually inspected to check tautomers, protonation states, rotatable hydrogen positions, and specific side chain orientations. This is necessary because x-ray protein crystallography, for example, is not generally high resolution enough to determine the protonation state of histidine (His) or to distinguish between rotamers of a glutamine (Gln) side chain that places the terminal NH and CO at opposite positions. When addressing this issue in the automated preparation script, the Reduce program is used for optimization of the orientations of OH, SH, NH_3^+, and methionine (Met) methyl hydrogens; asparagnine (Asn) and Gln sidechain amides; and His rings [38]. Finally, hydrogen positions are minimized using the MMFF94s force field (and default parameters in MOE, specifically 100 steps of Steepest Descent with a gradient test of 1000 kcal/mol·Å followed by 100 steps of Conjugate Gradient with a gradient test of 100 kcal/mol·Å followed by 200 steps of truncated Newton with a gradient test of 0.01 kcal/mol·Å). In the script, the option exists to minimize the hydrogens for the entire complex together or to minimize hydrogens on the protein separately. It may also help to minimize the entire complex structure, subject to harmonic constraints, to relieve initial bad contacts prior to the virtual screen.

13.4.2 SITE POINT SELECTION

A chemically labeled site point set can be generated in a fully automated manner using the script MCSS2SPTS [22] as is described above. In the docking step, site points labeled as acceptor can be matched to pharmacophore points (in the case of PhDock or ligand atom points in the case of DOCK) labeled as acceptor or dual (atoms capable of both donating and accepting hydrogen bonds); donors can match to donor or dual; duals match to donor, acceptor, or dual; centroids match to centroid; and neutrals match to any type. Alternatively, site points could be generated using the DOCK SPHGEN utility and then labeled using grid values calculated with the program GRID [39]. Furthermore, experimentally known ligand atom positions or water positions, typically determined in a x-ray crystallographic protein–ligand complex structure, can be used alone or to augment or validate a site point set calculated with a method such as MCSS2SPTS. In addition to being chemically labeled, DOCK site points can also be identified as being part of a critical cluster, requiring that any acceptable orientation match at least one site point in each defined critical cluster.

13.4.3 SCORING FUNCTIONS AVAILABLE

PhDOCK uses the standard DOCK contact and energy scoring functions. DOCK also has a chemical score available but we have limited experience with it. For each function, the score values are precalculated on a grid within the user specified box surrounding the binding site, which greatly speeds up the scoring event. Typically the grid spacing used is between 0.2 and 0.3 Å. When a docked ligand is scored, the grid value at the position of each ligand atom is used to determine the overall score. The contact score is a count of the number of protein atoms within a certain distance (e.g., 4 Å) of each ligand atom. The energy score is the AMBER all-atom force field [40], typically calculated with a distant dependent dielectric constant and a nonbonded term cutoff of 12 Å. These are the only functions currently available for scoring the orientations during the actual docking run. Other functions can be used later to rescore the best poses for each molecule in a stand-alone manner or, alternatively, a consensus can be taken.

13.4.4 DOCKING RUN PARAMETERS

For PhDock, only manual matching is allowed and typically we specify a minimum of 4 nodes and a maximum of 15. A minimum of 3 nodes can be specified but with only 3 matches the mirror image of each pose must be scored. During the docking, initial attempts at matching use the maximum number of nodes to span the largest extent of the ligand molecule possible. Typically the maximum number of orientations to try for each ligand is set to between 2500 and 5000 and the distance tolerance for matching is between 0.5 and 1.0 Å. With PhDock, because pharmacophore points are matched and not ligand atom positions, the maximum number of orientations is rarely exceeded.

13.4.5 Postprocessing Docking Output

Once a molecular docking run is complete, the output can be processed and analyzed in any of a number of different ways to select hits for testing. This postprocessing can involve taking a consensus score or rescoring with a more rigorous scoring function, additionally including full minimization of the ligand or protein–ligand complex. The approach taken depends on the initial scoring functions used in the docking, the level of computer power available, and, ultimately, the nature of the target binding site. A restrictive site, for example, may require using more ligand minimization with a molecular-mechanics force field for any hits to score favorably. That is, if the vdW term for the initial docked pose is high, which is more likely to be the case for a restrictive site, additional steps of minimization may be required. A highly charged site may require that solvation be considered so that specific charge–charge interactions do not overly dominate the score.

For a given PhDock run, poses can be scored using the contact or energy functions. If the orientations are scored using both contact and energy, separate hit lists are maintained for each. One can select a consensus consisting of molecules with representation on both energy and contact lists. This kind of consensus scoring attempts to leverage the complementary filtering properties of the contact and energy scoring functions. Representation on both lists suggests both good charge and shape matching to the site. The consensus option, however, can sometimes eliminate true hits, by misevaluating specific ligands due to limitations in one of the scoring functions. Contact scoring, in particular, is significantly influenced by the size of the candidate ligand. Thus, we have observed a highly charged, relatively small, weak-binding ligand, known to bind experimentally, ranked near the top of the energy list, but near 3000 on the contact list. Membership on both the contact and energy lists does not ensure that the same pose exists in both, therefore a better alternative may be rescoring the poses in the contact list with energy scoring and vice versa and selecting poses scoring better than a threshold value by both scoring functions. Another option is to score orientations for the entire database using only the contact function or some other equally fast scoring filter, saving multiple poses for each ligand. By using a scoring function, such as contact, that is relatively insensitive to exact positioning and as such does not require ligand minimization, several hundred thousand molecules can be rapidly screened. In contrast with the energy scoring function, due in particular to the steepness in the vdW term, rigid-body minimization of each orientation must be performed to obtain meaningful results, thereby significantly increasing the run time.

Once the entire database is screened with a fast filter, multiple poses for a few thousand of the best scoring molecules can then be rescored using a more physically realistic function. In this case, the initial filter only needs to be accurate enough to pass along the correct pose in say the top ten for that molecule. One obvious choice for rescoring is to use a molecular-mechanics force field, such as AMBER, with full minimization of the ligand in the fixed (or mainly fixed) protein site. In an example, discussed below in Section 13.6, we used the program FLO, which employs AMBER [40] on a grid with short nonbond cutoffs to rescore hits. With a program such as FLO [6], it is also possible to allow a few side chains or a region of the protein around the ligand to be flexible during the minimization. Including solvation also should improve hit selection. We have automated procedures (written as SYBYL Programming Language scripts) for rescoring using Poisson–Boltzmann (PB) electrostatics [41]. In addition, using a related approach, we have an automated SVL MOE script that takes as input protein–ligand complexes and calculated molecular-mechanics energy terms, ligand strain terms, PB electrostatics, and various physical properties (including MW, clogP, rotatable bonds, etc.). In this way, protein–ligand complexes from docking experiments, crystal structures, and *de novo* design can be similarly evaluated. Furthermore, a Microsoft® Excel spreadsheet containing ligand structures and their corresponding calculated energy terms and properties can be automatically generated.

13.5 HIV-1 PROTEASE TEST CASE (A CROSS-DOCKING EXAMPLE)

As a test of MCSS2SPTS and the quality of the site points it generates and of the possibility in general for docking ligands to a single protein structure, a cross-docking experiment can be carried out whereby the bound conformations for a series of ligands are docked into a single protein structure. Taking HIV-1 protease as a test system [22], 24 known inhibitors (taken from PDB complex structures) were docked into the binding site (of the 1DIF protein structure from the PDB) using a MCSS2SPTS generated site point set and PhDock. Before docking, the site point set (Figure 13.4) was compared to the 24 ligand structures (superimposing protein main chain atoms) to determine how well the known ligand pharmacophores were reproduced. After superimposing the other HIV-1 protease structures with the 1DIF structure, 22 out of 24 had at least 4 features within 1 Å of an appropriate site point. This suggests that the docking algorithm should appropriately sample orientations near those of the experimentally determined complexes for at least 22 of the ligands. The ligand structures were then docked using PhDock and energy scoring with 25 steps of rigid-body simplex minimization for each acceptable orientation. The ligand database (containing the 24 molecules in their x-ray conformations) consisted of 28 pharmacophores each with one associated conformer (4 of the 24 complex structures give 2 alternative conformers for the bound ligand). The docked poses were compared to the corresponding complex structures of the ligands. In the PhDock run using the 1DIF site point set, 20 out of 24 ligands were docked, and 16 out of the 24 ligands were correctly docked. That is, using a distance tolerance of 0.75 Å for the matching (distances between site points to distances between pharmacophore

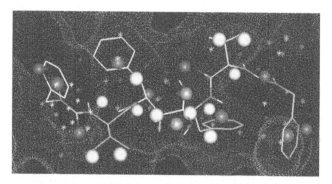

FIGURE 13.4 A set of site points generated with MCSS2SPTS for the HIV-1 protease structure. Spheres indicate site points that overlay with similar features in A79285 bound to HIV-1 protease. A79285 is shown in stick rendering and a molecular surface for HIV-1 protease is in white.

points), an acceptable orientation could not be found for 4 of the ligands. Two of the misdocked ligands (those that docked, but not in the correct orientation) were in the binding site in a flipped orientation that might be expected given that the binding site and the ligands are fairly symmetric. The other two misdocked ligands could not fit into the 1DIF protein structure without a side chain or two shifting somewhat; that is, the 1DIF protein structure is not competent to bind these ligands without undergoing an induced fit. Thus, even without accounting for protein flexibility about two-thirds (75% if you count the 2 that were flipped) of the ligands could be correctly docked.

13.6 DHFR ENRICHMENT STUDIES

13.6.1 DHFR Actives Database for Enrichment-Factor Calculations

The structures of 12 known binders to *Lactobacillus casei* dihydrofolate reductase (DHFR) were conformationally expanded and organized into pharmacophoric ensembles as described above for the other databases. This set of binders is composed of the ligands from the following protein–ligand complex structures: 3dfr, 7dfr, 1rf7, 1dyh, 1jol, 1dis, 1bzf, 1dyj, 1boz, 1dyr, as well as brodimoprim and piritrexim. Each of these ligands is known to bind to this strain of DHFR [42–50].

13.6.2 Decoy Database

Enrichment studies with sets of known actives require a set of random or presumed inactive compounds. For initial enrichment studies, different ACD database partitions, representing different levels of molecular complexity, were selected for screening. For the enrichment studies examining various scoring schemes, a set of 2000 molecules selected randomly from the filtered ACD database (see Section 13.3 above) was treated as described above to generate a pharmacophoric ensemble database.

13.6.3 DOCK 4.0 vs. PHDOCK ENRICHMENT FACTORS

The DHFR actives database was screened for binding to DHFR using both PhDock and DOCK 4.0 [16]. Thereby, enrichment rates for the knowns relative to ACD database partitions were determined. For the DOCK 4.0 calculations, the pharmacophores were removed from the database and each conformer was separately docked and scored. Overall, we found a greater enrichment factor with PhDock relative to DOCK 4.0. In fact, the partitions that showed the most dramatic speedup (up to 26-fold faster) using PhDock compared to DOCK 4.0 also showed the most significant relative enrichment of DHFR knowns (see Figure 13.5A for an example). This result suggests that the poses selected with DOCK 4.0 (although having equivalent or better contact scores than the top hits selected by PhDock) are less chemically relevant.

As a further illustration of the ability of PhDock to select more relevant poses, for a given database partition the scores for the top 100 hits selected by DOCK 4.0 and PhDock were examined. The docking runs were carried out using the contact score, and often the top DOCK 4.0 hits had better contact scores than the top PhDock hits (Figure 13.5B shows an example for medium-sized molecules with an average of 8.1 pharmacophore centers per molecule). When those top poses selected by DOCK 4.0 and PhDock, respectively, were rescored using the energy function, a more chemically aware score, the PhDock hits score significantly better (Figure 13.5C).

13.6.4 DHFR ACTIVES SEEDED INTO ENTIRE ACD DATABASE

The DHFR actives database was also seeded into the entire processed ACD database of over 110,000 molecules [16]. The search took an average of 0.03 sec/conformer on a SGI R12000 400 MHZ processor; 42% of the knowns are found in the top 1% of the ranked database, 58% in the top 2%, and 75% in the top 5%.

13.6.5 EVALUATION OF VARIOUS SCORING SCHEMES

As an initial study of the use of various scoring schemes with PhDock, the DHFR actives database was seeded into the 2000 decoys database. The resulting database was screened for binding to DHFR using the standard contact and energy scores, respectively. For energy scoring, 15 steps of rigid-body simplex minimization is carried out for each acceptable orientation. For the contact scoring run, the top 20 poses for each molecule were saved. Rescore of the best contact pose with rigid-body energy minimization in DOCK and with full minimization of the ligand in FLO [6], and similar rescore of the top 20 contact poses for each molecule was also performed. As described above in Section 13.4.5, FLO uses a modified AMBER force field [40] that is therefore similar to the DOCK energy function. Both scores are calculated on a grid.

In the examples in the other sections above we plotted the enrichment simply based upon the rank of the known relative to the decoys (a standard enrichment plot). Here, because we are exploring the use of various scoring schemes, we plot the enrichment only for correctly posed knowns. Because x-ray complex structures

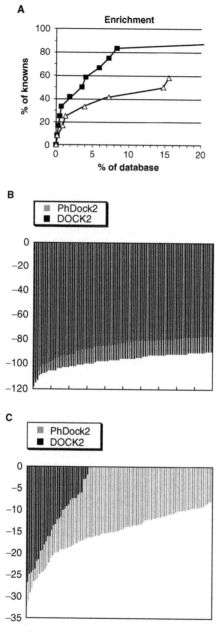

FIGURE 13.5 Enrichment of 12 DHFR knowns seeded into a random ACD database partition of over 500 unique molecules. In the plot of % knowns vs. % ranked database, the enrichment obtained using PhDock is shown in the curve with black squares while that obtained using DOCK 4.0 is shown in the curve with the white triangles (A). The contact scores for the top 100 ranked hits are shown in (B), in grey for PhDock and black for DOCK 4.0. The energy scores after rescore of the top 100 contact poses with the energy function are shown in (C), in grey for PhDock, and black for DOCK 4.0.

do not exist for all of the knowns, a pose is considered to be correct if the pteridine-like group and the acid overlap with the position of the corresponding groups in methotrexate in the 3DFR PDB structure (the structure used for the calculations). An enrichment plot is shown in Figure 13.6 for the various scoring schemes.

Using contact scoring, the correct pose is selected for 6 out of the 12 known ligands. At 2% of the ranked hits, the enrichment factor is approximately 8 (Figure 13.6B, line with white triangles), with 16% of the correctly docked knowns in the top 2% of the entire ranked database. The enrichment factor is similar at 2% of the database ranked by energy scoring; however, only two of the knowns are docked correctly using the energy score (Figure 13.6B, line with black squares). The energy force field scoring does so poorly here due to the sensitivity of the vdW term and the inability of 15 steps of rigid-body minimization to eliminate bad contacts in a fairly restrictive binding site. Taking the top contact pose of each molecule and rescoring by the standard DOCK energy function, the enrichment drops off quite significantly (Figure 13.6A, dashed grey line). Rescoring the top 20 contact poses for each known improves the enrichment, but it is overall slightly worse than with the original contact scoring (Figure 13.6A, solid grey line). For the energy rescoring, 50 steps of rigid-body simplex minimization was done for each pose.

If the best contact poses are rescored using FLO minimization, which allows for flexibility of the ligand, the enrichment is also somewhat worse than for the original contact scoring (Figure 13.6A, dashed black line). However, if the top 20 contact poses are rescored using FLO minimization, the enrichment does improve significantly (Figure 13.6A, solid black line). The rigid-body ligand minimization with DOCK is fast, but does not allow the conformation of the ligand to adjust. The FLO minimization allows for full minimization of the ligand. Because even correct poses (e.g., within say 1.5 Å of a x-ray structure) selected by the relatively soft contact function may have unfavorable vdW interactions with the protein, the ligand must be allowed to flex to alleviate the bad contacts and obtain favorable energy scores. Rigid-body minimization does not appear to be sufficient.

Although full minimization of the ligand with FLO in a rescore improves the enrichment, still only 45% of the knowns are found in the top 2% of the database (Figure 13.6A, solid black line). Furthermore, these plots were generated rescore 20 poses for each of the knowns and only one pose for each decoy. When the calculations are repeated, rescoring 20 poses for each decoy also, the enrichment falls off somewhat to 35% at 2% (see Figure 13.7, dashed black line). When the x-ray poses are included for 10 of the 12 ligands, however, the enrichment improves significantly to 50% at 2% even with 20 poses rescored for each decoy (Figure 13.7, solid black line). The latter result indicates that part of the reason some knowns are not docked correctly is that the correct bioactive conformer does not exist in the database. Again, because the DHFR binding site is fairly restrictive, even with contact scoring, a conformer close to the bioactive one needs to exist for the database molecule to score well. With a more open site, small changes in ligand position may compensate for inaccuracies in a conformer.

The rescoring schemes presented represent initial attempts at a two-tiered scoring scheme using a fast filter (e.g., the contact score) as the first tier, and a more rigorous scoring function for the second tier. The results show that using a contact score with

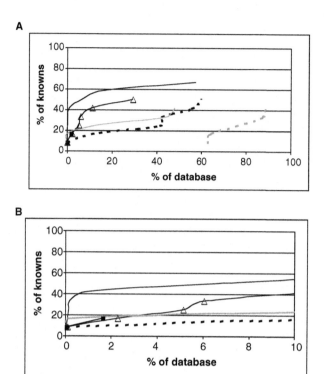

FIGURE 13.6 Enrichment of 12 DHFR knowns seeded into the 2000 decoys database. The plot of % knowns vs. % ranked database is shown in the curve with white triangles for contact scoring, the curve with black squares for energy scoring, dashed grey for rescore of top contact poses with energy, solid grey for rescore of top 20 contact poses with energy, dashed black for rescore of top contact poses with FLO minimization, solid black for rescore of top 20 contact poses with FLO minimization. An enlargement of the plot up to 10% of the ranked database is shown in (B).

the pharmacophoric matching is a reasonable starting point. Including solvation in the rescore is likely to yield even better results.

13.7 CONCLUSIONS

PhDock is fast and successful at placing molecules into the chemically relevant space of a target active site. Protein-derived pharmacophores, generated or augmented with the use of MCSS2SPTS, can be utilized for more efficient sampling. Furthermore, the efficiency of the database generation as well as the overall docking process has been significantly enhanced. With PhDock, as with all other docking methods, careful consideration should be given to protein preparation, site point generation, scoring function and run parameters, and postprocessing of the VS results in as automated a manner as possible. The use of a fast scoring filter, such as the

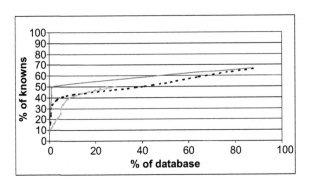

FIGURE 13.7 Enrichment of 12 DHFR knowns seeded into the 2000 decoys database. The plot of % knowns vs. % ranked database is shown in grey for contact scoring, in dashed grey for contact scoring including x-ray conformers for most of the knowns, in dashed black for FLO rescoring of the top 20 poses for the knowns and decoys, in solid black for FLO rescoring of the top 20 poses for the knowns and decoys including x-ray conformers for most of the knowns.

DOCK contact function, followed by a more rigorous scoring function to rank selected hits has been shown to enrich hit lists. A multitiered docking strategy that combines the speed of PhDOCK and the rigorousness of a PB scoring function may be capable of further enriching hit lists [41].

ACKNOWLEDGMENTS

We thank Thomas S. Rush, III, for his efforts on rescoring automation, providing some of the database statistics, as well as many helpful discussions. Also, Christian Kastrup and Timothy Hattori both helped develop the APOP script.

REFERENCES

[1]   O.F. Guner, History and evolution of the pharmacophore concept in computer-aided drug design, *Curr. Top. Med. Chem.* 2:1321–1332, 2002.

[2]   J.S. Mason, A.C. Good, and E.J. Martin, 3-D pharmacophores in drug discovery, *Curr. Pharm. Des.* 7:567–597, 2001.

[3]   J. van Drie, Strategies for the determination of pharmacophoric 3D database queries, *J. Computer-Aided Mol. Des.* 11:39–52, 1997.

[4]   Y.C. Martin, Computer design of bioactive compounds based on 3-D properties of ligands, *NIDA Res. Mono.* 134:84–102, 1993.

[5]   J.A. Grant, M.A. Gallardo, and B.T. Pickup, A fast method of molecular shape comparison: a simple application of a Gaussian description of molecular shape, *J. Comput. Chem.* 17:1653–1666, 1996.

[6]   C. McMartin and R.S. Bohacek, QXP: powerful, rapid computer algorithms for structure-based drug design, *J. Computer-Aided Mol. Des.* 411:333–344, 1997.

[7]   G. Jones, P. Willett, R.C. Glen, A.R. Leach, and R. Taylor, Development and validation of a genetic algorithm for flexible docking, *J. Mol. Biol.* 267(3):27–748, 1997.

[8]   D.K. Gehlhaar, G.M. Verkhivker, P.A. Rejto, C.J. Sherman, D.B. Fogel, L.J. Fogel, and S.T. Freer, Molecular recognition of the inhibitor AG-1343 by HIV-1 protease: conformationally flexible docking by evolutionary programming, *Chem. & Biol.* 2:317–324, 1995.

[9]   G.M. Morris, D.S. Goodsell, R.S. Halliday, R. Huey, W.E. Hart, R.K. Belew, and A.J. Olson, Automated docking using a lamarckian genetic algorithm and an empirical binding free energy function, *J. Comput. Chem.* 19:1639–1662, 1998.

[10]   G.M. Morris, D.S. Goodsell, R. Huey, and A.J. Olson, Distributed automated docking of flexible ligands to proteins: parallel applications of AutoDock 2.4, *J. Computer-Aided Mol. Des.* 10(4):293–304, 1996.

[11]   B. Kramer, M. Rarey, and T. Lengauer, CASP2 experiences with docking flexible ligands using FLEXX, *Proteins* Suppl. 1:221–225, 1997.

[12]   M. Rarey, S. Wefing, and T. Lengauer, Placement of medium-sized molecular fragments into active sites of proteins, *J. Computer-Aided Mol. Des.* 10(1):41–54, 1996.

[13]   S. Makino and I.D. Kuntz, Automated flexible ligand docking method and its application for database search, *J. Comput. Chem.* 18(14):1812–1825, 1997.

[14]   S.K. Kearsley, D.J. Underwood, and M.D. Miller, Flexibases: a way to enhance the use of molecular docking methods, *J. Computer-Aided Mol. Des.* 8:565, 1994.

[15]   D.M. Lorber and B.K. Shoichet, Flexible ligand docking using conformational ensembles, *Protein Sci.* 7(4):938–950, 1998.

[16]   D. Joseph-McCarthy, B.E. Thomas, IV, M. Belmarsh, D. Moustakas, and J.C. Alvarez, Pharmacophore-based molecular docking to account for ligand flexibility, *Proteins* 51:172–188, 2003.

[17]   A.I. Su, D.M. Lorber, G.S. Weston, W.A. Baase, B.W. Matthews, and B.K. Shoichet, Docking molecules by families to increase the diversity of hits in database screens: computational strategy and experimental evaluation, *Proteins* 42:279–293, 2001.

[18]   R.M.A. Knegtel, D.M. Bayada, R.A. Engh, W. von der Saal, V.J. van Geerestein, and P.D.J. Grootenhuis, Comparison of two implementations of the incremental construction algorithm in flexible docking of thrombin inhibitors, *J. Computer-Aided Mol. Des.* 13:167–183, 1999.

[19]   V. Schnecke and L.A. Kuhn, Virtual screening with solvation and ligand-induced complementarity, *Perspect. Drug Discov. Des.* 20:171–190, 2000.

[20]   I.D. Kuntz, J.M. Blaney, S.J. Oarley, R. Langridge, and T.E. Ferrin, A geometric approach to macromolecule-ligand interactions, *J. Mol. Biol.* 161:269–288, 1982.

[21]   T. Ewing, S. Makino, A. Skillman, and I. Kuntz, DOCK 4.0: Search strategies for automated molecular docking of flexible molecule databases, *J. Computer-Aided Mol. Des.* 15:411–428, 2001.

[22]   D. Joseph-McCarthy and J.C. Alvarez, Automated generation of MCSS-derived pharmacophoric DOCK site points for searching multiconformation databases, *Proteins* 51:189–202, 2003.

[23]   A. Miranker and M. Karplus, Functionality maps of binding sites: a multiple copy simultaneous search method, *Proteins* 11:29–34, 1991.

[24]   E. Evensen, D. Joseph-McCarthy, and M. Karplus, *MCSSv2,* Cambridge, MA: Harvard University Press, 1997.

[25]   D. Joseph-McCarthy, Structure-based combinatorial library design and screening: application of the multiple copy simultaneous search method. In A.K. Ghose and V.N. Viswanadhan, Eds., *Combinatorial Library Design and Evaluation for Drug Discovery: Principles, Methods, Software Tools and Applications,* New York: Marcel Dekker, 2000.

[26] C.A. Lipinski, F. Lombardo, B.W. Dominy, and P.J. Feeney, Experimental and computational approaches to estimate solubility and permeability in drug discovery and development settings, *Adv. Drug. Delivery Rev.* 23:3–25, 1997.

[27] T.I. Oprea, Property distribution of drug-related chemical databases, *J. Computer-Aided Mol. Des.* 14:251–264, 2000.

[28] T.I. Oprea, A.M. Davis, S.J. Teague, and P.D. Leeson, Is there a difference between leads and drugs? A historical perspective, *J. Chem. Info. Comput. Sci.* 41:1308–1315, 2001.

[29] D.F. Veber, S.R. Johnson, H.-Y. Cheng, B.R. Smith, K.W. Ward, and K.D. Kopple, Molecular properties that influence the oral bioavailability of drug candidates, *J. Med. Chem.* 45:2615–2623, 2002.

[30] OpenEye Scientific Software, 2001: www.eyesopen.com.

[31] S.L. Mayo, B.D. Olafson, and W.A. Goddard, III, DREIDING: a generic force field for molecular simulations, *J. Phys. Chem.* 94:8897, 1990.

[32] J. Bostrom, Reproducing the conformations of protein-bound ligands: a critical evaluation of several popular conformational searching tools, *J. Computer-Aided Mol. Des.* 15(12):1137–1152, 2002.

[33] J. Bostrom, J.R. Greenwood, and J. Gottfries, Assessing the performance of OMEGA with respect to retrieving bioactive conformations, *J. Mol. Graphics & Modelling* 21(5):449–462, 2003.

[34] MDL Information Systems Inc. 2000: www.mdl.com.

[35] T.A. Halgren, MMFF VI. MMFF94s option for energy minimization studies, *J. Comput. Chem.* 20(7):720–729, 1999.

[36] Chemical Computing Group Inc.: www.chemcomp.com.

[37] A. Yamaguchi, K. Iida, N. Matsuri, S. Tomoda, K. Yura, and M. Go, HET-PDB Navi.: a database for protein-small molecule interactions, *J. Biochem.* 135:79–84, 2004: http://daisy.nagahama-i-bio.ac.jp/golab/hetpdbnavi.html.

[38] J.M. Word, S.C. Lovell, J.S. Richardson, and D.C. Richardson, Asparagine and glutamine: using hydrogen atom contacts in the choice of sidechain amide orientation, *J. Mol. Biol.* 285:1733–1745, 1999.

[39] P. Goodford, A computational procedure for determining energetically favorable binding sites on biologically important macromolecules, *J. Med. Chem.* 28:849–857, 1985.

[40] W.D. Cornell, P. Cieplak, C.I. Bayly, I.R. Gould, K.M. Merz, D.M. Ferguson, D.C. Spellmeyer, T. Fox, J.W. Caldwell, and P.A. Kollman, A 2nd generation force-field for the simulation of proteins, nucleic-acids, and organic-molecules, *J. Am. Chem. Soc.* 117(19):5179–5197, 1995.

[41] T. Rush, III, E.S. Manas, G.J. Tawa, and J.C. Alvarez, Solvation-based scoring for high throughput docking. In J.C. Alvarez and B.K. Shoichet, Eds., *Virtual Screening in Drug Discovery*, New York: Marcel Dekker, 2005.

[42] J.T. Bolin, D.J. Filman, D.A. Matthews, R.C. Hamlin, and J. Kraut, Crystal structures of *Escherichia coli* and *Lactobacillus casei* dihydrofolate reductase refined at 1.7 A resolution: I. General features and binding of methotrexate, *J. Biol. Chem.* 257:13650, 1982.

[43] J. Basran, M.G. Casarotto, I.L. Barsukov, and G.C.K. Roberts, Role of the active site carboxylate in dihydrofolate reductase: kinetic and spectroscopic studies of the Aspartate 25 -> Asparagine mutant of the *Lactobacillus casei* enzyme, *Biochem.* 34:2872–2882, 1995.

[44] B. Birdsall, A.S. Burgen, E.L. Hyde, G.C. Roberts, and J. Feeney, Negative cooperativity between folinic acid and coenzyme in their binding to *Lactobacillus casei* dihydrofolate reductase, *Biochem.* 20:7186–7195, 1981.

[45]   W.D. Morgan, B. Birdsall, V.I. Polshakov, D. Sali, I. Kompis, and J. Feeney, Solution structure of a Brodimoprim analogue in its complex with *Lactobacillus casei* dihydrofolate reductase, *Biochem.* 34:11690–11702, 1995.

[46]   G. Martorell, M.J. Gradwell, B. Birdsall, C.J. Bauer, T.A. Frenkiel, H.T. Cheung, V.I. Polshakov, L. Kuyper, and J. Feeney, Solution structure of bound trimethoprim in its complex with *Lactobacillus casei* dihydrofolate reductase, *Biochem.* 33:12416–12426, 1994.

[47]   R.L. Then and F. Hermann, Properties of Brodimoprim as an inhibitor of DHFR, *Chemotherapy* 30:18–25, 1984.

[48]   V.M. Reyes, M.R. Sawaya, K.A. Brown, and J. Kraut, Isomorphous crystal strutures of *Escherichia coli* dihydrofolate reductase complexed with folate, 5-deazafolate, and 5,10-dideazatetrahydrofolate: mechanistic implications, *Biochem.* 34:2710–2723, 1995.

[49]   J.L.J. Woolley, J.L. Ringstad, and C.W. Sigel, Competitive protein binding assay for piritrexim, *J. Pharm. Sci.* 78:749–752, 1989.

[50]   A. Gangjee, A.P. Vidwans, S.F. Queener, and R.L. Kisliuk, 2,4-diamino-5-deaza-6-substituted pyrido[2,3-d]pyrimidine antifolates as potent and selective non-classical inhibitors of dihydrofolate reductases, *J. Med. Chem.* 39:1438–1446, 1996.

14 Fragment-Based High Throughput Docking

Peter Kolb, Marco Cecchini, Danzhi Huang, and Amedeo Caflisch

14.1 INTRODUCTION

Structural genomics programs around the world are delivering an abundance of three-dimensional (3-D) structures of proteins, some of which are pharmacologically highly relevant. Hence, computer programs for automatic docking of libraries of compounds are being developed further and applied to design drugs against a plethora of diseases including AIDS, Alzheimer's disease, cancer, malaria, and sleeping sickness. In this chapter, we first review the most common approaches for structure-based flexible ligand docking. Some technical improvements for more efficient sampling and more appropriate scoring functions are then presented. Finally, a number of practical suggestions are given for high throughput docking (HTD) with special emphasis on our fragment-based approach.

14.2 OVERVIEW

The basic strategy of any docking approach is to generate a conformation of a putative ligand, which is then placed (or *docked*) in the binding site of a protein target (also referred to as *receptor*). The result of these two operations is usually called a *pose*. A *score* has to be assigned to each pose, thus producing a *ranking*, with the correct pose (i.e., the natural binding mode) at the first rank or at least as close as possible to it.

14.2.1 DEFINING THE BINDING SITE

Prior to any attempt of docking, the approximate location of the binding site needs to be defined. It is easiest for the case in which the crystal structures of the receptor in complex with some ligands are already known. Usually, the binding site is then defined as the residues lying within a certain cutoff from the ligands.

A greater challenge is presented when only the 3-D structure of the protein is known. In that case, profound knowledge of the function of the protein is necessary. There are programs that analyze the protein surface and provide quantitative information on it, among them GRASP (Graphical Representation and Analysis of Structural Properties) [1] and HYDROMAP [2], which calculate the electrostatic potential and hydrophobicity map, respectively. Alternatively, some programs use so-called "flood filling" algorithms that attempt to identify cavities on the protein surface. Basically, they fill the space that is not occupied by the protein with points and then roll a large "eraser" over the surface of the protein. All remaining points are said to be in protein pockets [3].

In general, the residues in the binding site are important because their interaction with the ligand is stronger and usually treated in more detail. The binding site residues are explicitly used during the computation of the score and they are sometimes also considered as entities providing anchor points for the positioning of a conformation. Therefore, they should be chosen according to the type and function of the receptor, as well as the program's strategy to determine ligand poses.

Recently, the program AutoDock [4,5] was tested on "blind" docking, that is without defining any selected portion of the protein as binding site [6]. Docking was successful for ligands with less than 10 rotatable bonds, but only at high computational cost (in the order of days). Hence, the definition of the binding site is necessary for virtual screening (VS) of large databases.

Another aspect is the selection of an appropriate protein (and thus binding site) conformation. McGovern and Shoichet have performed a comparative study [7], using the x-ray structures of the complexed and uncomplexed protein as well as conformations obtained by homology modeling of 10 different proteins. The highest enrichment of known ligands in a database was in most cases achieved with the complexed structure. Using a conformation from a complex introduces a bias toward known inhibitors, however, and should thus be complemented by other protein structures in a screening project.

14.2.2 GENERATING A POSE

Two main types of approaches to obtain a ligand pose have to be distinguished: the ones that use only the complete structure of the ligand and those that follow an incremental strategy. Section 14.2.2.1 and Section 14.2.2.2 refer to the first type; the incremental methods are described in the Section 14.2.2.3.

14.2.2.1 Generation of Ligand Conformations

Typically, docking programs modify only the torsional degrees of freedom of rotatable bonds to produce different ligand conformations. It is important to at least modify the torsional angles of groups carrying hydrogen bond donors (HDO) to allow optimization of this type of interaction. Torsional angles of bonds in rings, double or triple bonds, or single bonds to symmetrical groups (like methyl) are normally kept fixed. In one study with the focus on protein flexibility, the backbone of peptidic inhibitors was considered as being rigid and only "sidechain" flexibility was allowed [5]. A rigorous test of a docking program should consider full flexibility, however [8,9]. An important exception is the docking of small fragments (like benzene or benzamidine), for whom the rigid body approximation is an appropriate description of their limited flexibility [10,11]. Some programs do not allow ligand flexibility, but the success rates in these cases are low if one does not use the conformation found in the crystal structure [12]. Clearly, such methods can hardly be used to predict the binding modes of "new" ligands. The program DOCK [13,14] also started as a rigid-body docking tool, but ligand flexibility was introduced in DOCK 4.0, using an exhaustive search and conformational refinement with the simplex method [15].

There are two common approaches for generating different ligand conformations:

1. In procedures that search the conformational space of the ligand outside of the binding site, a pool of relevant conformations with low internal energy is generated, and they are subsequently docked rigidly. The sampling of the ligand conformational space can be done exhaustively, modifying each torsional angle in discrete steps [16,17]. Alternatively, the procedure can employ rotamer libraries which assign the most probable values to torsions depending on the atom types [9,15,18].
2. The conformations can be subject to an optimization algorithm, where the torsional angles correspond to the variables of the optimizer. One can further distinguish between two optimizer types: Monte Carlo (MC) searches [3] (also used for *de novo* design by DeWitte et al. [19,20]) and genetic algorithms and other evolutionary approaches [4,8,21–23]. MC approaches use a single conformation that is randomly perturbed and improved. Genetic algorithms (GAs) employ a multitude of information-containing chromosomes (usually referred to as the *population*), which interact with each other and evolve to better solutions. These algorithms are more promising for docking [4], because the energy surfaces to be searched are rugged. MC methods tend to be rather slow, which is a disadvantage for large-scale library screening. Furthermore, if one uses MC–simulated annealing approaches, the additional problem of choosing an appropriate initial temperature and a cooling schedule arises.

14.2.2.2 Defining Ligand Positions

There are several strategies to position and orient the ligand in the binding site:

- The translational degrees of freedom can be encoded in an optimizer.
- The position can be determined by matching the shape of the ligand to the binding site.
- The conformation can be superimposed on a set of points that contain information about the binding site (for references see below).

As an example of approaches that follow the first strategy, the chromosomes in a GA can additionally carry genes for the translational degrees of freedom of the ligand and three (in the case of Euler angles) or four (when quarternions are used [4,24]) variables specifying the ligand orientation.

In approaches that follow the second strategy, the surface of the binding site is compared to the solvent accessible surface of the current ligand conformation. An optimal position is found based on some measure of similarity between those two. LigandFit [3] uses an algorithm developed by Oldfield [25,26], which treats both the binding site and the ligand as a collection of grid points. The shape of such a collection is characterized by a matrix. From the eigenvalues of these matrices, the shape discrepancy can be computed and used to assign a score to each conformation. FRED [17] employs a bump map, which is a Boolean grid representing the receptor, with true values where ligand atoms can potentially be placed. After this initial filtering step, several other scoring functions can be applied, among them Gaussian shape fitting. This function has favorable values when the ligand and the protein have high surface contact and little volume overlap.

DOCK [14,15] follows the third strategy by first filling the binding site with spheres of different sizes. The centers of these spheres are considered as anchors for atoms of the ligand. Variations of this approach at different levels of sophistication include the use of HDOs and HACs (hydrogen bond acceptors) as well as hydrophobic surface points as anchors [27,28]. An example of this is SEED, which was developed to dock small molecules with solvation [10,11]. It uses anchors on the surface of the receptor and performs an exhaustive search on a discrete space by matching donor and acceptor vectors (or vectors of hydrophobic interaction centers) and rotating the ligand around these axes. Other programs use information from the placement of predefined small molecular fragments to match their positions to similar entities in the ligand [16]. The Fragment-based Flexible Ligand Docking (FFLD) program utilizes the results from the docking of small and mainly rigid molecules that have been specifically chosen to match chemical moieties actually present in the ligand [8]. The underlying assumption for all these methods is that the interaction between a protein and a ligand is dominated by some key groups of the ligand. Hence, if the positions of these groups are determined correctly, the rest of the ligand will almost inevitably assume the correct pose.

14.2.2.3 Incremental Methods

Programs like FlexE [9] (an advanced version of FlexX [18]), SLIDE [28], or DOCK 4.0 [15] also try to optimize the interactions of the key groups, but do this individually for each group. The ligand is first split into several units (fragments), the first of which is placed as a seed. Usually, the determination of the pose of the first fragment is done with high accuracy. Sequentially, all the other fragments are connected in their due order, whereby each position is optimized, often exhaustively. At every step, the highest ranking solutions are retained and the next fragment is connected to each of them. It is important to carefully select only a small number of candidate solutions at every step (pruning) to control the exponential increase of possible solutions.

14.2.3 RANKING THE POSES

At the beginning of this chapter, we distinguished between exhaustive searches and optimization techniques. The latter minimize an objective function that is usually computationally not too expensive, because it has to be called quite frequently, and a force-field-based binding energy is evaluated for the final ranking. Exhaustive searches use only one energy function.

14.2.3.1 Objective Function

The objective function approximates the interaction energy between ligand and receptor and the internal strains of the ligand and the protein, if the latter is also flexible. Typical components are the intermolecular van der Waals (vdW) and Coulombic energy, and sometimes a term for hydrogen bonds. The internal strain is usually estimated by the intraligand vdW energy and sometimes the dihedral energy. Most objective functions do not take into account terms for bond, angle, and torsional strains. It has been proposed to increase the chances of the optimizer by smoothing the energy landscape. Whitfield et al. [29] introduced a gravitational force that dominates all other forces in the initial steps of the search and then decreases over time. It is assumed that the position of the global optimum does not change due to the smoothing and that only the well depth is modified. Hansmann and Wille [30] developed energy landscape paving, which penalizes scores that are found repetitively. Searches can thus escape local minima and go into regions of different energy.

Most of the docking programs that use physics-based functions (like DOCK [13–15], AutoDock [4,5], and FFLD [8]) employ a grid-based approach for efficiency reasons. These grids contain the Coulombic potential and vdW potential of the protein and avoid the need for recalculating the full energy for every pose during a database screen. Trilinear interpolation [31] is often used to compute the interaction energies from the grid values of the potential.

Empirical-based functions (such as the one used in FlexX [18] and FlexE [9]) use additive approximations to estimate the binding free energy. They contain several terms corresponding to hydrogen bonding, hydrophobic interactions, entropic changes, and sometimes, interactions with metal ions. The coefficients of each term in the sum are obtained from a fit to known experimental binding energies for various protein–ligand complexes [32,33].

14.2.3.2 Binding Energy Function and Postprocessing

After a docking run, the best poses of the ligand can be reranked using a more accurate force field [34,35]. This often contains the same terms as the objective function, but takes longer ranging interactions and ligand and receptor desolvation into account. Sometimes, the ligand pose is also minimized within the receptor using a molecular mechanics force field [36,37]. In our group, ligand poses are normally minimized with CHARMm [36] using the CHARMm22 force field (Accelrys, Inc.), and often also with the TAFF-force-field (Tripos). Additionally, the score and rank of each pose can be redetermined using more accurate energy functions that include electrostatic solvation like the one in SEED [10,11] or knowledge-based interaction fields like SuperStar [38], potential of mean force (PMF) [39], Small Molecule Growth (SMoG) [40], and DrugScore [41]. The energy rankings produced by the different scoring functions are usually compared, as a number of studies suggest that consensus scoring improves the chance of finding a true hit [42,43].

14.2.3.3 Solvation

The effects of solvation play a key role in molecular recognition events. To calculate the electrostatic contribution to solvation in the continuum dielectric approximation, one could solve the finite-difference Poisson–Boltzmann (PB) equation [44–47] for every new position of the ligand molecule. Considering the current computer power, this would be forbiddingly expensive, especially for HTS. Therefore, only a few docking programs take into account electrostatic solvation effects. The continuum dielectric approximation and the generalized Born (GB) approach [48,49] are used in SEED [10,11], Program to Engineer Peptides (PEP) [50,51], and DOCK [52]. Fairly recently, Arora and Bashford have presented a modified GB approach that estimates desolvation by an integral over the occluded volume [53].

Some docking programs treat solvation effects just with respect to the presence or absence of conserved water molecules that form interactions that are either essential for the protein conformation or necessary to mediate interactions between ligand and protein. Clearly, this approximation completely neglects the bulk properties of water (e.g., dielectric screening). Österberg et al. use grids that have been derived by averaging over several crystal structures, some of which can contain water molecules [5]. Although the method has mainly been developed to incorporate protein flexibility, heterogeneities in the presence of water molecules can be taken into account as well. Schnecke et al. consider water explicitly and have a term penalizing the replacement of water molecules by a hydrophobic group of the ligand [28]. Finally, Rarey et al. have described a method to precompute positions of water molecules and place them if they can form hydrogen bonds with the (partial) ligand during the incremental construction in FlexX [54].

14.2.4 PROTEIN FLEXIBILITY

In principle, it would be ideal to allow full flexibility for the protein to model large displacements upon ligand binding. Such studies have already been undertaken [55], but because the computational time was in the order of days for a single ligand, this can clearly not be applied to the screening of large libraries of compounds. As a consequence, flexibility of the protein, if any, is mostly limited to the binding site and its vicinity. Three different approaches shall be highlighted here.

AutoDock [5] incorporates both protein mobility and structural water heterogeneity. It first generates the energy grids for a number of different protein structures. The program then offers several ways to combine these grids into a single grid. It either computes simple point-by-point averages or weights the different grid points according to their energies and physico-chemical characteristics. This mean grid approach has the advantage that one can still dock to one rigid structure, which facilitates the analysis of the results compared to docking to several distinct conformations. On the other hand, it can only be used to approximate minor displacements. Moreover, the mean grid structure is the product of an averaging scheme and thus might not be observable in reality. Another drawback is the fact that no protein structure is present, but only its representation as a grid. One could thus not follow a multiple step approach (See Section 14.3.4) and do minimization with CHARMm [36], for example.

FlexE [9] is based on a so-called united protein description [56], which is derived from superimposing the backbones of an ensemble of different crystal structures. Variations of the structure in the binding site region are either maintained as distinct possibilities or are combined to one structure in case they are similar. During the incremental construction algorithm, the ligand is placed fragment by fragment into the active site of the united protein description. After each construction step, all possible interactions between the (partially) placed ligand and all instances of the united protein description are determined. The score is then assigned for the (partial) ligand in the best instance.

SLIDE [28] goes one step further and first docks a rigid scaffold into a rigid binding site. Gradually, the other parts of the ligand are attached to the scaffold. Clashes between the ligand and the protein are resolved by allowing rotations of bonds (both in the ligand and the protein) that have been defined as flexible beforehand. The bonds that should be rotated are determined with mean-field theory, which is capable of finding the minimum amount of rotations necessary to resolve all clashes [57–59]. Although flexibility is limited to the binding site residues, this approach comes close to an induced fit.

One of the most thorough approaches besides [55] has been undertaken by Lin et al. [60]. For their relaxed complex method, first long molecular dynamics (MD) simulations of 2 ns were conducted, with snapshots taken every 10 picoseconds (ps). Two candidate compounds were then docked to the ensemble of MD conformations. This technique recognizes the fact that ligands may bind tightly to conformations that appear only infrequently in the dynamics of a protein. However, every molecule has to be docked to a large number of different protein structure which strongly limits the size of the library.

14.3 TECHNICAL IMPROVEMENTS

14.3.1 CURRENT LIMITATIONS

As mentioned above, docking approaches can be described as a combination of two components—the search strategy and the scoring function. Because in most cases the objective function (See Section 14.2.3.1) is also used as the binding energy function (See Section 14.2.3.2), in the following, the term *scoring function* will be employed. The critical element of the search procedure is the amount of time required to effectively sample the relevant conformational space. The scoring function has to be fast enough to allow its application to a large number of potential solutions and, in principle, be able to effectively distinguish the experimentally observed binding mode from all others explored in the search. Consequently, the scoring function should include and appropriately weight just the energetic contributions that are relevant in the binding process. Nevertheless, an accurate scoring function will generally be computationally expensive and so the function's complexity is often reduced at the expense of a loss in accuracy.

The proper combination of an effective search algorithm and an adequate scoring function, whose global minimum corresponds to the biologically relevant complex, will solve the docking problem in a reasonable amount of time. However, because the approaches published up to date can fail, especially in cross-docking, this ideal combination has obviously not been found yet. Therefore, improvements in the efficiency of the search strategy and the accuracy of the scoring function are required as they will increase the reliability of the docking predictions and reduce the computational requirements, which is important for screening large libraries.

Docking predictions are still prone to fail and often the proposed binding modes do not reproduce the crystal structure of the protein–ligand complex [6,9,35,61]. In case of failure, the predicted binding mode can have a worse or a better score than the x-ray structure of the ligand. In the first case, the search strategy adopted in the docking approach could have been not effective enough. The search algorithm was thus not able to generate a pose sufficiently close to the experimental binding mode. In the second case, the failure might arise from an inadequate scoring function that allows more favorable binding modes than the one in the crystal structure. In the first case, one should focus on the improvement of the search procedure; in the second case, one should concentrate on the optimization of the scoring function.

Unfortunately the situation is much more complicated because the components of a docking protocol are not separate entities and as such they should be improved together. In the first scenario, for example, the scoring function could have played an important role because the resulting energy landscape was not smooth enough to allow the search to proceed efficiently while avoiding premature convergence. Although the scoring function described the protein–ligand interactions well, it was not suitable for the applied search strategy. In the second scenario, it could have happened that the experimentally determined structure was not close to a minimum of the scoring function. In this case, any energy comparison is much less meaningful. Although a proper combination of an efficient search algorithm and an accurate scoring function are the keys for a successful docking protocol, it is certainly not clear what "proper," "efficient," and "accurate" mean. In Section 14.3.2. and Section 14.3.3, we describe some important requirements for both the search strategy and the scoring function and how they are embedded in our docking approach.

14.3.2 SEARCH STRATEGY

Docking procedures belong to the category of global optimization techniques where the aim is finding the global minimum of the scoring function. A rigorous search algorithm would exhaustively investigate all possible binding modes between the ligand and the receptor. The degrees of translational and rotational freedom of the ligand would be explored along with the internal conformational degrees of freedom of both the ligand and the receptor. However, this is impractical because of the size of the search space, even when considering a rigid protein. Only a small amount of the total conformational space can be sampled and a balance must be reached between the computational expense and the amount of search space examined. A wide range of global optimization algorithms are currently available, but not all of them are suitable for docking. Most optimization algorithms for docking fall into one of three classes—gradient-based algorithms, combinatorial algorithms, and stochastic algorithms [62].

The strength of gradient-based methods is that they efficiently find a local minimum close to the initial conformation. Because gradient-based methods do not allow the system to escape from local minima they have to be combined with other search strategies, such as cycles of MC perturbations and gradient minimizations [63]. Moreover, most scoring functions do not have an analytical gradient.

Combinatorial algorithms have the potential advantage of being extremely fast and effective. The most successful combinatorial algorithms used for molecular docking [10,11,18,64,65] have set themselves apart in their ability to dock libraries of small molecules in a reasonable amount of time. Unfortunately, increasing the number of conformational degrees of freedom leads to an explosion of the dimension of the search space. To be able to sample such large spaces, the computational expense is usually controlled by a discretization of the space, which can restrict the effectiveness of the algorithm.

Stochastic algorithms have the advantage that, irrespective of the dimensionality of the problem and given enough time, they get arbitrarily close to the global minimum. On the other hand, they have the disadvantage that they require a large amount of central processing unit (CPU) time to achieve an acceptable degree of reliability [4,62]. Although computationally expensive, stochastic optimization algorithms seem to be the most suitable for flexible docking. In fact, the dimensionality of the search space and the ruggedness of the binding energy landscape make both gradient-based and combinatorial methods less effective. GAs are stochastic optimization methods that mimic the process of natural evolution by manipulating a population of data structures called chromosomes [66,67]. Although requiring rather large amounts of CPU time, GAs have been shown to effectively explore rough energy surfaces and to be suitable as search strategies for docking [4,8,21–23,68,69]. A GA was chosen as the search strategy for the original version of FFLD [8], the docking protocol developed in our group. During the FFLD evolution, a loop over generations is performed until the maximum number of steps is reached. Starting from an initial random population of chromosomes containing the dihedral angles of the ligand as genes, the GA repeatedly applies two mutually exclusive evolutionary operators—one-point crossover and mutation. This yields new chromosomes (children) that replace appropriate members (parents) of the population. These non-linear genetic operators help to overcome the barriers of the binding energy landscape and the search can proceed efficiently. Throughout the simulation, a constant evolutionary pressure is kept by selecting parent chromosomes with a bias toward the fittest. This pressure moves the population toward conformations related to the global minimum and increases the fitness of the individuals. The selection of the members of the population that should be replaced by new chromosomes is a crucial step. To avoid premature convergence, it is important to keep structural diversity. In the search strategy used in FFLD [8], both the energy difference and the conformational similarity are taken into account to determine if a given member of the population should be replaced by a new chromosome. At the end of each GA step, every new chromosome is compared with the old population by the following procedure: if a similar chromosome is found in the old population, it is replaced by the new chromosome only if the energy of the new one is more favorable; otherwise, the new chromosome is discarded. The similarity test significantly improves the efficiency of the search strategy and avoids premature convergence [50].

Following a comparative study of several search engines in AutoDock [4], a hybrid search procedure was introduced in the latest version of FFLD [35]. The hybrid search combines a global optimization procedure based on a GA with a local minimization algorithm to improve exploration of regions within energy basins. Local optimization has been shown to dramatically improve the success rate of the GA search without any loss in efficiency [4]. For the best 10% of the new individuals, a local optimization is performed to improve the ligand fitness before performing the similarity test. To evaluate the performance of the hybrid search procedure implemented in FFLD, it was compared with the GA of the original version [8]. The simulations showed that the hybrid search is more efficient than the canonical GA as it always reached a conformation with lower energy. The results of two docking experiments carried out with both search methods are presented in Figure 14.1. The first experiment, in which a ligand with 10 rotatable bonds was docked in human-immunodeficiency virus type 1 (HIV-1) protease (Figure 14.1, top), shows that the hybrid search procedure is more efficient than the genetic algorithm especially at the beginning of the simulation where the energy gap is large. At about 60% of the evolution the gap decreases and the performance of the two methods is comparable. Docking a ligand with 21 rotatable bonds in HIV-1 protease (Figure 14.1, bottom) shows that the hybrid search procedure performs better during the entire simulation and the energy gap increases until the end. Moreover, the standard deviation of the hybrid search evolutions (shown as error bars in Figure 14.1, bottom) is larger, indicating that it is less prone to converge prematurely. This comparison shows that the local search improves the quality of the docking predictions in case the conformational space of the ligand is large. This is mainly due to the fact that the random perturbations of binary strings performed by the GA during the evolution correspond to radical jumps in the energy landscape and may be too large. On the contrary, the local optimizer is able to refine the large perturbations due to crossover

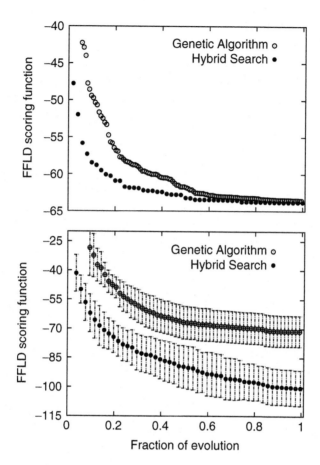

FIGURE 14.1 Evolution of the best individual of the population averaged over 10 docking runs for two different experiments. Empty and filled bullets indicate evolutions performed by GA and hybrid search procedure, respectively. Docking of HIV-1 protease ligands with 10 and 21 rotatable bonds are shown from top to bottom, respectively. In the bottom plot, the vertical bars show the standard deviation computed over 10 docking runs.

and mutations and leads to a better investigation of the energy landscape. The results of this docking study [35] suggests, in agreement with previous studies [4], that hybrid search methods should be preferred to canonical GAs.

The similarity test and the hybrid search procedure are just examples of possible means one can adopt in a protocol to increase the efficiency and accuracy of the search algorithm. However, the study clearly indicates that there is still room for improvement and that novel concepts can be effective. It is worth stressing again that the search algorithm is only half of the docking problem; the other factor to be incorporated into a successful protocol is the scoring function. In Section 14.3.3, the requirements for a scoring function that are suitable for docking are discussed.

14.3.3 SCORING FUNCTION

Underlying any docking approach is a model of ligand–protein interactions describing molecular recognition. In principle, a complete thermodynamic description of this process involves contributions from several balancing factors, including solvent reorganization, conformational entropy, and vdW and electrostatic interaction energies. For biomolecular systems, it is difficult to evaluate these terms with sufficient accuracy to permit quantitative predictions. Moreover, the complete energy function necessary for prediction of accurate binding affinities may not be suitable for docking simulations. The scoring function used in docking simulations should be a simple model of ligand–protein interactions rather than an estimation of the free energy of binding. It must be simple enough to permit a rapid evaluation and, more importantly, the resulting energy landscape must be smooth enough to allow the search to proceed efficiently without getting trapped in local minima. Nevertheless, a scoring function that is suitable for docking needs to be accurate, because it must be able to distinguish the experimental binding mode from all the other modes explored by the search algorithm.

With respect to this point, Verkhivker et al. [69] suggested that such an energy function should fulfill both a thermodynamic and a kinetic requirement. In other words, the energy related to the crystallographic structure of the ligand in the complex must be the global minimum of the binding energy landscape (*thermodynamic requirement*), but at the same time this conformation must be accessible during the search (*kinetic requirement*). The complexity of a complete and accurate force field that describes the binding process precisely, although it would fulfill the thermodynamic requirement, typically results in a rugged energy landscape and thus does not meet the kinetic criterion of the docking problem. The multitude of energetically similar but structurally different local minima inevitably leads to kinetic bottlenecks that dramatically reduce the frequency of successful structure predictions. This is the case for standard molecular mechanics force fields [36,37], because they have not been designed to reduce the ruggedness of the energy landscape. One of the critical factors that determines the success rate in predicting the structure of ligand–protein complexes is the roughness of the binding energy landscape [68,69]. Consequently, the applicability of standard force fields in docking is limited and simpler molecular recognition models that fulfill both the thermodynamic and kinetic requirements are to be designed and developed.

A fundamental component of models for molecular recognition is the steric energy function, which is based on surface complementarity. However, this term alone is not sufficient to distinguish effectively between alternative binding modes. Electrostatic interactions may provide additional specificity to discriminate between true and false solutions and they should be embedded in the scoring function. Finally, an intraligand energy term is also required; it largely reduces the conformational space to be investigated by preventing strained dihedrals and steric clashes among atoms of the ligand. Hence, the three key elements of a scoring function necessary for robust structural assessment during docking are:

1. Ligand–protein steric interactions
2. A simple description of ligand–protein electrostatics
3. An intraligand strain

In the FFLD docking approach developed in our group [8], the scoring function is

$$\Delta E_{total} = E_{dihedral}^{ligand} + E_{vdW}^{ligand} + E_{vdW}^{inter} + E_{polar}^{inter} \tag{14.1}$$

The dihedral energy of the ligand $\left(E_{dihedral}^{ligand}\right)$ has recently been implemented in FFLD (D. Huang, unpublished results) using the lowest order terms of a cosine expansion for each torsion. The second $\left(E_{vdW}^{ligand}\right)$ and the third $\left(E_{vdW}^{inter}\right)$ terms of Equation 14.1 are intraligand and ligand–receptor vdW energies, respectively. Both terms are described as the sum of an attractive dispersion and a steep repulsion term by the 6-12 Lennard-Jones potential. The last term in Equation 14.1 is the protein–ligand polar interaction energy $\left(E_{polar}^{inter}\right)$. The intermolecular polar term approximates electrostatic interactions and includes hydrogen bonds (HB) and unfavorable polar contacts (UP), namely two HAC (or HDO) atoms close to each other. Hence

$$E_{polar}^{inter} = \sum_{i=1}^{N_{HB}} E_i^{HB} + \sum_{i=1}^{N_{UP}} E_i^{UP} \tag{14.2}$$

where N_{HB} and N_{UP} are the number of hydrogen bonds and the number of unfavorable polar contacts, respectively. The energies E_{HB} and E_{UP} are approximated by constant values [35]. Distance- and angle-dependent criteria are considered for the definition of a hydrogen bond, but only a distance dependence is applied for unfavorable polar contacts. Originally, the distance dependence of both terms in Equation 14.2 and the directionality of the hydrogen bonds follow simple step functions (Figure 14.2, top left and top right, dashed lines) that are efficiently evaluated [8]. The steep repulsive part of the Lennard-Jones potential directly affects the height of the energy barriers and generates a rough energy surface. To reduce the steepness of this energy component, an intermolecular soft-core vdW term was implemented [8]. Following previous studies by Gehlhaar et al. [68], the repulsive part of the Lennard-Jones potential was linearized in FFLD, such that the functional form has a finite value when the interatomic distance approaches zero (Figure 14.2, bottom). The intermolecular soft-core vdW does not penalize binding modes with small atomic interpenetrations of the ligand with the protein and permits the formation of unphysical states that could open multiple pathways leading to the crystal structure. These states, otherwise forbidden by the presence of realistic energy barriers in standard force fields, may provide kinetically accessible routes to the global minimum.

In a recent study [35], a significant improvement with respect to the original version of our docking approach [8] has been observed by replacing the step

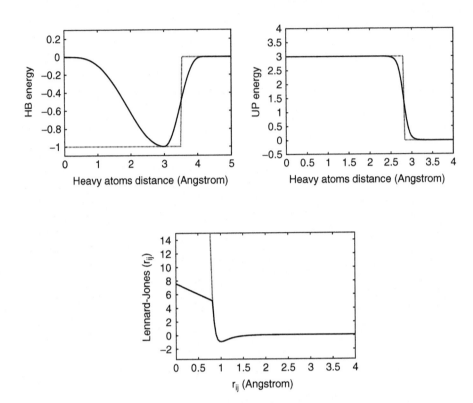

FIGURE 14.2 The distance dependence of hydrogen bonds (HB), unfavorable polar contacts (UP) and ligand-receptor vdW interactions is presented from left to right, respectively. The smooth functions (solid lines) [35] used for replacing the original stepwise functions (dashed lines) in the intermolecular polar interaction term are shown. On the bottom, the intermolecular soft-core vdW (solid line) [8] is compared with the 6-12 Lennard-Jones potential (dashed line). Values are in kcal/mol.

functions in the ligand–receptor polar interaction term $\left(E_{polar}^{inter} \right)$ with *smooth* functions. Smooth functions allow the optimization of the hydrogen bonding pattern avoiding discontinuities on the energy landscape. The continuous gradient can guide the search algorithm toward lower energy conformations at every point. In the latest version of the FFLD docking program, a sigmoidal function was used to describe the unfavorable polar contacts and bathtub-shaped functions were used for the distance dependence and the directionality of the hydrogen bonds (Figure 14.2, top left and top right, solid lines). Furthermore, it was observed that the distance- and angle-dependence in the polar term significantly reduced the noise arising from the energy degeneration of structurally different ligand conformations and improved the convergence of the docking runs [35].

Previous works by Gehlhaar and Verkhivker [68,69] suggested that a dynamical modification of the scoring function is helpful. In their docking experiments, an adaptive scoring function based on a piecewise linear potential was used. During docking, the height of the energy barriers had been continuously augmented by increasing the repulsive term of the potential. Thus, in the later stages of the simulation, this adaptive procedure narrowed the search to only a few energetically favorable binding modes, funneling the algorithm to the global minimum. According to the authors, the adaptive softness of the energy function facilitated the conformational search both by promoting escape from local minima and by destabilizing alternative solutions. Increasing the repulsive term of the potential yields a rougher energy landscape, but the energy function becomes more and more accurate and leads the search to the global minimum. Similar dynamical modifications of the energy function have been adopted to mimic the docking funnel [29,30,55]. Although the essential idea is rather simple, no general rules for adapting the potential are available and the optimal way for scaling the barriers may be strictly dependent on the system explored. Moreover, if the scaling is not accomplished in a proper way, the adaptive scoring function might not fulfill the kinetic requirement. Because of these limitations, we and others [35,70–73] have chosen an alternative approach. This is described in Section 14.3.4.

14.3.4 MULTIPLE-STEP DOCKING

Combining different scoring schemes into a single docking approach is a useful method to increase the effectiveness of a docking protocol. A two-step strategy makes use of a simple molecular recognition model based on the minimal frustration principle [68,69], followed by a more accurate energy evaluation to rank the docking solutions. When using multiple-step procedures, there is a clear distinction between the objective function, which is fast but approximative, and the binding energy function (See Section 14.2.3.1 and Section 14.2.3.2).

The basic assumption behind multiple-step approaches is that there is at least one low-lying minimum of the objective function inside the global minimum basin of the binding energy. The fast objective function is then thought of as a coarse-grained description of the more accurate binding energy function. The first step intends to overcome the kinetic bottlenecks of the accurate energy function by using a simpler and much less frustrated energy model. After the first step of the procedure, the final set of ligand conformations can undergo a gradient-based minimization with a standard force field. The minimized conformations are then ranked according to their energy. Multiple-step docking approaches are widely used and have been published [70–73]. A multiple-step procedure was also applied in the most recent version of our docking approach [35]. The results of FFLD [8] were postprocessed by CHARMm minimization [36] of the flexible ligand in the rigid receptor. The docking study showed the effectiveness of a multiple-step strategy. It was possible to correctly reproduce the binding mode of highly flexible inhibitors (up to 22 rotatable bonds) of HIV-1 protease, if the strain in their covalent geometry upon binding was not too large. Moreover, it was observed that the postprocessing step led to more reliable predictions and improved the success rate of the docking experiments [35].

14.4 PROTOCOLS

In this section, we will explain the use of our docking approach. However, many of the guidelines and recommendations introduced here will also hold true when using other docking programs.

14.4.1 OUR DOCKING APPROACH

The SEED/FFLD approach uses a GA to optimize ligand conformations and previously docked fragments to place the ligand in the binding site. It relies on the assumption that the most significant interactions with the protein are formed by three or more fragments of the ligand. Hence, it should be possible to first investigate the binding modes of the fragments and then use this information to place the whole molecule. This docking approach consists of four separate steps, the principles of which shall be described below. A more detailed protocol can be found in the following subsections and the original articles [8,10,11,35].

At first, those parts of the ligand that are supposed to account for most of the interactions (the fragments, Figure 14.3) have to be defined. This choice is rather important, for example, fragments that are too small will yield anchor positions that cannot discriminate the physicochemical characteristics of the binding site. A computer program has been developed to automatically choose at least three fragments (P. Kolb et al., unpublished), because the matching algorithm employed in the last step uses triangles. In the second step, the selected fragments are minimized with a force field to obtain low energy conformations. Subsequently, they are docked as rigid molecules with SEED [10,11] (Figure 14.4). As described before, SEED uses polar and hydrophobic vectors as anchors. The polar vectors are distributed around HDOs and HACs, whereas apolar vectors are used to mark hydrophobic regions. The latter are obtained by placing a low dielectric sphere (methane) at equal intervals on the solvent accessible surface of the protein. Points that have a favorable interaction energy are retained and the vectors are defined by joining each point with the corresponding atom center. During docking, every vector is matched to the complementary vectors on the fragments and the fragments are rotated exhaustively around these vector-defined axes. For each fragment position on each SEED point, a binding energy, which

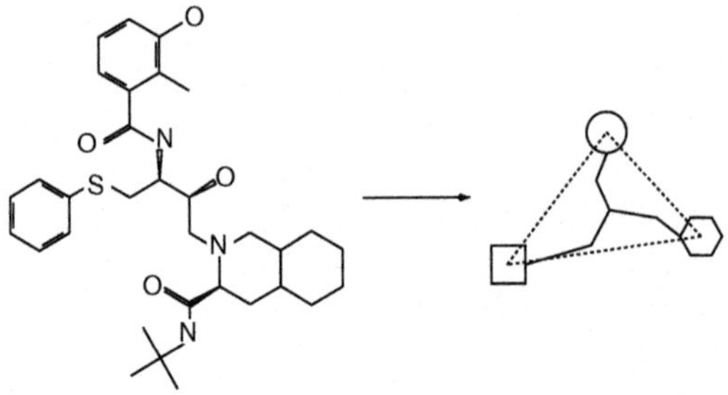

FIGURE 14.3 Schematic depiction of the fragment selection process. The molecule is Viracept (Agouron/Pfizer).

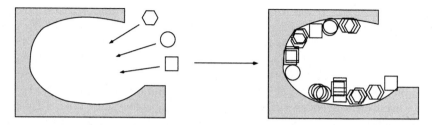

FIGURE 14.4 Schematic depiction of the docking process of the small fragments.

includes electrostatic solvation, is evaluated. Thus, if the fragments chosen are rigid (which is the case for small molecules and aromatic systems), the ranking is determined with high reliability. The information obtained from SEED consists of the 3-D coordinates of the geometrical centers of the fragment poses as well as their binding energies. Each fragment pose is one possible corner point of the placement triangle used in the last step. On average, a SEED run yields up to 100 poses per fragment type.

In the third step, this number is reduced to obtain a manageable number of possible triangle combinations. In practice, we reduce it to 20, using a clustering method which is based both on geometric proximity and the value of the binding energy for each pose [35]. For each fragment, the 20 points define a map that contains the important information from SEED and is still diverse enough to offer useful anchor points (Figure 14.5). Diversity is especially important because using only the top-ranked poses of the fragments does not always lead to the solution. This is due to the fact that the binding mode of the entire ligand is a compromise that tries to satisfy most of the fragments.

The fourth and last step is the docking of the complete putative ligand. This is done with the program FFLD [8], which uses a scoring function consisting of ligand dihedral and vdW energy, and protein–ligand polar and vdW contributions (See Section 14.3.3). Ligand conformations are generated and optimized by a GA, which encodes the torsional angle values of the rotatable bonds. For each conformation, the geometrical centers of the three fragments define a triangle. Based on the side lengths of the ligand triangle, FFLD finds those SEED points that form triangles of approximately the same shape. It then

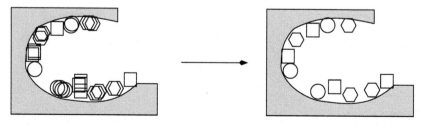

FIGURE 14.5 Schematic depiction of the clustering procedure. The different fragment types are shown for clarity.

tries to match the ligand triangle with each of the possible SEED triangles using a least-squares-fitting method (a variant of the Kabsch algorithm [74]) and assigns the score of the best placement to this conformation (Figure 14.6). The output of FFLD consists of the final poses for all conformations, usually 100 to 200 in total. It is worth noting that, because every conformation yields multiple poses at each step, FFLD will not only find the best binding mode, but also a number of alternative binding modes of comparable score. The alternative binding modes, in fact, can be used as a starting point for further postprocessing with more accurate energy functions.

14.4.2 PREPARATION OF THE LIBRARY OF COMPOUNDS

The first and most basic requirement is that the ligand is a chemically complete molecule (i.e., all valences must be satisfied). Special care must be taken to specify the correct bond types, because this will be the basis for the definition of the bonds that are rotatable. Another main concern is the correct assignment of the partial charges. These are needed for the calculation of the interaction energy in SEED and the postprocessing step. We use the modified partial equalization of orbital electronegativity (MPEOE) method developed by No et al. [75,76] as implemented in WITNOTP (A. Widmer, Novartis Pharma AG, Basel, unpublished), which yields partial charges consistent with those of the protein atoms in the CHARMm22 force field. Other implementations should also give reliable partial charges, but we have not tested them.

As a prerequisite to docking, one has to consider the state of ionizable groups in the protein (see below) and the ligand. Because the physiological conditions for protein–ligand complexes are in most cases close to pH 7, acidic groups are usually in a deprotonated and basic groups in a protonated state. A pK_a calculation could be done with a finite-difference Poisson solver in case of uncertainties. For a heterogenous library of compounds, it is much more difficult to assign formal charges. We usually check for groups where the assignment is evident (e.g., primary, secondary or tertiary amines, which are positively charged). Afterward, atom types for the CHARMm22 force field have to be assigned. Any ligand should furthermore be minimized with an accurate force field to obtain a low-energy conformation.

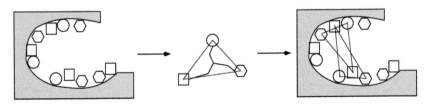

FIGURE 14.6 The docking process: FFLD tries to place the triangle defined by a conformation of the ligand (generated with the GA) on the anchor points computed by SEED.

14.4.3 Fragment Choice

The decomposition of a ligand into fragments and the choice of the anchor fragments have been automatized recently (P. Kolb et al., unpublished). We will list the major rules here as they can be of general interest. The decomposition is guided by the fact that SEED treats all molecules as rigid. Hence, preference is given to aromatic rings and other small rings and molecules that contain several amidic, double, or triple bonds. The fact that nonaromatic ring systems might have several distinct conformations can be accounted for by the ability of SEED to dock multiple (predefined) conformations at the same time. If one of these conformations can be docked with a lower binding energy than the others, it will automatically be chosen in the subsequent steps, because it will receive higher ranks.

The selection follows a few simple rules:

1. All atoms in a fragment must be connected by rigid or terminal bonds (for the definition of rigid bonds see above).
2. Large fragments are preferred because there are more steric constraints for large entities, as a consequence these should be positioned first.
3. Cyclic fragments are preferred because they usually are more rigid than acyclic moieties.
4. Because the fragments should be involved in the most significant interactions, those that contain HDOs and HACs are selected. Charged groups usually do not make such good anchors, because they tend to be positioned at the borders of the binding site, which are more exposed to the solvent. (However, there are exceptions as in the case of thrombin, where a favorable electrostatic interaction is provided by a charged aspartic acid in the specificity pocket [8].
5. Fragments that are close to the center of the molecule are omitted, especially if they have a high number of substituent groups. Such central or scaffold fragments will hardly ever form specific interactions.
6. Finally, fragments should not overlap (i.e., one atom should not be part of two fragments), because this would mean that there are no rotatable bonds in between, so their relative position cannot be changed.

These rules can be exemplified with the molecule XK263 (DuPont Merck, Figure 14.7). In principle, there are three fragment types that could be chosen—naphthalene, benzene, and the cyclic urea in the center. The largest fragment would be the cyclic urea. According to Rule 5, this is not a good choice as it is the core fragment and has four substituents. Furthermore, it is the most flexible of the three types, which is another point against its choice according to Rule 2. The remaining two types are aromatic and thus a recommended choice (Rule 1). Finally, it is better to select the two naphthalenes, because they are larger than the benzenes (Rule 2).

A more difficult choice is presented by acetyl-pepstatin (Figure 14.8), because it has no rings and almost no rigid bonds. All the fragments obtained by the application of Rule 1 are therefore small. All the larger fragments with a rigid

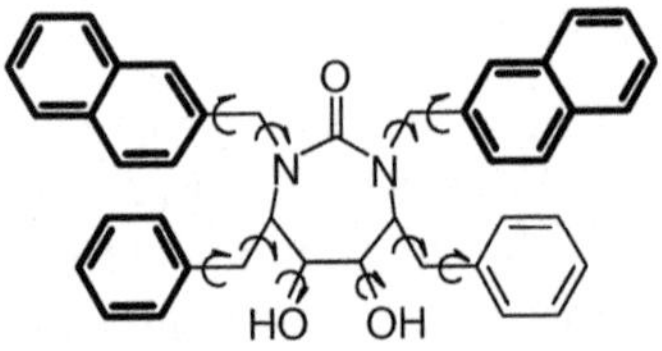

FIGURE 14.7 XK263 (Dupont Merck) is a nanomolar inhibitor of HIV-1 aspartic protease (PDB accession code of the complex: 1HVR). Selected fragments are bold. Curly arrows denote rotatable bonds.

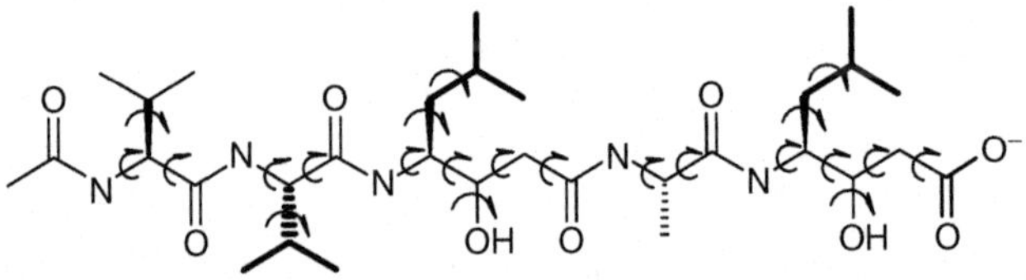

FIGURE 14.8 Acetyl-pepstatin is a micromolar inhibitor of HIV-1 aspartic protease (PDB accession code of the complex: 5HVP). Selected fragments are bold. Curly arrows denote rotatable bonds.

bond (the amide groups) are located in the backbone and will not make good anchors (Rule 5). One of the few choices remaining is to select three i-butanes (the sidechains), which are preferable with respect to the terminal carboxylic group, because this group is charged (Rule 4).

14.4.4 PROTEIN PREPARATION

It has to be emphasized that the preparation of the protein is a crucial step in the protocol and should be done carefully. It is not advisable to use automatic methods, as they cannot take into account all eventualities and special cases.

14.4.4.1 First Checks

The attention of the experimenter should be turned to all specific and unusual details, like nonstandard amino acids (e.g., cysteine-sulfonic acid, selenomethionine, etc.). Furthermore, the protein can contain prosthetic groups, cofactors, or other small molecules. Prosthetic groups should be kept for the docking run in all cases, because they will most probably be present in the protein in its native environment. Whether or not cofactors should be considered, depends on the system. Most probably, they can be removed, since they will not compete with an inhibitor, unless they have a strong affinity to the protein by themselves and will be present in the binding site most of the time. In general, small molecules (such as polyethylene glycol) are due to the crystallization conditions and should be removed. The final decision, however, has to be taken *ad hoc* for every system.

In any case, one should check in the pdb-file that no atoms are missing in the aforementioned residues and molecules, because most structure manipulation programs do not check on nonstandard residues automatically. Quite frequently, crystal structures will lack even whole parts of the protein due to poor electron density in disordered regions. This fact is usually commented on in the pdb-file or in the paper. It is then up to the researcher to decide if this is negligible or not. Judging from our experience, in the majority of cases, these incomplete regions are far away from the binding site. Thus, they will not have a great influence on the binding energy evaluation. Unless there are only one or two amino acids missing, it is not advisable to rebuild the protein in those regions. The error introduced by guessing the conformation without proper equilibration will probably be larger than the error due to the absence of the residues.

Another special case are ions. Those that are required for the stability of the protein should be kept, especially if they are close to the binding site. An ion in the binding site should always make a favorable interaction with an oppositely charged group in the ligand. It is advisable to determine the charged warhead for the candidate ligands *a priori* and discuss the simpleness of synthesis of the resulting compounds with a medicinal chemist.

Lastly, the presence of disulfide bonds has to be investigated. Information whether or not there are any should be listed in the pdb-file in a line commencing with "SSBOND." However, it is safer to visualize all cysteine residues. If the sulfur–sulfur distance between two cysteine residues is around 2 Å and the relative geometry is right, they will most likely form a disulfide bond.

14.4.4.2 Charged Residues

Special care should be exercised when treating residues with ionizable groups. The most sophisticated approach is to solve the finite-difference Poisson equation to calculate the pK_a of all titratable groups. If the *in vitro* tests are done at physiological pH, we normally assume both basic and acidic sidechains as well as the terminal carboxyl and amino group as ionized.

The situation for histidine residues is more complicated. First, one has to select a protonation state and then, in the case of monoprotonation, also which nitrogen (δ or ε) should be protonated. To properly assign the protonation state of the histidines, it is important to consider the local environment of these residues in the folded structure of the protein. At low pH (pH≤6), a diprotonated state should be assigned to histidines partially or fully exposed to the solvent. For calculations at physiological pH, a monoprotonated state is commonly preferred and we assign a monoprotonated state to the histidines irrespective of their position. If the environment does not indicate a clear preference for one of the two variants because of potential HACs or steric hindrance, we arbitrarily choose the δ-protonated variant.

Related to the issue of the charged residues is the choice of the interior dielectric constant of the protein, which is necessary for SEED. The value of this constant influences the strength of the coulombic interactions and can lead to significantly different results, as model calculations have shown (Majeux et al., unpublished results). Previously, values ranging from 1 to 4 have been used [10,11]. It is useful to perform preliminary docking runs with interior dielectric values of 1, 2, and 4 and compare the results with available crystal structures.

14.4.4.3 Adding Hydrogens

It is necessary to add hydrogens, because files in the Protein Data Bank (PDB) usually do not contain any. This should be done with a program like CHARMm [36] using the HBUILD module, which first places those hydrogens whose positions can be determined unambiguously, such as hydrogens connected to a peptidic nitrogen, and afterwards performs exhaustive searches to place hydroxyl hydrogens on serine, threonine, and tyrosine. To assign atom types, we use the atom type definition of the CHARMm22 force field. It has proven useful to recheck on all nonstandard residues to verify the correct assignment. Finally, the hydrogens should be minimized with an appropriate force field while keeping the protein backbone rigid.

14.4.4.4 Binding Site Definition

As mentioned above, this step is of high importance. To begin with, one should have a look at the publication describing the crystal structure and the interactions. The basis for the selection of the residues belonging to the binding site will most often be the pose of a known ligand. If such information is not available, one has to select the binding site by hand. In that case, in-depth knowledge of the function of the protein or crystal structures of closely related proteins of the same family are necessary.

We select the binding site by first determining all protein atoms that are within a cutoff radius of 5 Å from any ligand atom. It is important that there is a clear inside and outside of the binding site to avoid the positioning of anchors in solvent-exposed regions of the protein. Hence, selecting residues whose sidechains point away from the binding site have to be avoided. To achieve this, only residues which have at least 50% of their atoms within the cutoff distance are marked as members of the binding site. The cutoff should not be too small, as the bias toward the binding mode of the known ligand would be too big and no alternative ones could be detected. On the other hand, because the binding site residues are providing the anchor points for SEED, the number of anchors correlates with the number of residues. Thus, docking would take increasingly long as the binding site becomes larger and would additionally yield too many solutions, which are then difficult to rank. If a large binding site is really needed, it is probably better to split it into several (overlapping) sectors. Sometimes, it is advisable to manually alter the definition until one is satisfied with the distribution and the number of the anchor points. In this case, one has to remember that the binding mode (and consequently the ranking of a library of compounds) might be affected by the human intervention, which is usually based on previous knowledge. This bias might preclude interesting surprises like alternative binding modes [77].

As was mentioned before, SEED puts anchor vectors on atoms of the binding site residues. Clearly, only vectors pointing inside the binding site should be used. For that reason, the latest version of SEED employs a cutoff based on the angle between the vector and predefined points in the binding site (usually the heavy atoms of a native ligand) for choosing the most suitable ones [35]. Using the atoms of a ligand from a known complex to define the binding site does not introduce a bias, though, and corresponds to the situation in an advanced drug design program, where one or more crystal structures of protein/ligand complexes have already been solved.

Another critical issue is the ionization state of groups in the binding site. This is probably best illustrated by the case of the aspartic proteases, which contain an aspartyl dyad in the cleavage site. Piana et al. [78] have shown that, besides the pH, the ligand has an influence, as it can stabilize either the neutral, negatively or dinegatively charged form of the dyad state. Consequently, the charge state of the dyad can influence the types of ligands that will receive a high ranking.

14.4.4.5 Conserved Water Molecules

In many proteins, water molecules located at distinct positions can play a crucial role because they provide important interactions with the ligand. Wrongly positioned water molecules, on the other hand, can impede docking and make the detection of the correct binding mode impossible. Deciding which water molecules to keep is not trivial. Evidence can come from multiple x-ray structures with different ligands. If a water molecule is repeatedly found at the same position and also forms hydrogen bonds with the ligand, it is likely to be conserved because of structural relevance. Additional help is offered by prediction programs such as ConSolv [79], which compares the ligand-free form of the protein with the complex.

Our example, HIV-1 protease, for which numerous x-ray structures are available, normally contains a water molecule bridging the two flaps and the inhibitor. This water is necessary if one wants to reproduce the binding mode of acetyl-pepstatin in its native protein structure, 5HVP. The structure of 1HVR, however, does not contain a water molecule at that specific position. During binding, the carbonyl group of the cyclic urea displaces this water and directly stabilizes the two flaps of the protease. Therefore, docking the ligand XK263 in 1HVR requires the water site to be empty. It is possible to reproduce its binding mode only after removal of the water. However, it is not possible to know this *a priori* for every molecule in a large database for screening. Hence, in the absence of further information, we suggest removing all water molecules from the binding site.

14.4.4.6 Reference Structure

For every new project, the setup of the approach chosen for docking should be validated. The most common way to do this is by redocking a ligand to the corresponding protein structure from the complex. However to judge the performance of the method, it is crucial not to use the exact pose of the ligand from the crystal structure. This pose is the time-average over the ligand poses during the collection of the diffraction data (as is the case for the conformation of the protein).

Thus, it is likely that, according to the parameters of the applied scoring function, some atom positions have clashes with the protein. This problem can be solved by minimizing the ligand within the binding site with a gradient-based method applying the same scoring function as will be used for docking, while keeping the protein rigid. The minimized ligand then offers an appropriate reference structure for redocking calculations.

The ligand conformation which is used as input structure for the docking experiments should have been minimized with a force field outside of the binding site to remove any geometrical bias. However, one has to bear in mind that the force field will not only modify the torsion angles, but also bond lengths and bond angles. If the strain in the ligand conformation is large upon binding, the minimization outside of the receptor might yield a covalent geometry that is not compatible with the binding site. Therefore, because in the docking search only torsional degrees of freedom are considered, the docking approach might not be able to reproduce the experimental binding mode [35].

14.4.5 RUNNING SEED

SEED provides the anchors for the final docking procedure. Thus, it is worth analyzing the SEED results in detail. One should have a close look at the binding site with a molecular viewer to see the distribution of the polar and apolar vectors used by SEED to dock the fragments. If a project is in an advanced stage and a considerable amount of structural information is available, the user should eventually change the number of the polar and apolar vectors as well as the definition of the binding site or the interior dielectric constant.

14.4.6 RUNNING FFLD

The only parameters that should be modified in FFLD are the input values for the hybrid search algorithm. It has to be emphasized that optimal input values depend on the shape of the energy hypersurface and can thus hardly be predicted. As the limiting factor rather is the computer power, the user might want to select fewer chromosomes or fewer steps (which results in fewer energy evaluations) or a smaller frequency for the local search.

It is important, however, to perform multiple runs with different seeds for the random generation of the initial population. As with any stochastic search method, the hybrid search can be trapped in local minima. This is only detectable by comparing the results of many runs, therefore we typically perform 10 runs with different random seed numbers per ligand. Moreover, to judge the quality of the predictions, it is important to have a look at the convergence rate (i.e., which percentage of the different runs reach a similar conformation) [35]. This finding was obtained in a cross-docking study (which corresponds to the situation in a screening project) on 5 complexes of HIV-1 protease. Each of the 5 ligands was docked into all protein structures except its native one, which resulted in a total number of 20 docking experiments. For each docking experiment, convergence toward the lowest energy conformation (which is not

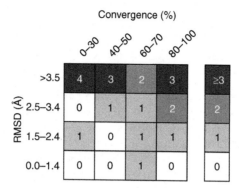

FIGURE 14.9 The density plot with the frequency of a certain rmsd from the experimentally determined structure for a given amount of convergence in 10 GA runs with different seeds. As an example, the "3" in the top right corner means that in 3 of the 20 docking experiments between 8 and 10 runs converged to the same conformation and this conformation has a rmsd larger than 3.5 Å from the experimental structure.

necessarily identical to the experimental structure) in 10 FFLD runs with different seeds was determined. The convergence values were then used to build a density plot that reports the frequency of finding a binding mode with a certain root-mean-square deviation (rmsd) from the experimental structure for a given amount of convergence (Figure 14.9). This density plot is almost upper triangular, which implies that experiments with less than 60% of convergence have probably failed to locate the global minimum. Consequently, these runs should not be relied on. On the other hand, a high convergence is no guarantee for successful docking, as is shown by the high number of runs that fully converged on a wrong structure (Figure 14.9, top right corner). The reason for this is probably to be searched for in the oversimplified nature of the energy function, which precludes an accurate detection of the solution. Taken together, these results suggest that a high convergence rate in multiple GA runs may be a necessary, although not sufficient, criterion for a good prediction.

ACKNOWLEDGMENTS

We thank Dr. Nicolas Majeux for interesting discussions. We also thank Fabian Dey for comments on this chapter. The development of our docking programs has been financially supported by Novartis, Aventis, and the Swiss National Center of Competence in Research (NCCR) in Structural Biology.

REFERENCES

[1]   A. Nicholls, K. Sharp, B. Honig, Protein folding and association: insights from the interfacial and thermodynamic properties of hydrocarbons, *Proteins: Structure, Function and Genetics* 11:281–296, 1991.

[2]   M. Scarsi, N. Majeux, and A. Caflisch, Hydrophobicity at the surface of proteins, *Proteins: Structure, Function and Genetics* 37:565–575, 1999.

[3]   C.M. Venkatachalam, X. Jiang, T. Oldfield, M. Waldman, LigandFit: a novel method for the shape-directed rapid docking of ligands to protein active sites, *J. Mol. Graphics Modelling* 21:289–307, 2003.

[4]   G.M. Morris, D.S. Goodsell, R.S. Halliday, R. Huey, W.E. Hart, R.K. Belew, and A.J. Olson, Automated docking using a Lamarckian Genetic Algorithm and an empirical binding free energy function, *J. Comput. Chem.* 19:1639–1662, 1998.

[5]   F. Österberg, G.M. Morris, M.F. Sanner, A.J. Olson, and D.S. Goodsell, Automated docking to multiple target structures: incorporation of protein mobility and structural water heterogeneity in AutoDock, *Proteins: Structure, Function, and Genetics* 46:34–40, 2002.

[6]   C. Hetényi, and D. van der Spoel, Efficient docking of peptides without prior knowledge of the binding site, *Protein Sci.* 11:1729–1737, 2002.

[7]   S.L. McGovern and B.K. Shoichet, Information decay in molecular docking screens against holo, apo, and modeled conformations of enzymes, *J. Med. Chem.* 46:2895–2907, 2003.

[8]   N. Budin, N. Majeux, and A. Caflisch, Fragment-based flexible ligand docking by evolutionary optimization, *Biol. & Chem.* 382:1365–1372, 2001.

[9]   H. Claussen, C. Buning, M. Rarey, and T. Lengauer, FlexE: Efficient molecular docking considering protein structure variations, *Algorithmica* 308:377–395, 2001.

[10]   N. Majeux, M. Scarsi, J. Apostolakis, C. Ehrhardt, and A. Caflisch, Exhaustive docking of molecular fragments with electrostatic solvation, *Proteins: Structure, Function, and Genetics* 37:88–105, 1999.

[11]   N. Majeux, M. Scarsi, and A. Caflisch, Efficient electrostatic solvation model for protein-docking, *Proteins: Structure, Function, and Genetics* 42:256–268, 2001.

[12]   A. Fahmy and G. Wagner, TreeDock: a tool for protein docking based on minimizing van der Waals energies, *J. Am. Chem. Soc.* 124:1241–1250, 2002.

[13]   E.C. Meng, B.K. Shoichet, and I.D. Kuntz, Automated docking with grid-based energy evaluation, *J. Comput. Chem.* 13:505–524, 1992.

[14]   I.D. Kuntz, E.C. Meng, S.J. Oatley, R. Langridge, and T.E. Ferrin, A geometric approach to macromolecule–ligand interactions, *J. Mol. Biol.* 161:269–288, 1982.

[15]   T.J.A. Ewing, S. Makino, A.G. Skillman, and I.D. Kuntz, DOCK 4.0: search strategies for automated molecular docking of flexible molecule databases, *J. Computer-Aided Mol. Design* 15:411–428, 2001.

[16]   A.N. Jain, Surflex: fully automatic flexible molecular docking using a molecular similarity-based search engine, *J. Med. Chem.* 46:499–511, 2003.

[17]   OpenEye Software, FRED, 2002. http://www.eyesopen.com/products/applications/fred.html.

[18]   M. Rarey, B. Kramer, T. Lengauer, and G. Klebe, A fast flexible docking method using an incremental construction algorithm, *J. Mol. Biol.* 261:470–489, 1996.

[19]   R. DeWitte, and E. Shakhnovich, SMoG: *de novo* design method based on simple, fast, and accurate free energy estimates. 1. Methodology and supporting evidence, *J. Am. Chem. Soc.* 118:11733–11744, 1996.

[20] R. DeWitte, A. Ishchenko, and E. Shakhnovich, SMoG: *de novo* design method based on simple, fast, and accurate free energy estimates: 2. Case studies in molecular design, *J. Am. Chem. Soc.* 119:4608–4617, 1997.

[21] M. Thormann and M. Pons, Massive docking of flexible ligands using environmental niches in parallelized genetic algorithms, *J. Comput. Chem.* 22:1971–1982, 2001.

[22] G. Jones, P. Willett, and R.C. Glen, Molecular recognition of receptor sites using a genetic algorithm with a description of desolvation, *J. Mol. Biol.* 245:43–53, 1995.

[23] C.M. Oshiro, I.D. Kuntz, and J.S. Dixon, Flexible ligand docking using a genetic algorithm, *J. Computer-Aided Mol. Design* 9:113–130, 1995.

[24] P.G. Maillot, *Graphics Gems*, London: Academic Press, p. 498, 1996.

[25] T.J. Oldfield, A number of real-space torsion-angle refinement techniques for proteins, nucleic acids, ligands and solvent. *Acta Cryst.* D57:82–94, 2001.

[26] T.J. Oldfield, X-ligand: an application for the automated addition of flexible ligands into electron density, *Acta Cryst.* D57:696–705, 2001.

[27] G. Jones, P. Willett, R.C. Glen, A.R. Leach, and R. Taylor, Development and validation of a genetic algorithm for flexible docking, *J. Mol. Biol.* 267:727–748, 1997.

[28] V. Schnecke and L. Kuhn, Virtual screening with solvation and ligand-induced complementarity, *Persp. Drug Discov. Des.* 20:171–190, 2000.

[29] T.W. Whitfield, and J.E. Straub, Gravitational smoothing as a global optimization strategy, *J. Comput. Chem.* 23:1100–1103, 2002.

[30] U.H.E. Hansmann and L.T. Wille, Global optimization by energy landscape paving, *Phys. Rev. Lett.* 88:068105, 2002.

[31] W.H. Press, S.A. Teukolsky, W.T. Vetterling, and B.P. Flannery, *Numerical Recipes in Fortran*, Cambridge, UK: Cambridge University Press, 1992.

[32] H.J. Bohm, The development of a simple empirical scoring function to estimate the binding constant for a protein–ligand complex of known three-dimensional structure, *J. Computer-Aided Mol. Design* 8:243–256, 1994.

[33] M.D. Eldridge, C.W. Murray, T.R. Auton, G.V. Paolini, and R.P. Mee, Empirical scoring functions: I. the development of a fast empirical scoring function to estimate the binding affinity of ligands in receptor complexes, *J. Computer-Aided Mol. Design* 11:425–445, 1997.

[34] G.M. Verkhivker, D. Bouzida, D.K. Gehlhaar, P.A. Rejto, S. Arthurs, A.B. Colson, S.T. Freer, V. Larson, B.A. Luty, T. Marrone, and P.W. Rose, Binding energy landscapes of ligand–protein complexes and molecular docking: principles, methods, and validation experiments. In A.K. Ghose, and V.N. Viswanadhan, Eds., *Combinatorial Library Design and Evaluation: Principles, Software, Tools, and Applications in Drug Discovery*, New York: Marcel Dekker, pp. 157–195, 2001.

[35] M. Cecchini, P. Kolb, N. Majeux, and A. Caflisch, Automated docking of highly flexible ligands by genetic algorithms: a critical assessment, *J. Comput. Chem.* 25:415–422, 2004.

[36] B.R. Brooks, R.E. Bruccoleri, B.D. Olafson, D.J. States, S. Swaminathan, and M. Karplus, CHARMm: a program for macromolecular energy, minimization, and dynamics calculations, *J. Comput. Chem.* 4:187–217, 1983.

[37] W. Cornell, P. Cieplak, C. Bayly, I. Gould, K. Merz, Jr., D. Ferguson, D. Spellmeyer, T. Fox, J. Caldwell, and P. Kollman, A second generation force field for the simulation of proteins, nucleic acids, and organic molecules, *J. Am. Chem. Soc.* 117:5179–5197, 1995.

[38] M.L. Verdonk, J.C. Cole, P. Watson, V.J. Gillet, and P. Willett, SuperStar: improved knowledge-based interaction fields for protein binding sites, *J. Mol. Biol.* 307:841–859, 2001.

[39] I. Muegge, A knowledge-based scoring function for protein–ligand interactions: probing the reference state, *Persp. Drug Discov. Des.* 20:99–114, 2000.

[40] A.V. Ishchenko and E.I. Shakhnovich, Small Molecule Growth 2001 (SMoG2001): an improved knowledge-based scoring function for protein–ligand interactions, *J. Med. Chem.* 45:2770–2780, 2002.

[41] H. Gohlke, M. Hendlich, and G. Klebe, Knowledge-based scoring function to predict protein–ligand interactions, *J. Mol. Biol.* 295:337–356, 2000.

[42] P.S. Charifson, J.J. Corkery, M.A. Murcko, and W.P. Walters, Consensus scoring: a method for obtaining improved hit rates from docking databases of three-dimensional structures into proteins, *J. Med. Chem.* 42:5100–5109, 1999.

[43] R.D. Clark, A. Strizhev, J.M. Leonard, J.F. Blake, and J.B. Matthew, Consensus scoring for ligand/protein interactions, *J. Mol. Graphics Modelling* 20:281–295, 2002.

[44] J. Warwicker and H.C. Watson, Calculation of the electric potential in the active site cleft due to α-helix dipoles, *J. Mol. Biol.* 157:671–679, 1982.

[45] M.K. Gilson and B.H. Honig, Energetics of charge-charge interactions in proteins, *Proteins: Structure, Function, and Genetics* 3:32–52, 1988.

[46] D. Bashford, and M. Karplus, pK_a's of ionizable groups in proteins: atomic detail from a continuum electrostatic model, *Biochem.* 29:10219–10225, 1990.

[47] M.E. Davis, J.D. Madura, B.A. Luty, and J.A. McCammon. Electrostatics and diffusion of molecules in solution: simulations with the University of Houston Brownian dynamics program, *Comput. Phys. Comm.* 62:187–197, 1991.

[48] W.C. Still, A. Tempczyk, R.C. Hawley, and T. Hendrickson, Semianalytical treatment of solvation for molecular mechanics and dynamics, *J. Am. Chem. Soc.* 112:6127–6129, 1990.

[49] M. Scarsi, J. Apostolakis, and A. Caflisch, Continuum electrostatic energies of macromolecules in aqueous solutions, *J. Phys. Chem.* A101:8098–8106, 1997.

[50] N. Budin, S. Ahmed, N. Majeux, and A. Caflisch. An evolutionary approach for structure-based design of natural and non-natural peptidic ligands, *Comb. Chem. High Throughput Screen.* 4:695–707, 2001.

[51] N. Budin, N. Majeux, C. Tenette-Souaille, and A. Caflisch, Structure-based ligand design by a build-up approach and genetic algorithm search in conformational space, *J. Comput. Chem.* 22:1956–1970, 2001.

[52] X. Zou, Y. Sun, and I.D. Kuntz, Inclusion of solvation in ligand binding free energy calculations using the generalized-Born model, *J. Am. Chem. Soc.* 121:8033–8043, 1999.

[53] N. Arora, and D. Bashford, Solvation energy density occlusion approximation for evaluation of desolvation penalties in biomolecular interactions, *Proteins: Structure, Function, and Genetics* 43:12–27, 2001.

[54] M. Rarey, B. Kramer, and T. Lengauer. The particle concept: placing discrete water molecules during protein–ligand docking predictions, *Proteins: Structure, Function, and Genetics* 34:17–28, 1999.

[55] J. Apostolakis, A. Plückthun, and A. Caflisch, Docking small ligands in flexible binding sites, *J. Comput. Chem.* 19:21–37, 1998.

[56] R.M.A. Knegtel, I.D. Kuntz, and C.M. Oshiro, Molecular docking to ensembles of protein structures, *J. Mol. Biol.* 266:424–440, 1997.

[57] R.M. Jackson, H.A. Gabb, and M.J.E. Sternberg, Rapid refinement of protein interfaces incorporating solvation: application to the docking problem, *J. Mol. Biol.* 276:265–285, 1998.

[58] P. Koehl and M. Delarue, Application of a self-consistent mean field theory to predict protein side-chains conformation and estimate their conformational entropy, *J. Mol. Biol.* 239:249-275, 1994.

[59] P. Koehl and M. Delarue, Mean-field minimization methods for biological macro-molecules, *Curr. Opin. Struct. Biol.* 6:222–226, 1996.

[60] J.H. Lin, A.L. Perryman, J.R. Schames, and J.A. McCammon, Computational drug design accommodating receptor flexibility: the relaxed complex scheme, *J. Am. Chem. Soc.* 124:5632–5633, 2002.

[61] E. Yuriev and P.A. Ramsland, Mcg light chain dimer as a model system for ligand design: a docking study, *J. Mol. Recognit.* 15:331–340, 2002.

[62] J.D. Diller and C.L.M.J. Verlinde, A critical evaluation of several global optimization algorithms for the purpose of molecular docking, *J. Comput. Chem.* 20:1740–1751, 1999.

[63] A. Caflisch, P. Niederer, and M. Anliker, Monte Carlo docking of oligopeptides to proteins, *Proteins: Structure, Function, and Genetics* 13:223–230, 1992.

[64] M. Miller, S.K. Kearsley, D.J. Underwood, and M.D. Sheridan, FLOG — a system to select quasi-flexible ligands complementary to a receptor of known 3-dimensional structure, *J. Computer-Aided Mol. Design* 8:153–174, 1994.

[65] S. Makino and I.D. Kuntz, Automated flexible ligand docking method and its applica-tion for database search, *J. Comput. Chem.* 18:1812–1825, 1997.

[66] D.E. Goldberg, *Genetic Algorithms in Search Optimization and Machine Learning*, Reading, MA: Addison-Wesley, 1989.

[67] L. Davis, Ed., *Handbook of Genetic Algorithms*, New York: Van Nostrand Reinhold, 1991.

[68] D.K. Gehlhaar, G.M. Verkhivker, P.A. Rejto, C.J. Sherman, D.B. Fogel, L.J. Fogel, and S.T. Freer, Molecular recognition of the inhibitor AG-1343 by HIV-1 protease: conformationally flexible docking by evolutionary programming, *Chem. Biol.* 2:317–324, 1995.

[69] G.M. Verkhivker, P.A. Rejto, D.K. Gehlhaar, and S.T. Freer, Exploring the energy landscape of molecular recognition by a genetic algorithm: analysis of the requirements for robust docking of HIV-1 protease and FKBP-12 complexes, *Proteins: Structure, Function, and Genetics* 25:342–353, 1996.

[70] L. Schaffer and G.M. Verkhivker, Predicting structural effects in HIV-1 protease mutant complexes with flexible ligand docking and protein side-chain optimization, *Proteins: Structure, Function, and Genetics* 33:295–310, 1998.

[71] D. Hoffman, B. Kramer, T. Washio, T. Steinmetzer, M. Rarey, and T. Lengauer, Two-stage method for protein–ligand docking, *J. Med. Chem.* 42:4422–4433, 1999.

[72] J. Wang, P.A. Kollman, and I.D. Kuntz, Flexible ligand docking: a multistep strategy approach, *Proteins: Structure, Function, and Genetics* 36:1–19, 1999.

[73] M.L.P. Price, and W.L.J. Jorgensen, Analysis of binding affinities for celecoxib analogues with COX-1 and COX-2 from combined docking and Monte Carlo simulations and insight into the COX-2/COX-1 selectivity, *J. Am. Chem. Soc.* 122:9455–9466, 2000.

[74] W. Kabsch, A solution for the best rotation to relate two sets of vectors, *Acta Cryst.* A32:922–923, 1976.

[75] K. No, J. Grant, and H. Scheraga, Determination of net atomic charges using a modified partial equalization of orbital electronegativity method: 1. Application to neutral molecules as models for polypeptides, *J. Phys. Chem.* 94:4732–4739, 1990.

[76] K. No, J. Grant, M. Jhon, and H. Scheraga. Determination of net atomic charges using a modified partial equalization of orbital electronegativity method: 2. Application to ionic and aromatic molecules as models for polypeptides, *J. Phys. Chem.* 94:4740–4746, 1990.

[77] K. Hilpert, J. Ackermann, D.W. Banner, A. Gast, K. Gubernator, P. Hadvary, L. Labler, K. Müller, G. Schmid, T. Tschopp, and H. van de Waterbeemd, Design and synthesis of potent and highly selective thrombin inhibitors, *J. Med. Chem.* 37:3889–3901, 1994.

[78] S. Piana, D. Sebastiani, P. Carloni, and M. Parrinello, *Ab initio* molecular dynamics-based assignment of the protonation state of pepstatin A/HIV-1 protease cleavage site, *J. Am. Chem. Soc.* 123:8730–8737, 2001.

[79] M.L. Raymer, P.C. Sanschagrin, W.F. Punch III, S. Venkataraman, E.D. Goodman, and L.A. Kuhn, Predicting conserved water and water-mediated ligand interactions in proteins using a k-nearest-neighbor genetic algorithm, *J. Mol. Biol.* 265:445–464, 1997.

15 Protein–Ligand Docking and Virtual Screening with GOLD

Jason C. Cole, J. Willem M. Nissink,
and Robin Taylor

15.1 INTRODUCTION

GOLD (Genetic Optimization for Ligand Docking) is a program for docking flexible ligands into protein binding sites [1,2]. It was originally written by Jones at the University of Sheffield, England. Since its release in 1998, it has been distributed, maintained, and improved by the Cambridge Crystallographic Data Centre (CCDC). Two other companies, Astex Technology, Ltd. and GlaxoSmithKline PLC, have contributed significantly to the program's development. This chapter describes the current version of GOLD (version 2.1). Although the focus is unashamedly on that program, much of the discussion is relevant to docking and virtual screening (VS) in general.

A docking program requires two basic abilities: a method of scoring any trial ligand pose and a search algorithm for finding the pose with the best score. In addition, two other components are essential although often taken for granted. First, the program must be sufficiently well tested that its reliability and limitations are clearly understood. Second, it must offer easy-to-use options for dealing with practical issues such as applying constraints, handling large volumes of input and output, customizing program settings for specific problems, distributing jobs over many processors, and so forth. GOLD is therefore described under the main headings Scoring Functions, Search Strategy, Prediction Reliability, and Program Infrastructure.

15.2 SCORING FUNCTIONS

The scoring function used to estimate the energy of a given protein–ligand arrangement is arguably the most important feature of a docking program. All scoring functions are compromises between the accuracy with which protein–ligand energy is estimated and the speed of calculation. The bias is invariably toward the latter, mainly because speed is a lot easier to achieve than accuracy. Because all functions are compromises, each has its own strengths and weaknesses (though there is rarely consensus on what these are). A common observation is that one scoring function works well on one protein but not on another, whereas a different function shows the opposite trend. For that reason, GOLD offers a choice of two scoring functions with different parameters and functional forms, together with an application programming interface (API) that can be used to implement other functions. This section describes the two built-in functions, GoldScore (the original and default GOLD scoring function) and ChemScore.

15.2.1 GOLDSCORE

GoldScore [1,2] comprises three main terms: protein–ligand hydrogen-bonding energy (*external H-bond*), protein–ligand van der Waals energy (*external vdW*), and ligand internal strain energy (*internal strain*). A fourth term, intramolecular ligand H-bond energy (*internal H-bond*), can be included optionally. All terms are computed from molecular-mechanics expressions using an all-atom force field. GoldScore does not use atomic formal or partial charges or bond dipoles and therefore relies on the use of atom types and molecular connectivity to infer the characteristics (e.g., H-bond donor, H-bond acceptor, hydrophobe) of atoms.

15.2.1.1 External H-Bond Energy

Any atom capable of H-bonding (Table 15.1) is assigned a H-bonding type, comprised of a base type, normally the atom's standard SYBYL atom type [3], followed by D (donor) or A (acceptor), such as O.3A (ether oxygen acceptor). Occasionally, SYBYL atom types are further subdivided; for example, amide carbonyl oxygen acceptor atoms are treated differently from other carbonyl oxygens. Carbon is not considered a H-bond donor, although there is evidence that CH...O H-bonding can contribute to the stability of protein–ligand complexes [4]. Halogen atoms and some oxygens in conjugated systems (e.g., furan) are not regarded as acceptors, which is consistent with published information [5,6]. All thionelike sp² sulfur atoms are deemed acceptors, although there is tentative evidence [5,7] to suggest that only activated sulfurs (such as in thioureas and thioamides, where an α–nitrogen increases the sulfur electron density) readily accept H-bonds.

The total external H-bond score is the sum of contributions from all protein–ligand donor–acceptor pairs within H-bonding distance. The contribution from a given pair depends on:

- The donor and acceptor atom types, which determine the energy that will be contributed if the H-bond geometry is ideal (the maximum H-bond energy for the donor–acceptor pair).
- A weight, used to attenuate the energy as the geometry is distorted from ideal.

The table of maximum H-bond energies for different donor–acceptor combinations was originally derived from gas-phase molecular-mechanics calculations on model molecules using PM3-derived Mulliken charges and a dielectric constant of 1 [1,2]. To correct for desolvation, the maximum H-bond energy for the pair D...A

TABLE 15.1

GOLD H-Bond Atom Types and Acceptor Directional Preferences Assumed when Using GoldScore

Atom Description	Base Type	Donor[1]	Acceptor	Acceptor Directionality
sp³ nitrogen	N.3	Yes	Yes	Along lone pairs
sp² nitrogen	N.2	Yes	Yes	Along lone pairs
sp nitrogen	N.1	No	Yes	Along lone pair
Acidic nitrogen[2]	N.acid[1]	Yes	Yes	Along lone pairs
Aromatic nitrogen	N.ar	No	Yes	Along lone pairs
Amide nitrogen	N.am	Yes	No	
Quaternary nitrogen	N.4	Yes	No	
Uncharged trigonal nitrogen [3]	N.pl3	Yes	No	
Charged trigonal nitrogen[4]	N.plc	Yes	No	
Hydroxyl oxygen	O.3	Yes	Yes	In plane of lone pairs
Ether oxygen	O.3	No	Yes	In plane of lone pairs
Carboxylate oxygen	O.co2	No	Yes	Along lone pairs
Carbonyl oxygen	O.2	No	Yes	In plane of lone pairs
Nitro oxygen	O.2	No	Yes	Along lone pairs
N-oxide oxygen	O.2	No	Yes	None
Amide oxygen	O.2	No	Yes	In plane of lone pairs
Neutral sulfur-bound oxygen[5]	O.2	No	Yes	None
Charged sulfur-bound oxygen[6]	O.co2	No	Yes	Cone
Phosphate oxygen	O.co2	No	Yes	Cone
Borate oxygen	O.co2	No	Yes	Cone
Other negatively charged oxygen	O.co2	No	Yes	None
Negatively charged sulfur	S.m	No	Yes	Along lone pairs
sp² sulfur	S.a	No	Yes	Along lone pairs

[1.] Provided at least one H-atom covalently bound.

[2.] An acidic nitrogen is a nitrogen bound by at most two single bonds.

[3.] As in an uncharged histidine residue, for example.

[4.] As in a guanidino residue, for example.

[5.] As in sulfonamides, sulfoxides, sulfones.

[6.] As in sulphate groups.

was taken as the sum of the optimized energies of the two H-bonds ($M_D...M_A$) and (water...water) minus the summed energies of the H-bonds ($M_D...$water) and ($M_A...$water), where M_D and M_A are the model molecules for D and A. This occasionally meant that the formation of a particular protein–ligand H-bond was regarded as unfavorable. Because the methodology was crude and higher quality intermolecular perturbation theory calculations often gave different results, the table of maximum H-bond energies was eventually simplified. Now, ion pair H-bonds are generally assigned energies of –10 kcal/mole, neutral H-bonds are typically assigned values of –2 or –4 kcal/mol, and H-bonds where just one of the interacting moieties is charged will typically be –6 kcal/mol. Adoption of this much simpler scheme did not reduce overall docking accuracy.

Energy attenuation as H-bonds are distorted from ideal geometry depends on both distance and angles. An ideal D...A distance is defined (2.9 Å for all types of H-bonds) and the energetic contribution from a H-bond whose length is within 0.5 Å of this value is given full weight. As the deviation between actual and ideal length increases beyond 0.5 Å, the bond energy is reduced linearly until it reaches zero at a specified point. The angular attenuation of H-bond energy depends on the nature of the acceptor group. Specifically, acceptors are divided into four types, according to how they prefer to form H-bonds:

1. Along lone-pair directions (e.g., carboxylate oxygens)
2. Within the lone-pair plane (e.g., ether oxygens)
3. In a cone, whose axis is the covalent bond between the acceptor atom and its neighbor (e.g., phosphate oxygens)
4. Show no directional preferences at all

Assignment of acceptors to these categories was based on H-bond geometries observed in small-molecule crystal structures [5]. Attenuation of H-bond energies is then based on an angular parameter that is defined differently for each category (Figure 15.1) but which, in each case, depends on the deviation of the D-H...A angle from linearity and the difference between the observed and preferred directions of approach of the hydrogen atom to the acceptor.

The H-bond energy contribution from a D...A pair is ignored if the protein atom involved in the contact is highly solvated in the absence of the ligand. This is to avoid ligands becoming docked at the mouths of binding sites. Docking at the periphery of the binding site is further discouraged by the imposition of an additional, large penalty term if any ligand atom is more than 10 Å from the nearest protein atom.

Coordination of ligand atoms to protein-bound metal ions (Mg, Ca, Mn, Fe, Co, Zn, Gd) is treated as a pseudo H-bonding interaction. Standard energies for the coordination of various types of electronegative atoms to each of these ions are available to the program. These energies (typically –15 or –10 kcal/mol) have limited theoretical justification, but give reasonable results in practice. All of the ions are allowed to form octahedral coordination geometries; Mg, Mn, Fe, and Zn are also allowed to be tetrahedral. Given the partial coordination sphere determined by interactions between the metal ions and coordinating protein atoms, the remaining available coordination positions are identified for use in ligand docking. Each position can be occupied by a single ligand atom or shared between a pair of electro-

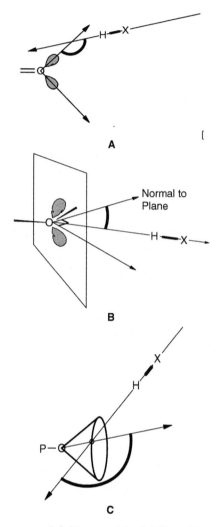

FIGURE 15.1 Angle parameter definitions used in GoldScore H-bond energy terms. A: For lone-pair directional acceptors, angles to all lone pair vectors are calculated. The largest angle is used. The angular energy weight is a maximum when the subtended angle is 180°. B: For planar acceptors, the angular energy weight is a maximum when the subtended angle is 90°. C: For cone acceptors, the angular energy weight is a maximum when the angle subtended is 180°.

negative ligand atoms that are bonded to a common atom (e.g., a bidentate carboxylate group). GOLD is thereby able to reproduce coordination numbers other than 4 or 6. Ideal metal coordination distances were estimated from surveying metal complexes in small molecule crystal structures [1,2]. The energy of interaction between a metal and a coordinating ligand atom is attenuated if the geometry departs from ideal, the weighting methodology being similar to that used for H-bonds. Thus, the directional preferences assumed for various electronegative atoms when they H-bond (see Table 15.1) will be assumed also when they coordinate to metals.

15.2.1.2   External vdW Energy

Protein–ligand vdW interaction energy is obtained by summing the individual atom-pair (ij) contributions

$$E_{ij} = A/d^8_{ij} - B/d^4_{ij} \qquad (15.1)$$

where the sum is over protein–ligand atom pairs closer than a specified maximum distance. The expression is softer than traditional exp-6 or 12-6 potentials and sometimes leads to dockings containing unrealistically short protein–ligand contacts. This is deliberate. In common with most other docking programs, GOLD assumes that the protein binding site is rigid (apart from H-bonding groups, see Section 15.3.1). This prohibits a protein moving slightly to accommodate a ligand that is just a little too large, although in reality such movement is highly likely. The soft potential compensates by allowing unusually short nonbonded contacts. The net effect is an improvement in overall docking prediction accuracy, although the presence of a few short contacts is occasionally alarming to uninitiated users. Even with the softer 8-4 potential, the energy of very short contacts can become extremely large and lead to instability in the search algorithm. A cutoff is therefore applied such that, at very short distances, the energy rises only linearly (Figure 15.2). The distance at which the cutoff is applied changes as the search algorithm proceeds (see Section 15.3.4). The vdW interaction energy of H-bonded H...A pairs is ignored.

In comparison with the Tripos force field [8], the GOLD potentials tend to be deeper for hydrophobic (e.g., carbon...carbon) pairs. This encourages hydrophobic protein–ligand contacts but, as was shown by early test runs in which hydrophobic ligands were poorly docked, not to a sufficient extent. The overall external vdW term in GoldScore is therefore artificially increased by an empirical factor (1.375) chosen to promote hydrophobic contact area without diminishing too much the importance of H-bonding interactions.

15.2.1.3   Internal Strain and H-Bond Energy

Internal ligand strain energy is estimated using the vdW and torsional parts of the Tripos force field [8]. The latter component is optional but switched on by default. Internal ligand H-bonding energy can be computed, using similar methodology to that used for external H-bonds, but this term is switched off by default.

15.2.1.4   Other Aspects of GoldScore

GoldScore is open to several criticisms. It is not trained against experimental binding affinities. It has no penalty term to account for loss of entropy when the ligand is frozen into a particular conformation upon binding. Because it does not use atomic partial charges, it cannot account for electrostatic interactions other than H-bonding. In common with most other scoring functions, it is poor at penalizing some unfavorable arrangements such as the occlusion of a H-bonding group by a hydrophobic group. Despite these limitations, GoldScore has proven useful in practice.

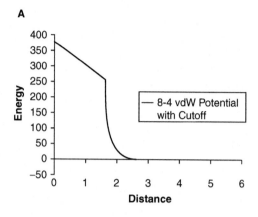

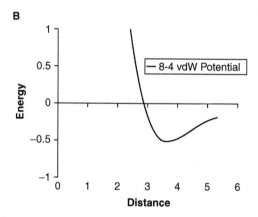

FIGURE 15.2 Example 8-4 vdW potential used in GOLD (energy in kcal/mol, distance in Å). A: Cutoff of GoldScore protein–ligand nonbonded repulsion energy at short distances. B: Same curve with larger energy scale to show showing attractive region at moderate distances.

The implementation of GoldScore within GOLD is highly configurable. All force field parameters (e.g., H-bond energies) are defined within an editable parameter file and hence can be changed. Other terms such as the overall weights applied to the external vdW and internal strain energy terms are also customizable. New atom types and torsion definitions can be added. Several switches are available, such as for switching between 8-4 and 12-6 nonbonded potentials, altering vdW cutoff distances, preferred H-bond geometries, etc. One or two of these options are experimental and little tested. For example, there is an experimental desolvation term and an option for discouraging close approach of ligand acceptor atoms to protein acceptor atoms and/or ligand donor hydrogens to protein donor hydrogens. As with all empirical scoring functions, the effect of changing parameters is often unpredictable. It is quite possible that the default GoldScore configuration used in GOLD could be further optimized.

15.2.2 ChemScore

The ChemScore scoring function [9] was originally derived for affinity prediction and, unlike GoldScore, was parameterized by regression against literature protein–ligand binding constants. Subsequently, it was adapted for use in docking in the program PRO_LEADS [10]. It approximates the free energy of binding ($\Delta G_{binding}$) of a ligand (in a given pose) to a protein binding site as

$$\Delta G_{binding} = \Delta G_0 + \Delta G_{hbond} + \Delta G_{metal} + \Delta G_{lipo} + \Delta G_{rot} \qquad (15.2)$$

Each term on the right-hand side estimates the free-energy contribution from a specific type of interaction or physical effect, viz.

ΔG_0 Baseline contribution (nonligand-dependent part of the free energy of binding)

ΔG_{hbond} Change of free energy due to forming ligand–protein H-bonds

ΔG_{metal} Change of free energy due to forming ligand–metal coordination contacts

ΔG_{lipo} Change of free energy due to forming lipophilic ligand–protein atom contacts

ΔG_{rot} Change of free energy due to freezing ligand rotatable bonds

Equation 15.2 can be rewritten in a form suitable for regression

$$\Delta G_{binding} = v_0 + v_{hbond} P_{hbond}() + v_{metal} P_{metal}() + v_{lipo} P_{lipo}() + v_{rot} P_{rot}() \qquad (15.3)$$

The P terms are functions that estimate the magnitudes of each type of interaction or physical effect for a given pose; the v terms are regression coefficients derived by fitting to a training set of 82 complexes of known affinity. Full details are given in the literature [9,11]. Here, we focus on the differences between the original ChemScore equation and that implemented in GOLD by Verdonk et al. [11].

15.2.2.1 Gaussian Smoothing

In the literature version of ChemScore, ΔG_{hbond}, ΔG_{lipo}, and ΔG_{metal} are estimated using convolutions of one or more block functions of the form

$$B(x, x_{ideal}, x_{max}) = \begin{cases} 1 \text{ if } x < x_{ideal} \\ 1 - \dfrac{x - x_{ideal}}{x_{max} - x_{ideal}} \text{ if } x_{ideal} \le x \le x_{max} \\ 0 \text{ if } x > x_{max} \end{cases} \qquad (15.4)$$

or its mirror image. An example of an unsmoothed angular block function is shown in Figure 15.3. In GOLD's implementation of ChemScore, the block functions are further convoluted with a Gaussian smoothing function

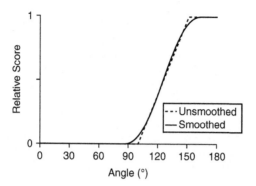

FIGURE 15.3 Unsmoothed and smoothed block functions of type used in ChemScore.

$$B'(x, x_{ideal}, x_{max}, \sigma) = \frac{\int_{-\infty}^{+\infty} B(x-u, x_{ideal}, x_{max})g(u,\sigma)du}{\int_{-\infty}^{+\infty} g(u,\sigma)du} \qquad (15.5)$$

where

$$g(u,\sigma) = e^{-u^2/2\sigma^2} \qquad (15.6)$$

σ values of 10° and 0.1 Å are used for angular and distance block functions, respectively. The effect of smoothing can be seen in Figure 15.3. Its intention is to reduce discontinuities in the ChemScore hypersurface; for example, the smoothed functions should be more tolerant of small inaccuracies in protein geometry than are the original block functions.

15.2.2.2 Metal Coordination Parameterization

ChemScore incorporates a metal coordination term

$$P_{metal} = \sum_{\substack{All\ ligand \\ acceptors}} \sum_{\substack{All\ protein \\ metals}} B'(r_{aM}, R_{ideal}, R_{max}, \sigma_{metal}) \qquad (15.7)$$

In the PRO_LEADS implementation of ChemScore [10], this term is calculated with R_{ideal} = 2.2 Å and R_{max} = 2.6 Å. In GOLD, a longer range term is used with R_{ideal} = 2.6 Å and R_{max} = 3.0 Å. This was necessary to prevent metal–ligand interactions from being missed.

15.2.2.3 Modification of ChemScore for Docking

The original ChemScore equation does not contain any terms that penalize bad contacts between the protein and the ligand or that represent internal ligand strain energy. Both terms (the latter comprising torsion-energy and nonbonded-clash components) were added to ChemScore for use in GOLD. The new terms follow those used in the PRO_LEADS implementation of ChemScore [10] but are modified for GOLD to account for the possibility of ring flexibility [11]. A further modification was made to improve docking speed. The rotatable bond freezing term from the original ChemScore [9] requires the set of frozen rotatable bonds to be identified for each trial pose. In GOLD, all bonds are regarded as frozen when a ligand is docked, which reduces the term to a constant for a given ligand.

15.3 SEARCH STRATEGY

15.3.1 Search Space; Ligand and Protein Flexibility

During docking, GOLD allows the torsion angles of all acyclic, rotatable bonds in the ligand to vary except for amide bonds and bonds to conjugated trigonal nitrogens. Amide linkages are set to trans planar and are not varied unless amide flipping has been turned on, in which case trans and cis planar geometries are allowed. Similarly, trigonal sp^2 nitrogen atoms (e.g., a R,R'-N-disubstituted aniline) can be made to flip in the search algorithm between the two alternative planar geometries. Ligand ring conformations may be altered by corner flipping [12]. Inversion of pyramidal nitrogen atoms is not allowed. The input stereochemistry of ligands is maintained, so it may be necessary to dock multiple stereoisomers for each ligand in a data set. Neither ligand nor protein bond lengths nor angles are allowed to change. The protein is held rigid except for H-bond donor groups (serine, threonine, and tyrosine –OH; lysine –NH$_3^+$), which are allowed to rotate unless involved in H-bonding to nearby protein acceptor groups. Water molecules can be included in the binding site, but their orientations are not optimized. Ligand movement is confined to a user-defined region of the binding site (see Section 15.5.2); this region should be large enough to include all realistic binding modes.

15.3.2 Genetic Algorithm Parameters

GOLD uses a genetic algorithm (GA) [13] to search for favorable ligand dockings. A population of chromosomes is manipulated during each GA run, each chromosome representing a trial docking. A chromosome contains all the information needed to completely define a trial ligand pose and is associated with a fitness value, computed from whichever scoring function is in use. The GA itself is controlled by a number of parameters. The population is distributed over a number of different islands, n_islands. The number of chromosomes on each island is pop_size. Three genetic operations are allowed, viz. mutation of a single chromosome, crossover between two chromosomes on the same island, and migration of a chromosome from one island to another. The total number of operations applied during the GA is determined by n_ops. Each time an operation is performed, the specific operator to be used is chosen by roulette-wheel selection based on operator weights *mutation, crossover,* and *migration.* By default, the first two of these are set so that mutations and crossovers are performed with equal frequency (though it is not clear whether this is optimum). The ratio of mutation to crossover has the largest influence on how much time the GA spends in exploring the search space (encouraged by high crossover weight) compared to how much time is spent exploring local minima (encouraged by high mutation weight). The migration weight is set low by default (there is no evidence that the use of islands in GOLD improves GA performance or evidence to the contrary). The parameter select_pressure represents the relative probability of selecting the most fit member of the current population for the next operation to be performed, compared to the probability of selecting an individual whose fitness is equal to the average fitness of all population members. Higher select_pressure settings force the algorithm to converge more quickly (hence, n_ops could be reduced), but at the cost of a less thorough search.

After each operation, the child chromosome will normally replace the least fit member of the population (hence, pop_size remains constant). However, niching is used to increase population diversity: the chromosome population is sorted into niches, each niche containing chromosomes that code for similar docking solutions. When a new child is generated, it is compared with every member of the existing population. If it is found to be similar to an existing chromosome, it replaces the least fit member of the niche to which that chromosome belongs. The maximum size of a niche is defined by the parameter niche_size. Even with niching, it is not clear whether there is much diversity within a GOLD GA population during the latter half of a typical run.

It is usual to perform several GA runs on a single protein–ligand complex. The maximum number of runs is determined by the parameter n_runs. Because GAs are nondeterministic, each run may find a different docking solution, so several possible binding modes are usually found. These are ranked by fitness score so that the top-ranked pose is the one with the best score. Before ranking, each solution is subjected to a final simplex optimization to take its score to the local optimum. It has been observed that, if the same solution is found several times in repeated docking runs on the same complex, it is often a correct solution. Thus, an early termination option is provided: as solutions are generated, they are compared with one another, and the program terminates — even if not all n_runs dockings have been performed — if a point is reached at which the current best three solutions are effectively identical.

15.3.3 Chromosome Composition and Ligand Placement

A chromosome encapsulates all the conformational and positional information required to define a ligand pose. Conformational information comprises the torsion angles around rotatable ligand bonds and bonds to rotatable protein H-bonding groups. Positional information is held in a rather unusual way: rather than store (say) the (x, y, z) coordinates of the ligand centroid and a set of Euler angles to define orientation, the chromosome stores a mapping of H-bond donor and acceptor ligand groups onto complementary H-bonding points on the protein, and of hydrophobic points on the ligand onto hydrophobic points on the protein. The ligand is then positioned in the binding site by a two-step process. An initial least-squares fit is performed of all mapped pairs of ligand and protein fitting points. Pairs that are not in close proximity after this fitting are eliminated and a second, final fit performed on the remaining pairs. In consequence, a given mapping of ligand and protein fitting points, together with the rotatable torsion values held in the chromosome, unambiguously define a specific ligand pose, as required. The reason for storing positional information indirectly as fitting-point mappings rather than directly as coordinates and Euler angles is that the mappings, being chemically favorable (H-bond donor onto acceptor; acceptor onto donor; hydrophobe onto hydrophobe), inherently bias the GA toward exploring likely binding modes.

Generation of H-bond fitting points is straightforward, being based on the observed geometrical preferences of H-bonds in small molecule crystal structures (see Section 15.2.1.1). Because hydrophobic interactions are much less directional, the positioning of hydrophobic fitting points is more problematic. By default, GOLD superimposes a grid on the binding site and evaluates at each grid position the vdW energy of a single carbon atom with the protein residues. The energy is evaluated by application of the user-specified nonbonded potential (i.e., the 12-6 or 8-4 potential terms); every point with an energy below −2.5 kcal/mol is added to the list of fitting points. In this way, a map is generated that contains points at which the placement of a hydrophobic ligand atom would be favorable. An individual chromosome will contain mappings between a subset of these points and a subset of the lipophilic atoms (carbons bound to at least one hydrogen) in the ligand. Alternatively, customized hydrophobic fitting points on the protein can be defined in cases where the set of points generated by GOLD is unsatisfactory; these, for example, can be derived from a hot-spot identification program such as SuperStar [14]. GOLD also uses special fitting points positioned above and below tryptophan ring systems, onto which ligand NMe_3^+ groups can be mapped; this is to reproduce interactions between the electron-rich tryptophan system and soft electropositive groups [15].

15.3.4 ANNEALING

The aim of annealing is to allow relatively unfavorable interactions to exist at the beginning of a GA run, it being assumed that they will improve as the run progresses. Two parameters are annealed when GoldScore is used. The first is the value at which the repulsive r^8 term in the external vdW energy component is replaced by a linear function (Figure 15.2A): this occurs at longer distances at the start of a GA run than at the end, allowing short contacts to be tolerated in the early stages of the search. Secondly, the maximum allowed deviation between a H-bond distance and its ideal value decreases as a run progresses. H-bond annealing is also performed when ChemScore is used. At the beginning of the run, the maximum H...acceptor distance is set to 3 Å and the minimum donor-H–acceptor angle to 70°. When 75% of the total number of operations has been carried out, the maximum distance becomes 2.5 Å and the minimum permissible donor-H–acceptor angle 100°.

15.3.5 TUNING THE GENETIC ALGORITHM

The default GA parameters originally used in GOLD resulted in slow, thorough searches, acceptable for docking an individual ligand but of limited use for high throughput VS. The current version of GOLD therefore offers a choice of parameter sets (Table 15.2), each of which gives a particular balance between speed and docking accuracy. Profiling of central processing unit (CPU) usage shows that GOLD spends most time in fitness-score calculation. The number of fitness-score calculations depends on the number of children generated by the GA, which, in turn, is almost entirely determined by the parameter n_ops. For a given fitness function, significant reductions in execution time can therefore only be achieved by reducing this parameter. The pop_size parameter is a secondary determinant of program speed. This is because niching requires each new child to be compared with the existing population (i.e., pop_size comparisons). No other parameters have a significant influence on execution time. Table 15.3 shows the effect on execution time of changing n_ops and pop_size. Times given ignore the fixed overhead of protein initialization, which can be significant but is performed only once per VS experiment. The table confirms that execution times can be very significantly decreased by reducing n_ops; further speed-up can be obtained by reducing pop_size.

Fast GA settings are only useful if the resulting dockings are reasonably reliable; some decrease in accuracy can be tolerated, but not too much. Validation results (see Section 15.4.3.1) show only modest deterioration in docking reliability for faster GA settings. Use of 10,000 to 30,000 operations is a good general purpose compromise. Remarkably, GOLD achieves some success even when n_ops is as small as 100 (however, this is not recommended). This suggests that much of its predictive ability comes from the way in which the initial population is set up. In particular, the mapping of H-bonding and hydrophobic points on the ligand onto complementary points on the protein is likely to increase the chance of the starting population containing some reasonable solutions.

15.3.6 USE OF TORSION DISTRIBUTIONS

Early GOLD development versions allowed all rotatable bonds to vary through a full 360° range. Searches of the Cambridge Structural Database (CSD) [16], however, show that the experimentally observed ranges of many common types of torsion angles are very restricted, either because of steric or conjugative factors [17]. A small library of experimentally observed torsion-angle distributions was therefore developed from the CSD [1,2]. During ligand initialization, GOLD attempts to match each rotatable bond with a torsion distribution in the library. If a match is found, the GA will no longer sample the full 360° range, but will constrain the angle to those ranges observed to be experimentally accessible. Values outside these ranges will never be generated and it is therefore only safe to include well-populated torsion distributions in the library.

TABLE 15.2
Standard GA Parameter Sets in GOLD

	Parameter Set I	Parameter Set II	Parameter Set III	Parameter Set IV	Parameter Set V
n_ops	100,000	50,000	30,000	10,000	1,000
n_islands	5	5	3	1	1
pop_size	100(500)	100(500)	100(300)	100(100)	50(50)
mutation	95	95	95	100	100
crossover	95	95	95	100	100
migration	10	10	10	0	0
select_ pressure	1.1	1.1	1.1	1.1	1.125

See text for explanation of symbols. Total population over all islands is given in parentheses.

TABLE 15.3
CPU Time (Secs) for One GA Run* for Different Combinations of Scoring Function, Number of Operations, and Population Size

Number of Operations	Total Population Size	Average Speed per GA Run		Parameter Set (Table 15.2)
		GoldScore	ChemScore	
100,000	500	97.8	46.9	I
100,000	250	82.1	41.3	n.a.
50,000	500	54.7	25.5	II
30,000	300	32.5	14.5	III
10,000	100	10.6	4.8	IV
1,000	100	4.1	3.4	n.a.
1,000	50	2.7	1.1	V

Times are averages over all 305 complexes in the GOLD validation set.

*Pentium 4 1.8 GHz processor under Debian Linux operating system.

Using the torsion library does not make the GA go any faster, because the same number of operations is still performed, but it focuses the search on lower-energy regions of ligand conformational space. The default torsion library can be extended, replaced by an alternative library, or ignored. The efficacy of using torsion libraries is unclear. They are unlikely to detract from predictive accuracy unless distributions based on an inadequate number of observations are used. They may not give as much advantage as might be anticipated, however, because unfavorable ligand conformations may well be rejected quickly by the GA anyway.

15.4 PREDICTION RELIABILITY

15.4.1 VALIDATION SETS FOR TESTING DOCKING PROGRAMS

Docking programs must be validated by extensive application to known protein–ligand complexes. The first GOLD validation set [2] was compiled to test a fundamental requirement for all docking programs, viz. the capacity to find and single out native ligand binding modes. It comprised 100 Protein Data Bank (PDB) entries (extended to 134 a short time later) and was the first docking test set to be made publicly available, hence becoming something of a benchmark. Other large test sets were subsequently collated: for example, 200 complexes were used to test FlexX [18], 154 for EUDOC [19], and 70 for PRO_LEADS [20]. Many docking tools, however, are still validated against distinctly smaller sets of protein–ligand complexes. Thus, a recent study describes DOCK 4.0 results for 12 complexes [21]; DARWIN results are shown for 3 examples [22]; Blom and Sygusch [23] report results for 7 complexes; 19 were used for MCDOCK [24]; and Wang et al. report results on 12 complexes for DOCK 3.5 [25].

A validation set should be large (so that results are statistically significant), diverse (to avoid biases toward particular protein families), and comprise accurate structures. As GOLD progressed, it became apparent that even the GOLD validation set of 134 entries was not good enough to assess reliably the effect of new developments and an extended test set was collated [26]. This added a further 171 complexes to the original 134. Of the original entries, 3 were updated (2ack, 3mth, and 4aah) and known errors were fixed. Forty-eight entries from the ChemScore test set [20] that were not in the original GOLD set were added. Selection of the remaining 123 new entries was influenced by several considerations. An attempt was made to cover pharmaceutically interesting protein families. The diversity of the set was increased by focussing on structurally different classes of proteins — specifically, by searching for entries from the 58 best-populated protein structural classes [27] and their homologues. Relibase+ [28] was used to search for and browse through families of homologous proteins and their ligands. With one or two exceptions, complexes were rejected if their crystallographic resolution was worse than 3.0 Å. In the case of disordered ligands, one of the orientations of the ligand was taken after visual inspection.

Protonation states within binding sites were inferred by inspecting the immediate H-bonding environment of ligand and protein groups, taking particular care in assigning protonation and tautomeric states of acidic groups and histidines. Filters were applied to identify and flag suspect entries. These covered:

- The presence of severe factual and structural errors in the original PDB file.
- Inconsistency between the electron density in the binding site and the reported position and conformation of the ligand.
- An unlikely ligand conformation.

- The presence of severe clashes between protein and ligand atoms.
- The necessity of including crystallographically related protein residues to describe the binding site correctly.

Of the 305 complexes in the new set, a diverse subset of entries not affected by any of these problems is recommended for validation purposes (Table 15.4).

TABLE 15.4
PDB Codes of the Clean Set, a Subset of the GOLD Validation Set Containing Diverse, Relatively Problem-Free Entries

1a28	1a42	1a4g	1a4q	1a6w	1a9u	1aaq	1abe	1abf	1acj
1acl	1acm	1aco	1aec	1ai5	1aoe	1apt	1apu	1aqw	1ase
1atl	1azm	1b58	1b59	1b9v	1baf	1bbp	1bgo	1bl7	1blh
1bma	1bmq	1byb	1byg	1c12	1c1e	1c5c	1c5x	1c83	1cbs
1cbx	1cdg	1cil	1ckp	1cle	1com	1coy	1cps	1cqp	1cvu
1cx2	1d0l	1d3h	1d4p	1dbb	1dbj	1dd7	1dg5	1dhf	1did
1dmp	1dog	1dr1	1dwb	1dwc	1dwd	1dy9	1eap	1ebg	1eed
1ei1	1ejn	1eoc	1epb	1epo	1eta	1etr	1ets	1ett	1f0r
1f0s	1f3d	1fax	1fen	1fgi	1fkg	1fki	1fl3	1flr	1frp
1glp	1glq	1hak	1hdc	1hfc	1hiv	1hos	1hpv	1hri	1hsb
1hsl	1htf	1hvr	1hyt	1ibg	1ida	1imb	1ivb	1ivq	1jap
1kel	1lah	1lcp	1ldm	1lic	1lna	1lpm	1lst	1lyb	1lyl
1mbi	1mcq	1mdr	1mld	1mmq	1mrg	1mrk	1mts	1mup	1nco
1ngp	1nis	1okl	1okm	1pbd	1pdz	1phd	1phg	1poc	1ppc
1pph	1ppi	1pso	1ptv	1qbr	1qbu	1qcf	1qpe	1qpq	1rds
1rne	1rnt	1rob	1rt2	1slt	1snc	1srj	1tdb	1tlp	1tmn
1tng	1tnh	1tni	1tnl	1tpp	1trk	1tyl	1ukz	1ulb	1uvs
1uvt	1vgc	1wap	1xid	1xie	1ydr	1ydt	1yee	25c8	2aad
2ack	2ada	2ak3	2cht	2cmd	2cpp	2ctc	2dbl	2fox	2gbp
2h4n	2ifb	2lgs	2mcp	2pcp	2phh	2pk4	2qwk	2r07	2tmn
2tsc	2yhx	2ypi	3cla	3cpa	3erd	3ert	3gpb	3hvt	3tpi
4aah	4cox	4cts	4dfr	4est	4fbp	4lbd	4phv	5abp	5cpp
5er1	6rnt	6rsa	7tim						

15.4.2 INTERPRETING VALIDATION RESULTS

Given a validation set of known protein–ligand complexes, the reliability of a docking program can be quantified by the percentage of ligands that it docks sufficiently closely to the experimental position. However, this headline figure is influenced by many factors, both systematic and random, that obscure the underlying performance of the docking algorithm. These should be eliminated or at least characterized. Systematic factors include:

- The nature of the test systems used — Large ligands with many rotatable bonds pose a tougher problem than small, rigid ones bound in a small cavity. To provide a useful assessment of docking performance, the validation set should cover a sufficiently broad range of ligand and cavity sizes. Results will be influenced by whether ligand rings are allowed to flex or not.
- Structural errors in the data — A test case may never be predicted properly if the experimental structure is severely in error. Erroneous entries should be excluded as far as possible (see Section 15.4.1). This also includes entries that display extensive crystal-packing effects in the vicinity of the binding site.
- Absence of water molecules — Waters may well play a structural role in the binding of certain ligands in the test set but are usually excluded from docking (only waters strongly coordinated to metal ions are retained in the GOLD validation set).
- The way in which information is passed on to the docking program — The binding site definition will influence results: in particular, confining the search to a small box around the observed ligand position presents an easier problem than if a large box is used. Also, the native ligand pose should not be provided as input as some algorithms may indirectly use this information and thereby obfuscate results.

Random influences on validation results include:

- The nature of the docking algorithm — The nondeterministic nature of methods like GOLD's GA will randomly influence results. This effect will be more pronounced for difficult cases, like large and flexible ligands.
- The validation set size — A small test set will give rise to large sampling errors (see later in this section).
- Data precision — Protein crystal structures may be of low resolution.

Systematic influences are often closely related to the actual contents of the data set and can be curbed by choosing the test systems carefully. Random effects are to some extent dependent on the set's makeup (structural resolution) and, to a large extent, the result of the validation set being a limited sample from a universe of good test systems. When using a validation set to determine docking success rates, we are actually estimating the success rate that would be obtained on the underlying, hypothetical population of all possible good test systems. The only way to obtain a precise estimate is by making the validation set as large as possible. The precision of the estimate can be determined by bootstrapping [26]. It is essential to know this precision to assess whether success rates for different docking methodologies are significantly different. Unfortunately, reported success rates seldom include such error estimates.

Most current validation sets assess the ability of the docking program to reproduce known binding poses of ligands within a protein cavity. As such, they do not account for the inherent flexibility of protein structures during binding. Attempts have been made to address protein flexibility, either during the docking process or by means of cross-docking experiments, but interpreting such results is difficult due to the validation sets being small [29].

15.4.3 GOLD Validation Results

15.4.3.1 Current Validation Results for Alternative Genetic Algorithm Settings

Docking validation results are usually quoted as success rates giving the percentage of ligands that are docked within a certain root-mean-square deviation (rmsd) of the experimentally observed pose. A rmsd threshold of 2 Å is commonly used to define success, although very small ligands probably merit tighter thresholds. In validating GOLD, only the top-ranked solution for a given ligand is used (e.g., a docking is not considered a success if the second-ranked solution has a rmsd of 1.5 Å but the top-ranked solution has a rmsd of 2.5 Å). Table 15.5 reports results for various subsets of the new GOLD validation set. These include the full set of 305 entries, the clean set of 224 reliable, diverse entries (Table 15.4), subsets of the clean set meeting various resolution criteria [26], and a subset of 139 complexes involving druglike ligands. The latter was suggested by Verdonk et al. [11] and derived by application of Veber's rules [30] for predicting bioavailability. In fact, this set contains many small, easy-to-dock structures, some of which are arguably not druglike but are at least leadlike. The full and clean sets contain ligands ranging from very small, rigid molecules to large, flexible compounds. Results are given for several of the GA parameter sets in Table 15.2.

GOLD yields better results for the clean set, from which doubtful structures have been removed, than for the full set. Defining a good solution as one with rmsd < 2.0 Å, the average success rate was 66(1.1)% for the full set and 70(1.3)% for the clean set (Parameter Set I, i.e., slow setting, standard deviation for 50 runs in parentheses). Marginally improved success rates are obtained for better resolution structures (71(1.5)% for resolution better than 2.5 Å, 180 entries; 76(2.2)% for resolution better than 2.0 Å, 92 entries). Faster, medium-speed settings (Parameter Set III) give only slightly lower success rates and an approximately three-fold speedup. Results for the druglike set are distinctly better than results for both clean and full sets, presumably because the former mainly comprises rather small ligands with few rotatable bonds. For both medium and slow settings, GoldScore outperforms ChemScore on both the clean and druglike sets. ChemScore, however, is up to three times faster [11]. As a compromise, it appears that consensus docking, in which dockings are performed using ChemScore and subsequently rescored using

TABLE 15.5
GOLD Success Rates for Different Subsets of the Validation Set

Parameter Set I

	$f_{RMS<0.5 Å}$		$f_{RMS<1.0 Å}$		$f_{RMS<1.5 Å}$		$f_{RMS<2.0 Å}$		$f_{RMS<2.5 Å}$		$f_{RMS<3.0 Å}$	
	GS	CS	GS	CS	GS	CS	GS	CS	GS	CS	GS	CS
All entries	12(1.0)	9(0.8)	39(1.2)	33(1.4)	57(1.2)	54(1.3)	66(1.1)	64(1.3)	73(1.2)	70(1.3)	78(1.1)	74(1.1)
Clean set	14(1.2)	10(1.0)	45(1.5)	37(1.7)	63(1.4)	59(1.6)	70(1.3)	68(1.6)	76(1.4)	73(1.2)	80(1.5)	77(1.1)
Clean set, R < 2.5Å	16(1.4)	11(1.2)	46(1.8)	38(1.8)	64(1.6)	60(1.8)	71(1.5)	69(1.7)	78(1.7)	74(1.5)	81(1.8)	78(1.4)
Clean set, R < 2.0Å	17(2.2)	13(1.7)	51(2.5)	44(2.4)	69(2.1)	68(2.2)	76(2.2)	75(2.0)	82(2.7)	79(1.8)	85(2.6)	81(1.7)
Druglike set	21(1.8)	15(1.4)	57(1.4)	48(2.3)	74(1.4)	72(1.8)	79(1.3)	78(1.5)	84(1.3)	82(1.3)	87(1.4)	86(1.2)

Parameter Set III

	$f_{RMS<0.5 Å}$		$f_{RMS<1.0 Å}$		$f_{RMS<1.5 Å}$		$f_{RMS<2.0 Å}$		$f_{RMS<2.5 Å}$		$f_{RMS<3.0 Å}$	
	GS	CS	GS	CS	GS	CS	GS	CS	GS	CS	GS	CS
All entries	12(1.1)	10(0.9)	39(1.5)	32(1.5)	57(1.4)	52(1.4)	65(1.5)	61(1.4)	72(1.5)	67(1.5)	77(1.5)	71(1.4)
Clean set	15(1.4)	11(1.1)	45(1.8)	37(1.7)	62(1.6)	57(1.5)	69(1.8)	65(1.4)	75(1.8)	70(1.6)	79(1.7)	75(1.7)
Clean set, R < 2.5Å	16(1.5)	11(1.3)	45(1.8)	37(1.7)	63(1.6)	58(1.7)	70(2.0)	66(1.5)	76(2.0)	71(1.9)	79(1.9)	75(1.9)
Clean set, R < 2.0Å	17(1.7)	13(1.6)	50(2.7)	43(2.4)	68(2.5)	66(2.2)	75(2.6)	72(2.0)	81(2.6)	76(2.2)	85(2.3)	79(2.2)
Druglike set	21(2.2)	16(1.6)	57(2.1)	49(2.1)	75(1.6)	71(1.7)	80(1.7)	77(1.7)	86(1.3)	81(1.5)	88(1.3)	84(1.6)

Parameter Set IV

	$f_{RMS<0.5 Å}$		$f_{RMS<1.0 Å}$		$f_{RMS<1.5 Å}$		$f_{RMS<2.0 Å}$		$f_{RMS<2.5 Å}$		$f_{RMS<3.0 Å}$	
	GS	CS	GS	CS	GS	CS	GS	CS	GS	CS	GS	CS
All entries	11(0.9)	9(1.1)	36(1.3)	30(1.6)	53(1.2)	48(1.5)	61(1.4)	57(1.5)	68(1.6)	63(1.6)	73(1.4)	68(1.5)
Clean set	14(1.2)	11(1.4)	42(1.5)	34(1.9)	59(1.5)	54(1.8)	66(1.7)	62(1.5)	72(1.9)	67(1.7)	75(1.7)	71(1.6)
Clean set, R < 2.5Å	16(1.5)	11(1.4)	43(1.7)	35(2.1)	60(1.7)	55(2.1)	67(1.8)	63(1.6)	73(1.9)	67(1.6)	76(1.6)	71(1.6)
Clean set, R < 2.0Å	16(2.3)	14(2.3)	47(2.6)	41(2.6)	65(2.7)	62(2.1)	72(2.5)	68(2.2)	78(2.2)	72(1.9)	81(1.9)	75(2.2)
Druglike set	21(2.0)	16(2.1)	55(2.3)	47(2.5)	73(2.0)	69(2.0)	79(2.0)	75(1.6)	84(2.0)	79(1.8)	86(1.7)	83(1.9)

Parameter Set V

	$f_{RMS<0.5 Å}$		$f_{RMS<1.0 Å}$		$f_{RMS<1.5 Å}$		$f_{RMS<2.0 Å}$		$f_{RMS<2.5 Å}$		$f_{RMS<3.0 Å}$	
	GS	CS	GS	CS	GS	CS	GS	CS	GS	CS	GS	CS
All entries	6(1.0)	6(1.2)	27(1.3)	23(1.3)	42(1.6)	39(1.6)	51(1.4)	48(1.8)	58(1.7)	55(1.7)	63(1.7)	60(1.8)
Clean set	8(1.4)	7(1.4)	31(1.7)	26(1.7)	48(1.7)	44(2.0)	57(1.8)	52(2.1)	63(2.1)	58(1.9)	67(2.0)	62(2.1)
Clean set, R < 2.5Å	9(1.5)	7(1.7)	32(2.1)	27(1.8)	49(1.8)	44(2.4)	58(1.8)	53(2.3)	63(2.1)	58(2.2)	67(2.0)	62(2.4)
Clean set, R < 2.0Å	8(2.1)	8(2.3)	33(2.8)	32(2.6)	51(3.1)	50(2.7)	60(2.8)	57(2.8)	65(3.0)	62(2.8)	69(2.9)	66(2.8)
Druglike set	12(2.0)	10(2.2)	44(2.7)	39(2.5)	65(2.4)	60(2.5)	72(2.4)	69(2.2)	78(2.5)	74(2.3)	81(2.2)	78(2.3)

Standard deviations in parentheses.

Success rates are percent of top-ranked solutions with rmsd from the experimental pose less than the indicated threshold; results are averaged over 50 runs; parameter sets refer to Table 15.2; GS = GoldScore, CS = ChemScore.

GoldScore, is a good option for VS applications in which small-to-average-sized compounds need to be processed quickly [11]. For average-to-large structures, docking using GoldScore seems a good approach. The efficiency of ChemScore for docking tails off quickly with increasing number of rotational bonds (Table 15.6).

Bootstrapping can be used to estimate the total standard deviation of a success rate, including not only the uncertainties due to the stochastic nature of the GA but also the sampling error that arises from selecting a validation set from an unknown universe of all possible good test complexes [26]. The bootstrapping procedure picks a large number of random sets of a given size from the available entries (full set, 305; clean set, 224; druglike set, 139), allowing entries to be picked more than once. Given that the validation set is a reasonably random sample from the universe of good test complexes, the variance of success rates observed for these artificial sets is indicative of the sampling error. Table 15.7 shows bootstrapped standard deviations for the success rates in Table 15.5; they amount to about 3 percentage units for the clean set and 4 to 5 units for smaller sets containing about 100 protein–ligand complexes.

15.4.3.2 Importance of Mediating Water Molecules

The removal of waters from test complexes is expected to detract from docking performance as they can mediate interactions that are essential for ligand binding. GOLD tests on 40 complexes with mediating waters (at least one water within 2.9 Å of both protein and ligand) and 55 complexes without show slightly worse results for the former (Table 15.8) [26]. In all cases, docking was performed with all waters removed. ChemScore and GoldScore perform similarly on entries in which binding is mediated by water, but the latter performs better for entries where water plays no mediating role (albeit, conclusions are tentative because these sets are small).

15.4.3.3 Performance on Different Classes of Proteins

Table 15.9 summarizes GoldScore and ChemScore success rates, using the medium-speed parameter set III, for different classes of protein. The small size of the sets causes standard errors to be large, which hampers analysis. Some tentative trends can be observed. Results are poor on aspartic proteases (these often involve large ligands that require slower GA settings), but possibly better than expected on lectins. ChemScore seems to perform better than GoldScore for kinases and lyases. The lyases are difficult as they have relatively shallow binding sites and polar ligands, sometimes partly solvent-exposed; crystal waters sometimes mediate binding. Most kinase entries feature quite hydrophobic ligands and binding sites. In addition, most of them contain weak N-heterocycle CH...O H-bonds that are not modeled explicitly by GOLD.

15.4.3.4 Prediction of Ligand Affinity; Virtual High Throughput Screening

For VS, a scoring function must be able to predict ligand binding affinities. This is related to, but not identical with, the aim of predicting binding poses. In predicting affinity, it is reasonable to assume the starting point of a well-docked ligand, so it

TABLE 15.6
GOLD Validation Results as a Function of Number of Rotational Bonds

Rotational Bond Range	Average Number of Bonds	$f_{RMS<1.5 Å}$		$f_{RMS<2.0 Å}$		$f_{RMS<2.5 Å}$	
Parameter Set I		GS	CS	GS	CS	GS	CS
0 to 4	2.7	82(1.0)	85(2.2)	84(0.9)	88(1.5)	88(1.2)	89(1.5)
5 to 9	6.6	62(2.6)	62(2.2)	70(2.3)	69(2.2)	77(2.5)	73(2.0)
10 to 14	11.3	68(4.0)	56(4.3)	75(4.0)	68(3.6)	80(4.0)	77(2.4)
15 to 19	15.8	40(9.6)	20(7.2)	54(7.9)	41(9.7)	60(8.6)	55(9.1)
20 to 29	25.5	20(8.3)	7(6.7)	33(11)	20(10)	42(9.2)	29(10)
30 to 39	35	10(20)	1(7)	16(24)	13(24)	23(25)	25(34)
Parameter Set III		GS	CS	GS	CS	GS	CS
0 to 4	2.7	82(1.5)	84(2.3)	85(1.6)	87(1.8)	89(1.7)	87(2.0)
5 to 9	6.6	64(2.8)	60(2.7)	72(2.7)	67(2.4)	79(2.1)	72(2.2)
10 to 14	11.3	68(4.1)	54(4.2)	75(4.1)	65(4.9)	80(4.3)	73(3.8)
15 to 19	15.8	33(7.2)	17(6.4)	45(7.6)	34(7.4)	52(8.6)	49(8.0)
20 to 29	25.5	7(7.5)	4(5.2)	16(9.1)	13(8.1)	27(9.8)	22(10)
30 to 39	35	1(7.1)	2(9.9)	6(16)	4(14)	11(21)	12(22)
Parameter Set IV		GS	CS	GS	CS	GS	CS
0 to 4	2.7	83(1.6)	82(2.5)	86(1.7)	86(2.2)	90(1.7)	87(2.2)
5 to 9	6.6	62(3.0)	56(3.2)	70(2.9)	65(2.3)	75(3.2)	69(2.6)
10 to 14	11.3	60(4.3)	47(4.4)	69(4.1)	57(4.5)	75(4.5)	66(4.4)
15 to 19	15.8	24(7.6)	17(7.0)	37(8.3)	32(8.0)	44(8.4)	42(7.5)
20 to 29	25.5	3(4.3)	3(4.0)	9(7.4)	9(7.4)	16(8.9)	16(9.1)
30 to 39	35	-(-)	-(-)	1(7.1)	-(-)	6(16.4)	6(16.4)

Standard deviations in parentheses. Entries from clean list.

Success rates are percent of top-ranked solutions with rmsd from the experimental pose less than the indicated threshold; results are averaged over 50 runs; parameter sets refer to Table 15.2; GS = GoldScore, CS = ChemScore.

is unnecessary for the scoring function to contain terms for penalizing bad clashes. Also, more time can be devoted to scoring the single, best docking from a GOLD run than can be spent on evaluating the fitness of each of thousands of trial dockings generated during the search for the best pose. On the other hand, a function for predicting affinity must be able to make comparisons between chemically unrelated ligands; whereas, all that is necessary for pose prediction is to discriminate between different poses of the same molecule. Such considerations suggest it is desirable to use different scoring functions for pose and affinity prediction and, indeed, GOLD has the ability to do this. However, much published work involves the use of the same function for both tasks.

Current scoring functions reproduce known binding affinities with a standard deviation of about 2 kcal/mol (approximately 1.5 orders of magnitude) for known

TABLE 15.7
Bootstrapped Standard Deviations (s) for Success Rates $f_{RMS<x\AA}$

Parameter Set I [a]	$f_{RMS<1.5\ \AA}$ s		$f_{RMS<2.0\ \AA}$ s		$f_{RMS<2.5\ \AA}$ s	
	GS	CS	GS	CS	GS	CS
All entries (305)	2.8	2.7	2.7	2.7	2.6	2.5
Clean set (224)	3.1	3.1	3.0	3.0	2.9	2.9
Druglike set (139)	4.3	4.2	4.1	4.0	3.8	3.8

[a] Reported bootstrap standard deviations are averaged bootstrap results for 5 different validation runs (performed using Parameter Set I). Standard deviations of these averages were all less than or equal to 0.1. Values obtained for different parameter sets were similar to those for Set I.

Obtained by bootstrapping single validation runs for the full set of 305 entries, the clean set (Table 15.4), and the list of druglike entries. Size of the validation set given in parentheses after set name. Parameter sets refer to Table 15.2; GS = GoldScore, CS = ChemScore.

TABLE 15.8
GOLD Docking Success Rates for Structures in which Binding Is Mediated by Water (but These Waters Are Omitted from the Calculation) Compared with Success Rates for Structures with No Water Mediation

Parameter Set III	$f_{RMS<1.5\ \AA}$		$f_{RMS<2.0\ \AA}$		$f_{RMS<2.5\ \AA}$	
	GS	CS	GS	CS	GS	CS
No mediating waters	70(3.3)	63(2.7)	75(2.9)	70(3.1)	78(2.3)	73(2.6)
Mediating waters present	63(2.7)	60(4.6)	70(3.1)	69(3.4)	73(2.6)	73(3.7)

Standard deviations in parentheses.

Success rates are percent of top-ranked solutions with rmsd from the experimental pose less than the indicated threshold; results are averaged over 50 runs; parameter sets refer to Table 15.2; GS = GoldScore, CS = ChemScore.

(i.e., experimentally observed) binding modes [9,31]. This accuracy drops quickly for binding modes generated by docking programs. Even modest accuracy, however, may often be sufficient to distinguish possible leads from nonbinders, which is the usual requirement. Choice of scoring function is clearly important. Verdonk et al. [11] found that GoldScore outperforms ChemScore in predicting affinities, especially at fast search settings (surprising, because GoldScore was not trained explicitly to reproduce affinities whereas ChemScore was). It is likely that the optimum choice of scoring function is protein dependent, and users should examine different scoring functions to find the one best suited to the system at hand.

TABLE 15.9
GOLD Docking Success Rates for Different Classes of Protein

	Success Rates					
	$f_{RMS<1.5\ Å}$		$f_{RMS<2.0\ Å}$		$f_{RMS<2.5\ Å}$	
Parameter Set III	GS	CS	GS	CS	GS	CS
Hydrolases (92)	56(3.0)	48(2.7)	65(3.4)	57(2.7)	71(3.5)	62(3.6)
Metalloproteases (14)	72(6.4)	66(7.4)	78(6.3)	72(7.5)	83(7.2)	81(6.8)
Aspartic proteases (19)	32(4.6)	30(4.6)	38(6.6)	35(5.9)	45(7.8)	41(7.3)
Serine proteases (25)	48(7.5)	51(5.9)	58(7.5)	57(6.4)	61(7.5)	63(6.1)
Glycosidases (10)	71(4.0)	47(9.8)	73(5.6)	55(7.6)	77(6.5)	56(7.5)
Transferases and kinases (26)	61(5.1)	68(4.9)	64(4.1)	78(4.0)	68(4.4)	81(4.0)
Kinases only (7)	76(9.8)	96(7.2)	85(3.4)	98(4.7)	86(0.0)	99(2.8)
Lyases (11)	43(8.4)	56(9.5)	47(9.2)	71(6.4)	68(9.2)	72(7.4)
Oxidoreductases (25)	66(4.8)	65(4.3)	75(3.3)	75(3.6)	78(2.8)	77(3.1)
Immunoglobulins (22)[a]	66(5.2)	62(6.3)	78(3.2)	71(5.0)	84(2.5)	81(3.7)
Isomerases (11)	83(6.6)	60(5.5)	87(5.8)	61(4.1)	96(5.6)	62(3.7)
Lectins (6)	97(7.4)	90(10)	97(7.0)	90(10)	97(7.0)	90(10)
Virus proteins (2)	67(34)	15(23)	87(24)	50(35)	96(14)	76(27)

[a] This set also contains catalytic antibodies.

Standard deviations in parentheses. Entries from clean set. The number of protein–ligand entries per class is shown in parentheses.

Success rates are percent of top-ranked solutions with rmsd from the experimental pose less than the indicated threshold; results are averaged over 50 runs; parameter sets refer to Table 15.2; GS = GoldScore, CS = ChemScore.

Several strategies have been reported for obtaining good correlations of fitness scores with experimental binding constants or IC_{50} (inhibitory concentration 50%) data. Verdonk et al. [11] found it necessary to disregard the intramolecular term in the GoldScore scoring function, presumably because it is relative to an arbitrary origin and also can be highly sensitive to the geometry of the ligand. Some GOLD users have achieved success by using the average scores of all the high-ranking docking poses for a molecule, rather than the score of the top-ranked pose alone. Presumably, if multiple well-fitting (and thus high-scoring) solutions are found, the compound is more likely to be a good binder. Barbanton et al. [32] augmented GoldScore by including additional information such as whether known key H-bonds were present. Using this approach, a discriminant equation was set up for a training set of 82 (44 active, 38 inactive) aldose reductase (AR) inhibitors docked into four different conformations of the protein. Prediction rates were 90% for the training set; for a test set of 680 compounds, 78% could be discriminated correctly as active or inactive.

15.4.4 GOLD IN THE LITERATURE

Large scale VS is widely used in industry but reported in the literature sparingly. Nonetheless, it appears to be useful: recently published examples of successful structure-based VS include work on deoxyribonucleic acid (DNA) gyrase (using Ludi) [33], retinoic acid receptor [34], nuclear hormone receptor [35], and thyroid hormone receptor [36] (with ICM), kinesin [37] and protein kinase CK2 [38] (DOCK), farnesyltransferase (EUDOC) [39], and the estrogen receptor (PRO_LEADS) [40]. GOLD has been reported to perform quite well for several protein targets. Bissantz et al. [41] applied GOLD, DOCK, and FlexX to thymidylate kinase and estrogen receptor targets and concluded that GOLD poses are the most suitable for use in VS. Their screening experiment utilized a pool of random compounds spiked with a small number of known binders, most of them micromolar. They observed that GOLD reproduces binding modes well. Verdonk et al. [42] reported screening experiments with GOLD on neuraminidase, ptp1b, cdk2, and the estrogen receptor and obtained enrichment rates of 10 to 60. Jenkins et al. [43] performed VS runs for angiogenin inhibitors on a total set of 18,000 compounds and used the results to direct selection of compounds for high throughput screening (HTS). They found that HTS hits ranked in the top 2% by GOLD included 42% of the true hits (all but one in the top 38%), but only 8% of the false positives, and concluded that the combination of VS and HTS is effective in selecting inhibitors with midmicromolar dissociation constants, typical of leads that are commonly obtained from primary screens.

McFadyen et al. [44] docked both morphine and a steroid into a 7TM domain model structure of the μ-opioid receptor. Binding of the structurally different steroid was unexpected. GOLD produced first-ranked binding modes for both ligands that were consistent with affinity data from protein mutation studies, showing that, in principle, VS can lead to structurally novel drug candidates. Afzelius et al. [45] docked a set of CYP2C9 inhibitors into a protein homology model based on the crystallized rabbit CYP2C5 structure. Suitable GOLD poses were selected and subsequently used to build a good-quality 3D-QSAR (quantitative structure–activity relationship) model from which the authors were able to predict affinities for external data sets within 0.5 logarithmic units of the experimental value.

Specific targets may require customized scoring functions. Logean et al. [46] compared GoldScore, ChemScore, DockScore, PMFScore, and Score to the Fresno scoring function for predicting the binding affinity of 26 peptide inhibitors of the Class I major histocompatibility protein HLA-B*2705. They concluded that Fresno outperforms all, but of the six, ChemScore, DockScore, and GoldScore show significant correlations with experimental affinities. Wang et al. [47] made a comparison of 11 scoring functions. GoldScore yielded a correlation with experimental affinity of 0.57 (X-Score, PLP (Piecewise Linear Potential), and DrugScore also showed correlations). Its success rate for ranking poses was reported to be low, but this might be due to the methodology used: ligand poses were generated using the AutoDock program, and resulting conformations were not optimized as a function of GoldScore (for example, torsion angles of ligand methyl groups are not optimized by AutoDock even though doing so is expected to lead to improved GoldScore values). Keseru [48] reported a VS study for high affinity cytochrome P450cam substrates. Of

GoldScore, PMFScore, DockScore, and FlexXScore, only the former correlated with experimental affinities for both experimental (x-ray) and docked ligand poses. Good results were obtained for discerning actives from inactives, using a combination of PMF and GoldScore.

Clark et al. [49] described the development of a consensus-scoring approach that uses GoldScore as one of its constituents. Consensus scoring was also used by Paul et al. [50] but these authors clustered ligand poses generated by DOCK, FlexX, and GOLD rather than using scoring functions directly. Terp et al. [51] used GOLD to dock a set of 40 inhibitors into 3 matrix metalloproteinases and subsequently derived a consensus score (MultiScore) based on 8 scoring functions, including GoldScore, and using GOLD's docked poses. A good correlation between experimental affinities and calculated scores was obtained. Van Hoorn [52] described docking of several steroids to the estrogen receptor α and β forms. Though GOLD's docking results were consistent with experimental data, the potential of mean force (PMF) scoring function was found to be superior to both the GOLD fitness score and SYBYL's CScore methods. Clemente et al. [53] used GOLD to design new ligands for leukocyte elastase and to evaluate their hypothesis of an inhibition mechanism.

In [54], Bissantz et al. investigated whether homology models of G-protein-coupled receptors can be used as targets for screening. DOCK, FlexX, and GOLD were used in combination with seven scoring functions. Results from very fast GOLD GA settings (Parameter Set V) were found to be roughly comparable with those of FlexX for antagonists, using consensus scoring schemes; for agonist screening, GOLD did not perform well, though it was noted that it was able to propose reliable poses for known actives. GOLD's better GA settings were not explored and it has been noted that rescoring poses with the GoldScore option of the CScore module [49] may be problematic.

15.5 PROGRAM INFRASTRUCTURE

15.5.1 PROTEIN AND LIGAND PREPARATION

Correct preparation of protein and ligand input structures is essential. Both protein and ligand should have reasonable bond lengths and angles, because these are held fixed during docking. It is arguable whether the starting ligand geometry should be minimized: this will ensure reasonable bond lengths and angles but might bias the docking toward the starting conformation, because this is the one for which the valence angles will have been optimized. This consideration aside, the input conformation of the ligand is irrelevant when using GOLD. That of the protein is critical, because it will not be varied (apart from rotation of H-bonding groups). Proteins exhibiting conformational flexibility can only be dealt with by docking separately into the different known conformations.

Geometry apart, the two key issues in setting up the protein and ligand are atom-type assignment and hydrogen atom placement.

15.5.1.1   Atom Typing

Each protein and ligand atom in GOLD must have an atom type (with few exceptions, SYBYL types are used as seen later in this section). These are used both in computing fitness scores and in identifying H-bond donors and acceptors and their properties. Consequently, the initial assignment of atom types is important: misassigned types may lead to over- or underestimation of an atom's contribution to scoring function terms or, worse, cause GOLD to fail to consider the particular mapping into the binding site that would yield the correct pose. Atom types do not have to be assigned manually (though they can be). For ligands and cofactors, GOLD provides an automated mechanism for deriving them from 2D connectivity, using a rule-based algorithm that is reliable provided that bond types are set correctly on input. The GOLD manual lists the required bond typing for groups that have more than one conventional valence-bond representation. The atom-typing algorithm is fairly robust and is recommended over manual atom-type assignment. Certain groups can cause problems but these are getting fewer as the program is improved.

For protein residues, atom-type assignment is trivial provided substructure information is provided (i.e., the amino-acid identity of each residue). Consequently, substructure information must be available in the input file for the peptide chains, even if MOL2 format is used rather than PDB. If PDB format is used, CONECT records must be present for any cofactors in the vicinity of the binding site; multiple CONECT records for the same bond can be used to specify double and triple bonds (but not, unfortunately, aromatic bonds: a better approach is to use MOL2 file format in this situation).

In addition to the standard SYBYL atom types, GOLD uses the four types N.acid (nitrogen acceptor forming only two bonds, both single), N.plc (positively charged planar nitrogen, e.g., guanidino), S.m (thiolate sulfur), and S.a (sp^2 sulfur double bonded to carbon). The newest version of GOLD uses the substituents at trigonal nitrogen atoms to establish if a nitrogen atom is sp^2 or sp^3 hybridized; only the latter is deemed to be able to act as a H-bond acceptor or coordinate metal atoms.

15.5.1.2   Hydrogen Atom Placement

GOLD uses an all-atom model for scoring and also requires hydrogen-atom positions for generation of the donor fitting points that are used to map ligands into binding sites. Hydrogen atoms are also used for atom-type elucidation. Consequently, it is necessary for users to ensure that hydrogens are added to the protein and ligand before input to GOLD. The exact orientation of the hydrogens is relatively unimportant for XH (X = N, O, S), because H-bonding groups on both protein and ligand will be rotated anyway, but it is essential that the correct numbers of hydrogens are placed on each heavy atom. In particular, the ionization and tautomeric states of both protein residues and ligand will be assumed by GOLD to be correct on input and are likely to have a big effect on docking accuracy if they are, in fact, wrong. In case of ambiguity, especially in the protein (e.g., for histidine residues), the only sensible strategy is to do multiple dockings with the various protonation-state possibilities.

15.5.2 Binding-Site Definition

The protein binding site can be defined in several ways in GOLD. The simplest approach is to define explicitly the list of atoms that are deemed to constitute the binding site (for example, all atoms within a certain distance of the ligand in a complex of known structure). Alternatively, a point in space can be specified together with a radius. Solvent-accessible atoms within the sphere thus defined are taken to constitute the binding site. A grid-based flood-fill algorithm is used to determine solvent accessibility: any atom that desolvates a point in the flood-fill grid is deemed to be solvent accessible. The default sphere radius in GOLD is 10.0 Å; this occasionally needs to be increased for large or ill-defined binding sites (slower GA settings may then be needed to properly explore the larger search space).

Both these approaches can be combined with cavity detection, which reduces the atoms held to be in the binding site by selecting only those that are on the surface of cavities. Two cavity-detection algorithms, due to Hendlich et al. [55] and Delaney [56], are available in GOLD; the former usually performs better (and more efficiently) and so is the default choice. Shallow cavities are sometimes poorly recognized by cavity-detection algorithms, in which case the cavity may need to be defined by hand, using an explicit atom list.

15.5.3 Constraints and Restraints

Constraints and restraints are available within GOLD to deal with covalently bound ligands and to encourage dockings containing features that are expected, such as by analogy with experimentally observed binding modes. Recent work indicates that use of constraints can be beneficial in virtual HTS [42,57]. GOLD offers several options.

15.5.3.1 Covalent Constraints

Covalently bound ligands can only be docked in GOLD by applying a constraint that forces a bond between an atom on the protein with an unsatisfied valence and a corresponding atom on the ligand. Each trial ligand position generated by the GA is initially translated so that a covalent contact (possibly of poor geometry) is formed. A rotation matrix, centered on the protein covalent link atom, is then calculated and used to rotate the ligand to improve the covalent geometry. A Tripos force field term is added to the fitness score to describe the covalent link energy. This ensures that only poses whose covalent link has reasonable bond length and angles are scored highly, and so poor linkages disappear from the population as the GA progresses.

15.5.3.2 Distance Restraints

Restraints are available for biasing searches toward dockings containing a contact between a particular protein–ligand atom pair. This is achieved by adding a quadratic penalty term to the scoring function that reduces fitness if the distance between the specified pair of atoms lies outside a desired range, that is

$$F_{restrained} = F_{unrestrained} - w.x^2 \tag{15.8}$$

where x is given by

$$x = \begin{cases} d - d_{max} \text{ if } d > d_{max} \\ d_{min} - d \text{ if } d < d_{min} \\ 0 \text{ if } d_{min} \leq d \leq d_{max} \end{cases} \tag{15.9}$$

Here, d is the distance between the protein and ligand atoms and d_{min} and d_{max} are the minimum and maximum values, respectively, of the required distance range. This approach does not guarantee that the desired contact will be present in all GA population members but increases its probability of occurrence.

If the ligand contains atoms that are topologically equivalent to the atom specified in a distance restraint, GOLD calculates the minimum possible penalty contribution by testing each topologically equivalent atom in turn. This is necessary to avoid bias in the docking procedure. For example, a sulfonamide is bound to two topologically equivalent, prochiral oxygen atoms. Regarding only one of them as the restraint atom would bias docking toward solutions in which this atom, not the other, satisfies the distance restraint. This is not usually what is desired.

It is possible to define a distance restraint by providing a substructure and nominating an atom, X_S, within that substructure. GOLD then matches the substructure onto each ligand in turn and identifies all atoms that match the atom X_S (a given substructure might occur in a ligand more than once). The constraint term is calculated to all atoms that match X_S, and the minimum of the penalty terms thus calculated is applied to the score.

15.5.3.3 Similarity-Based (Pharmacophore) Restraints

Similarity-based restraints are used to encourage dockings in which the ligand overlaps well with a molecular fragment (molecule x) positioned within the binding site. The effect is to bias dockings toward solutions that express a particular pharmacophore or occupy a particular region in the binding site. The approach used is related to that of Fradera [58]. Gaussian functions are placed at the location of (optionally) all atoms, donor atoms or acceptor atoms of molecule x and the overlap [59] of these functions with a corresponding set of functions centered on ligand atoms of the same type is computed for the pose under consideration, y. The overlap is calculated from

$$O(x,y) = \frac{f(x,y)}{f(x,x)} \tag{15.10}$$

where

$$f(x,y) = \sum_{i=1}^{natoms_x} \sum_{j=1}^{natoms_y} \left(\frac{\pi}{\sigma_i + \sigma_j} \right)^{3/2} e^{\frac{\sigma_i \sigma_j d_{ij}^2}{\sigma_i + \sigma_j}} \tag{15.11}$$

Here, d_{ij} is the separation distance between the ith atom in x and the jth atom in y. σ_i and σ_j are the sigma values in the exponents of the Gaussian functions (in GOLD, hard coded to $0.64 \times$ vdW radius of the atom). The overlap $O(x,y)$, multiplied by a user-specified weight, is then added to the fitness score.

Two carbonic anhydrase complexes — PDB entries 2h4n (complex with acetazolamide) and 1h4n (complex with tromethamine) [60] — serve to illustrate the technique. Entry 2h4n is in the GOLD validation set and is one of the minority of structures that GOLD generally gets wrong: the thiadiazole ring is rarely placed forming a H-bond to the OH of threonine-200 (Thr200). By applying a similarity restraint, dockings can be biased toward the pose of tromethamine. Figure 15.4 shows the top-ranked pose of acetazolamide obtained when a similarity-based restraint is applied using the observed position of tromethamine as the template and applying nonzero weights to acceptor atoms only. A weight of 10.0 (the default) was used for the restraint term. The tromethamine position is shown in the figure for reference. The new pose generated for acetazolamide has its acceptor atoms in positions similar to those observed in tromethamine. The fitness scores provide an insight into the overall effect of the restraint. The top-ranked docking without a restraint had a fitness of 51.1 fitness units. In the restrained run, the total fitness was higher, at 56.8 fitness units. However, a large component of this was the overlap score (9.3 units), so, in effect, the restraint is promoting solutions that are less well ranked by GoldScore, but exhibit the desired interactions with the protein.

15.5.3.4 H-Bonding Restraints

Two types of restraints are provided to force the formation of particular H-bonds. First, it is possible to specify a donatable hydrogen or an acceptor atom in the protein and a complementary atom in the ligand. GOLD then ensures that fitting points from these two atoms are mapped onto each other with a relatively high weight each time a trial ligand pose is generated by least-squares fitting of the ligand into the binding site. In the second type of restraint (protein H-bond restraint), the user can specify that a particular protein atom must form a H-bond, but without having to specify which ligand atom is to receive the H-bond. GOLD ensures that the protein atom is mapped to a suitable complementary atom in the ligand. This fitting point is then up-weighted in the least-squares fit of the ligand into the binding site. This, in itself, does not guarantee that a H-bond is present in the resultant pose. A penalty is therefore added to the scoring function to penalize solutions that lack the desired H-bond. The penalty is simply the number of H-bonds that were requested, but are missing, multiplied by a user-specified weight.

Protein H-bond restraints were validated on AR complexes. Inhibitors of this enzyme form a specific set of H-bonds to tyrosine-48 (Tyr48), histidine-110 (His110), and tryptophan-111 (Trp111). A library of 1500 compounds was docked

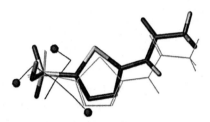

FIGURE 15.4 Top-ranked pose of acetazolamide (stick representation) in carbonic anhydrase, obtained with a similarity-based restraint using the observed position of tromethamine (black wireframe) as the template and applying nonzero weights to acceptor atoms only (shown as spheres). The restraint moves the thiadiazole group closer to the observed position (in gray). In the unrestrained docking, the molecule is misplaced completely, with the heterocyclic ring rotated by 180°.

into the enzyme using data from PDB entry 1frb [61]. The docking was run initially without restraints. Then, a protein H-bond restraint was specified with the hydroxyl hydrogen of Tyr48 as the nominated atom. The run was repeated with a variety of differing restraint weights. Figure 15.5 shows the number of solutions containing a H-bond to the specified atom plotted against unrestrained fitness (i.e., the fitness score with the restraint term removed). Compared with the unrestrained run, considerably more solutions containing the desired H-bond are seen in the higher fitness range.

Further analysis was performed to ascertain whether the restraint merely prevents solutions without the H-bond from being generated, rather than promoting the formation of new poses that conform to the restraint. The docking solutions for each ligand were clustered to identify the different poses that had been found. Because 10 different GA runs were performed on each ligand, the number of different poses found has to lie between 1 (same pose found 10 times) and 10 (each run produced a different pose). Figure 15.6 is a histogram showing the distribution of the *number of different poses per ligand* for the unrestrained docking run and for docking runs using four protein H-bond restraint weights. As the restraint is imposed more heavily, the average number of poses per ligand reduces, but not to a large degree, suggesting that new poses containing the desired H-bond are being found.

15.5.4 Dealing with Large Numbers of Ligands

15.5.4.1 Parallel Processing

Even at fast GA settings, high throughput VS benefits by use of parallel processing. GOLD has a parallel implementation that uses third-party software [62] to distribute dockings across a network of processors. Parallelization is performed at the ligand level. Each subprocess, in effect, runs GOLD independently of the scheduler, which merely instructs each subprocess as to which ligand to analyze and then harvests the results. Specialist software houses also support versions of GOLD customized to operate on a grid of personal computers. The potential benefits are massive given that, in a grid environment, GOLD can run on hundreds of desktop systems at once.

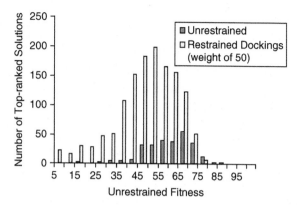

FIGURE 15.5 Results of docking a library of 1500 compounds into AR using a protein H-bond restraint. Fitness score, with restraint term removed, is plotted against the number of solutions containing a H-bond to the specified protein atom. Results from unrestrained docking are shown for comparison.

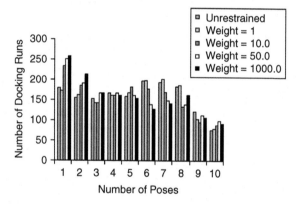

FIGURE 15.6 Histogram showing the distribution of number of different poses per ligand for an unrestrained docking run and for docking runs using four protein H-bond restraint weights.

Furthermore, by such an approach, organizations can use existing infrastructure to increase docking throughput.

15.5.4.2 Selection of the Best Results

In VS, users may dock hundreds of thousands of ligands, but will only really be interested in analyzing the best solutions from a run. It is therefore possible in GOLD to save only the best n solutions from a run, erasing the rest, or the best m poses for each ligand. This makes it relatively easy to minimize the volume of saved output for a given run. Solutions can be saved to a single SD format file [63] containing fitness component tags that can then be transferred to a spreadsheet for postdocking analysis.

15.5.5 CUSTOMIZATION

The latest release of GOLD (version 2.1) offers an API [64], developed in collaboration with Astex Technology, Ltd., which allows users to implement their own scoring functions and which, by exposing several of GOLD's C routines and structures, makes possible many other customizations. The API is based on the program architecture shown in Figure 15.7. The main GOLD shared object (or dynamically loadable library) is responsible for most of the program's operations, but delegates all scoring function calculations to a second shared object, libfitfunc.so. The signatures of the C functions from which this latter library is built are documented in the API user manual, and it is possible to code customized versions of one or more of the functions. These may then be used to build a modified version of the shared object that can be used in place of the standard GoldScore implementation. (In fact, GOLD switches from using GoldScore to ChemScore by this mechanism, that is, by loading a ChemScore version of libfitfunc.so in place of the default GoldScore implementation.)

Apart from functions and structures allowing access to essential data such as atomic coordinates, connectivity information, etc., the main components of the API are:

- Functions for evaluating fitness scores
- Functions for copying a fitness score from one individual to another
- Functions for writing out results
- Annealing functions

Examples of customizations that can be implemented by modifying these functions include:

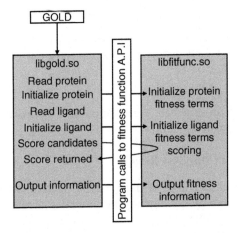

FIGURE 15.7 GOLD program architecture, showing interactions between main and scoring function shared objects.

- Calculating and writing out extra information characterizing each docking solution (e.g., whether a particular protein pocket is occupied by ligand atoms).
- Adding an extra term to the GoldScore function (e.g., to penalize contacts between protein hydrophobic atoms and ligand H-bond donor or acceptor atoms).
- Rescoring docked solutions with a second scoring function (to be meaningful, this would include simplex reoptimization of the docked ligand to reach the local optimum of the new scoring function).
- Implementing a completely new scoring function and using it to drive the GA search (the standard GoldScore mechanism could still be used for implementing covalent, distance, and other constraints).

The disadvantage of the API is that it requires low-level coding in C. Its advantage is that it affords a high degree of flexibility in how GOLD may be customized.

15.6 FUTURE WORK

The docking problem is only partially solved. State-of-the-art programs like GOLD will correctly dock typically 60 to 80% of ligands, depending on the circumstances, but much remains to be done. There is a need for additional filters for screening out dockings that contain unlikely interactions, better treatment of water molecules and protein flexibility, improved infrastructure for postdocking analysis, further optimization of search strategies, better scoring functions, particularly for affinity prediction, and so on. Research in many of these areas is ongoing at CCDC and our intention is to continue to develop GOLD actively for the foreseeable future.

ACKNOWLEDGMENTS

The original version of GOLD was written by Gareth Jones. Contributions have also been made by Peter Willett (University of Sheffield); Bobby Glen and Andrew Leach (GlaxoWellcome and GlaxoSmithKline PLC); and Mike Hartshorn, Chris Murray, Richard Taylor, and Marcel Verdonk (Astex Technology, Ltd.). The staff members of the CCDC are thanked for their work on the technical and administrative support of GOLD.

REFERENCES

[1]   (A) G. Jones, P. Willett, and R.C. Glen, A genetic algorithm for flexible molecular overlay and pharmacophore elucidation, *J. Computer-Aided Mol. Des.* 9:532–549, 1995; (B) G. Jones, P. Willett, and R.C. Glen, Molecular recognition of receptor sites using a genetic algorithm with a description of desolvation, *J. Mol. Biol.* 245:43–53, 1995; (C) G. Jones, P. Willett, R.C. Glen, A.R. Leach, and R. Taylor, Further development of a genetic algorithm for ligand docking and its application to screening combinatorial libraries. In A.L. Parrill and M.R. Reddy, Eds., *Rational Drug Design,* ACS Symposium Series, Vol. 719, Washington, D.C.: American Chemical Society, pp. 217–291, 1999.

[2] G. Jones, P. Willett, R.C. Glen, A.R. Leach, and R. Taylor, Development and validation of a genetic algorithm for flexible docking, *J. Mol. Biol.* 267:727–748, 1997.

[3] SYBYL, Tripos, Inc.: www.tripos.com.

[4] (A) A.C. Pierce, K.L. Sandretto, and G.W. Bemis, Kinase inhibitors and the case for CH...O hydrogen bonds in protein–ligand binding, *Proteins* 49:567–576, 2002; (B) B. Loll, G. Raszewski, W. Saenger, and J. Biesiadka, Functional role of C(alpha)-H...O hydrogen bonds between transmembrane alpha-helices in photosystem, *J. Mol. Biol.* 328:737–747, 2003.

[5] I.J. Bruno, J.C. Cole, J.P. Lommerse, R.S. Rowland, R. Taylor, and M.L. Verdonk, IsoStar: a library of information about nonbonded interactions, *J. Computer-Aided Mol. Des.* 11:525–537, 1997.

[6] (A) H.J. Böhm, S. Brode, U. Hesse, and G. Klebe, Oxygen and nitrogen in competitive situations: which is the hydrogen-bond acceptor? *Chem. Eur. J.* 2:1509–1513, 1996; (B) J.D. Dunitz and R. Taylor, Organic fluorine hardly ever accepts hydrogen bonds, *Chem. Eur. J.* 3:89–98, 1997.

[7] F.H. Allen, C.M. Bird, R.S. Rowland, and P.R. Raithby, Resonance-induced hydrogen bonding at sulfur acceptors in $R_1,R_2C=S$ and R_1CS_2–Systems, *Acta. Cryst.* B53:696–701, 1997.

[8] M. Clark, R.D. Cramer, III, and N. van Opdenbosch, Validation of the general purpose Tripos 5.2 force field, *J. Computer-Aided Mol. Des.* 10:982–1012, 1989.

[9] M.D. Eldridge, C.W. Murray, T.R. Auton, G.V. Paolini, and R.P. Mee, Empirical scoring functions: I. The development of a fast empirical scoring function to estimate the binding affinity of ligands in receptor complexes, *J. Computer-Aided Mol. Des.* 11:425–445, 1997.

[10] C.A. Baxter, C.W. Murray, D.E. Clark, D.R. Westhead, and M.D. Eldridge, Flexible docking using Tabu search and an empirical estimate of binding affinity, *Proteins* 33:367–382, 1998.

[11] M.L. Verdonk, J.C. Cole, M.J. Hartshorn, C.W. Murray, and R.D. Taylor, Improved protein–ligand docking using GOLD, *Proteins*, 52:609–623, 2003.

[12] A.W. Payne and R.C. Glen, Molecular recognition using a binary genetic search algorithm, *J. Mol. Graph.* 11:74–91, 121–123, 1993.

[13] D.E. Goldberg, *Genetic Algorithms in Search, Optimization and Machine Learning,* Reading, MA: Addison-Wesley, 1989.

[14] (A) M.L. Verdonk, J.C. Cole, and R. Taylor, SuperStar: a knowledge-based approach for identifying interaction sites in proteins, *J. Mol. Biol.* 289:1093–1108, 1999; (B) M.L. Verdonk, J.C. Cole, P. Watson, V. Gillet, and P. Willett, SuperStar: improved knowledge-based interaction fields for protein binding sites, *J. Mol. Biol.* 307:841–859, 2001; (C) D.R. Boer, J. Kroon, J.C. Cole, B. Smith, and M.L. Verdonk, SuperStar: comparison of CSD and PDB-based interaction fields as a basis for the prediction of protein–ligand interactions, *J. Mol. Biol.* 312:275–287, 2001.

[15] J.P. Gallivan and D.A. Dougherty, Cation-pi interactions in structural biology, *Proc. Natl. Acad. Sci. USA* 96:9459–9464, 1999.

[16] F.H. Allen, The Cambridge Structural Database: a quarter of a million crystal structures and rising, *Acta. Cryst.* B58:380–388, 2002.

[17] G. Klebe and T. Mietzner, A fast and efficient method to generate biologically relevant conformations, *J. Computer-Aided Mol. Des.* 8:583–606, 1994.

[18] B. Kramer, M. Rarey, and T. Lengauer, Evaluation of the FlexX incremental construction algorithm for protein–ligand docking, *Proteins* 37:228–241, 1999.

[19] Y.P. Pang, E. Perola, K. Xu, and F.G. Prendergast, EUDOC: a computer program for identification of drug interaction sites in macromolecules and drug leads from chemical databases, *J. Comput. Chem.* 22:1750–1771, 2001.

[20] C.A. Baxter, C.W. Murray, B. Waszkowycz, J. Li, R.A. Sykes, R.G. Bone, T.D. Perkins, and W. Wylie, New approach to molecular docking and its application to virtual screening of chemical databases, *J. Chem. Info. Comput. Sci.* 40:254–262, 2000.

[21] T.J. Ewing, S. Makino, A.G. Skillman, and I.D. Kuntz, DOCK 4.0: search strategies for automated molecular docking of flexible molecule databases, *J. Computer-Aided Mol. Des.* 15:411–428, 2001.

[22] J.S. Taylor and R.M. Burnett, DARWIN: a program for docking flexible molecules, *Proteins* 41:173–191, 2000.

[23] N.S. Blom and J. Sygusch, High resolution fast quantitative docking using Fourier domain correlation techniques, *Proteins* 27:493–506, 1997.

[24] M. Liu and S. Wang, MCDOCK: a Monte Carlo simulation approach to the molecular docking problem, *J. Computer-Aided Mol. Des.* 13:435–451, 1999.

[25] J. Wang, P.A. Kollman, and I.D. Kuntz, Flexible ligand docking: a multistep strategy approach, *Proteins* 36:1–19, 1999.

[26] J.W. Nissink, C. Murray, M. Hartshorn, M.L. Verdonk, J.C. Cole, and R. Taylor, A new test set for validating predictions of protein–ligand interaction, *Proteins* 49:457–471, 2002.

[27] I.N. Shindyalov and P.E. Bourne, An alternative view of protein fold space, *Proteins* 38:247–260, 2000.

[28] A. Bergner, J. Gunther, M. Hendlich, G. Klebe, and M. Verdonk, Use of Relibase for retrieving complex three-dimensional interaction patterns including crystallographic packing effects, *Biopolymers* 61:99–110, 2001.

[29] (A) H.B. Broughton, A method for including protein flexibility in protein–ligand docking: improving tools for database mining and virtual screening, *J. Mol. Graphics & Modelling* 18:247–257, 302–304, 2000; (B) R. Najmanovich, J. Kuttner, V. Sobolev, and M. Edelman, Side-chain flexibility in proteins upon ligand binding, *Proteins* 39:261–268, 2000; (C) H. Claussen, C. Buning, M. Rarey, and T. Lengauer, FlexE: efficient molecular docking considering protein structure variations, *J. Mol. Biol.* 308:377–395, 2001; (D) C.H. Li, X.H. Ma, W.Z. Chen, C.X. Wang, A soft docking algorithm for predicting the structure of antibody-antigen complexes, *Proteins* 52:47–50, 2003; (E) C.W. Murray, C.A. Baxter, and A.D. Frenkel, The sensitivity of the results of molecular docking to induced fit effects: application to thrombin, thermolysin and neuraminidase, *J. Computer-Aided Mol. Des.* 13:547–562, 1999.

[30] D.F. Veber, S.R. Johnson, H.Y. Cheng, B.R. Smith, K.W. Ward, and K.D. Kopple, Molecular properties that influence the oral bioavailability of drug candidates, *J. Med. Chem.* 45:2615–2623, 2002.

[31] H.J. Böhm, The development of a simple empirical scoring function to estimate the binding constant for a protein–ligand complex of known three-dimensional structure, *J. Computer-Aided Mol. Des.* 8:243–256, 1994.

[32] J. Barbanton, G. Perrot, and S. Salisbury, Multi-processor molecular docking with GOLD, Dechema Assessment Report, 1999.

[33] H.J. Böhm, M. Boehringer, D. Bur, H. Gmuender, W. Huber, W. Klaus, D. Kostrewa, H. Kuehne, T. Luebbers, N. Meunier-Keller, and F. Mueller, Novel inhibitors of DNA gyrase: 3D structure based biased needle screening, hit validation by biophysical methods, and 3D guided optimization: a promising alternative to random screening, *J. Med. Chem.* 43:2664–2674, 2000.

[34]   M. Schapira, B.M. Raaka, H.H. Samuels, and R. Abagyan, Rational discovery of novel nuclear hormone receptor antagonists, *Proc. Natl. Acad. Sci. USA* 97:1008–1013, 2000.

[35]   M. Schapira, R. Abagyan, and M. Totrov, Nuclear hormone receptor targeted virtual screening, *J. Med. Chem.*, 46:3045–3059, 2003.

[36]   M. Schapira, B.M. Raaka, S. Das, L. Fan, M. Totrov, Z. Zhou, S.R. Wilson, R. Abagyan, and H.H. Samuels, Discovery of diverse thyroid hormone receptor antagonists by high-throughput docking, *Proc. Natl. Acad. Sci. USA* 100:7354–7359, 2003.

[37]   S.C. Hopkins, R.D. Vale, and I.D. Kuntz, Inhibitors of kinesin activity from structure-based computer screening, *Biochem.* 39:2805–2814, 2000.

[38]   E. Vangrevelinghe, K. Zimmermann, J. Schoepfer, R. Portmann, D. Fabbro, and P. Furet, Discovery of a potent and selective protein kinase CK2 inhibitor by high-throughput docking, *J. Med. Chem.* 46:2656–2662, 2003.

[39]   E. Perola, K. Xu, T.M. Kollmeyer, S.H. Kaufmann, F.G. Prendergast, and Y.P. Pang, Successful virtual screening of a chemical database for farnesyltransferase inhibitor leads, *J. Med. Chem.* 43:401–408, 2000.

[40]   B. Waskowycz, T.D.J. Perkins, R.A. Sykes, and J. Li, Large-scale virtual screening for discovering leads in the postgenomic era, *IBM Systems* 40:360–376, 2001.

[41]   C. Bissantz, G. Folkers, and D. Rognan, Protein-based virtual screening of chemical databases: 1. Evaluation of different docking/scoring combinations, *J. Med. Chem.* 43:4759–4767, 2000.

[42]   M.L. Verdonk, V. Berdini, M.J. Hartshorn, W.T.M. Mooij, C.W. Murray, R.D. Taylor, and P. Watson, Virtual screening using protein–ligand docking: avoiding artificial enrichment, *J. Chem. Info. Comput. Sci.* 44:793–806, 2004.

[43]   J.L. Jenkins, R.Y. Kao, and R. Shapiro, Virtual screening to enrich hit lists from high-throughput screening: a case study on small-molecule inhibitors of angiogenin, *Proteins* 50:81–93, 2003.

[44]   I.J. McFadyen, H. Houshyar, L.Y. Liu-Chen, J.H. Woods, and J.R. Traynor, The steroid 17alpha-acetoxy-6-dimethylaminomethyl-21-fluoro-3-ethoxy-pregna-3, 5-dien-20-one (SC17599) is a selective mu-opioid agonist: implications for the mu-opioid pharmacophore, *Mol. Pharmacol.* 58:669–676, 2000.

[45]   L. Afzelius, I. Zamora, M. Ridderstrom, T.B. Andersson, A. Karlen, and C.M. Masimirembwa, Competitive CYP2C9 inhibitors: enzyme inhibition studies, protein homology modeling, and three-dimensional quantitative structure-activity relationship analysis, *Mol. Pharmacol.* 59:909–919, 2001.

[46]   A. Logean, A. Sette, and D. Rognan, Customized versus universal scoring functions: application to class I MHC-peptide binding free energy predictions, *Bioorg. Med. Chem. Lett.* 11:675–679, 2001.

[47]   R. Wang, Y. Lu, and S. Wang, Comparative evaluation of 11 scoring functions for molecular docking, *J. Med. Chem.* 46:2287–2303, 2003.

[48]   G.M. Keseru, A virtual high throughput screen for high affinity cytochrome P450cam substrates: implications for *in silico* prediction of drug metabolism, *J. Computer-Aided Mol. Des.* 15:649–657, 2001.

[49]   R.D. Clark, A. Strizhev, J.M. Leonard, J.F. Blake, and J.B. Matthew, Consensus scoring for ligand/protein interactions, *J. Mol. Graphics & Modelling* 20:281–295, 2002.

[50]   N. Paul and D. Rognan, ConsDock: A new program for the consensus analysis of protein–ligand interactions, *Proteins* 47:521–533, 2002.

[51]   G.E. Terp, B.N. Johansen, I.T. Christensen, and F.S. Jorgensen, A new concept for multidimensional selection of ligand conformations (MultiSelect) and multidimensional scoring (MultiScore) of protein–ligand binding affinities, *J. Med. Chem.* 44:2333–2343, 2001.

[52] W.P. van Hoorn, Identification of a second binding site in the estrogen receptor, *J. Med. Chem.* 45:584–589, 2002.

[53] A. Clemente, A. Domingos, A.P. Grancho, J. Iley, R. Moreira, J. Neres, N. Palma, A.B. Santana, and E. Valente, Design, synthesis and stability of N-acyloxymethyl- and N-aminocarbonyloxymethyl-2-azetidinones as human leukocyte elastase inhibitors, *Bioorg. Med. Chem. Lett.* 11:1065–1068, 2001.

[54] C. Bissantz, P. Bernard, M. Hibert, and D. Rognan, Protein-based virtual screening of chemical databases: II. Are homology models of G-Protein Coupled Receptors suitable targets? *Proteins* 50:5–25, 2003.

[55] M. Hendlich, F. Rippmann, and G. Barnickel, LIGSITE: automatic and efficient detection of potential small molecule-binding sites in proteins, *J. Mol. Graphics & Modelling* 15:359–363, 389, 1997.

[56] J.S. Delaney, Finding and filling protein cavities using cellular logic operations, *J. Mol. Graph.* 10:174–177, 163, 1992.

[57] S.A. Hindle, M. Rarey, C. Buning, and T. Lengauer, Flexible docking under pharmacophore type constraints, *J. Computer-Aided Mol. Des.* 16:129–149, 2002.

[58] X. Fradera, R.M. Knegtel, and J. Mestres, Similarity-driven flexible ligand docking, *Proteins* 40:623–636, 2000.

[59] A.C. Good, E.E. Hodgkin, and W.G. Richards, Utilization of Gaussian functions for the rapid evaluation of molecular similarity, *J. Chem. Info. Comput. Sci.* 32:188–191, 1992.

[60] C.A. Lesburg, C. Huang, D.W. Christianson, and C.A. Fierke, Histidine --> carboxamide ligand substitutions in the zinc binding site of carbonic anhydrase II alter metal coordination geometry but retain catalytic activity, *Biochem.* 36:15780–15791, 1997.

[61] D.K. Wilson, T. Nakano, J.M. Petrash, and F.A. Quiocho, 1.7 A structure of FR-1, a fibroblast growth factorinduced member of the aldo-keto reductase family, complexed with coenzyme and inhibitor, *Biochem.* 34:14323–14330, 1995.

[62] A. Geist, A. Beguelin, J. Dongarra, J. Weicheng, R. Manchek, and V. Sunderam. In J. Kowalik, Ed., *PVM: Parallel Virtual Machine. A User's Guide and Tutorial for Networked Parallel Computing*, Cambridge, MA: MIT Press, 1994.

[63] CTfile Formats. In MDL Information Systems: MDL User Documentation, MDL, 2001.

[64] J.C. Cole, J.W.M. Nissink, and R. Taylor, Gold Scoring-Function A.P.I. Documentation. In Gold Online Documentation v2-1, CCDC, 2003.

16 A Brief History of Glide: A New Paradigm for Docking and Scoring in Virtual Screening

Thomas A. Halgren, Robert B. Murphy,
and Richard A. Friesner

16.1 INTRODUCTION

A wide variety of docking programs are currently employed in the pharmaceutical and biotechnology industries to advance lead discovery and lead optimization projects, the most widely used of which have been GOLD [1], FlexX [2], and DOCK [3]. Over the past several years, considerable success has been reported for these programs in virtual screening (VS) applications [4–6]. However, neither these programs nor others that are available can be viewed as offering a robust and accurate solution to the docking problem, even in the context of a rigid protein receptor. Moreover, although many methods readily recognize favorable protein–ligand interactions, current-generation scoring functions do not adequately (if at all) penalize nonphysical interactions that oppose binding, and hence cannot efficiently rule out false positives. We and other developers of Glide are working to enable Glide, in reasonable computational time, to find the correctly docked structure in the vast majority of cases and to distinguish reliably between active ligands and ligands that could not plausibly bind to a given receptor. In this chapter, we will summarize the progress we have made to date in developing Glide 2.7 and will describe what users need to do or to know to employ Glide to best advantage in VS applications.

16.2 COMPUTATIONAL METHODOLOGY

To flexibly dock a ligand into a protein site, Glide uses a series of hierarchical filters to search for possible locations of the ligand in the active site region of the receptor. The shape and properties of the receptor are represented on a grid by several different sets of fields that provide progressively more accurate scoring of the ligand pose. (By *pose*, we mean a complete specification of the ligand: position and orientation relative to the receptor, core conformation, and rotamer-group conformations.) These fields are generated as preprocessing steps in the calculation and hence need to be computed only once for each receptor.

Once the various field representations of the receptor have been generated, the next step is to generate a set of initial ligand conformations. These conformations are selected from an exhaustive enumeration of the minima in the ligand torsion-angle space and are represented in a compact combinatorial form. They are not necessarily the final ones that will be chosen, however, as flexible geometry optimization in torsion space is carried out in the receptor field at a later stage of the calculation. Furthermore, the torsional optimization is supplemented by a conformational exploration of nearby torsional minima for a small number of promising candidates.

Given the initial set of ligand conformations, a series of initial screens are performed over the entire phase space available to the ligand to locate promising ligand poses. These initial screens are computationally inexpensive and do not incorporate any ligand geometry optimization. Rather, they seek to locate promising points in phase space from which to begin more accurate calculations. Prescreening of this type drastically reduces the region of phase space over which computationally expensive energy and gradient evaluations need to be performed, at the same time avoiding the use of stochastic methods; such methods can miss key phase–space regions a certain fraction of the time, thus precluding development of a truly robust algorithm. To our knowledge, Glide is unique in its reliance on the techniques of exhaustive systematic search, though clearly it requires approximations and truncations to achieve acceptable computational speed.

Starting from the poses selected by the initial screening protocols, the ligand is then minimized in the field of the receptor using a standard molecular-mechanics energy function (in this case, that of the OPLS-AA (Optimized Potentials for Liquid Simulations — all atoms) force field [7]) in conjunction with a distance-dependent dielectric model. Finally, the 3 to 5 lowest energy poses obtained in this fashion are subjected to a Monte Carlo procedure in which nearby torsional minima are examined. This procedure is needed in some cases to properly orient peripheral groups and occasionally alters internal torsion angles.

16.3 PROTEIN AND LIGAND PREPARATION

Our philosophy in the application of rigid docking methods to VS is that, if possible, information based on existing cocrystallized structures for the receptor should be exploited. Our protein-preparation methodology is based on this idea. Obviously, there will be situations in which no experimentally determined structure exists, such as when dealing with genomic targets. In this case, a variety of strategies are possible, for example the use of homology modeling to utilize cocrystallized structures for homologous proteins. Schrödinger's protein structure prediction package, Prime (Protein Integrated Modelling Environment), includes new technology that can help to build high-quality homology models that are suitable for use in docking [8]. We focus here, however, on the case in which an experimental structure is available.

The procedure normally starts with a protein and a cocrystallized ligand. It finishes with a partially optimized protein–ligand complex to which hydrogens have been added, subject to adjustment of protonation states for ionizable residues, modification of tautomeric forms for histidine residues, and reorientation of hydroxyl and sulfhydryl hydrogens. The procedure accomplishes the latter, and relieves unacceptable steric clashes, by carrying out a series of restrained molecular-mechanics minimizations of the complex in which progressively smaller restraint force constants are used until the cumulative root-mean-square deviation (rmsd) from the input protein structure reaches a target value the user sets (default, 0.3 Å). We perform these minimizations with either the Impact™ [9] or MacroModel [10] protein modeling codes.

Control of protonation states for residues far from the active site seems unlikely to be a critical factor for Glide, as the distance-dependent dielectric employed in the molecular-mechanics force field renders long-range electrostatics relatively unimportant. However, we have nevertheless implemented a standard protocol that takes into account dielectric screening by the solvent and by counterions that would normally neutralize the charges of ionized surface residues. In this protocol, residues that would be charged at physiological pH are generally treated as neutral unless they participate in a salt bridge or are relatively close to the ligand binding site. Protein sites prepared using this procedure are also employed in detailed binding affinity studies using Liaison™ and in mixed quantum mechanical/molecular mechanical (QM/MM) calculations using QSite™ [11], where the neutralization of distant residues, or of the total charge on the protein–ligand complex [12], may be of greater importance.

Our experience is that proper protein preparation is important for attaining accurate docking with Glide. It is essential that the physically untenable steric clashes often found in crystallographically determined protein sites be annealed away, so that the native ligand (and similar ligands) can yield favorable vdW interaction energies for properly docked structures. It is also important that protonation states and hydrogen (H)-bonding patterns be correct. See Section 16.8.1 for a further discussion.

It is also important that ligands have good initial geometries and be protonated correctly. See Section 16.8.2 for a discussion.

16.4 SCORING IN GLIDE

Glide 2.7 employs two forms of GlideScore, the principal scoring method used for predicting ligand binding affinities and for comparing the binding of diverse ligands:

1. GlideScore 2.7 SP, used by Standard-Precision Glide
2. GlideScore 2.7 XP, used in Extra-Precision mode

These scoring functions use similar terms but are formulated with different objectives in mind. Specifically, GlideScore 2.7 SP is a softer, more forgiving function that is adept at identifying ligands that have a reasonable propensity to bind, even in cases in which the Glide pose has significant imperfections. This version is attuned to minimizing false negatives and is appropriate for many database screening applications. In contrast, GlideScore 2.7 XP is a harder function that exacts severe penalties for poses that violate established physical chemistry principles, such as that charged and strongly polar groups make an appropriate complement of H-bonds or be adequately exposed to solvent. This version of GlideScore is more adept at minimizing false positives and can be especially useful in lead-optimization or other studies in which the number of compounds that will be considered experimentally is limited and each compound identified computationally needs to be as high in quality as possible.

GlideScore 2.7 modifies and extends the ChemScore function [13] as follows

$$\Delta G_{bind} = C_{lipo-lipo} \, \Sigma f(r_{lr}) + C_{hbond-neut-neut} \, \Sigma g(\Delta r) \, h(\Delta \alpha) + C_{hbond-neut-charged} \, \Sigma g(\Delta r) \, h(\Delta \alpha)$$

$$(16.1)$$

$$+ \, C_{hbond-charged-charged} \, \Sigma g(\Delta r) \, h(\Delta \alpha) + C_{max-metal-ion} \, \Sigma f(r_{lm}) + C_{rotb} \, H_{rotb}$$

$$+ \, C_{polar-phob} \, V_{polar-phob} + C_{coul} \, E_{coul} + C_{vdW} \, E_{vdW} + Solvation \; Terms$$

GlideScore 2.7 SP uses the same functional form and the same weighting of the contributing terms as did GlideScore 2.5 SP. Results obtained with Glide 2.7, however, differ because the parameterization of the underlying OPLS-AA force field has changed to some degree as a result of major improvements made to the atom typing mechanism. The lipophilic–lipophilic term and the H-bonding term use the same functional form as in ChemScore; the latter, however, is separated into differently weighted components that depend on whether the donor and acceptor are neutral, one is neutral and the other is charged, or both are charged. In the optimized scoring function, the first of these contributions is found to be the most stabilizing, and the last, the charged-charged term, is the least important. The metal–ligand interaction term (the fifth term) also uses the same functional form as is employed in ChemScore, but varies in three principal ways. First, Glide considers only interactions with anionic acceptor atoms (such as either of the two oxygens of a carboxylate group). This change allows Glide to recognize the evident strong preference for coordination of anionic ligand functionality to metal centers in metalloproteases [14,15]. Second, GlideScore 2.7 counts just the single best interaction when two or more metal ligations are found. We set the coefficient to -2.0 kcal/mol, a value we believe to be reasonable, though the parameter refinement would have preferred an even more strongly negative value. Third, Glide assesses the net charge on the metal ion in the unligated, apoprotein (generally straightforward via examination of the coordinated protein sidechains). If the net charge is positive, the preference for an anionic ligand is incorporated into the scoring function. On the other hand, if the ion is net neutral (as it is, for example, in the case of the zinc metalloprotein farnesyl protein transferase, which accepts neutral ligands such as substituted imidazoles [16]), the preference is suppressed. The sixth term penalizes rotatable bonds in the ligand that are restricted in the docked pose, and the seventh term, from Maestro™'s Active Site Mapping facility, rewards instances in which a polar but non-H-bonding atom (as classified by ChemScore) is found in a hydrophobic region.

GlideScore 2.7 also incorporates contributions from the Coulomb and van der Waals (vdW) interaction energies between the ligand and the receptor. To make the Coulomb interaction energy a better predictor of binding (and a better contributor to a composite scoring function), we reduce, by approximately 50%, the net ionic charge on centers such as Zn^{2+} and other protein metals and on formally charged groups such as carboxylates and guanidiniums; we also reduce the vdW interaction energies for the atoms involved to keep intermolecular geometries from being greatly perturbed. As we have shown elsewhere [17], the wide disparities in the original interaction energies are greatly reduced, though charged-charged interactions are still favored over charged-neutral or neutral-neutral interactions to some extent. The Coulomb–vdW energies used in GlideScore (but not those used in EModel) employ these reductions in net ionic charge except in the case of anionic ligand–metal interactions; to be consistent with the preference of metal centers such as Zn^{2+} for anionic ligands, the full interaction energy is used in these cases.

The third major component is the introduction of a solvation model. The use of a gas-phase potential-energy function is a principal flaw in earlier versions of Glide-Score and in other empirical scoring functions. Scoring functions that ignore solvation do not take into account the severe restrictions on possible ligand poses that arise from the requirement that charged and polar groups of both the ligand and protein be adequately solvated. Although GlideScore 1.8 and 2.0 sought to penalize poses that fail to satisfy this requirement, the terms they used for this purpose did not properly address the consequences of burying charged groups or strongly polar groups in hydrophobic regions of the protein. Charged groups, in particular, require careful assessment of their access to solvent. Finally, water molecules may be trapped in hydrophobic pockets by the ligand, also an unfavorable situation. The solvation terms play only a minor role in Glide SP, but in Glide XP they aggressively penalize poses that are judged to make inappropriate physical interactions.

16.5 DOCKING ACCURACY

Table 16.1 summarizes the rms errors obtained as a function of the flexibility of the ligand for docking into 282 PDB-derived complexes, starting from conformationally optimized geometries for the cocrystallized ligands. As expected, rms errors and central processing unit (CPU) times increase with ligand flexibility. Both, however, are quite modest for sets of ligands having 0 to 8 or 0 to 10 rotatable bonds such as are often employed in database screens carried out to find new leads. In general, Glide gives a reasonable docking performance over a wide range of rotatable bonds and chemical functionality.

Detailed docking accuracy results for GOLD and FlexX posted on the GOLD [18] and FlexX [19] Web sites have enabled us to make the head-to-head comparisons shown in Table 16.2 and Table 16.3. These tables compare rms deviations (Å) given by Glide and GOLD and by Glide and FlexX for common sets of noncovalently bound ligands having up to 10 and up to 20 rotatable bonds; comparisons are also presented for all ligands Glide can handle (i.e., up to 35 rotatable bonds). The Glide calculations use the conformationally optimized versions of the native ligands. The comparison for ligands having up to 10 rotatable bonds seems to us to be the most

TABLE 16.1
Average RMSDs for Flexible Docking on 282 PDB Complexes

Number of Rotatable Bonds	Number of Cases	Average RMS Top Ranked Pose (Å)	Average CPU Time (Min.)
0–3	51	1.06	0.2
4–6	92	1.47	0.6
7–10	48	1.66	1.6
0–8	164	1.35	0.5
0–10	191	1.41	0.7
0-20	263	1.84	2.0

Times are AMD Athlon MP 1800+ CPU minutes.

TABLE 16.2
Comparison of RMSDs (Å) for Flexible Docking by Glide and GOLD

Method	≤ 10 Rotbonds (72 Cases)		≤ 20 Rotbonds (86 Cases)		All Ligands (93 Cases)	
	Average RMSD	Maximum RMSD	Average RMSD	Maximum RMSD	Average RMSD	Maximum RMSD
Glide	1.42	8.5	1.62	8.5	2.05	13.7
GOLD	2.56	14.0	2.92	14.0	3.06	14.0

TABLE 16.3
Comparison of RMSDs (Å) for Flexible Docking by Glide and FlexX

Method	≤ 10 Rotbonds (133 Cases)		≤ 20 Rotbonds (175 Cases)		All Ligands (189 Cases)	
	Average RMSD	Maximum RMSD	Average RMSD	Maximum RMSD	Average RMSD	Maximum RMSD
Glide	1.42	8.5	1.81	9.2	2.14	13.7
FlexX	2.99	12.6	3.48	13.4	3.72	15.5

relevant to database screening applications, given the priority usually placed on finding active ligands that are relatively rigid. On average, Glide gives rms deviations that are about 60% of those given by GOLD and half those given by FlexX.

16.6 ACCURACY IN VIRTUAL SCREENING

To characterize the ability of Glide to identify active compounds, we summarize below results for database screening obtained for the following nine receptors, five of which are represented by two or more alternative cocrystallized receptor sites:

1. Thymidine kinase (1kim)
2. Estrogen receptor (3ert; 1err)
3. CDK-2 kinase (1dm2; 1aq1)
4. p38 MAP kinase (1au9; 1bl7; 1kv2)
5. HIV protease (1hpx)
6. Thrombin (1dwc; 1ett)
7. Thermolysin (1tmn)
8. Cox-2 (1cx2)
9. HIV-RT (1vrt; 1ht1)

These receptors cover a wide range of receptor types and therefore constitute a stringent test of the docking and scoring algorithms. Each screen employs 1000 database ligands and between 7 and 33 known actives. The known binders for the first two systems were specified by Bissantz et al. [5], and ligands for CDK-2 kinase and for p38 MAP (mitogen-activated protein) kinase were provided by pharmaceutical and biotech collaborators. For thrombin, 12 of the 16 known binders were taken from the studies by Engh et al. [20] and by von der Saal et al. [21]. Others are ligands for the same target protein taken from our docking-accuracy test set or were developed from multiple sources in the literature. Omitting the large HIV (human immunodeficiency virus) protease ligands, the active ligands average approximately 410 in molecular weight (MW).

Most of these screens employed a set of presumed inactives (decoys) that averaged 400 in MW. These compounds were selected from a large database of purchasable compounds that had been processed by FirstDiscovery™'s ionizer utility to protonate or deprotonate ionizable functional groups (subject to limits of ±2 on the net charge and of a total of no more than 4 charged groups) so as to yield ionic states likely to be present in measurable concentration. (This is to allow for shifts in pK_a induced by the protein site.) The selection process was configured to ensure that the average numbers of quantities such as rotatable bonds, rings, heteroaromatic rings, charged and neutral H-bond donors and acceptors, second- and higher-row atoms, divalent oxygens, and amide hydrogens match the profile given by a set of World Drug Index compounds to which we had access. For thymidine kinase (tk), which has a relatively small active site, we used a similar (but in this case more competitive) set of average MW 360. All compounds considered have 20 or fewer rotatable bonds and 100 or fewer atoms. We believe these compounds to be representative of the chemical sample collections of pharmaceutical and biotechnology companies.

By using MMFF94s (Merck Molecular Force Field) [22] with a "4r" distance-dependent dielectric, FirstDiscovery's `premin` utility was employed to minimize the ligands before docking, starting from CORINA [23] geometries. The known actives (some taken from cocrystrallized ligands) were also MMFF94s-optimized, but used unbiased input geometries obtained via a short MacroModel conformational search.

Glide docking runs typically use reduced atomic vdW radii for nonpolar atoms to mimic minor readjustments of the protein site upon ligand binding. The ligand and protein vdW scale factors can be optimized for each particular receptor, but the present results all use the default 1.0 protein scaling (which means that the OPLS-AA vdW radii for the protein are not changed) and 0.8 ligand scaling. Thus, the performance described here reflects what can be expected when Glide is run out-of-the-box; in some cases, to be sure, improved results can be obtained by optimizing the scale factors (see Section 16.8.3).

We report enrichment factors in graphical and tabular form and present accumulation curves that show how the fraction of actives recovered varies with the percent of the database screened. Following Pearlman and Charifson [24], the enrichment factor can be written as

$$EF = \{Hits_{sampled}/N_{sampled}\} \, / \, \{Hits_{total}/N_{total}\} \qquad (16.2)$$

Equivalently, this can be written as

$$EF = \{N_{total} \, / \, N_{sampled}\} * \{Hits_{sampled} \, / \, Hits_{total}\} \qquad (16.3)$$

Thus, if only 10% of the scored and ranked database (i.e., $N_{total}/N_{sampled} = 10$) needs to be assayed to recover all of the $Hits_{total}$ actives, the enrichment factor would be 10. But, if only half of the total number of known actives are found in this first 10% (i.e., if $Hits_{sampled}/Hits_{total} = 0.5$), the actual enrichment factor would be 5.

When the objective is to quantify performance for recovering a substantial fraction of the active ligands, we modify the definition of the EF as follows

$$EF' = \{50\% \, / \, APR_{sampled}\} * \{Hits_{sampled} \, / \, Hits_{total}\} \qquad (16.4)$$

In this equation, $APR_{sampled}$ is the average percentile rank of the $Hits_{sampled}$ known actives. Intuitively, this makes sense: if the actives are uniformly distributed over the entire ranked database, the average percentile rank for an active would be 50% and the enrichment factor would be 1. This formula considers the rank of each of the $Hits_{sampled}$ known actives, not just the rank of the last active found (which is what $N_{sampled}$ is likely to be). As a result, the enrichment factor will be larger than the value computed from Equation 16.2 or Equation 16.3 if the actives are concentrated toward the beginning of the $N_{sampled}$ ranked positions, but will be smaller if the actives are grouped toward the end of this list. This is appropriate, as a key objective is to find active compounds as early as possible in the ranked database. The modified definition is better at indicating when this is happening.

TABLE 16.4
Comparison of Enrichment Factors for Glide 1.8, Glide 2.0, and Glide 2.7

| | | EF' (Equation 16.4) | | | EF (Equation 16.3)[a] | | | EF (Equation 16.3)[a] | |
| | | 70% Recovery | | | 2% of Database | | | 5% | 10% |
Screen	Site	GS 1.8	GS 2.0	GS 2.7	GS 1.8	GS 2.0	GS 2.7	GS 2.7	GS 2.7
Thymidine kinase	1kim	4.2	7.6	20.8	0.0	10.0	20.0	16.0	10.0
Tk-pyrimidine ligands	1kim	4.5	6.7	18.3	0.0	7.1	21.4	14.3	10.0
Estrogen receptor	3ert	88.4	79.8	29.1	35.0	35.0	30.0	12.0	7.0
Estrogen receptor	1err	88.4	37.5	45.8	35.0	30.0	35.0	14.0	7.0
CDK-2 kinase	1dm2	3.6	3.9	4.7	5.0	10.0	15.0	10.0	5.0
CDK-2 kinase	1aq1	2.1	3.8	3.8	5.0	15.0	5.0	6.0	4.0
p38 MAP kinase	1a9u	2.0	1.8	2.7	0.0	2.9	2.9	2.4	2.9
p38 MAP kinase	1bl7	1.8	2.9	4.7	2.9	5.9	8.8	5.9	5.3
p38 MAP kinase	1kv2	4.5	2.9	5.0	8.8	11.8	14.7	8.2	4.7
HIV protease	1hpx	10.8	7.8	38.6	30.0	13.3	33.3	16.0	9.4
Thrombin	1dwc	5.8	2.8	17.7	6.2	3.1	18.8	13.8	7.5
Thrombin	1ett	5.2	5.6	45.7	12.5	3.1	34.4	18.8	10.0
Thermolysin	1tmn	1.6	15.2	52.6	5.0	15.0	40.0	18.0	9.0
Cox-2	1cx2	3.6[b]	3.4[b]	4.2[b]	7.6	3.0	13.6	7.9	4.5
Cox-2 (site 1 ligands)	1cx2	5.7	5.7	12.3	10.9	4.3	19.6	11.3	6.5
HIV-RT	1rt1	4.6	3.2	8.5	3.0	0.0	12.1	9.1	7.0
HIV-RT	1vrt	1.8	2.0	5.6	0.0	0.0	6.1	7.3	5.8
Average enrichment factor		5.4	6.4	14.8	6.0	7.1	18.0	11.2	6.8

[a.] Maximum possible enrichment factors are 50 for 2% sampling, 20 for 5% sampling, and 10 for 10% sampling.

[b.] EF'(60) value.

Table 16.4 compares the performance of Glide 1.8, 2.0, and 2.7 using enrichment factors EF'(70), which reports the enrichment for recovering 70% of the known actives (a measure of global enrichment), and EF(2%), which measures enrichment for assaying just the top 2% of the ranked database (early enrichment). We also report enrichment factors EF(5%) and EF(10%) for Glide 2.7. Though 70% recovery is arbitrary, we feel this is a realistic standard for docking into a rigid protein site, given that such a site is unlikely to be properly shaped to house all the known actives when the site is relatively plastic. Also listed are average enrichment factors computed using a generalized geometric mean that weights the smaller enrichment factors more heavily.[1]

The table shows that Glide 2.7 performs much better than these two earlier versions for both global and early enrichment (we show elsewhere that it is comparable to its immediate predecessor, Glide 2.5 [25]). Because of improvements to the more difficult screens, both global and early enrichment on average have tripled since Glide 1.8, the first widely distributed commercial release. The CDK-2 and p38 screens are still problematic, but thymidine kinase now does well, and thrombin, HIV, thermolysin, cox-2, and HIV-RT all have improved substantially. For the first listing line for cox-2, the EF' value is shown is $EF'(60)$ because fewer than 70% of the 33 known actives dock into the primary 1cx2 binding site with a negative Coulomb–vdW interaction energy. The second line for cox-2 gives results based on the 23 ligands that can do so (only 21 of which dock successfully with the default vdW scale factors used here).

Figure 16.1 summarizes Glide's ability to rank active ligands in the first 2%, 5%, and 10% of the scored and ranked database. For EF(2%), $N_{total}/N_{sampled} = 50$. As the fraction of the known actives ($Hits_{sampled}/Hits_{total}$) found in this portion of the database cannot be greater than 1, it follows that the maximum attainable value for EF(2%) is 50. Similarly, maximum enrichment factors are 20 for EF(5%) and 10 for EF(10%). The enrichment factors show that Glide 2.7 performs exceedingly well for many of the targets and reasonably well for most of the others. More detailed discussions may be found elsewhere [25].

For the same set of screens, Figure 16.2 compares the ability of Glide 2.7 to identify active ligands to two earlier versions of Glide and to random chance. This figure shows the progress Schrödinger has been making in virtual screening efficiency.

16.7 EXTRA-PRECISION GLIDE

Within the limitations of docking to a single rigid protein structure, it is, in our view, simply not possible to design a fast empirical scoring function that for all possible receptors of pharmaceutical interest avoids penalizing any arbitrary active compounds that are docked and simultaneously is effective at screening out false positives. A soft potential can avoid penalizing active compounds, but by the same token will fail to eliminate many inactives. Conversely, a hard potential will prevent active compounds from docking if these compounds are not compatible with the particular conformation of the receptor that is being used in the docking study. Consequently, our recommended approach for using XP scoring is, when possible, to dock into multiple receptor conformations and to then combine the results of the individual docking runs. Multiple conformations of the receptor can be obtained if several cocrystallized complexes are available. Alternatively, the Schrödinger protein modeling methodology, Prime [8], can generate low-energy receptor conformations that can be used for VS. We present results below for the first approach; we are at present working on the second and expect to have illustrative examples available shortly.

To achieve the higher accuracy in docked structures needed for XP scoring, it is necessary to expend a larger amount of CPU time than in normal Glide docking. The present version of XP docking and scoring is typically 10 to 15 times more expensive than a Glide SP calculation on the same molecule when the default of five recycling iterations is selected. However, our suggested mode of operation is to run Glide SP first, retain only the top 10 to 30% of the ranked poses, and to run XP docking on this subset after minimizing the ligands to remove any bad steric clashes.

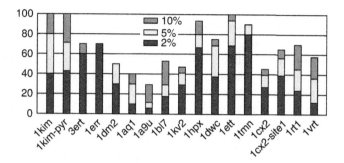

FIGURE 16.1 Percent of actives recovered with Glide 2.7 for assaying 2%, 5%, and 10% of the ranked database for the screens considered in this chapter.

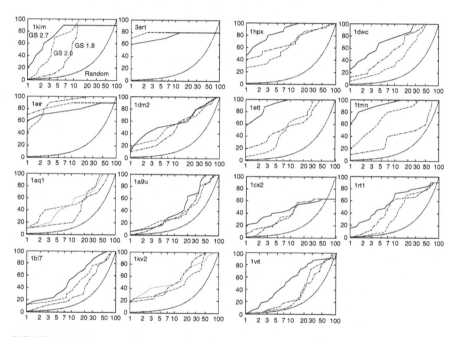

FIGURE 16.2 Percent of known actives found (y axis) vs. percent of the ranked database screened (x axis) for Glide 2.7 (solid line), Glide 2.0 (dashed line), and Glide 1.8 (dot-dashed line). Black dotted lines shows results expected by chance.

When deployed in this fashion, SP plus XP docking typically requires 2 to 5 times as much CPU time for VS of a library as does SP docking alone.

As previously described, GlideScore includes a number of terms that are aimed at assessing various contributions to the binding affinity. These terms include hydrophobic atom–atom pair interactions, H-bond interactions, the gas-phase Coulomb and vdW protein–ligand interaction energies, and a rotatable bond term. All of these terms are important in detecting poses that have the potential for high binding affinity. However, they do not take into account the severe restrictions on possible ligand poses that arise from requirements such as that charged and polar groups of both the ligand and protein be adequately solvated (or be located in an otherwise appropriate electrostatic environment).

To model solvation in Glide, explicit water molecules are docked into a list of protein–ligand complexes that otherwise receive good GlideScores, and descriptors based on the interaction of these water molecules with various charged and polar groups of the ligand and protein are used as a measure of whether the complex is physically realistic. Penalties are assigned to structures where statistical results indicate that one or more groups is inadequately solvated. A large database of cocrystallized structures, as well as the database screens described herein, have been used to optimize the parameters associated with the penalty terms. The explicit-water technology and descriptors are also used in Glide SP scoring. However, the improved sampling allows XP docking to assign substantially higher penalties to serious violations of physical principles; it is not possible to just score SP poses or poses generated by other docking methods with XP scoring.

GlideScore 2.7 XP also includes a term that specifically rewards occupancy of well-defined hydrophobic pockets by hydrophobic ligand groups. Hydrophobic reward terms are employed in empirical scoring functions such as ChemScore and the SP version of GlideScore in the form of lipophilic–lipophilic pair terms; other empirical scoring functions use lipophilic surface area contact terms for this purpose. Our investigations, however, have convinced us that simple pair terms underestimate hydrophobic effects in certain cases. We developed the new hydrophobic term to offset this underestimation. The term can confer up to several kcal/mol of additional binding energy in favorable cases. Its effect is to substantially improve enrichment factors, particularly for early enrichment, for screens in our test set that are dominated by hydrophobic binding, such as p38 MAP kinase, CDK-2, HIV-RT, and cox-2. That these are also the receptors that have been most poorly handled by previous versions of Glide is consistent with our view that conventional hydrophobic pair terms miss an important aspect of the underlying physics.

Achievement of the higher docking accuracy needed for XP scoring naturally requires the expenditure of greater computation time for sampling. In particular, we generate an enhanced ensemble of poses in the important regions of phase space via a recycling technology. An initial Glide run is used to identify a small subset of poses with the best scores; then, additional sampling, via repetition of the Glide protocol constrained to the relevant phase-space regions, is carried out. This protocol is successful in enabling properly docked active compounds to avoid penalties for all but a small subset of our test cases. In cases in which the ligand in principle can fit, the current XP sampling procedure encounters difficulties when the original Glide SP run fails to locate the correct basin of attraction (as opposed to the detailed pose within that basin); this in turn is typically due to challenging cross-docking situations in which the fit of the ligand into the target rigid receptor is marginal (i.e., tight). We are currently developing a fundamentally improved sampling algorithm that will more effectively address problems of this type.

Detailed results for XP docking have been published for the set of database screens used above for characterizing SP docking [26]. As previously indicated, we first screen the database (including seeded actives) using Glide SP; then, we take a fraction of the top-scoring results, minimize the ligands to remove any bad steric clashes, and carry out XP calculations. The fraction we have selected varies between 10 and 30% for different screens and corresponds roughly to our perception of the difficulty of the receptor for SP docking (the more problems SP docking has, the larger the fraction of the database that should be passed on to XP). We would not expect the results to differ greatly if XP docking were used for all ligands.

We will consider here only a few of the applications of XP docking we have made — those for thymidine kinase, CDK-2 kinase, and p38 MAP kinase. All XP calculations used the default setting of 5 for the number of redocking cycles. Five cycles is unnecessarily expensive, however, for smaller active sites such as thymidine kinase, CDK-2, HIV-RT, and cox-2, where the number of rotatable bonds for known actives is on average five. For smaller active sites as these, three cycles appear to be adequate.

16.7.1 Thymidine Kinase (1kim)

Thymidine kinase (tk) is an enzyme that recognizes and acts on nucleic acid bases. Both purines and pyrimidines fit into the active site, but as discussed elsewhere [25], the protein conformation required for each series to fit is different. In our present testing of XP scoring, we consider only the docking of the seven pyrimidines [5]. The cavity is small and mostly hydrophilic, though there are some hydrophobic regions. As is shown in Table 16.5 and in Figure 16.3, the results for XP scoring are excellent and represent a quantitative improvement over SP docking. The horizontal offset between the two accumulation curves in the figure shows that a significantly smaller fraction of the ranked database needs to be assayed to recover a given number of active compounds when XP scoring is used.

16.7.2 CDK-2 (1dm2; 1aq1)

CDK-2 has a largely hydrophobic active site that can adopt a number of conformations, though the differences for the receptor variants considered here are not as dramatic as those seen in p38 MAP kinase. We have examined two receptor struc-

TABLE 16.5
Glide 2.7 XP and SP Rankings for Pyrimidine-Based Actives Docked into the 1kim Thymidine Kinase Receptor

Ligand	XP Rank	SP Rank
idu	1	7
ahiu	2	26
dT	3	1
mct	4	21
dhbt	5	38
hpt	21	40

vdW scalings are 1.0 protein/0.9 ligand.

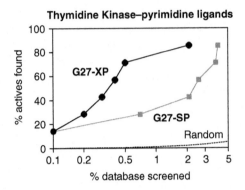

FIGURE 16.3 Pyrimidine-based actives recovered as a function of the percent of the ranked database screened for thymidine kinase (1kim receptor). Squares: standard-precision Glide 2.7 results; circles: extra-precision Glide 2.7 results.

tures, 1dm2 and 1aq1, the native ligands for which are hymenialdisine and staurosporine. Docking results are summarized in Table 16.6.

Figure 16.4 shows that XP scoring for the 1aq1 site of CDK-2 represents a substantial improvement over Glide 2.7 SP or Glide 2.0 scoring (also standard precision). This improvement is not seen for the 1dm2 receptor (Figure 16.5), but we anticipate that further enhancements in the XP sampling protocols currently being developed will yield substantially improved results.

TABLE 16.6
Glide 2.7 XP and SP Rankings for CDK-2 Actives Docked into the 1aq1 and 1dm2 Receptors

	1aq1 Receptor		1dm2 Receptor	
Ligand	XP Rank	SP Rank	XP Rank	SP Rank
Staurosporine	1	1	---[a]	---[b]
Indolin-2-one	2	34	4	40
Cgp60474	5	42	141	12
Hymenialdisene	7	291	7	16
PurB	15	118	30	23
Indolin-2-one-2	18	55	28	4
Ag12073	59	108	135	292
Pyrpyr	---[a]	---[b]	185	194
4aquin	---[a]	---[b]	231	236

[a] Did not make SP cutoff.

[b] Not in top 303 ranked positions (30%).

vdW Scalings are 1.0 protein/0.8 ligand.

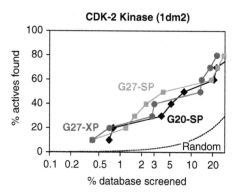

FIGURE 16.4 CDK-2 actives recovered for the 1dm2 receptor site as a function of the percent of the ranked database for Glide 2.0 scoring and for Glide 2.7 SP and XP scoring.

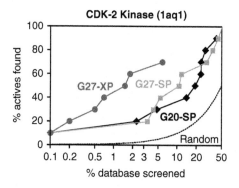

FIGURE 16.5 CDK-2 actives recovered for the 1aq1 receptor site as a function of the percent of the ranked database for Glide 2.0 scoring and for Glide 2.7 SP and XP scoring.

16.7.3 P38 MAP KINASE (1A9U; 1BL7; 1KV2)

The p38 MAP kinase has a large, complex active site that has been used for many docking studies in a pharmaceutical context. The enzyme exhibits a particularly wide range of conformational changes, involving, in some cases, large movements of loops and sidechains. In this situation, our expectation is that there will be a wide range of variability in the ability to dock active compounds into different conformations of the receptor. This expectation is borne out by the current study.

Our SP study [25] presented results for docking our screening database into three receptor structures — 1a9u, 1bl7, and 1kv2. As active compounds, we used 14 p38 ligands obtained from a colleague in the biopharmaceutical industry plus the three cocrystallized ligands. The 1kv2 receptor appears to be the most appropriate one for a majority of the active compounds we consider here. In what follows, we discuss the 1kv2 and 1a9u results, eliminating 1bl7 for purposes of XP scoring.

Table 16.7 and Table 16.8 summarize the docking results obtained using two choices of vdW scalings. Comparison of the results for the two vdW scalings indicates that limited experiments with the scalings can help to obtain optimal ranks.

TABLE 16.7
Glide 2.7 XP and SP Rankings for p38 Actives Docked into the 1kv2 and 1a9u Receptors

Ligand	1kv2 Receptor		1a9u Receptor	
	XP Rank	SP Rank	XP Rank	SP Rank
1kv2_lig	1	1	1	4
p38_12	6	94	226	183
p38_9	7	5	293	200
p38_4	10	162	---[a]	---[b]
p38_11	11	157	33	166
p38_6	28	17	138	30
p38_1	38	24	244	73
p38_10	41	12	5	172
p38_8	42	2	12	198
1a9u_lig	43	31	67	56
p38_14	80	204	---[a]	---[b]
p38_13	167	151	93	60
p38_5	268	228	---[a]	---[b]
1bl7_lig	---[a]	---[b]	6	122

[a.] Did not make SP cutoff.

[b.] Not in top 305 ranked positions (30%).

vdW scalings are 1.0 protein/0.8 ligand.

Docking into 1kv2 yields good early enrichment for both SP and XP docking, with particularly good XP results (4 actives in the top 10 positions in Table 16.7). A few of the actives that do not dock well into 1kv2, however, are ranked reasonably well in 1a9u. Thus, a composite XP screen with these two receptors extends the enhanced enrichment further down into the active list, as can be seen by comparing the results for the individual screens (Figure 16.6 and Figure 16.7) with the result for the composite screen (Figure 16.8). Overall, the XP results improve substantially on SP docking with either Glide 2.7 SP or Glide 2.0.

The composite screen was generated by keeping the best scoring instance for each individual ligand in the separate 1kv2 and 1a9u screens, after first adjusting the 1kv2 GlideScore values to compensate for the fact that 1kv2 typically yields a much lower docking score for its well-docked ligands than does 1a9u. As further discussed in Section 16.8.6, these lower scores appear to reflect enhanced binding opportunities in a higher energy form of the receptor that results from a significant conformational change; the offset values take this protein-preparation energy into account. Specific offsets used were +4 kcal/mol for the 1kv2 XP dockings and +2 kcal/mol for the Glide 2.7 and 2.0 SP dockings. That the XP and SP offset values are not the same indicates that the offsets must also in part account for differences

TABLE 16.8
Glide 2.7 XP and SP Rankings for p38 Actives Docked into the 1kv2 and 1a9u Receptors

Ligand	1kv2 Receptor		1a9u Receptor	
	XP Rank	SP Rank	XP Rank	SP Rank
1kv2_lig	1	1	9	5
p38_5	6	192	---[a]	---[b]
p38_8	8	6	56	193
p38_12	11	153	120	146
p38_1	12	9	13	219
p38_9	14	7	300	80
p38_10	15	15	6	114
p38_6	16	25	220	43
p38_4	18	274	---[a]	---[b]
1a9u_lig	38	41	17	60
p38_14	73	256	---[a]	---[b]
p38_11	144	147	83	142
p38_13	210	197	297	52
1bl7_lig	---[a]	---[b]	4	131
p38_7	---[a]	---[b]	18	127
p38_3	---[a]	---[b]	247	220

[a.] Did not make SP cutoff.

[b.] Not in top 305 ranked positions (30%).

vdW scalings are 0.9 protein/0.9 ligand.

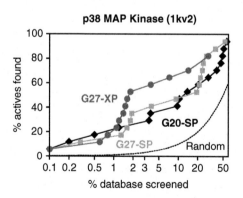

FIGURE 16.6 The p38 MAP kinase actives recovered as a function of the percent of the ranked database for the 1kv2 receptor site using 0.9 protein/0.9 ligand scaling. Results are shown for Glide 2.0 scoring and for Glide 2.7 SP and XP scoring.

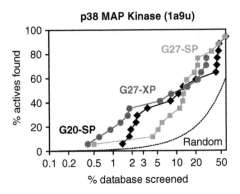

FIGURE 16.7 The p38 MAP kinase actives recovered as a function of the percent of the ranked database for the 1a9u receptor site using 0.9 protein/0.9 ligand scaling.

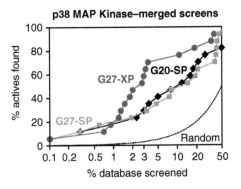

FIGURE 16.8 Number of p38 actives recovered as a function of the percentage of the ranked database after merging by offset the results obtained in separate screens using the 1kv2 and 1a9u receptor sites with 0.9 protein/0.9 ligand scaling.

in the scale of predicted binding affinities for the two methods; as discussed earlier, the large contributions from the new hydrophobic term, which apply in many cases to 1kv2-docked ligands but do not comparably affect 1a9u ligands, distort to some degree the correspondence between predicted binding affinities and what would be expected to be observed experimentally. As described in Section 16.8.6, the separate dockings can also be combined by using the ligand ranks as the docking scores; see that discussion for a full accounting of the present results.

An overall comparison to SP Glide is summarized in Table 16.9, which compiles results for 21 SP and XP screens consisting of the screens discussed above or in the full study [26] (except for thermolysin, a metalloprotein for which XP scoring has not yet been configured) plus variants of those screens that use the default 1.0 protein/0.8 ligand vdW scaling (where different). The table lists counts for the total number of actives found in the top 5, 10, or 20 ranked positions (out of totals of 105, 210, or 420 positions), together with average enrichment factors computed using the weighted geometric mean approach employed for Table 16.4. The com-

TABLE 16.9

Comparison of Ranks and Early Enrichment Factors for Glide 2.5 SP and Glide 2.5 XP Dockings

Method	Number of Top-Ranked Actives			EF′(40)	EF′(50)
	Top 5	Top 10	Top 20		
Glide 2.7 SP	37	59	90	20.7	17.2
Glide 2.7 XP	44	76	110	32.3	29.2

parison shows that XP places substantially more actives in the top-ranked positions and achieves enrichment factors, EF′ (computed as shown in Equation 16.4), for recovering 40 or 50% of the actives (a reasonable percentage to expect to dock correctly into a single rigid protein site) that are about 50% higher than are found for SP docking. On these systems, therefore, XP docking performs significantly better than does SP docking; based on ongoing work, we expect this margin to widen in future releases.

16.8 OPTIMIZING GLIDE'S PERFORMANCE

Glide can usually be used to good effect with default settings and procedures. However, taking some degree of care at critical junctures can help to get optimal results. It is particularly important that the protein and the ligands being screened be prepared properly, and the results can sometimes be improved by adjusting the vdW scaling factors used to regulate the effective size of the protein site. In addition, the user should know how to run Glide to screen large databases, how to use XP docking in conjunction with SP docking, and how to take account of receptor flexibility by combining results from dockings into distinct receptor conformations. These matters are discussed in this section.

16.8.1 PREPARING THE PROTEIN CORRECTLY FOR GLIDE

16.8.1.1 Choosing the Protein Site or Sites

If two or more cocrystallized complexes are available, a first decision that needs to be made is whether to select a single protein site or to choose two or more sites for use in independent docking experiments. A single site should suffice for a rigid protein, as some proteins are known to be. To determine whether a single site is likely to be sufficient in other cases, it helps to transpose the known protein–ligand complexes (or a subset chosen to represent the diversity of the cocrystallized ligand structures) into the coordinate frame of a reference complex using a modeling package that can transform the protein and its bound ligand as a single entity; we use Maestro for this purpose. The objective is to judge whether the reference site appears to be compatible with all the cocrystallized ligands or, if not, whether another site appears to be more suitable or whether two or more sites should be chosen as independent docking targets.

For example, an initial screening of the CDK-2 kinase receptor used 1hck as the docking site, the cocrystallized ligand for which is ATP (adenosine triphosphate). Other known cocrystallized complexes include 1aq1, 1di8, 1dm2, 1fvt, and 1fvv. Superimposing these five complexes onto 1hck using all Cα atoms in common revealed that at least four of the five ligands (all active binders) cannot fit into the 1hck site because its active site channel is too short. The reason for this is that ATP and a Mg²⁺ ion bound to its terminal phosphate group pull glutamate and lysine sidechains more deeply into the 1hck cavity, where they form an ion pair that closes off the cleft. Based on this visual examination, we chose to use 1dm2 instead of 1hck in our initial docking experiments. This site is considerably more open than the 1hck site, though somewhat less so than the 1aq1 or 1fvv sites. Glide was far more successful in docking the known binders into the 1dm2 site than into 1hck.

16.8.1.2 Preparing the Site or Sites

Schrödinger's protein preparation procedure was summarized in Section 16.3 and is described in detail elsewhere [27]. As previously noted, this procedure adjusts protonation states and then performs a series of restrained minimizations to improve hydrogen orientations and relax unphysical steric clashes in the protein–ligand complex. Our experience is that a properly relaxed site and proper protonation states are important for the correct operation of Glide.

Although many site preparation choices are correctly made by the preparation procedure, the user ultimately must take responsibility for seeing that protonation states are assigned correctly. In particular, the user should check for inconsistencies in H-bonding to see whether a misprotonation of the cocrystallized ligand or the protein might have left two acceptor groups close to one another without an intervening H-bond. For example, the protein preparation procedure leaves both active site aspartates negatively charged in HIV protease (which means that one carboxylate would clash with the oxygen atom of a hydroxyl-based ligand), and in thermolysin leaves a carboxylate oxygen of a negatively charged glutamic acid-143 (Glu143) residue and an imidazole nitrogen of a neutral histidine-231 (His231) within about 3 Å of one of the zinc-bound oxygens of a carboxylate or PO_2^- group. If abutting acceptor atoms are found, one of them should usually be protonated. It is normally sufficient to just add the proton and perform approximately 50 steps of steepest-descent minimization to correct the nearby bond lengths and angles, as this optimizer will not make large-scale changes that could alter the protein or ligand conformation significantly. However, if comparison to the original complex shows that the electrostatic mismatch due to the misprotonation has appreciably changed the positions of the ligand or protein atoms during the restrained minimization, it is best to reprotonate the original structure and redo the restrained minimization of the complex.

16.8.1.3 Making Sure the Site Properly Accommodates the Cocrystallized Ligand

To be sure that the prepared complex does not contain unresolved steric clashes, Glide's score-in-place facility can be used to score the cocrystallized ligand in the (restraint optimized) geometry obtained from the structure preparation, and the Coulomb and vdW interaction energies and the H-bond and (where appropriate) metal ligation components of GlideScore can be examined. A reasonable range for the vdW interaction energy is from $-N/2$ to $-N$ in kcal/mol, where N is the number of ligand atoms; enclosed sites tend to have more negative interaction energies for a given N because a tight-binding ligand can contact the protein on all sides. If the calculated vdW energy is appreciably less negative than $-N/2$ (or is positive), the prepared complex may contain unresolved steric clashes that will prevent Glide from properly docking even the native ligand. Steric clashes can also be detected by displaying the ligand and protein in Maestro and using the Measurements > Contacts folder to visualize bad or ugly contacts. It should be noted, however, that Maestro currently defines bad contacts purely on the basis of the ratio of the interatomic distance to the sum of the vdW radii it assigns. As a result, normal H-bonds will show up as bad contacts. We are not talking about these.

If steric clashes are found, the user should repeat the restrained-minimization portion of the protein preparation procedure, but allow a greater rmsd from the starting heavy atom coordinates (the default is 0.3 Å). Alternatively, an additional series of restrained optimizations can be applied to the already prepared ligand–protein complex to allow the site to relax further from its current geometry.

Finally, the Coulombic interaction energy and the H-bond or metal-ligation components of GlideScore should be examined. If the Coulomb energy is positive, or is less negative than about 3 times the H-bond score (assuming use of the default 2r distance-dependent dielectric), there may be a problem with the charge representation of the site. In a few cases, we have found that a nonpolar ligand can have an unfavorable Coulomb interaction energy, even though the ligand and the protein form one or more H-bonds. Nonetheless, poor Coulomb interaction energies need to be looked into. If changes in protonation state are needed, a local correction can be made (see above) or a portion of the restrained optimization can be repeated.

16.8.1.4 Choosing the Enclosing Box

In setting up the docking grids, it is important to use knowledge of the protein and of the candidate ligands to minimize the size of the Enclosing Box (the grid box within which all the ligand atoms must lie) to the degree possible. In particular, a relatively small Enclosing Box should be chosen in most cases, not the large default-sized box that allows ligands with up to 100 atoms to dock. This will speed up the calculation and will weed out ligands that cannot fit; such ligands either will find no viable position within the box or will be given unfavorable docking scores because of unresolved steric clashes. It also keeps available "slots" in the initial, rough-scoring stages of the docking for ligand poses that potentially could prove to be of interest in the more detailed stages that follow. Use of a relatively small Enclosing Box may also prevent Glide from finding positions outside of the active site (e.g., positions on the protein surface) that can compete with the scores of true actives. If Glide's scoring function were perfect, this would not be a problem. But at present, it helps to give Glide a hint, much as one would do in using Glide's constraints facility [28]. Problems of this type are most likely to occur when ligand binding, as assessed by Glide, is relatively weak and when the scores of properly docked actives, for example, are only in the range 7 to 9 kcal/mol; a receptor that yields GlideScore values of 12 kcal/mol is unlikely to find extraneous positions that can give competitive GlideScores.

16.8.2 Preparing the Ligands Correctly for Glide

To give best results, ligands need to have good initial geometries and be properly protonated. The initial geometries are important because docking only modifies the torsional (and rigid body) coordinates of the ligand; all conformers employ the bond lengths and angles found in the input structure. Schrödinger has recently developed its own LigPrep procedure for generating three-dimensional (3D) structures from initial 2D models. However, the work reported in this chapter used CORINA [23] for this purpose. Table 16.10 shows how the initial ligand geometry affects docking accuracy for the 282 cocrystallized complexes considered in Section 16.5. As can be seen, optimization of the CORINA-derived geometries (with MMFF94s [20]) produces a systematic improvement. With this in mind, we strongly recommend that ligand databases be preminimized prior to being docked by Glide. For customers who have a license to MacroModel, Schrödinger provides a convenient premin facility that requires less than half a second per ligand on a modern Linux processor. This time is insignificant in comparison to times required for docking; moreover, preminimization only needs to be performed once for a given ligand dataset.

Database ligands also need to have proper ionization states. That is, carboxylic acids should be deprotonated if the physiological state of interest is near pH 7, and aliphatic amines should be protonated. Otherwise, for example, a neutral aliphatic amine could improperly act as a H-bond acceptor in the docking calculations or could occupy a hydrophobic region without incurring the large desolvation penalty that Glide XP docking would have assessed if the amine had been properly protonated.

Protonation states are particularly crucial when the receptor site is a metalloprotein such as thermolysin or a matrix metalloprotease (MMP). Glide normally assigns a special stability to ligands in which anionic functionality coordinates to the metal center; to benefit, groups such as carboxylates, hydroxamates, and thiolates need to be anionic. Moreover, as discussed in Section 16.8.1.2, the protein residues

TABLE 16.10
Effect of Input Ligand Geometry on Docking Accuracy for Glide Using the GOLD Test Set

Ligand Set	Average RMS (Å)
MMFF94s-optimized native ligands	2.02
MMFF94s-conformational search	2.05
CORINA	2.34
MMFF94s-optimized CORINA	1.96

that line the approach to the metal center (such as Glu143 and His231 in thermolysin) need to be protonated in a manner compatible with the coordination of an anionic ligand such as a carboxylate or hydroxamate. Anionic ligands may also be satisfactory for a site such as that in farnesyl protein transferase [16] or in a hydrogen deacetylase, where the metal center (Zn^{2+} in this case) and its directly coordinated protein residues are net neutral. However, Glide does not give preference to anionic ligands in such a case, and neutral ligands may also dock and score well. The point is that the cocrystallized complex needs to be examined to determine how the protein and the ligands should be protonated. In some cases, two or more protomeric forms of the protein may need to be used in independent docking experiments to cover the range of physically reasonable ligand dockings. Section 16.8.6 describes how to combine the results of such dockings.

16.8.3 OPTIMIZING THE VDW SCALE FACTORS

Glide recognizes specific interactions at a high level of geometric detail through its use of hard interaction energetics on a Coulomb-vdW grid. This attention to geometric detail is one of the factors that allows Glide to do as well as it does at finding ligands that could reasonably bind and rejecting those that clearly could not. But this also means that how the protein is prepared is important (*vide supra*) and that fine-tuning the fit of known ligands to the receptor site may improve the results. Methods that are more forgiving of geometry errors may require less careful protein preparation and less fine-tuning. But approaches that too readily allow known actives to bind computationally to sites that are not geometrically correct for them inevitably will also give good scores to ligands that could not bind to the physical receptor, increasing the number of false positives.

As it happens, Glide does provide a mechanism for adjusting the interaction with the protein site relative to what the full vdW potential would produce. This mechanism works by scaling down the vdW radii of nonpolar protein or ligand atoms. Cross-docking tests in which known ligands are docked into a single instance of a common receptor have consistently shown that it is important to modify the final vdW surface in this manner to allow breathing room for known ligands that are slightly larger than the native ligand for the cocrystallized complex. The default in Glide 2.7 is not to scale the protein radii, but to scale the radii of nonpolar ligand atoms by 0.8. This (1.0/0.8) scaling usually works quite well and is suitable for routine use, but better results can often be obtained by determining what is right for a particular case.

Table 16.11 shows how the choice of scaling factors effects the enrichment obtained in the Glide 2.7 SP database screens described in Section 16.6. Thermolysin is an example of a rigid, open site, but the p38 MAP kinase site is highly mobile and a key hydrophobic region of the estrogen-receptor site is tightly enclosed. Listed are enrichment factors for assaying 2%, 5%, and 10% of the scored and ranked database (computed as shown in Equation 16.3) and for recovering 70% of the known actives (computed as shown in Equation 16.4). Alternative scalings shown in the first listing lines for tk, tk-pyr, 3ert, 1err, 1dwc, 1hpx, 1tmn, and 1rtl are cases in which the preferred scaling for Glide 2.0 identified in the *FirstDiscovery 2.0*

TABLE 16.11
Sensitivity of Glide 2.7 SP Enrichment Factors to vdW Scaling Factors

Screen	Site	vdW Scaling		Enrichment Factor			
		Protein	Ligand	EF(2%)	EF(5%)	EF(10%)	EF'(70)
Tk	1kim	1.0	0.9	20.0	18.0	9.0	22.9
		0.9	0.8	20.0	14.0	9.0	19.8
		1.0	0.8	20.0	16.0	10.0	20.8
Tk-pyr[a]	1kim	1.0	0.9	14.3	17.1	8.6	19.3
		0.9	0.8	28.6	14.3	8.6	23.4
		1.0	0.8	21.4	14.3	10.0	18.3
Estrogen receptor	3ert	0.9	0.8	40.0	16.0	9.0	70.7
		1.0	0.8	30.0	12.0	7.0	29.1
	1err	0.9	0.8	35.0	18.0	9.0	72.8
		1.0	0.8	35.0	14.0	7.0	45.8
CDK-2	1aq1	0.9	0.8	5.0	4.0	5.0	4.1
		1.0	0.8	5.0	6.0	4.0	3.8
	1dm2	0.9	0.8	0.0	6.0	5.0	3.7
		1.0	0.8	15.0	10.0	5.0	4.7
p38	1a9u	0.9	0.9	2.9	2.4	2.9	3.3
		0.9	0.8	2.9	2.4	3.5	3.3
		1.0	0.8	2.9	2.4	2.9	2.7
	1bl7	0.9	0.9	8.8	5.9	5.3	5.5
		0.9	0.8	8.8	7.1	5.3	4.6
		1.0	0.8	8.8	5.9	5.3	4.7
	1kv2	0.9	0.9	17.6	8.2	4.7	4.0
		0.9	0.8	14.7	8.2	4.1	4.1
		1.0	0.8	14.7	8.2	4.7	5.0
Thrombin	1dwc	1.0	1.0	12.5	12.5	8.1	12.7
		0.9	0.8	15.6	10.0	6.9	10.7
		1.0	0.8	18.8	13.8	7.5	17.7
	1ett	0.9	0.8	34.4	17.5	9.4	37.3
		1.0	0.8	34.4	18.8	10.0	45.7
HIV protease	1hpx	0.9	0.8	33.3	16.0	8.7	38.3
		1.0	0.8	33.3	16.0	9.3	38.6
Thermolysin	1tmn	1.0	1.0	35.0	18.0	10.0	41.9
		0.9	0.8	35.0	16.0	8.0	38.7
		1.0	0.8	40.0	18.0	9.0	52.6
Cox-2	1cx2	0.9	0.8	10.6	6.1	4.5	4.7[b]
		1.0	0.8	13.6	7.9	4.5	4.2[b]
	1cx2-site1	0.9	0.8	15.2	8.7	6.5	8.1
		1.0	0.8	19.6	11.3	6.5	12.3
HIV-RT	1rt1	0.9	0.8	6.1	7.3	6.4	7.2
		1.0	1.0	16.7	8.5	5.8	3.8
		1.0	0.8	12.1	9.1	7.0	8.5
	1vrt	0.9	0.8	1.5	4.2	4.8	4.4
		1.0	0.8	6.1	7.3	5.8	5.6

[a.] Screen uses only the seven pyrimidine-based actives. (See C. Bissantz et al., *J. Med. Chem.* 43:4759–4767, 2000.)

[b.] EF'(60) value.

In each case, the preferential scaling model used with Glide 2.0 is listed first.

Technical Notes differs from the current default of (1.0/0.8) scaling. The table shows that the new default scaling works better than the previously identified preferential scaling in three of the eight cases and does about equally well in three others. The original scaling does give substantially better $EF'(70)$ or $EF(2\%)$ enrichment factors for 3ert and 1err. However, the enrichment factors are high in these cases and the default enrichments are also reasonably good.

Also shown in Table 16.11 for each of the screens are enrichment factors for (0.9/0.8) scaling, the Glide 2.0 default. These results show that the stronger scaling provides clearly superior results only for the 3ert and 1err cases already cited. For most of the other cases, the new default scaling gives better results. As an overall measure, enrichment factor averages for (0.9/0.8) scaling computed as in Table 16.4 are 14.3 for $EF'(70)$ and 14.0 for $EF(2\%)$. For comparison, the average values are 14.8 and 18.0 for (1.0/0.8) scaling. Thus, (0.9/0.8) scaling performs comparably well for global enrichment, but the current default is the better choice for early enrichment. However, a few cases — evidently, those with particularly tight active sites — profit substantially from the use of the more generous (0.9/0.8) scaling.

Table 16.12 shows that default scaling allows almost all of the known binders to dock successfully, but that (0.9/0.8) scaling occasionally allows one or two additional actives to dock. Moreover, the more generous scaling usually produces a significantly lower rank for the last common active found (e.g., the 8th for 3ert, the 9th for 1err, or the 21st for 1cx2). This, too, indicates that (0.9/0.8) scaling produces a better physical model when the fit is tight. The 1rtl site is an exception because (0.9/0.8) scaling is not in fact the optimal scaling, as Table 16.11 shows.

The conclusion we draw from these comparisons is that use of optimal scaling factors can improve the results and should be considered for high performance screens. If active ligands are unavailable or will not be used to determine the scaling factors and if a single scaling factor will be employed, the current default should normally be used. However, if the protein heavy atom coordinates are taken directly

---

TABLE 16.12

Number of Known Actives Docked with Negative Coulomb-vdW Interaction Energies as a Function of the Protein and Ligand vdW Scale Factors for Nonpolar Atoms

Screen	Site	Number of Actives	Number Docked		Rank of Last Common Active	
			(1.0/0.8)	**(0.9/0.8)**	**(1.0/0.8)**	**(0.9/0.8)**
Estrogen receptor	3ert	10	8	9	130	8
Estrogen receptor	1err	10	9	9	465	51
HIV protease	1hpx	15	15	15	150	369
Cox-2 (site 1)	1cx2	33	21	23	512	351
HIV-RT	1rtl	33	30	30	661	624

from the x-ray structure without opportunity for relaxation of steric clashes, it may be better to use (0.9/0.8) scaling to reduce the effect of unattended steric clashes. The more generous (0.9/0.8) scaling should also be used in cases in which it is known that the active site region is tight and enclosed (an example being the hydrophobic channel of the estrogen receptor), as it will be difficult in such cases for many active ligands to avoid serious steric clashes with the rigid site. Conversely, a lesser degree of scaling might be tried if the site is open and is known to be relatively rigid.

To use active ligands to select the scaling factors, the most direct and powerful approach is to dock some or all of the known binders together with a subset of the ligands to be used in the full database screen, using combinations of protein and ligand scaling factors ranging from 0.8 to 1.0. The model that yields the highest enrichment factors, while maximizing the number of known actives recovered, would then be chosen for use in the full database screen. A less direct approach would be to dock and score just the known binders, or a subset of them, and examine the output data. Obviously, scaling factors that produce strongly negative Coulomb-vdW energies for as many known binders as possible should be chosen, and H-bond (and, where appropriate, metal-ligation) scores should be as favorable as possible. Comparing GlideScore values, however, is more problematic. This is because in Glide 2.7 protein scaling (but not ligand scaling) affects the vdW radii used to calculate lipophilic–lipophilic terms (Equation 16.1) that usually contribute strongly to the computed GlideScore values. The retraction of the protein surface means that favorable GlideScore values will get progressively less negative on going from 1.0 to 0.9 and then to 0.8 protein scaling. Thus, basing the choice of scaling model on computed GlideScore values will tend to choose a model that involves a lesser degree of protein scaling. As minimizing the scaling should reduce the number of false positives, this may in general be a good thing, but is something the user should be aware of.

16.8.4 USING GLIDE TO SCREEN LARGE DATABASES

16.8.4.1 Dividing the Screen Over Multiple Processors

When the number of ligands is small (e.g., approximately 1000), Maestro can be used to submit a job that runs on a single processor. For larger jobs, the FirstDiscovery para_glide facility can be employed to divide the screen over several processors or machines if the licensing allows multiple copies of Glide to run simultaneously. The Maestro graphical user interface (GUI) should first be employed to write the template <inp-file> that para_glide needs. In doing so, it can be helpful to elect to save Ligand Library files rather than Pose Viewer files to avoid saving multiple copies of the receptor. A Pose Viewer file can always be reconstituted later by appending all but the first five lines of a Ligand Library file to a copy of the receptor file using a command like:

```
tail +6 lig_lib.mae >> copy_of_receptor.mae
```

How to run para_glide is indicated below:

```
Usage: $SCHRODINGER/utilities/para_glide -i <inp-
file> [<options>]
```

Options:

-n	`<njobs>`	number of subjobs to prepare
-f	`<firstlig>`	first ligand to include
-l	`<lastlig>`	last ligand to include
-j	`<jobnum>`	subjob number to prepare
-x		launch jobs after writing input files
-s		split input ligand file by subjob
-o		have log file written directly to output dir(only meaningful if -x option also used.)

For example, to divide the screen into 10 equal parts based on the template file my-screen.inp, a command like the following would be given:

```
$SCHRODINGER/utilities/para_glide -i my-screen.inp -n
10 -x -s -HOST <hostname>
```

In this case, para_glide will count the number of ligands in the input Maestro or Molecular Design Limited (MDL) SD structure-data file specified in my-screen.inp and assign `<firstlig>` and `<lastlig>` values that divide the ligands into 10 equal segments. To keep the jobs separate, para_glide creates a series of subdirectories named my-screen_<firstlig>_<lastlig> if they do not already exist. In addition, it names the cloned *.inp files in a similar manner and modifies the names of the output files it finds in the template file by inserting the string _<firstlig>_<lastlig>. The para_glide script also produces a my-screen_M_N_report.sh file that can be used to concatenate the output structure files from the individual jobs and produce a unified report of the docking scores, where M is `<firstlig>` from the first of the spawned jobs and N is `<lastlig>` from the last such job.

Other capabilities are indicated in the synopsis or in the example command shown above. For example, jobs can be submitted to a different machine or to a queue by including the -HOST `<hostname>` option, and `<firstlig>` or `<lastlig>` can be specified explicitly. After the job is submitted, the my-screen_M_N_status.sh script also written by para_glide can be executed to report the current status of each subjob (e.g., submitted, running, or finished); error conditions are also reported (see Section 16.8.4.3).

16.8.4.2 Using **glide_sort** to Work Up the Results

Combining the individual job outputs from a screen subdivided into segments by para_glide is facilitated by the FirstDiscovery glide_sort script, which can be applied to one or a series of Glide output structure files. These files can be final Pose Viewer or Ligand Library files or, for flexible docking, can be the `*.ext` files Glide writes while it processes the ligands; indeed, each Glide subjob generates its output report and structure files by applying `glide_sort` to its own `*.ext` file when it has finished processing the ligands. In part, the synopsis for `glide_sort` reads as follows:

```
Usage: $SCHRODINGER/utilities/glide_sort
<output_options> [<options>] <pv-or-lib-files>

Output options: (at least one is required)

  -o  <output-file> write the best-scoring poses to
                    <output-file>

  -r  <report-file> create a report of the best
                    scores in <report-file>

  -R                write a report of the best scores
                    to standard output

Options:

  -n  <nreport>     retain only the <nreport> lowest-
                    scoring poses

  -norecep          don't look for receptor structure
                    in the input file(s)

  -best-by-title    keep only the single best pose
                    for each ligand with a given
                    title
```

By setting the $-n$ option, `glide_sort` can also be used to extract a specified number of top-ranking hits for visual assessment or further processing.

16.8.4.3   Dealing with a Subjob that Fails

Glide is fairly stable at this point in its development. However, we cannot rule out the possibility that an unanticipated condition or a remaining error in the program code might lead to an abnormal temination. More commonly, a subjob that does not complete will fail because a CPU time limit has been exceeded or because a network or hardware problem has occurred. In such (hopefully rare) cases, the user will need to manually complete the subjob and combine the output from the original and resubmitted runs to generate the `*_lib.mae` or `*_pv.mae` file that `glide_sort` needs to process the completed screen.

Indication that a subjob has failed may result from running the `*_status.sh` script that `para_glide` writes and finding that a subjob has died (the job crashed), was killed (the CPU time may have expired), was stopped by the user, became unreachable because of network problems, or was stranded (i.e., terminated abnormally, such that output could not be returned). It could also arise by executing the `*_report.sh` script and finding that one or more `*_lib.mae` or `*_pv.mae` files are listed as being missing.

If a subjob failed or did not produce the expected structure output file, the output files in the associated subdirectory (the name will be indicated in the `*_status.sh` or `*_report.sh` output) will need to be examined to determine what the problem might be:

- If the `*.log` file ends by listing the CPU time used but there is no `*_lib.mae` or `*_pv.mae` file, the output structure file may not have been successfully written. In this case, `glide_sort` can be run on the `*.ext` file, with the `-o` option set appropriately, to produce the needed `*_lib.mae` or `*_pv.mae` file. Alternatively, in rare cases the subjob may have run correctly but not found any qualifying ligands (i.e., docked poses with negative Coulomb-vdW interaction energies); when this happens, the last few lines of the `*.log` file will so indicate.
- If the subjob terminated abnormally, the last ligand processed needs to be determined so that a Glide job that starts with the next ligand can be submitted. To determine which ligand this is, the following command can be given:

```
egrep "DOCKING RESULTS FOR LIGAND|skip" *.log | tail
```

The screen output will list the numbers of the last few ligands successfully processed. The ligand following the last one successfully docked (or the last one skipped for having too many atoms or rotatable bonds, etc.) would be the one that failed. Thus, the first ligand for the Glide resubmission would be the last ligand successfully docked plus (at least) 2. The job can be resubmitted by using the `para_glide` script with the `-n` option set to 1 (or to a larger value) and with `<firstlig>` set by using the `-f` option. If the original `para_glide` submisssion employed the `-s` option, the subjob directory will include a structure input file containing just the ligands needed for the subjob; in this case, nothing more is needed. If the `-s` option was not used, however, `<lastlig>` will need to be specified by using the `-l` option.

When the resubmission finishes, the `*_report.sh` script written by `para_glide` can be used to bring the output `*_lib.mae` or `*_pv.mae` file back into the current directory, and `glide_sort` can be used to combine this file with the `*.ext` file from the incomplete subjob and to write the `*_lib.mae` or `*_pv.mae` file that the `*_report.sh` script written by `para_glide` expects to find.

16.8.5 Using Glide XP to Improve Pose Quality or Enhance Early Enrichment

Except in the case of metalloproteins, for which XP scoring has not yet been defined [26], we recommend that Glide XP be used for high performance lead-discovery screens and for lead-optimization dockings in which the best possible accounting of pose geometry or binding affinity is wanted. As Table 16.9 shows, Glide XP assigns higher ranks to known actives and achieves better early enrichment. Our recommended procedure is to first carry out a SP screen and to then select the top 10 to 30% of high scoring ligands for XP docking. Because Glide SP intentionally allows relatively close intraligand contacts to implicitly allow for the breathing the actual protein site might undergo as a larger ligand docks, the ligands from the Glide SP run should first be minimized to remove any bad steric clashes (e.g., using the premin utility discussed in Section 16.8.2). The Glide XP run can then be submitted in the normal manner, apart from selecting XP mode in the Maestro GUI. The para_glide utility should again be used, but the user should keep in mind that individual XP dockings can be 10 to 15 times slower than SP dockings; thus, the -n option may need to be set to a greater value than that used in the SP dockings.

Good XP results can be obtained under favorable conditions by docking into a single conformation of the protein. However, two or more receptor conformations may need to be used when the protein site is plastic. In which case, results from the individual dockings will need to be combined in the manner described in Section 16.8.6.

16.8.6 USING MULTIPLE RECEPTOR SITES TO DEAL WITH RECEPTOR FLEXIBILITY

A key limitation for Glide is the restriction to docking to a rigid receptor site. This restriction can be ameliorated to some degree by scaling down the vdW radii of nonpolar ligand or protein atoms. Such an approach can be effective in allowing for subtle deformations of a protein site. But when discrete conformational changes occur in the receptor upon ligand binding, a better approach is to carry out flexible ligand docking into multiple rigid protein structures. We will illustrate this approach in the context of Glide XP for the p38 MAP kinase screen, which uses the 1kv2 and 1a9u sites. This is a case in which the site is known to be highly variable.

One way is to merge the docking results based on the ranks of the compounds within each screen. That is, a compound is assigned the highest rank it achieves over all of the screens, and this number is used to rerank the database. Ties in rank could be broken on the basis of which receptor has the most favorable total score for the best scoring compound, on the basis of which of the equally ranked ligands has the more negative GlideScore, or simply on the basis of which comes first in the input stream. All are somewhat arbitrary, but have a minimal effect on enrichment factors and other quantitative measures of performance. We adopt here the last of the three options.

A second approach for combining multiple receptor dockings is to merge the data by selecting the instance of each ligand that has the lowest GlideScore value. An important refinement is to assign a constant offset to the scores obtained in a given receptor and then to retain the highest adjusted score for each compound. The physical justification is that some receptor conformations have an intrinsically higher energy than others, and this promotion energy must be added to the computed binding affinity to correspond to what would be observed experimentally. Use of this approach requires some way of calibrating the offset values. If there are known active compounds, these can be used as a training set, and the offsets can be optimized to yield the best overall enrichment factors. But our experience is that this approach does not always give a physically justifiable value for the promotion energy. In principle, one could construct the offset based on the total energies of the different protein conformations. Computing sufficiently reliable total energies for this purpose, however, is a daunting task and it remains to be seen whether this can be successfully done in practice. A more straightforward approach is to examine the GlideScore values for the top-ranking N ligands for the alternative receptor sites and to presume that the intrinsically higher energy protein conformer will pay for its higher energy cost by producing intrinsically more negative computed binding affinities. In the case of the 1kv2 site for p38, a stronger interaction with a novel bound ligand that occupies a newly revealed hydrophobic pocket appears to be the factor that drives the population of the higher energy protein state [29].

This second approach is explored in Table 16.13, which lists average GlideScore values computed using the `glide_rescore` utility for the first 10, 20, 50, or 100 ligands for the 1a9u and 1kv2 sites of p38. The table shows that the average GlideScore value is consistently more negative for 1kv2, but by lesser extents as the number of ligands increases. Given that the objective is to find high ranking actives, use of an offset based on one of the first listing rows would seem appropriate — e.g., of perhaps +4 kcal/mol for 1kv2.

Table 16.14 compares computed EF(2%), EF(5%), EF(10%), and EF'(70) enrichment factors for the individual screens and for composite screens obtained by merging the individual screens based on ranks, on computed GlideScores, and on GlideScores offset by adding +1, +2, +3, +4, and +5 kcal/mol to the 1kv2 GlideScore values. Merging on the raw GlideScores corresponds to use of an offset of 0; in this case, because the 1kv2 GlideScore values are much more favorable, the combined screen in effect reverts to the pure 1kv2 results. Best results overall are obtained by merging with offsets of +3 or +4 kcal/mol and by merging on ranks, each of which increases the EF'(70) value substantially and increases the EF(5%) and EF(10%) values slightly; the merge illustrated in Section 16.7.3 used an offset of +4 kcal/mol. Also evident is that the EF'(70) value begins to decrease as the 1kv2 offset is further increased; application of a sufficiently large positive offset would simply revert to the (inferior) 1a9u results.

TABLE 16.13

Average GlideScore Values for Top-Ranking Ligands for Alternative Protein Conformations for p38 MAP Kinase (1a9u and 1kv2 Sites)

Number of Ligands	1a9u Site	1kv2 Site	Difference
10	-9.06	-14.13	5.07
20	-8.78	-13.43	4.65
50	-8.24	-12.00	3.76
100	-7.60	-10.79	3.19

TABLE 16.14

Computed Enrichment Factors for Individual and Composite Glide 2.7 XP Screens Using 0.9 Protein/0.9 Ligand Scaling

Method	EF(2%)	EF(5%)	EF(10%)	EF'(70)
1a9u screen	17.6	7.1	4.7	3.8
1kv2 screen	26.5	11.8	6.5	12.1
Merge GlideScores	26.5	11.8	6.5	12.1
Merge by rank	17.6	14.1	7.1	18.1
Offset by +1 kcal	26.5	11.8	7.6	15.3
Offset by +2 kcal	26.5	12.9	7.1	17.2
Offset by +3 kcal	26.5	14.1	7.1	20.4
Offset by +4 kcal	23.5	14.1	7.1	19.4
Offset by +5 kcal	14.7	11.8	7.6	13.3

To extract the best pose for each ligand, the most straightforward approach is to make use of the composite structure file created for each receptor screen by running the *_report.sh script that para_glide writes. The glide_rescore utility would then be employed in -rank or -offset mode to write modified structure files in which rank orders or offset GlideScore values have been inserted into the docking_score property variables in place of the normal GlideScore values (this does not destroy the GlideScore values, which are also stored separately). The glide_rescore synopsis is shown in part below:

```
Usage   : $SCHRODINGER/utilities/glide_rescore
[<options>] <pv-or-lib-files>

Purpose: Replace the "docking score" properties in
Glide pose output files with different values, so
that glide_sort's "best-by-title" option

can be used to combine different screens.

Options:

-rank               Replace "docking_score" with
                    ligand rank (default mode)

-offset <value>     Replace "docking_score" with
                    GlideScore plus this offset

-average            Calculate the average GlideScore
                    over all the poses; no output
                    besides this average is produced

-top <number>       Average only the top <number>
                    poses

-every <number>     Print running averages every
                    multiple of <number> poses
```

```
-o <output-file>Output to this filename, instead
              of default name (<input-
              file>.rank.mae or <input-
              file>.offset.mae)
```

Finally, glide_sort would be used with the -best-by-title option to combine the output structure files containing the desired docking_score values and to write the report and structure files that describe the composite screen.

16.9 CONCLUSIONS

This chapter has presented results for docking accuracy and for database screens covering a wide variety of receptor types. The results show that Glide docks ligands more accurately than either GOLD or FlexX for the test sets devised for these methods. We are not yet in a position to provide comparable comparisons for database enrichment. However, it is clear that Glide 2.7 does far better than early versions of the program for many of the more difficult screens; this qualitative improvement should be borne in mind when assessing comparative studies based on Glide 1.8 or 2.0 [30]. One theme that runs consistently through the database enrichment results is that Glide does best when the active ligands make multiple H-bonds to the receptor and worst when the site is hydrophobic and offers few such opportunities. From what we have seen in the literature, this behavior is not unique to Glide. One of the key problems in database screening — one on which we have made considerable progress in ongoing work [31] — is how to properly model binding when it is mainly hydrophobic in character.

The results presented in this chapter show that Glide XP docking and scoring can provide enhanced results for both binding mode and binding affinity prediction. Moreover, the greatest degrees of improvement have been found for CDK-2 and p38, the screens on which Glide SP performs the most poorly. In the latter case, the improvements arise partly from the use of two different receptor conformations. Better results have also been observed for HIV-RT [26], another relatively difficult screen for Glide SP, and for thymidine kinase. Visual examination of poorly scoring XP compounds typically reveals regions of poor complementarity to the receptor site. The key point about XP docking is that the correlation between a good docking score and a good quality pose is significantly enhanced by the aggressive (penalty wise) and detailed nature of the scoring.

The multiple-receptor p38 dockings presented here used alternative cocrystallized forms of the receptor. In many real-world applications, such alternative receptor shapes will be available. In other cases, accurate molecular-mechanics modeling of protein structure will be required to enumerate the variations in active site geometry that can be accessed at relatively low energies; calculations along these lines, if successful, would obviate the need for cocrystallized examples. Although modeling of this type is clearly quite difficult at present, methods using continuum solvation models like those developed by Schrödinger can in principle address this problem effectively. If this can be accomplished, it would greatly enhance the effectiveness of any docking methodology in a wide range of practical applications. Efforts along these lines are currently underway at Schrödinger and have yielded promising early results.

NOTE

1. In computing these averages, any enrichment factor of less than 1 was replaced by 1. This was done so that a single bad result (e.g., an enrichment factor of 0) would not overly influence the computed average value.

REFERENCES

[1]   G. Jones, P. Wilett, R.C. Glen, A.R. Leach, and R. Taylor, Development and validation of a generic algorithm and an empirical binding free energy function, *J. Mol. Biol.* 267:727–748, 1997.

[2]   M. Rarey, B. Kramer, T. Lengauer, and G.A. Klebe, A fast flexible docking method using an incremental construction algorithm, *Chem. Biol.* 261:470–489, 1996.

[3]   T.J.A. Ewing and I.D. Kuntz, Critical evaluation of search algorithms for automated molecular docking and database screening, *J. Comput. Chem.* 18:1175–1189, 1997.

[4]   P.S. Charifson, J.J. Corkery, M.A. Murcko, and W.P Walters, Consensus scoring: a method of obtaining improved hit rates from docking databases of three-dimensional structures into proteins, *J. Med. Chem.* 42:5100–5109, 1999.

[5]   C. Bissantz, G. Folkers, and D. Rognan, Protein-based virtual screening of chemical databases: 1. Evaluation of different docking/scoring combinations, *J. Med. Chem.* 43:4759–4767, 2000.

[6]   M. Stahl and M. Rarey, Detailed analysis of scoring functions for virtual screening, *J. Med. Chem.* 44:1035–1042, 2001.

[7]   W.L. Jorgensen, D.S. Maxwell, and J. Tirado-Rives, Development and testing of the OPLS all-atom force field on conformational energetics and properties of organic liquids, *J. Am. Chem. Soc.* 118:11225–11236, 1996.

[8]   Prime (Protein Integrated Modelling Environment) is scheduled for release by Schrödinger, LLC, in Fall 2003: www.schrodinger.com.

[9]   Impact is the computational engine of Schrödinger's FirstDiscovery Suite. Impact was developed in the laboratories of Prof. Ronald Levy (Rutgers University). Impact and the FirstDiscovery suite are available from Schrödinger, LLC, www.schrodinger.com.

[10] MacroModel (formerly known as BatchMin) is a general-purpose molecular-mechanics program available from Schrödinger, LLC, www.schrodinger.com. MacroModel was developed in the laboratories of Prof. Clark Still (Columbia).

[11] Together with Glide, Liaison, and QSite are modules in the FirstDiscovery suite available from Schrödinger, LLC: www.schrodinger.com.

[12] T. Hansson and J. Åqvist, Estimation of binding free energies for HIV proteinase inhibitors by molecular dynamics simulations, *Protein Eng.* 8:1137–1144, 1995.

[13] M.D. Eldridge, C.W. Murray, T.R. Auton, G.V. Paolini, and R.P. Mee, Empirical scoring functions: I. The development of a fast empirical scoring function to estimate the binding affinity of ligands in receptor complexes, *J. Computer-Aided Mol. Des.* 11:425–445, 1997.

[14] R.E. Babine and S.L. Bender, Molecular recognition of protein–ligand complexes: applications to drug design, *Chem. Rev.* 97:1359–1472, 1997.

[15] M. Whittaker, C.D. Floyd, P. Brown, and A.J.H. Gearing, Design and therapeutic application of matrix metalloproteinase inhibitors, *Chem. Rev.* 99:2735–2776, 1999.

[16] I.M. Bell, S.N. Gallicchio, M. Abrams, L.S. Beese, D.C. Beshore, H. Bhimnathwala, M.J. Bogusky, C.A. Buser, J.C. Culberson, J. Davide, M. Ellis-Hutchings, C. Fernandes, J.B. Gibbs, S.L. Graham, K.A. Hamilton, G.D. Hartman, D.C. Heimbrook, C.F. Homnick, H.E. Huber, J.R. Huff, and K.S.K. Kassahun, 3-aminopyrrolidinone farnesyltransferase inhibitors: design of macrocyclic compounds with improved pharmacokinetics and excellent cell potency, *J. Med. Chem.* 45:2388–2409, 2002.

[17] R.A. Friesner, J.L. Banks, R.B. Murphy, T.A. Halgren, J.J. Klicic, D.M. Mainz, M.P. Repasky, E.H. Knoll, M. Shelley, J.K. Perry, D.E. Shaw, L.C. Sander, P. Francis, and P.S. Shenkin, Glide: a new paradigm for rapid, accurate docking and scoring. I. Method and assessment of docking accuracy, *J. Med Chem.* 47:1739–1749, 2004.

[18] The results for the GOLD test set are available at http://www.ccdc.cam.ac.uk/products/life_sciences/validate/gold-validation/ccdc-test-html.

[19] The results for the FlexX test set are available at http://www.biosolveit.de/software/flexx/html/flexx-eval.html.

[20] R.A. Engh, H. Brandstetter, G. Sucher, A. Eichinger, U. Bauman, W. Bode, R. Huber, T. Poll, R. Rudolph, and W. van der Saal, Enzyme flexibility, solvent, and "weak" interactions characterize thrombin-ligand interactions: implications for drug design, *Structure* 4:1353–1362, 1996.

[21] W. von der Saal, R. Kucznierz, H. Leinart, and R.A. Engh, Derivatives of 4-aminopyridine as selective thrombin inhibitors, *Bioorg. Med. Chem. Lett.* 7:1283–1288, 1997.

[22] T.A. Halgren, MMFF VI. MMFF94s option for energy minimization studies, *J. Comput. Chem.* 20:720–729, 1999.

[23] J. Gasteiger, C. Rudolph, and J. Sadowski, Automated generation of 3D atomic coordinates for organic molecules, *Tetrahedron Comput. Methodol.* 3:537–547, 1990.

[24] D.A. Pearlman and P.S. Charifson, Improved scoring of ligand–protein interactions using OWFEG free energy grids, *J. Med. Chem.* 44:502–511, 2001.

[25] *FirstDiscovery Technical Notes*, version 2.7, Chapter 6, Schrödinger, LLC: www.schrodinger.com.

[26] *FirstDiscovery Technical Notes*, version 2.7, Chapter 7, Schrödinger, LLC: www.schrodinger.com.

[27] *FirstDiscovery User Manual*, Schrödinger, LLC, www.schrodinger.com.

[28] *FirstDiscovery Technical Notes*, version 2.7, Chapter 8, Schrödinger, LLC: www.schrodinger.com.

[29] C. Pargellis, L. Tong, L. Churchill, P.F. Cirillo, T. Gilmore, A.G. Graham, P.M. Grob, E.R. Hickey, N. Moss, S. Pav, and J. Regan, Inhibition of P38 map kinase by utilizing a novel allosteric binding site, *Nat. Struct. Biol.* 9:268, 2002.

[30] T. Schulz-Gäsch and M. Stahl, Binding site characteristics in structure-based virtual screening: evaluation of current docking tools, *J. Mol. Mod.* 9:47–57, 2003.

[31] R.A. Friesner and R.B. Murphy, unpublished research.

Index

A

Absolute binding affinity, 231–235
  binding constant, determination, 233
  PMF, 232
  standard state, 234
ACD. *See* Available Chemicals Directory
Aggregate-forming screening hits, promiscuous, 107–124
  air bubbles, 122
  albumin, decreased potency in presence of, 118
  compound concentration
    high, 122
    low, 122
  conceptual advances, 110–111
  data analysis, 120–122
    particles, 121
    scattered light, average intensity of, 122
    size of particles, 121–122
  denaturation, 108–109
  dissimilar enzymes, inhibition of, 115–116
  dust scattering, 122
  dynamic light scattering, 119–120
    laser, wavelength of, 120
    measurement, sample volume required for, 120
    signal to noise ratio, 120
    suppliers, 119
  enzyme concentration, sensitivity to, 116–117
  frequent hitters, neural net to identify, 110–111
  incubation effect, 117–118
    assay for, 118
  ionic strength, decreasing potency with, 118–119
  kinetic assays, 112–113
    buffer, 112
    components, 112
    enzyme concentration, 112
    inhibitor concentration, 113
    substrate concentration, 112
  light scattering, 119
  methods, 120
  noncompetitive inhibition, 119
  problems, 122
  protocols, 111–122
  recursive partitioning model, 110
  reversible adsorption, 109–110
  sensitivity to, detergent, 113–115
  steep inhibition curves, 119
  substrate sequestration, 108
  Triton X-100, 113–114
  troubleshooting, 122
Albumin, decreased potency in presence of, 118
Alignment, docking, comparison, interaction models, 29
Alignment-based screening, CDK2 inhibitors, 41–42
α4β1 antagonists (Biogen), 191–193
  inflammatory diseases, 191
  success screening of virtual library, 193
Anchor sensitivity, molecular docking, 329
Anticompensation, phenomenon of, 236
AutoDock, 257–258
Available Chemicals Directory (ACD), 93

B

Base placement, 31–32
Base selection, 31–32
Binding affinity, estimation of, 18
Binding free energy terms
  decomposability, 238–239
  separability of, 238–239
Binding mode, prediction of, 37–39
Binding pocket
  defined, 56
  solvent exposure of, 61–62
Binding region, defined, 56
Binding site, defining, 57, 349–350
Binding-site conformers, consideration of, 11
Biological activity, score, correlations between, 70
Biomolecular Ligand Energy Evaluation Protocol (BLEEP), 259
Bitmap size, virtual high throughput screening, pharmacophore multiplet fingerprints, 212
BLEEP. *See* Biomolecular Ligand Energy Evaluation Protocol
Bohm's function, 255–256
Born-based implicit solvent methods, free energy-based scoring functions, 240
Breast cancer, osteoporosis, estrogen ERα receptor (Organon), 193–194